BEYOND THE BODY PROPER

BODY, COMMODITY, TEXT

Studies of Objectifying Practice

A SERIES EDITED BY ARJUN APPADURAI,

JEAN COMAROFF, AND JUDITH FARQUHAR

BEYOND THE BODY PROPER

PROPER

Reading the Anthropology of Material Life

EDITED BY **MARGARET LOCK**

AND **JUDITH FARQUHAR**

Duke University Press

Durham and London 2007

© 2007 DUKE UNIVERSITY PRESS

ALL RIGHTS RESERVED

PRINTED IN THE UNITED STATES OF

AMERICA ON ACID-FREE PAPER ∞

DESIGNED BY AMY RUTH BUCHANAN

TYPESET IN MINION BY KEYSTONE

TYPESETTING, INC.

LIBRARY OF CONGRESS CATALOGING-

IN-PUBLICATION DATA APPEAR ON THE

LAST PRINTED PAGE OF THIS BOOK.

CONTENTS

NOTE ON THE FORMAT OF THE BOOK

Despite the cultural and historical range of the readings included in this collection, many bodies have not found their way into this book. Similarly, important parts of careful arguments have had to be excised from the bodies of some of the essays we did include. Where deletions have been made, we have used an ellipsis to so indicate.

References to literature not directly connected to the argument or the aims of this book have been deleted to save space; references to the sources of direct quotes from scholarly literatures have been maintained, however. Citation style remains consistent with the original usage for each article: endnotes including full citations appear in some articles, and in-text author and date references, with a list of sources at the end of the article, are used in others. A list of citations for the forty-seven articles in the book has been supplied to facilitate readers' efforts to locate the originals.

The introductions to each part are meant to help orient readers to the articles and extracts that follow. References cited in these part introductions will be found in the general bibliography. In addition to the works cited in the general introduction and in the part introductions, a few suggestions for additional reading are included at the end of some part introductions. These latter sources do not appear in the general bibliography.

ACKNOWLEDGMENTS

The project of assembling and bringing to press a collection like *Beyond the Body Proper* is mostly a matter of detailed and repetitive labor. In this task we have had much willing and intelligent assistance over a long period of time. For their clerical and editing help, we especially thank Monica Adamowicz, Jessica Ash, Wee-Teng Soh, and Virginia Sprague. Kelly Raspberry made many excellent suggestions for shortening articles near the beginning of this process, and Kathryn Goldfarb kept us organized and legal as we gathered permissions to reprint them at the end of it. The ever-helpful staff of our offices at McGill University, the University of Chicago, and the University of North Carolina provided material and moral support throughout.

INTRODUCTION

JUDITH FARQUHAR

MARGARET LOCK

In the last decades of the twentieth century, the topic of the body became almost conventional in the human sciences. In fields ranging from anthropology to literary studies, history to political science, researchers expanded the classical social science concern with either minds or bodies, meanings or behaviors, individual bodies or the body of the social to focus on a new hybrid terrain, that of the lived body. Seen as contingent formations of space, time, and materiality, lived bodies have begun to be comprehended as assemblages of practices, discourses, images, institutional arrangements, and specific places and projects. There has been a proliferation of fascinating empirical studies multiplying the kinds of bodies that can be perceived and widening the scholarly vision of human capacities.[1]

This emergence of the lived body was not entirely unprecedented. Phenomenological philosophy had long explored problems of embodiment—a term that emphasizes process and contingency—and offered a dynamic understanding of human being as inseparable from a universal human physicality existing within complex fields of influence.[2] The related philosophical traditions of vitalism and pragmatism also sought to understand material life beyond the dualities of mind and body, and some of the key works in these orientations are now being productively reread.[3] These intellectual movements resisted modernist tendencies to excise subjectivity and experience from material bodies and worlds, idealizing them for the disciplinary attention of humanists and social scientists.

Even so, most of social science continued for a long time to treat bodies as the naturalized, essentially passive atoms or building blocks of society. The body offered to social thought by nineteenth- and twentieth-century biomedical specialties, though a complex materiality in its own right, was easily appropriated in social thought as a capsule of nature that could be inhabited, but not altered, by culture (see part I). The classic problematic of the relations between individual and society that still provides analytic tools for most of the social sciences seems to require a "proper" body as the unit of individuality. This body proper, the unit that supports the individual from which societies are apparently assembled, has been treated as a skin-bounded, rights-bearing, communicating, experience-collecting, biomechanical entity. Our common sense has attributed basic needs to this discrete body along with fixed gender characteristics. In law it has been seen as the only possible basis for the citizen's responsibility to act and to choose. In the humanities it was long treated as the locus of an originary consciousness that is expressed in voice, image, and action.[4] However contradictory this complex hybrid body may seem, its naturalness and normality tend to be reinforced by the operations of common knowledge and standard operating procedure in many contemporary spheres of activity.

Recent scholarship in the human sciences, led perhaps by gender, ethnic, and rights activism in postmodern popular culture, has turned away from the commonsense body, however, learning to perceive more dynamic, intersubjective, and plural human experiences of carnality that can no longer be referenced by the singular term *the body*. It is difficult, however, to characterize one object of knowledge around which new research has been organized. What has emerged from interest in the human body, as it is lived, is a multiplicity of bodies, inviting a great many disciplinary points of view and modes of interpretation. If bodies and lives are historically contingent, deeply informed by culture, discourse, and the political, then they cannot be summed up in any one kind of narrative. There is no clear common ground, no simple foundation in physical human nature. One thing is clear, however: this is no longer the body that stands in a tidy contrastive relationship with the mind. Even if a clean distinction between body and mind was ever possible (and thoughtful rereading of the scholarly record suggests that it was not)[5], what is meant by *the body* today is historically variable, suffused with discourses, thoroughly mindful in its practice.[6] This is not a natural self-contained entity organized by mechanically functioning internal organs; it is not the site of will and personality; it is not the source of needs for food, clothing, and shelter. Or at least, it is not solely or completely any of these things. What the body is, however, cannot be simply stated or presumed in theory on the basis of our own historically located and spatially restricted experience. To make bodies a topic for anthropological, humanistic, sociological, and historical research is to ask how human life can be and has been constructed, imagined, subjectively known—in short, lived.

This book introduces a stellar group of writings that exemplify these changes. Ranging from classic works of social theory, history, and ethnography to more recent explorations of historical and cultural variations in lived embodiment, forty-seven articles and book excerpts are presented here for (re)reading. Each of these works in some way challenges the taken-for-granted situations in which moderns of the Eurocentric world have lived, in certain respects unthinkingly, for a long time. Both in this introduction and in prefatory notes for each of the book's nine parts, we draw out the potential of these humanities and social science approaches to embodiment for expanding understanding of human experience. Here the materialism of Karl Marx, with his emphasis on actual practical activity, joins with a structuralist reading of bodily symbols by Terence Turner in "The Social Skin." Maurice Merleau-Ponty's philosophical insights on the "lived-through" world of the body, beyond the reductive constructs of formal knowledge, inform the comparative history of the senses offered by Shigehisa Kuriyama in his article on Greek and Chinese pulse taking. The temporal bodies of Nuer pastoralists are compared with those of early modern English workers, and Chinese bodies are rendered processual through attention to breath and flavor. Bodies engaged in walking, birthing, art making, sexual foreplay, confessing fantasies, dressing, healing, reading, displaying themselves and being displayed are addressed in these readings. The articles and excerpts included here speak to each other and to the most influential trends in the human sciences in many surprising ways. Taken together, they lead us far beyond the individualist, positivist, and utilitarian presumptions that have thus far dominated our capacity to think about embodiment.

This scholarship challenges a number of classic postulates about human nature. Even putative universal needs for sex and food can be recast as contingent and unpredictable forms of desire, as Judith Farquhar's study of bodily life in postsocialist China has shown.[7] The traditional fixities of social science —invariant symptoms of illness,[8] rational self-interest, the priority of the economic in the structuring of motives[9]—have returned as problems to be explained rather than starting points to be assumed. As anthropology and other human sciences have challenged sociological and biological universals, new "local" bodies have become visible, offering a diverse history and geography of human material Being. Some of these conceptions are already influential in anthropology and science studies: Donna Haraway's widely adopted notion of the cyborg has introduced a local body that is familiar in our lives but transgressive of our linguistic distinctions between machines and humans, objects and subjects.[10] A hybrid of human and machine, flesh and information, the cyborg is a desiring and displaying creature of science fiction who eventually comes to stand for the normal body in the new millennium. Margaret Lock's notion of local biologies, moreover, an idea that demanded attention in her study of experiences of midlife in North America and Japan, has proven productive in medical anthropology. She and other researchers have begun to investigate whole fields of material practice—eating, aging, medical procedures, daily exertions, modes of attention, forms of subjectivity—as producing contingent forms of embodiment. Cyborgs and local biologies are neither purely mental nor purely physical. Rather they are evolving and historical forms of life that are multiple and material at once, refusing all biological reductions and proposing a new politics that seeks solidarity among bodies while refusing to resort to commonsense presumptions about universal bodies or human nature.

The relatively recent field of science studies has also concerned itself with multiple embodiments, finding cultural significance in high-tech practices that are situated near the heart of social life in contemporary society. Emerging technologies ranging from assisted reproduction to organ transplantation produce bodies that are both novel and "normal." Among the most powerful of such technologies are those of molecular genetics, a now-dominant paradigm that has deeply influenced our senses of ourselves in both biomedicine and the popular imagination.[11] One consequence of genetic testing and screening is that individual genomes are transformed into omens of future ill health, resulting in a new category of people, the presymptomatically ill. This is a diagnosis that may eventually include us all. The lived body is made, in effect, into a ticking time bomb. Popular appropriations of genetics thus increasingly find hope and fatality in a coded and mapped body. This is an occasion for all manner of medical and lay interpretations, as essays included in part IX of this volume attest. The body coded and decoded by genetics becomes one of many examples in this book in which embodiment must be seen as not just structural but temporal, not just an objective presence but a moment in a process that is thoroughly social and historical and hence diverse.

THE CATEGORY OF THE BODY IN THE
TWENTIETH-CENTURY HUMAN SCIENCES

A perusal of the literature of the humanities and social sciences published before the latter part of the twentieth century reveals that discussion about the lived body was scant.[12] This was due in large part to an unquestioned acceptance of the body proper. The bounded physical body was usually bracketed and set aside because it fell "naturally" into the domain of the biological sciences. Nevertheless, as the extract by Friedrich Engels that we have selected shows, social thinkers influenced by Charles Darwin were interested in the practical imperatives implied by the evolution of human anatomy. Nineteenth-century biology in general is a rich source of descriptions of the proper body in action. Research of the kind undertaken by Darwin on the expression of emotions and by Cesare Lombroso on criminal types, for example, reveals both the natural science interest in physical variation at the time and the universalizing and normalizing powers of an increasingly hegemonic biomedical worldview.[13] The most extreme form of this natural science of body variations culminated in the eugenic practices of the early twentieth century;[14] arguably social science has continued to collaborate with biomedicine to smuggle normative concerns with race, intelligence, and beauty into policy, clinical practice, and, of course, the desires of biomedicine's so-called consumers.

In contrast to these nineteenth- and twentieth-century biologizing projects, the selections that appear in part I of this book under the title "An Emergent Canon," all of which have become classics in anthropology, show how embodied individuals are thoroughly social. These works tend to draw from a Durkheimian tradition in the social sciences, in that they emphasize the irreducibility of social phenomena to the ends pursued by individuals. In this tradition the body is understood as the "first and most natural tool of man," a carnal template that furnishes the mind with marvelous, irresistible objects and relations, a source of endless symbolic and metaphorical analogies.[15] The most extended social reading of the natural body has been the widely influential work of Mary Douglas.[16] She shows how bodily analogies are utilized in both secular life and sacred events to make the dominant ordering of the social world appear to be natural. The body–society analogy, drawn on in virtually every culture it would seem, provides building blocks for enforcing moral order, as the selections by Robert Hertz, Marcel Granet, and Victor Turner in part I make clear. By focusing on bodies as active, intentional, and signifying forms of cultural life, Emile Durkheim, Marcel Mauss, and their anthropological descendants opened the construct of the natural individual to sociological analysis.

Analyses of metaphoric uses of natural symbols in producing and reproducing the social order have resulted in a substantial literature on the homologous relationships commonly found among physical topography, domestic architecture, social arrangements, deportment, parts of the body, and sexual behavior. Although at first glance it appears that such cultural homologies emphasize holism, inclusion, and unity, closer scrutiny reveals that more often than not hierarchies, difference, and exclusion are commonly naturalized as part of what comes to be understood as the only just and appropriate way to conduct daily life. The essay by Terence Turner, for example, is one of a genre that shows how social categories are literally inscribed on and into the body through precepts about the treatment of body fluids and the appropriate use of hairstyles, ornamentation, cosmetics, clothing, and so on. An oft-cited structural analysis by Pierre Bourdieu of the Kabyle (North Africa) house demonstrates the extent to which gender difference and asymmetry can penetrate every aspect of quotidian space and time. And Victor Turner's analyses of "social dramas" and their bodily correlates in healing have given anthropology some of its best examples of chronic local-level conflicts.[17] However, symbolic and structuralist anthropological methods tend to see culturally specific forms of daily life as affecting rather invariant or even abstract bodies. Though mundane practices are often noted, the multiple, variable lived body does not form part of the discussion—arguments move from the material body as being good to think with, a resource for symbolic language, directly to the social and moral order. Mauss and Victor Turner are partial exceptions in that they rely on a psychologized individual as a vehicle for mediation between the body and society, thus opening a door to inner worlds of experience. Yet they too presume a somewhat ethnocentric "mind" or pattern of needs and motives inhabiting the body they describe.

In contrast to writers who follow the Durkheimian tradition and who are occupied above all with social worlds and moral order, the phenomenological movement that took root in postwar France strives to overcome mind/body and subject/object dualisms. The work of Merleau-Ponty has become emblematic in this respect. (See the excerpt from *The Phenomenology of Perception* in this volume.) Others as well have been influential in Continental philosophy, maintaining a materialist tradition somewhat distinct from twentieth-century debates among Marxists and political economists.[18] In his book *Le Mystère de l'être* (1997), for example, Gabriel Marcel argues that the body, "my body," is always immediately present in experience.[19] To *have* a body means inevitably that one is embodied; consciousness can exist only as mediated through experienced embodiment. The body is never, therefore, simply a physical object but rather an embodiment of consciousness and the site where intention, meaning, and all practice originate. This phenomenological approach tends to center being on a kind of presence (and temporal present) that can be made conscious but is most often taken for granted.[20] Embodiment therefore seems to escape, at least in part, symbolization and discourse. As a result, the lived body as understood by Merleau-Ponty and other phenomenologists stands in strong contrast to the approach of writers who work in the Durkheimian vein. Because its focus is on a body presumed to be universal and individual, depicted from the point of view of the subject, embodiment in this tradition can lack both historical depth and sociological content. The invocation of phenomenology in history and ethnography is thus as often a limitation as an interpretive opportunity for the human sciences.

Bourdieu drew in part on phenomenology in the early 1970s to counter what he understood as a misplaced objectivity on the part of French structuralists, in particular Claude Lévi-Strauss, who was concerned above all with modeling the mental representations that he believed informed, cemented, and ordered social institutions.[21] Bourdieu by contrast showed that practical activity, material objects, and daily life could not be understood simply as reflections or expressions of structures of mind. Using a renovated understanding of Mauss's concept of *habitus*, Bourdieu's approach focuses on relatively inarticulate bodily practices, but he moves away from phenomenology as he engages explicitly with everyday life as it is acted out socially. Along with Michel de Certeau and Norbert Elias, Bourdieu has had a pervasive influence on social scientists as they struggle to show the extent to which embodiment is itself social.[22] His insistence on making practice—a thoroughly temporal and dynamic category—the foundation of his sociological analysis is a continuing inspiration to the anthropology of embodiment.[23] Bourdieu's turn from the body–mind divide, which kept biology and anatomy distinct from symbolism and cultural form, to empirical fields of practice has opened up a new arena for research on social life. For an anthropology of practice, the smallest gestures and the most taken-for-granted circumstances are infused with a historical and cultural significance.

Anthropology has a long tradition of drawing inspiration from neighboring fields, especially history and sociology. Contemporary social-cultural anthropology is also much indebted to recent theoretical advances in feminism as well as in literary and media studies. Perhaps most important, the anthropology of the body has taken inspiration from the work of Michel Foucault, focusing especially on his histories of madness, medicine, and imprisonment.[24] Foucault's accounts of the particular ways of speaking about and of disposing bodies within the institutions, built spaces, and communities of Europe in the eighteenth and nineteenth centuries have provided a carnal and material dimension to the genesis of the modern individual. Foucault argues that the commonsense individual (discussed above) did not simply emerge by throwing off the shackles of a confining medieval past, learning after centuries of darkness to express its essential autonomy and originality; rather our very modern sense that the individual is a natural linkage of material body and "soul" came to be hegemonic only quite recently through the practical work and writing of institutions of medicine and population management. Rather than reading *through* discourses to find underlying meanings, causes that can be posited, or essences that are expressed, Foucault demonstrated in his analysis of medical and welfare discourses that it is possible to learn from the explicit surfaces of the archive how social life changed its forms. His readings of often obscure historical documents show how both subjective and objective discursive practices changed in ways that are deeply consequential for the life of bodies. Foucault's method turned away from the earlier materialist histories (of, for example, the Annales school of historians) that privileged supposedly objective structures over discourse itself, and he also eschewed the idealist intellectual histories that privileged canonical theories and debates. In other words, insofar as the disciplines of the human sciences had organized their methods as either more mental or more bodily, he crafted a new, post-Cartesian historical field that no longer maintained a mind–body divide. His plane of analysis is precisely that which is now being sought out by the anthropology of the body. This is the domain of the taken for granted, the mundane records and routines that fill everyday life, the disciplinary protocols that quietly maintain the (historically contingent) normal (see especially parts II, IV, VII, and IX below). Contemporary research that goes beyond the body proper is thoroughly grounded in Foucault, and readers who do not yet know his work will do well to seek it out separately.

Cultural history has also been much influenced by Foucault, and there has been fruitful communication between cultural anthropology and cultural history. In the latter field, reflection on the methods that can open the lived past to historical analysis has led to innovative ways of reading archival materials. Many of the historical articles included in this volume sensitively explore embodied life beyond the traces found in surviving written and pictorial materials. (See Caroline Walker Bynum, Gregory Pflugfelder, Peter Stallybrass and Allon White, Barbara Duden, and Shigehisa Kuriyama in this volume, for example.) This general approach became important as scholars sought to learn more about the history of difference—feminist history, history of sexualities, subaltern history—through an archive that had mostly represented the voice and the conditions of the literate and powerful. The gender historian Joan Scott has addressed the historiographic challenges posed by the new cultural history of differences in several discussions of the notion of experience.[25] She has shown that experience is a crucial category for researchers who go beyond normative history to document the occluded worlds of the relatively voiceless (women, workers, the colonized), yet she argues that experience has been evoked by historians in naive ways. Even many cultural historians, she suggests, have treated a simple idea of experience as a foundation for empiricist histories, accumulating the evidence of the presumed experience of naturalized historical actors as additions to a growing archive of unproblematic facts. Scott counters this tendency by detailing the arguments of social and linguistic theorists that subjectivity—and thus experience and narrative accounts that claim the authority of experience—is always constructed in social practice. She shows that "experience is at once always already an interpretation *and* something that needs to be interpreted." As she returns the category of experience from the speculations of philosophy to the work of history, Scott also reminds us of the important contributions of feminism to social theory and human sciences scholarship. It still bears pointing out that everywhere bodies are somehow gendered. Families, communities, and societies are crossed by inequalities that are often taken to be rooted in forms of embodiment; thus it was feminist scholars who perhaps most powerfully in the twentieth century forced a political anthropology of the body.[26]

Scott's recasting of the problem of experience has important methodological implications for the human sciences. If we can find no natural invariants in bodily experience, we must continue to explore the shifting terrains presented by many modes of material life. Rather than taking (apparently) empirical bodies and (apparently) interior experience as a starting point (this is, we think, the central limitation of many phenomenological approaches), comparative scholarship in anthropology, history, and the humanities shows that the problem of the body can be read from many kinds of discourses, mundane practices, technologies, and relational networks. The bodies that come into being within these collective formations are social, political, subjective, objective, discursive, narrative, and material all at once. They are also culturally and historically specific, while at the same time mutable, offering many challenges to both scholarship and the everyday politics of a world compressed in time and space. As Haraway has persistently argued, all manner of hopeful alliances are possible once we have learned to see our worldly comrades with an expanded objectivity.[27]

A turn toward embodied lifeworlds in some medical anthropology[28] as well as in feminist and political ethnography[29] has begun to generate important empirical research that adds to our perceptions of human possibilities. An "anthropology of the senses" has emerged to challenge a mentalist bias in symbolic anthropology, for example.[30] Literary attention to "the materiality of the text" and explorations of embodied practices of reading have supplemented more classically humanist forms of reading.[31] Collections of "illness narratives" have appeared which mobilize narrative theories from philosophy and literary criticism.[32] Literary scholarship has thus made it increasingly implausible to separate the silent body from the expressive and articulate consciousness. Further, historical anthropologists have established important connections between political and bodily regimes[33] (see part V), while studies of popular culture have emphasized the extent to which bodily practices are contingent on particular gender and economic inequalities.[34] Perhaps most effectively of all, historians have demonstrated that the twentieth-century Euro-American body (diverse enough when many versions are attended to) is far from natural (see especially part III). The surprise or repulsion with which readers respond to descriptions of some medieval and early modern body practices and imagery makes this point clearly.[35] The body proper—that discrete, structured, individual myth of a European modernity—begins to disappear, to be replaced by an indeterminate site of natural–cultural processes that is full of possibilities and impossible to finally delimit. Not only is the body not singular, it is not very proper either.

A MATERIALISM OF LIVED BODIES

Perhaps what most distinguishes the anthropology of embodiment, and this collection of readings as well, from a more classic cultural anthropology is its commitment to materialist assumptions and methods in the study of cultural form. Yet we speak of materialism in a sense that is different from that of many other anthropologists. For some cultural anthropologists, *materialist* has been a negative epithet to be used against colleagues oriented to natural science methods and the positivist social sciences. Deep divides within anthropology and between its subdisciplines too often have sorted themselves out as debates between materialist and idealist habits of thought. Perhaps only Marxists have crossed this great divide with any success, but it is the rare Marxist anthropologist who departs from the economic sphere to consider other forms of materiality.

Some medical anthropologists use the term *materialism* negatively to express their discomfort with what they see as a biomedical reduction of human distress to structural-anatomic changes. They charge that medical materialism ignores the experience or even the humanity of patients, and they are often joined in this critique by health care workers. (Note especially the humanistic orientations of the "ethic of care" that has developed in nursing, itself an interesting site for locating nondualistic bodies.) But these critiques remain Cartesian in the sense that they tend to portray human ideals—mental phenomena—as violently reduced to the simplistic material level of a structural or mechanical body. By distinguishing human and subjective experiences from material and objective things, they continue to found their critique on a modernist humanism that fails to capture the life of bodies or even, we feel, the complexity of biomedical practice.

Symbolic anthropologists also remain Cartesian in relation to materialism. Their interpretations of ritual and cultural texts tend to dissolve the material world as they decode the concrete, seeking out underlying abstractions. Observable, material signifiers serve only to indicate abstract, ideal signifieds. This is another kind of ideal/material dualism, one which (as Jacques Derrida has shown) insistently privileges the ideal side of the semiotic dyad.[36] Though symbolic anthropological writing is replete with fascinating objects—from jaguars to wooden saints to red and white body paint—the analysis always leads us on to abstractions like social structure, cosmology, or the unconscious. These insistent efforts to stay on the culture side of the nature–culture divide, including approaches that emphasize the cultural constructedness of nature, have made it difficult to think about concrete existence and carnal life in any but reductionist terms.

The newer scholarship on the body in the human sciences, we believe, has advanced a new materialism for anthropology. With the assumptions of Cartesian common sense about bodies and minds, matter and spirit cleared away, it has been possible to approach actual forms of lived embodiment in the fields of practice in which they take form. Ethnographic and historical projects that read and delineate specific material-cultural (bodily) formations do much more than simply relativize cultures. The task only begins with a denaturalizing or "social constructionist" critique. Rather, this new empirical research opens a domain of human experience to the imagination that is at once subjective and objective, carnal and conscious, observable and legible. The problematic of perceiving bodily life in its actual empirical and material forms invites scholars to see social multiplicity more clearly and to adjust our actions more sensitively to the depths at which human Being varies. To make a topic of the body is to study cultural, natural, and historical variation in whole worlds.

Thus an anthropology and geography of space and place has combined environmental awareness with critical attention to the structured and structuring powers of built worlds over and within the bodies that live in them.[37] An ethnographically rich medical anthropology has had to address the messy and very concrete lives of sick people and their caretakers, critiquing the logic of both medicine and sociology with an insistence on understanding embodied practice itself.[38] Sociologists of bioscience and medical systems have also been very effective in showing how objects of medical concern—microbes, organs, diseases—have their own genealogies and a linked array of powers that loop back into reality.[39] As we argued above, a literary criticism that attends to the materiality of texts, their circulation, and their embodied reception has embarked on exploratory and speculative technologies of reading, stepping back from the hermeneutic quest for ultimate or hidden meanings.[40] With actor-network theory in science studies, an understanding of networks made up of both human and nonhuman "actants" has emerged, presenting new questions for the human sciences about the concrete material linkages among bodies, texts, and things.[41] All of these recent efforts could be said to be seeking a new style of materialism, neither reductive and economistic nor sealed off from the traditional humanistic concerns of signification, subjectivity, and ethics.

A materialist anthropology of embodiment is not really reinventing anthropology. It seeks only to indulge a widened curiosity about arenas of life that have previously been kept in the dark. It does not aim to displace either political economy or biological anthropology, nor does it seek to banish anthropologies of consciousness and meaning. The socially constructed formations and experiences it would describe are very real, and we predict that embodied readers will find much to recognize in the bodily lives of even those who are quite remote from them in time or space. However common the "bodies" of this anthropology may turn out to be, they cannot be seen as universal. By presenting them here, we hope to expand the ways we humans can imagine ourselves.

An expanded anthropology of embodiment, one with room for desires and microbes, significance and the taken-for-granted habitus, local biologies and transnational plagues, is needed and pertinent to contemporary scholarship and practical life. There is no shortage of sensitive and theoretically powerful writing that has already begun the task. A critical and ambitious rereading is now in order. Such a reading will not discard earlier concerns about bodily distinctions based on gender, ethnicity, and class, nor will it sideline the materiality of bodies; rather it can build on and further nuance the existing literature by challenging the givenness of many received categories, among which *the body* and *the mind* have too long held pride of place.

PATHS BEYOND THE BODY PROPER

The editors of this volume both came to read and write the anthropology of embodiment through their ethnographic and critical research in medical anthropology.

Judith Farquhar's first major project on the logic of practice in Chinese medicine, based on field and library research in a "traditional" medical college in Guangzhou, China, in the 1980s, required her to acknowledge that a different body seemed to be at issue in Chinese medicine from the one presumed by modern biomedicine.[42] In ancient and recrafted theories of traditional medicine as well as in the reported bodily experiences of patients and dinner companions discourses on embodiment proved easy to grasp and identify with—especially when speaking Chinese—yet they contained almost no reference to anatomical structures or scientifically verified substances. *Qi*, wind, and flavors were more salient to this body than muscles, intestines, and active ingredients in drugs. In later work, which moved outward from the world of medicine to that of popular culture, Farquhar continued to look for a processual body open to the constantly changing world as she explored the everyday life of urbanites in Beijing.[43] Even in this cosmopolitan city, where health education promotes all manner of globally recognized medical information, there are approaches to lived materiality that diverge deeply from what North Americans generally take for granted. In contemporary Chinese society the long, connected history of East Asian civilization is a constantly shifting presence, and all manner of published and broadcast materials offer resources for embodiment that challenge global common sense and cosmopolitan medical information. These resources flow through the experience of modern Beijingers in unique ways. Farquhar's current research on the way of cultivating life (*yangsheng*) in a Beijing neighborhood follows the routes of both traditional and global forms of embodiment and seeks to discover the historicity of bodies in several modes. Embodied memories of a dramatic modern history of revolution, reform, and globalization as well as practical appropriations of classical dance, martial arts, calligraphy, and medicine contribute to lives that combine many threads of history and discourse. By beginning with the everyday life of bodies, this is an anthropology that can lead anywhere except perhaps to the proper body of bourgeois common sense. Working for more than twenty years in China and reflecting for even longer on the foundations of her own experience as an American academic, Farquhar no longer believes that the body proper has ever existed anywhere.

Trained originally as a basic scientist, Margaret Lock had experiences in Japan while doing research in preparation for a Ph.D. in medical anthropology that were not dissimilar to those of Farquhar. While collecting ethnographic data in the 1970s in Kyoto clinics where East Asian medicine is practiced, she was struck by the facility with which practitioners and patients communicated in a discourse that made liberal use of such concepts as *ki* and blocked energy flows while at the same time drawing on the language of biomedicine. Many of the practitioners in these clinics are medical doctors who, like their patients, themselves made use of biomedical services at times.[44] Without apparent conflict, practitioners and patients are able to conceptualize more than one kind of body, and pluralism in medical thought and practices was and is commonplace in Japan and, for that matter, in most parts of the world.[45] Lock's later work, a comparative project carried out in North America and Japan, focused on menopause. The differences in symptom reporting between Japanese and North American women could not be accounted for simply by resorting to an argument for historical and cultural construction. Recognition of the coproduction and interdependence of biology and culture to embodiment was key to this research, although neither biology nor culture was essentialized; both are fluid in time and through space.[46] In recent years Lock has carried out research on death, notably the condition legalized as brain death in order that organs for transplant can be procured from such living/dead entities. Once again comparative research in Japan and North America proved very useful in bringing to the fore the unexamined assumptions about these practices that are present in the dominant thinking in North America.[47] In her current research into molecular genetics and complex disease, in which Alzheimer's disease is her primary (elusive) object of study, Lock joins a rising number of social scientists who are observing the way in which genomics is steadily bringing about an end to simple deterministic arguments so often associated with biomedicine, in particular genetics, while at the same time resisting any critical decentering of the body proper. We have entered an era when we can no longer deny biological variation, as many social scientists have continued to do. Now, more than ever before, it is crucial to pay serious attention to the lived body in its infinite variety.

Together, we venture to claim another characteristic that qualifies us to critically assemble and evaluate the literature on the body: like our readers, we too are embodied. This truism highlights one of the virtues of building an emphasis on embodiment within anthropology: embodiment has the potential to unite readers and writers, anthropologists and informants, doctors and patients, teachers and students. It does not require orientalist distinctions between East and West or developmental differences between North and South. At the level of embodiment, we are all "primitives," all excellent informants for a global anthropology. The commonalities of the carnal, while they cannot be presumed, often go unacknowledged. The extent to which the embodied existence of one party is obscured can be linked to the discomfort or indignity of the other, more obviously embodied person. Doctors hide inside white coats to examine disrobed patients; writers become distant authorities while readers question their own ability to understand the prose; anthro gists have the privilege of mobility and can remove their bodies from "und developed" environments that still threaten the health and livelihood of their research subjects. Still, one cannot really gaze disembodied upon another body; when we see or read about another's pain, we are likely to experience vicarious discomforts; when we read about sex or food, we are likely to experience desire or disgust.

Thus we hope a more focused attention to an anthropology of embodiment will "cut both ways," challenging privilege and its idealization while dignifying bodily existence and chartering a new materialist anthropology. A turn to the body in anthropology has the potential to do more than simply add a topic to the study of Man or append a footnote to our accumulated knowledge about human nature. Rather, as anthropology has moved beyond the body proper it has opened up a new stratum of social existence, one that offers a broad terrain for research between the impossible poles of a Cartesian social science. This is the domain of neither a cultural mind nor a biological body, but of a lively carnality suffused with words, images, senses, desires, and powers.

NOTES

1. Some frequently cited examples are Stallybrass and White 1986; Comaroff 1985; Laqueur 1990; Bynum 1987; Martin 1987; Foucault 1977.

2. See Csordas 1994 for a discussion of the term embodiment.

3. Massumi 2002.

4. The contradictions and historical contingency of this bourgeois individual have been critiqued by Lukes 1979; Mauss in Carrithers et al., eds., 1985; MacPherson 1962; and Lowe 1982, among others.

5. This point is parallel with Bruno Latour's critique of "the modern constitution" in *We Have Never Been Modern* (1993).

6. Lock and Scheper-Hughes 1987.

7. Farquhar 2002.

8. Lock 1993.

9. Stallybrass and White 1986.

10. Haraway 1985.

11. Keller 1992; Konrad 2005; Nelkin and Lindee 1995.

12. See Lock 1993 for a review article and extensive bibliography relating to the production of the social body. See also Turner 1984; Polhemus 1978; Blacking 1977.

13. Darwin 1899; Lombroso and Ferrero, 2004 (1893). See also relevant cultural history by Gould 1981; Gilman (1985, 1988); and Poovey (1995).

14. Proctor 1988.

15. Mauss 1973 (1934): 75.

16. Douglas 1966, 1970.

17. Turner 1957.

18. See, e.g., Gil 1998.

19. Marcel 1997.

20. But note Derrida's critique of the very possibility of presence (1976).

21. Lévi-Strauss 1969a, 1969b.

22. Certeau 1984; Elias 1982.

23. Comaroff 1985; Weiss 1996; B. Turner, 1984; Csordas 1994; Farquhar 2002; Wacquant 2004.

24. Foucault, 1965, 1973, 1977.

25. See, for example, Scott 1991.

26. Boston Women's Health Book Collective 1976; hooks 1981; Irigaray 1985; Kristeva 1982; Rich 1979.

27. Haraway 1991.

28. Good 1995; Csordas 1993.

29. Feldman 1991; Seremetakis 1991; Pandolfi, in this volume; Scarry 1985.

30. Classen 1993; Howes 1991; Stoller 1989.

31. Miller, in this volume; Boyarin 1993; de Man 1986; Derrida 1996.

32. Kleinman 1988; Mattingly and Garro, eds., 2000. See also Good 1994.

33. Comaroff 1985; Stoler 1995.

34. Hebdige 1979; de Lauretis 1987.

35. Bakhtin 1968; Bynum 1989; Stallybrass and White 1986.; Strathern 1992.

36. Derrida 1976.

37. Lefebvre 1991; Bachelard 1964; Low and Lawrence-Zuñiga 2003.

38. Kaufman 2005; Lock 2002;

39. Latour 1979 (1986) and in this volume; Hacking, in this volume; Mol, 2002.

40. See, for example, Liu 1995, 1999; Barker 1984.

41. Latour 1988; Callon 1986; and Law and Hassard 1999. See also C. Thompson, in this volume.

42. Farquhar 1994.

43. Farquhar 2002.

44. Lock 1980.

45. Nichter and Lock 2002; Scheid 2002.

46. Lock 1993.

47. Lock 2002.

PART I

An Emergent Canon, or Putting Bodies

on the Scholarly Agenda

It has been a long struggle in the human sciences to rehabilitate the body as something other than an appendage to mind. In medieval times debate about the body—its relation to soul, whether or not it has transcendental qualities, its incipient fecundity during putrefaction and in death, and its worth once fragmented—was ubiquitous. From the late sixteenth century, however, with the beginnings of modernity, the body was gradually bifurcated. Its earthiness, sensuality, and inspiration for aesthetic and religious expression were set off from the body proper, the object body that became a subject for systematic investigation by the natural sciences. Competing theories existed in various geographical locations, but natural scientists everywhere agreed that the human body must be understood as part of the natural, as opposed to cultural, order, subject to the laws of nature, not history.

René Descartes and like-minded thinkers of his day were in the forefront of this transition, and their debates enabled a style of reasoning that came to full flower in the writings of Julien Offray de la Mettrie, a French philosopher and physician who conceptualized the human body as fully machinelike. It was not until Darwin postulated the theory of biological evolution that bodily knowledge was once again put into motion and thus modified profoundly. Machinelike models were by no means totally displaced, but dimensions of time and space and images of change and progress central to evolutionary theory demanded that the human body be understood as being subject to change—it must evolve, making machine metaphors insufficient.

Both Karl Marx and Friedrich Engels were strong supporters of evolutionary theory, and in his essay "On the Part Played by Labor in the Transition from Ape to Man," written in 1925 and published posthumously, Engels dwells on the importance to evolution of the morphology and uses of the human hand. The freeing up of the hand is, for Engels, the "decisive step in the transition from ape to man," permitting the production of ever more sophisticated tools. But Engels follows Darwin closely and recognizes that the hand is only one part of a complex organism. He makes an association between the biological evolution of humans and changes in the articulation of the hand, upright posture, the development of speech, and changes in the size of the brain. Significantly, Engels goes further and emphasizes the way in which the labor of humans is a social endeavor that permits technological advances critical to human life as a whole. Transformations in individual human bodies and in the "evolution" of societies are interdependent.

Engels's essay has worn well with the passing years. We know now, as Engels did not, that other animals make tools and consequently that this skill is not unique to humans, but the opposable human thumb remains a cornerstone of human evolutionary history. It is currently postulated that this unique anatomical change may have taken place well before an increase in brain size, thus elevating its significance vis-à-vis human evolution. Of course, we no longer believe with Engels that societies as such evolve, but the dominant belief system of our time continues to be one of progress facilitated through technological innovation. Effects of the global transfer of technologies, documented by medical anthropologists, are showing clearly how intimate is the relationship between human biology and societal and technological change, but this relationship is complex and notoriously unpredictable. The Nigerian anthropologist Tola Olu Pearce has shown, for example, the enormous impact of contraceptive devices including birth control pills, injectables, implants, and sterilization practices on population growth as well as on received wisdom about the female body and sexuality in Nigeria. These transformations can be understood as being advantageous by some and damaging by others (1995; see also Bledsoe 2002). A more recent example, the long-lasting side effects of technology gone wrong, is documented by Adriana Petryna. The nuclear disaster at Chernobyl in the Ukraine transformed forever the bodies of more than 3.5 million living individuals as well as those of many more as yet unborn; at the same time it brought about a new form of biological citizenship created out of the need to prove eligibility for compensation (2002).

Writing in a very different vein from that of Engels and his materialist successors, Robert Hertz was a star pupil of Emile Durkheim. His essay "The Pre-eminence of the Right Hand" was published in 1909 in a journal entitled *Revue Philosophique* and has since become an anthropological classic. He follows Durkheim's lead by assuming that "man is double" and that a useful distinction can be made between the universal physical body and the "higher," morally imbued socialized body (Durkheim 1961). Hertz's article was perhaps the first to demonstrate how the body is "good to think with," and it set the stage for numerous subsequent articles that for more than half a century took an essentially similar approach in anthropology. Hertz states emphatically that "organic asymmetry in man is at once a fact and an ideal," but, in elaborating on this claim, he diminishes the biological contribution to hand usage and pursues a discussion that borders on social or even cultural determinism. Hertz argues that culturally informed manual practices and associated intellectual and moral representations about the right and left hands, respectively, are "anterior to all individual experience" and, following Durkheim, are products of the structure of social thought. Perhaps Hertz's most important insight is that even though social hierarchies *claim* to be founded on natural facts, they are nevertheless socially produced and therefore are not immutable. He makes hand usage into a remarkably rich illustrative example of the power of discourses that are legitimized through reference to the natural world, understood as the way things "should" be.

In the 1930s, Marcel Granet, writing about right and left symbolism in China, returned to Hertz's essay and challenged the stark dichotomy that Hertz had created when he stressed that the right hand is universally associated with the sacred and the left with the profane. No such black-and-white distinction exists in China, asserted Granet, who goes on to outline the "anatomy of the world" as it was understood in the Chinese literate tradition. Granet sets out the structural correlations that were made between the order of the universe, society, and the human body, characteristic of much Chinese thinking over the centuries. He focuses in particular on the effects of such thinking on the practice of etiquette in daily life, demonstrating the flexibility and even reversibility of the values associated with left and right and other entrenched dualities.

Numerous essays similar to those of Granet and Hertz have shown how natural symbols are incorporated into systems of thought. In such usages homologies between the natural and social worlds create a moral landscape for bodily behavior that is thoroughly practical and can be mapped through time and space. By reference to such classificatory systems, the social order is legitimized and reproduced, and individual behavior can be judged as morally appropriate or as wanting (see, for example, Griaule 1965; Needham 1973; Bourdieu 1970).

In keeping with this tradition, Terence Turner examines how cosmetics and body adornment among the Amazonian Kayapo, ranging from body painting to penis sheaths, enact a powerfully generative symbolic language. Deploying a rigorous structural analysis, he develops the concept of a "social skin" to describe the "socialization" of the naked human body through adornment. Turner's argument, similar to that of Hertz and Granet, emphasizes what he terms the subordination of the physical body to shared social values and recognized behavioral styles. Body decoration, including the use of symbolically loaded colors, reinforces the normative social order; while there is room for aesthetic creativity, the social hierarchies figured in body aesthetics are left intact.

In all of the above essays the reader is presented with apparently timeless cosmological orders, subject neither to change, threat from the outside, nor radical transformation from within. These authors would write differently today, but the rich data they have left us about classificatory systems and the widespread associations made by premodern peoples between order in the natural and social worlds are invaluable.

Marcel Mauss's essay "Techniques of the Body" returns us briefly to the 1930s to track a radically different approach to the study of embodiment from that of his teacher Durkheim. Mauss was the originator of the idea expressed in the essay included here that the body is the first and most natural tool of man. In using this idiom, Mauss is making self-conscious use of a double entendre in which he refers to use of the body and body parts as classificatory devices that inform the social order. At the same time it is through the body that, as individuals, we experience and transform the world around us. But he goes further than this. Mauss, who will later profoundly influence Pierre Bourdieu, is explicit that it is essential to move from the concrete to the abstract, and not the other way round. His is an early call to pay attention to practice as the main source of knowledge—and it is for this reason, perhaps more than any other, that this essay has become a classic.

There is another reason this essay is so important to an anthropology of the body. Mauss argues explicitly, in contrast to earlier writers, for recognition of a "triple man," one who is not only biological and social but in whom the psychological acts as mediator, bringing about the "total man." In his inimitable way Mauss "parade[s]" before his lecture audience an array of techniques of the body to demonstrate that everywhere we participate in "physio-psycho-sociological assemblages of series of actions." People internalize bodily behavior that then becomes part of the habitus of each individual, but this process of internalization must be set in motion through education in movement that is integral to human socialization in any given society. Here Mauss is explicitly breaking with the form of reasoning passed along from Auguste Comte to Durkheim. Unlike them, Mauss posits the psychological as a mediating factor between the social and the biological, thus creating a small space for the embodied individual as a willful actor with both the potential to resist the normative social order under certain conditions and the capacity to display an irreducibly social character. (See also parts V and VII.)

With the essay by Victor Turner we turn to one of the most influential anthropologists from the British tradition to write about ritual. Like Mauss, Turner is concerned with the relationship between bodily practices and social values, and, again like Mauss, he posits a psychologized individual as a mediating entity—"ritual adapts and periodically readapts the biopsychical individual to the basic conditions and axiomatic values of human social life." In collaboration with Ndembu (East Africa) ritual experts, whose fascination with symbolism he shared, Turner developed an empirically rich symbolic anthropology that drew upon bodily, social, natural, and cosmological domains of life.

Turner's ethnography traces the dense polysemic associations made between natural objects, human body parts, blood and effluvia, and the social status of individuals in Ndembu society. The strength of his essay lies not only in demonstrating these emotionally laden associations, but also in showing their flexibility and selective use depending upon the age, gender, and status of the individual or group of people who are the key participants in specific rituals. For Turner these symbols are "social facts" that bind individuals into the social order, and as such they assist individuals in transcending the contradictions that inevitably exist between their aspirations and desires and the demands of the social groups in which they must participate. The reader is left wondering how individuals who do not conform are dealt with in a society as apparently cohesive as that of the Ndembu. It is also tempting to ask what, if anything, might substitute for this immensely complex system of shared meanings in other societies in which the explicit elaboration of symbol systems is much less evident, as is the case even among neighboring peoples in East Africa.

The authors of these essays share the goal of attempting to create an approach in which the body is, above all, made social. They do not deny the physicality of the body proper, but, with the exception of Engels, they put this fact to one side. These essays set out a position in which it is shown how effectively the body—here still discrete and individual—is co-opted and subordinated to human intentions and to the demands of the social collectivity, as people everywhere seek to create a moral order for daily life and construct meaning out of the worlds in which they live. These goals are achieved, often very successfully, by objectifying the body, but, in contrast to the essays that make up part II, the lived body is not the primary consideration for these canonical authors.

ADDITIONAL READINGS

Descartes, René. 1951. *A Discourse on Method and Selected Writings.* Translated by John
Veitch. New York: E. P. Dutton.

Douglas, Mary. 1970. "The Two Bodies." In *Natural Symbols: Explorations in Cosmology.*
New York: Pantheon.

Durkheim, Emile. 1995 (1912). *The Elementary Forms of the Religious Life.* Translated by
Karen E. Fields. New York: Free Press.

ON THE PART PLAYED BY LABOR IN

THE TRANSITION FROM APE TO MAN

FRIEDRICH ENGELS

Labor is the source of all wealth, the economists assert. It is this—next to nature, which supplies it with the material that it converts into wealth. But it is also infinitely more than this. It is the primary basic condition for all human existence, and this to such an extent that, in a sense, we have to say that labor created man himself.

Many hundreds of thousands of years ago, during an epoch not yet definitely determined, of that period of the earth's history which geologists call the Tertiary period, most likely towards the end of it, a specially highly-developed race of anthropoid apes lived somewhere in the tropical zone—probably on a great continent that has now sunk to the bottom of the Indian Ocean. Darwin has given us an approximate description of these ancestors of ours. They were completely covered with hair, they had beards and pointed ears, and they lived in bands in the trees.

Almost certainly as an immediate consequence of their mode of life, for in climbing the hands fulfill quite different functions from the feet, these apes when moving on level ground began to drop the habit of using their hands and to adopt a more and more erect posture in walking. This was *the decisive step in the transition from ape to man.*

All anthropoid apes of the present day can stand erect and move about on their feet alone, but only in case of need and in a very clumsy way. Their natural gait is in a half-erect posture and includes the use of the hands. The majority rest the knuckles of the fist on the ground and, with legs drawn up, swing the body through their long arms, much as a cripple moves with the aid of crutches. In general, we can to-day still observe among apes all the transition stages from walking on all fours to walking on two legs. But for none of them has the latter method become more than a makeshift.

For erect gait among our hairy ancestors to have become first the rule and in time a necessity pre-supposes that in the meantime the hands became more and more devoted to other functions. Even among the apes there already prevails a certain separation in the employment of the hands and feet. As already mentioned, in climbing the hands are used differently from the feet. The former serve primarily for collecting and holding food, as already occurs in the use of the fore paws among lower mammals. Many monkeys use their hands to build nests for themselves in the trees or even, like the chimpanzee, to construct roofs between the branches for protection against the weather. With their hands they seize hold of clubs to defend themselves against enemies, or bombard the latter with fruits and stones. In captivity, they carry out with their hands a number of simple operations copied from human beings. But it is just here that one sees how great is the gulf between the undeveloped hand of even the most anthropoid of apes and the human hand that has been highly perfected by the labor of hundreds of thousands of years. The number and general arrangement of the bones and muscles are the same in both; but the hand of the lowest savage can perform hundreds of operations that no monkey's hand can imitate. No simian hand has ever fashioned even the crudest stone knife.

At first, therefore, the operations, for which our ancestors gradually learned to adapt their hands during the many thousands of years of transition from ape to man, could only have been very simple. The lowest savages, even those in whom a regression to a more animal-like condition, with a simultaneous physical degeneration, can be assumed to have occurred, are nevertheless far superior to these transitional beings. Before the first flint could be fashioned into a knife by human hands, a period of time must probably have elapsed in comparison with which the historical period known to us appears insignificant. But the decisive step was taken: *the hand became free* and could henceforth attain ever greater dexterity and skill, and the greater flexibility thus acquired was inherited and increased from generation to generation.

Thus the hand is not only the organ of labor, *it is also the product of labor.* Only by labor, by adaptation to ever new operations, by inheritance of the resulting special development of muscles, ligaments, and, over longer periods of time, bones as well, and by the ever-renewed employment of these inherited improvements in new, more and more complicated operations, has the human hand attained the high degree of perfection that has enabled it to conjure into being the pictures of Raphael, the statues of Thorwaldsen, the music of Paganini.

But the hand did not exist by itself. It was only one member of an entire, highly complex organism. And what benefited the hand, benefited also the whole body it served; and this in two ways.

In the first place, the body benefited in consequence of the law of correlation of growth, as Darwin called it. According to this law, particular forms of the individual parts of an organic being are always bound up with certain forms of other parts that apparently have no connection with the first. Thus all animals that have red blood cells without a cell nucleus, and in which the neck is connected to the first vertebra by means of a double articulation (condyles), also without exception possess lacteal glands for suckling their young. Similarly cloven hooves in mammals are regularly associated with the possession of a multiple stomach for rumination. Changes in certain forms involve changes in the form of other parts of the body, although we cannot explain this connection. Perfectly white cats with blue eyes are always, or almost always, deaf. The gradual perfecting of the human hand, and the development that keeps pace with it in the adaptation of the feet for erect gait, has undoubtedly also, by virtue of such correlation, reacted on other parts of the organism. However, this action has as yet been much too little investigated for us to be able to do more here than to state the fact in general terms.

Much more important is the direct, demonstrable reaction of the development of the hand on the rest of the organism. As already said, our simian ancestors were gregarious; it is obviously impossible to seek the derivation of man, the most social of all animals, from non-gregarious immediate ancestors. The mastery over nature, which begins with the development of the hand, with labor, widened man's horizon at every new advance. He was continually discovering new, hitherto unknown, properties of natural objects. On the other hand, the development of labor necessarily helped to bring the members of society closer together by multiplying cases of mutual support, joint activity, and by making clear the advantage of this joint activity to each individual. In short, men in the making arrived at the point where *they had something to say* to one another. The need led to the creation of its organ; the undeveloped larynx of the ape was slowly but surely transformed by means of gradually increased modulation, and the organs of the mouth gradually learned to pronounce one articulate letter after another. . . .

First comes labor, after it, and then side by side with it, articulate speech—these were the two most essential stimuli under the influence of which the brain of the ape gradually changed into that of man, which for all its similarity to the former is far larger and more perfect. Hand in hand with the development of the brain went the development of its most immediate instruments—the sense organs. Just as the gradual development of speech is inevitably accompanied by a corresponding refinement of the organ of hearing, so the development of the brain as a whole is accompanied by a refinement of all the senses. The eagle sees much farther than man, but the human eye sees considerably more in things than does the eye of the eagle. The dog has a far keener sense of smell than man, but it does not distinguish a hundredth part of the odours that for man are definite features of different things. And the sense of touch, which the ape hardly possesses in its crudest initial form, has been developed side by side with the development of the human hand itself, through the medium of labor.

The reaction on labor and speech of the development of the brain and its attendant senses, of the increasing clarity of consciousness, power of abstraction and of judgment, gave an ever-renewed impulse to the further development of both labor and speech. This further development did not reach its conclusion when man finally became distinct from the monkey, but, on the whole, continued to make powerful progress, varying in degree and direction among different peoples and at different times, and here and there even interrupted by a local or temporary regression. This further development has been strongly urged forward, on the one hand, and has been guided along more definite directions on the other hand, owing to a new element which came into play with the appearance of fully fledged man, viz. *society*. . . .

By the co-operation of hands, organs of speech, and brain, not only in each individual, but also in society, human beings became capable of executing more and more complicated operations, and of setting themselves, and achieving, higher and higher aims. With each generation, labor itself became different, more perfect, more diversified. Agriculture was added to hunting and cattle-breeding, then spinning, weaving, metal-working, pottery, and navigation. Along with trade and industry, there appeared finally art and science. From tribes there developed nations and states. Law and politics arose, and with them the fantastic reflection of human things in the human mind: religion. In the face of all these creations, which appeared in the first place to be products of the mind, and which seemed to dominate human society, the more modest productions of the working hand retreated into the background, the more so since the mind that plans the labor process already at a very early stage of development of society (e.g., already in the simple family), was able to have the labor that had been planned carried out by other hands than its own. All merit for the swift advance of civilization was ascribed to the mind, to the development and activity of the brain. Men became accustomed to explain their actions from their thoughts, instead of from their needs—(which in any case are reflected and come to consciousness in the mind)—and so there arose in the course of time that idealistic outlook on the world which, especially since the decline of the ancient world, has dominated men's minds. It still rules them to such a degree that even the most materialistic natural scientists of the Darwinian school are still unable to form any clear idea of the origin of man, because under this ideological influence they do not recognize the part that has been played therein by labor.

THE PRE-EMINENCE OF THE RIGHT HAND:

A STUDY IN RELIGIOUS POLARITY

ROBERT HERTZ

What resemblance more perfect than that between our two hands! And yet what a striking inequality there is!

To the right hand go honors, flattering designations, prerogatives: it acts, orders, and *takes*. The left hand, on the contrary, is despised and reduced to the role of a humble auxiliary: by itself it can do nothing: it helps, it supports, it *holds*.

The right hand is the symbol and model of all aristocracies, the left hand of all plebeians.

What are the titles of nobility of the right hand? And whence comes the servitude of the left?

ORGANIC ASYMMETRY

Every social hierarchy claims to be founded on the nature of things, *physei, ou nomo*: it thus accords itself eternity, it escapes change and the attacks of innovators. Aristotle justified slavery by the ethnic superiority of the Greeks over barbarians; and today the man who is annoyed by feminist claims alleges that woman is *naturally* inferior. Similarly, according to common opinion, the preeminence of the right hand results directly from the organism and owes nothing to convention or to men's changing beliefs. But in spite of appearances the testimony of nature is no more clear or decisive when it is a question of ascribing attributes to the two hands, than in the conflict of races or the sexes.

It is not that attempts have been lacking to assign an anatomical cause to right-handedness. Of all the hypotheses advanced only one seems to have stood up to factual test: that which links the preponderance of the right hand to the greater development in man of the left cerebral hemisphere, which as we know, innervates the muscles of the opposite side. Just as the center for articulate speech is found in this part of the brain, so the centers which govern voluntary movements are held to be also mainly there. As Broca says, "We are right-handed because we are left-brained." The prerogative of the right hand would then be founded on the asymmetric structure of the nervous centers, of which the cause, whatever it may be, is evidently organic.

It is not to be doubted that a regular connection exists between the predominance of the right hand and the superior development of the left part of the brain. But of these two phenomena which is the cause and which the effect? What is there to prevent us turning Broca's proposition round and saying, "We are left-brained because we are right-handed?" It is a known fact that the exercise of an organ leads to the greater nourishment and consequent growth of that organ. The greater activity of the right hand, which involves more intensive work for the left nervous centers, has the necessary effect of favoring its development. If we abstract the effects produced by exercise and acquired habits, the physiological superiority of the left hemisphere is reduced to so little that it can at the most determine a slight preference in favor of the right side.

The difficulty that is experienced in assigning a certain and adequate organic cause to the asymmetry of the upper limbs, joined to the fact that the animals most closely related to man are ambidextrous, has led some authors to deny any anatomical basis for the privilege of the right hand. This privilege would not then be inherent in the structure of *genus homo* but would owe its origin exclusively to conditions external to the organism.

This radical denial is at least bold. The organic cause of right-handedness is dubious and insufficient, and difficult to distinguish from influences which act on the individual from outside and shape him; but this is no reason for dogmatically denying the action of the physical factor. Moreover, in some cases, where external influence and organic tendency are in conflict, it is possible to affirm that the unequal skill of the hands is connected with an anatomical cause. In spite of the forcible and sometimes cruel pressure which society exerts from their childhood on people who are left-handed, they retain all their lives an instinctive preference for the use of the left hand. If we are forced to recognize here the presence of a congenital position to asymmetry we must admit that, inversely, for a certain number of people, the preponderant use of the right hand results from a bodily disposition. The most probable view may be expressed, though not very rigorously, in mathematical form; in a hundred individuals there are about two who are naturally left-handed, resistant to any contrary influence; a considerably larger proportion are right-handed by heredity; while between these two extremes oscillate the mass of people, who if left to themselves would be able to use either hand equally, with (in general) a slight preference in favor of the right. There is thus no need to deny the existence of organic tendencies towards asymmetry; but apart from some exceptional cases the vague disposition to right-handedness, which seems to be spread throughout the human species, would not be enough to bring about the absolute preponderance of the right hand if this were not reinforced and fixed by influences extraneous to the organism.

But even if it were established that the right hand surpassed the left, by a gift of nature, in tactile sensibility, strength, and aptitude, there would still remain to be explained why a humanly instituted privilege should be added to this natural superiority, why only the better-endowed hand is exercised and trained. Would not reason advise the attempt to correct by education the weakness of the less favored member? Quite on the contrary, the left hand is repressed and kept inactive, its development methodically thwarted. Dr. Jacobs tells us that in the course of his tours of medical inspection in the Netherlands Indies he often observed that native children had the left arm completely bound; it was to teach them *not to use it*. We have abolished the material bonds, but that is all. One of the signs which distinguishes a well-brought-up child is that its left hand has become incapable of any independent action.

Can it be said that any effort to develop the aptitude of the left hand is doomed to failure in advance? Experience shows the contrary. In the rare cases in which the left hand is properly exercised and trained, because of technical necessity, it is just about as useful as the right: for example, in playing the piano or violin, or in surgery. If an accident deprives a man of his right hand, the left acquires after some time the strength and skill that it lacked. The example of people who are left-handed is even more conclusive, since this time education struggles against the instinctive tendency to "unidexterity" instead of following and exaggerating it. The consequence is that left-handers are generally ambidextrous and are often noted for their skill. This result would be attained, with even greater reason, by the majority of people, who have no irresistible preference for one side or the other and whose left hand asks only to be used. The methods of bimanual education, which have been applied for some years, particularly in English and American schools, have already shown conclusive results: there is nothing against the left hand receiving an artistic and technical training similar to that which has up to now been the monopoly of the right.

So it is not because the left hand is weak and powerless that it is neglected: the contrary is true. This hand is subjected to a veritable mutilation, which is none the less marked because it affects the function and not the outer form of the organ, because it is physiological and not anatomical. The feelings of a left-hander in a backward society are analogous to those of an uncircumcised man in countries where circumcision is law. The fact is that right-handedness is not simply accepted, submitted to, like a natural necessity; it is an ideal to which everybody must conform and which society forces us to respect by positive sanctions. The child which actively uses its left hand is reprimanded, when it is not slapped on the over bold hand; similarly the fact of being left-handed is an offense which draws ridicule on the offender and a more or less explicit social reproof.

Organic asymmetry in man is at once a fact and an ideal. Anatomy accounts for the fact to the extent that it results from the structure of the organism; but however strong a determinant one may suppose it to be, it is incapable of explaining the origin of the ideal or the reason for its existence.

RELIGIOUS POLARITY

The preponderance of the right hand is obligatory, imposed by coercion, and guaranteed by sanctions; contrarily, a veritable prohibition weighs on the left hand and paralyzes it. The difference in value and function between the two sides of our body possesses therefore in an extreme degree the characteristics of a social institution; and a study which tries to account for it belongs to sociology. More precisely, it is a matter of tracing the genesis of an imperative which is half aesthetic, half moral. Now the secularized ideas which still dominate our conduct were born in a mystical form, in the realm of religious beliefs and emotions. We have therefore to seek the explanation of the preference for the right hand in a comparative study of collective representations.

One fundamental opposition dominates the spiritual world of primitive men—that between the sacred and the profane. Certain beings or objects, by virtue of their nature or by the performance of rites, are as it were impregnated with a special essence which consecrates them, sets them apart and bestows extraordinary powers on them, but which then subjects them to a set of rules and narrow restrictions. Things and persons which are denied this mystical quality have no power, no dignity: they are common and, except for the absolute interdiction on coming into contact with what is sacred, free. Any contact or confusion of beings and things belonging to the opposed classes would be baneful to both. Hence the multitude of prohibitions and taboos which, by keeping them separate, protect both worlds at once.

The significance of the antithesis between profane and sacred varies according to the position in the religious sphere of the mind which classifies beings and evaluates them. Supernatural powers are not all of the same order: some work in harmony with the nature of things, and inspire veneration and confidence by their regularity and majesty; others, on the contrary, violate and disturb the order of the universe, and the respect they impose is founded chiefly on aversion and fear. All these powers have in common the character of being opposed to the profane, to which they are all equally dangerous and forbidden. . . .

Dualism, which is essential to the thought of primitives, dominates their social organization. The two moieties or phratries which constitute the tribe are reciprocally opposed as sacred and profane. Everything that exists within my own phratry is sacred and forbidden to me; this is why I cannot eat my totem, or spill the blood of a member of my phratry, or even touch his corpse, or marry in my clan. Contrarily, the opposite moiety is profane to me; the clans which compose it supply me with provisions, wives, and human sacrificial victims, bury my dead, and prepare my sacred ceremonies. Given the religious character with which a primitive community feels itself invested, the existence of an opposed and complementary segment of the same tribe, which can freely carry out functions which are forbidden to members of the former group, is a necessary condition of social life. The evolution of society replaces this reversible dualism with a rigid hierarchical structure: instead of separate and equivalent clans there appear classes or castes, of which one, at the summit, is essentially sacred, noble, and devoted to superior works, while another, at the bottom, is profane or unclean and engaged in base tasks. The principle by which men are assigned rank and function remains the same: social polarity is still a reflection and a consequence of religious polarity.

The whole universe is divided into two spheres: things, beings, and powers attract or repel each other, implicate or exclude each other, according to whether they gravitate towards the one or the other of the two poles.

Powers which maintain and increase life, which give health, social preeminence, courage in war, and skill in work. All reside in the sacred principle. Contrarily, the profane (in so far as it infringes on the sacred sphere) and the impure are essentially weakening and deadly; the baleful influences which oppress, diminish, and harm individuals come from this side. So on one side there is the pole of strength, good, and life; while on the other there is the pole of weakness, evil and death. Or, if more recent terminology is preferred, on one side gods, on the other, demons.

All the oppositions presented by nature exhibit this fundamental dualism. Light and dark, day and night, east and south in opposition to west and north, represent in imagery and localize in space the two contrary classes of supernatural powers: on one side life shines forth and rises, on the other it descends and is extinguished. There is the same contrast between high and low, sky and earth: on high, the sacred residence of the gods and the stars which know no death: here below, the profane region of mortals whom the earth engulfs: and, lower still, the dark places where lurk serpents and the host of demons.

Primitive thought attributes a sex to all beings in the universe and even to inanimate objects: all of them are divided into two immense classes according to whether they are considered as male or as female. Among the Maori the expression *tama tane*, "male side," designates the most diverse things: men's virility, descent in the paternal line, the east, creative force, offensive magic, and so on; while the expression *tama wahine*, "female side," covers everything that is contrary of these. This cosmic distinction rests on a primordial religious antithesis. In general man is sacred, woman is profane; excluded from religious ceremonies, she is admitted to them only for a function characteristic of her status, when a taboo is to be lifted, i.e., to bring about an intended profanation. But if woman is powerless and passive in the religious order, she has her revenge in the domain of magic: she is particularly fitted for works of sorcery. "All evils, misery, and death," says a Maori proverb, "come from the female element." Thus the two sexes correspond to the sacred and to the profane (or impure), to life and to death. An abyss separates them, and a rigorous division of labor apportions activities between men and women in such a way that there can never be mixing or confusion. . . .

How could man's body, the microcosm, escape the law of polarity which governs everything? Society and the whole universe have a side which is sacred, noble, and precious, and another which is profane and common: a male side, strong and active, and another, female, weak and passive; or, in two words, a right side and a left side—and yet should the human organism alone be symmetrical? A moment's reflection shows us that this is an impossibility. Such an exception would not only be an inexplicable anomaly, it would *ruin* the entire economy of the spiritual world. For man is at the center of creation: it is for him to manipulate and direct for the better the redoubtable forces which bring life and death. Is it conceivable that all these things and these powers, which are separated and contrasted and are mutually exclusive, should be confounded abominably in the hand of the priest or the artisan? It is a vital necessity that neither of the two hands shall know what the other doeth: the evangelical precept merely applies to a particular situation this law of the incompatibility of opposites, which is valid for the whole world of religion.

If organic asymmetry had not existed, it would have had to be invented. . . .

A no less significant concordance links the sides of the body to regions in space. The right represents what is high, the upper world, and the sky; while the left is connected with the underworld and the earth. It is not by chance that in pictures of the Last Judgment it is the Lord's raised right hand that indicates to the elect their sublime abode, while his lowered left hand shows the damned the gaping jaws of Hell ready to swallow them. The relation uniting the right to the east or south and the left to the north or west is even more direct and constant, to the extent that in many languages the same words denote the sides of the body and the cardinal points. The axis which divides the world into two halves, the one radiant and the other dark, also cuts through the human body and divides it between the empire of light and that of darkness. Right and left transcend the limits of our body to embrace the universe.

According to a very widespread idea, at least in the Indo-European area, the community forms a closed circle at the center of which is the altar, the Ark of the Covenant, where the gods descend and from which place divine aid radiates. Within the enclosure reign order and harmony, while outside it extends a vast night, limitless and lawless, full of impure germs and traversed by chaotic forces. On the periphery of the sacred space the worshippers make a ritual circuit round the divine center, their right shoulders turned towards it. They have everything to hope for from one side, everything to fear from the other. The right is the *inside*, the finite, assured well-being, and certain peace; the left is the *outside*, the infinite, hostile, and the perpetual menace of evil.

The above equivalents would in themselves allow us to presume that the right side and the male element are of the same nature, and likewise the left side and the female element, but we are not reduced to simple conjecture on this point. The Maori apply the terms *tama lane* and *tama wahine* to the two sides of the body, terms whose almost universal extension we have already noted; man is compounded of two natures, masculine and feminine; the former is attributed to the right side, the latter to the left. Among the Wulwanga tribe of Australia two sticks are used to mark the beat during ceremonies: one is called the man and is held in the right hand, while the other, the woman, is held in the left. Naturally, it is always the "man" which strikes and the "woman" which receives the blows; the right which acts, the left which submits. Here we find intimately combined the privilege of the strong sex and that of the strong side. Undoubtedly God took one of Adam's left ribs to create Eve, for one and the same essence characterizes woman and the left side of the body—two parts of a weak and defenseless being, somewhat ambiguous and disquieting, destined by nature to a passive and receptive role and to a subordinate condition.

Thus the opposition of right and left has the same meaning and application as the series of contrasts, very different but reducible to common principles, presented by the universe. Sacred power, source of life, truth, beauty, virtue, the rising sun, the male sex, and—I can add—the right side: all these terms are interchangeable, as are their contraries; they designate under many aspects a single category of things, a common nature, the same orientation towards one of the two poles of the mystical world. Is it believable that a slight difference of degree in the physical strength of the two hands should be enough to account for such a trenchant and profound heterogeneity? . . .

THE FUNCTIONS OF THE TWO HANDS

The different characteristics of the right hand and the left determine the difference in rank and functions which exists between the two hands. . . . It is clear that there is no question here of strength or weakness, of skill or clumsiness, but of different and incompatible functions linked to contrary natures. If the left hand is despised and humiliated in the world of the gods and of the living, it has its domain where it commands and from which the right hand is excluded; but this is a dark and ill-famed region. The power of the left hand is always somewhat occult and illegitimate; it inspires terror and revulsion. Its movements are suspect: we should like it to remain quiet and discreet, hidden if possible in the folds of the garment, so that its corruptive influence will not spread. As people in mourning, whom death has enveloped, have to veil themselves, neglect their bodies, and let their hair and nails grow, so it would be out of place to take too much care of the bad hand: the nails are not cut and it is washed less than the other. Thus the belief in a profound disparity between the two hands sometimes goes so far as to produce a visible bodily asymmetry. Even if it is not betrayed by its appearance, the hand of sorcery is always the cursed hand. A left hand that is too gifted and too agile is the sign of a nature contrary to right order, of a perverse and devilish disposition: every left-handed person is a possible sorcerer, properly to be distrusted. To the contrary, the exclusive preponderance of the right, and a repugnance for requiring anything of the left, are the marks of a soul unusually associated with the divine and immune to what is profane or impure: such are the Christian saints who in their cradle were pious to the extent of refusing the left breast of their mother. This is why social selection favors right-handers and why education is directed to paralyzing the left hand while developing the right.

Life in society involves a large number of practices which, without being integrally part of religion, are closely connected with it. If it is the right hands that are joined in a marriage, if the right hand takes the oath, concludes contracts, takes possession, and lends assistance, it is because it is in man's right side that lie the powers and authority which give weight to the gestures, the force by which it exercises its hold on things. How could the left hand conclude valid acts since it is deprived of prestige and spiritual power, since it has strength only for destruction and evil? Marriage contracted with the left hand is a clandestine and irregular union from which only bastards can issue. The left is the hand of perjury, treachery, and fraud. As with jural formalities, so also the rules of etiquette derive directly from worship: the gestures with which we adore the gods serve also to express the feelings of respect and affectionate esteem that we have for one another. In greeting and in friendship we offer the best we have, our right. The king bears the emblems of his sovereignty on his right side; he places at his right those whom he judges most worthy to receive, without polluting them, the precious emanations from his right side. It is because the right and the left are really of different value and dignity that it means so much to present the one or the other to our guests, according to their position in the social hierarchy. All these usages, which today seem to be pure conventions, are explained and acquire meaning if they are related to the beliefs which gave birth to them. . . .

Thus, from one end to the other of the world of humanity, in the sacred places where the worshipper meets his god, in the cursed places where devilish pacts are made, on the throne as well as in the witness-box, on the battlefield and in the peaceful workroom of the weaver, everywhere one unchangeable law governs the functions of the two hands. No more than the profane is allowed to mix with the sacred is the left allowed to trespass on the right. A preponderant activity of the bad hand could only be illegitimate or exceptional; for it would be the end of man and everything else if the profane were ever allowed to prevail over the sacred and death over life. The supremacy of the right hand is at once an effect and a necessary condition of the order which governs and maintains the universe.

CONCLUSION

Analysis of the characteristics of the right and the left, and of the functions attributed to them, has confirmed the thesis of which deduction gave us a glimpse. The obligatory differentiation between the sides of the body is a particular case and a consequence of the dualism which is inherent in primitive thought. But the religious necessities which make the pre-eminence of one of the hands inevitable do not determine which of them will be preferred. How is it that the sacred side should invariably be the right and the profane the left? According to some authors the differentiation of right and left is completely explained by the rules of religious orientation and sun-worship. The position of man in space is neither indifferent nor arbitrary. In his prayers and ceremonies the worshipper looks naturally to the region where the sun rises, the source of all life. Most sacred buildings, in different religions, are turned towards the east. Given this direction, the parts of the body are assigned accordingly to the cardinal points: west is behind, south to the right, and north to the left. Consequently the characteristics of the heavenly regions are reflected in the human body. The full sunlight of the south shines on our right side, while the sinister shade of the north is projected on to our left. The spectacle of nature, the contrast of daylight and darkness, of heat and cold, are held to have taught man to distinguish and to oppose his right and his left.

This explanation rests on outmoded naturalistic conceptions. The external world, with its light and shade, enriches and gives precision to religious notions which issue from the depths of the collective consciousness; but it does not create them. It would be easy to formulate the same hypothesis in more correct terms and to restrict its application to the point that we are concerned with; but it would still run up against contrary facts of a decisive nature. In fact, there is nothing to allow us to assert that the distinctions applied to space are anterior to those that concern man's body. They all have one and the same origin, the opposition of the sacred and the profane: therefore they are usually concordant and support each other; but they are nonetheless independent. We are thus forced to seek in the structure of the organism the dividing line which directs the beneficent flow of supernatural favors towards the right side.

This ultimate recourse to anatomy should not be seen as a contradiction or a concession. It is one thing to explain the nature and origin of a force, it is another to determine the point at which it is applied. The slight physiological advantages possessed by the right hand are merely the occasion of a qualitative differentiation the cause of which lies beyond the individual, in the constitution of the collective consciousness. An almost insignificant bodily asymmetry is enough to turn in one direction and the other contrary representations which are already completely formed. Thereafter, thanks to the plasticity of the organism, social constraint adds to the opposed members, and incorporates in them those qualities of strength and weakness, dexterity and clumsiness [gaucherie], which in the adult appear to spring spontaneously from nature.

The exclusive development of the right hand has sometimes been seen as a characteristic attribute of man and a sign of his moral pre-eminence. In a sense this is true. For centuries the systematic paralyzation of the left arm has, like other mutilations, expressed the will animating man to make the sacred predominate over the profane, to sacrifice the desires and the interest of the individual to the demands felt by the collective consciousness, and to spiritualize the body itself by marking upon it the opposition of values and the violent contrasts of the world of morality. It is because man is a double being— *homo duplex*—that he possesses a right and a left hand that are profoundly differentiated.

RIGHT AND LEFT IN CHINA

MARCEL GRANET

I must begin by offering my excuses to those who may have been misled by the title of my paper and who will be disappointed by it. It is not about right and left in the political senses of these words. I shall not take it upon myself, thankfully, to explain Chinese politics. There is more than enough to do in expounding the mythology of Right and Left in China.

The facts that I shall have to set out are fairly complicated, and I should offer my apologies on this score also. I have chosen this topic partly because Henri Levy-Bruhl asked me for a paper on it, at rather short order, and partly because it provides an occasion for me to recall the fine work of the late and lamented Robert Hertz on the pre-eminence of the right hand.

In the latter study, Hertz maintained two theses. One, which was concerned with physiology, tried to explain the pre-eminence of the right hand, from a physiological point of view, by reasons of a social kind; the other, which was more general, dealt with the classification of religious facts. Hertz postulated an absolute opposition between left and right analogous to that between pure and impure, an opposition which he considered to be more essential than that between sacred and profane. Left and right are opposed absolutely as that which is *right* to that which is *sinister*, as good is to bad. i.e., in a diametrical opposition. Hertz was therefore led to speak, in this connection, of religious polarity.

The Chinese facts have no bearing of interest upon the physiological part of Hertz's thesis, and I shall leave this to one side. At most, I might be able to tell you that if the Chinese are right-handed this has to do with the very reason that Hertz gave: the Chinese are *obligatorily* right-handed—at least in certain respects.

But what may give their case a special interest is that *whereas the Chinese are right-handed, the honorable side for them is the left*. Hertz mentioned this difficulty in his work. Without going into this very complicated question, he simply suggested the idea that since China had an agricultural civilization the reason that the Chinese, though right-handed, preferred the left might perhaps be sought in the techniques of agriculture.

In fact, if there are technical reasons for the preference (which is limited in any case) of the Chinese for the left, these reasons might perhaps better be sought not in the techniques of agriculture but in military techniques.

The point on which the Chinese case presents a particular interest is related to the mythology of right and left: the latter is preferred to the right, but the right is not absolutely inauspicious, nor is the left always auspicious. The diametrical opposition or polarity of which Hertz spoke is not found in China.

The Chinese attribute values to left and right which are unequal and relative to the circumstances, but are always comparable. There is never question of an absolute pre-eminence, but rather of an alternation. This has to do with a number of characteristics of Chinese civilization and thought. There is nothing abstract in Chinese categories, and it would be vain to look among them for diametrical oppositions such as Being and Non-being. Space and time are conceived as a collection of domains, each with its own conventions; instead of absolute oppositions there are only correlations, so that formal indications or counter-indications are recognized, no absolute obligations, no strict taboos. *Everything is a matter of convention, because everything is a matter of what is fitting.*

The problem of right and left has to do with the very general issue of *Etiquette.* In China, etiquette governs both cosmography and physiology. It expresses the structure of the world—and this is no different from the structure of the individual; the architecture of the universe and that of the individual rest on exactly the same principles. It is therefore the anatomy of the world which will explain the alternate pre-eminence of the right and of the left.

I shall start with the most simple facts. The representations concerning the right and the left are strictly obligatory. They have to do with rites, which means that they are part of a body of rules by which attitudes, gestures, modes of existence, and conduct are imposed upon individuals.

The first of these rules or ritual attitudes favors the right. The right hand is the hand for eating with. Let us note at this point an indication in support of Hertz's thesis: as soon as children are capable of picking up food, they *must* be taught to eat with the right hand. The right is thus educated in a way which is intended to give it a certain pre-eminence. This education bears in the first place on activities relating to food. The fact that the right hand is the hand for eating with is well confirmed by the name that the Chinese give to the index finger: this is not the finger for pointing with (something which is dangerous and forbidden); it is the finger for eating with. In order to taste a sauce one dips the index finger into it and sucks this finger.

But next there is a fact which seems to point in the opposite direction (we are in the domain of etiquette, i.e., of complication). When children grow up, they are taught how to greet people. The ritual of salutation is different for boys and for girls. Boys cover the right hand with the left when they bow: they *conceal* the right and present the left. Girls, on the contrary, and no less obligatorily, pay their respects by covering the left hand with the right.

It will already be seen that left and right form part of that great system of bipartite classification, the classification by Yang and Yin. The left is *yang*, it belongs to the male: the right is *yin*, it belongs to the female. *Yang* and the left are male, *yin* and the right are female.

But the ritual of greeting leads us to a new complication. In time of mourning, men make their salutations as though they were women, i.e., in this case they do not present the left hand but the right; the right hand must then cover and conceal the left. An inversion is produced by which the left is associated with the auspicious and the right with the inauspicious. Another rite of greeting and respect consists in uncovering the shoulders, or more exactly in uncovering one shoulder. When one is to be punished one uncovers the right shoulder, and when one attends a joyful ceremony one uncovers the left shoulder. Here again, and in a number of other cases of the same kind, the left is the auspicious side whereas the right is the inauspicious side.

Let us now pass on to another ritual, one of the present day. At this point things begin to get complicated. In general, one gives to the left and one takes on the right. Hence a juridical custom: when two persons make a contract, they divide a slip, a cutting; the left half is kept by the one who has the advantage of the other, i.e., by the creditor, and the right half is kept by the one in the inferior position, i.e. by the debtor. Here again the left is pre-eminent. When it is a question of presents consisting of living things, the ritual is very complicated and there is a tendency to explain it by reasons of convenience which are not in fact at work. Sheep, horses, dogs, and prisoners of war must be presented on a leash and by giving this leash (for it is by giving the leash that possession is transferred). For sheep and horses, the giver has to hold the leash in his right hand, because (it is explained) sheep and horses are inoffensive animals. In order to give a dog, which may bite, the right hand must be free, ready for defense, so the leash is held with the left hand. The reason is the same, it is said, in the case of prisoners of war. But it is hard to believe that the alleged motive of convenience is the true reason. As a matter of fact, prisoners of war whom one gives while holding them by the left hand are persons from whom one has cut, or from whom one is about to cut, the left ear. Presumably, therefore, the ritual is inspired by complex reasons governed by certain religious conceptions.

Some of the most important facts about the etiquette of left and right are to be found in the ritual of oath-taking, of which there are two forms. First there is the oath that is concluded by clasping the right hands (and which is symbolized in writing by the image of two *right* hands). The right, in this case, seems to predominate, and it might be thought that it predominates because it is auspicious, but the facts demand a closer inspection. To swear by gripping in this way is an oath of conjugal or military companionship. It corresponds to a peace-making after a feud. For this reason it seems most often to be completed by the conclusion of a blood-pact, the *blood* being taken from the *right* arm. But in the second type of oath it is the left that prevails. When a solemn oath, one that is binding in law, is sworn under the eyes of the gods, a little blood has to be taken from the victim. This blood must be taken from close to one of the ears, which is obligatorily the *left* ear. The blood is used to anoint the lips, and it is sniffed (sometimes, instead of being taken from the ear of the victim, it is made to flow from the nose of the swearer). The facts differ from those concerning the oath by gripping, in that it is not the *blood* that is the essential thing but the *breath*. What counts is the animation of the oath-taker's word by the breath taken from the victim through the mediation of its blood. This is why the blood is taken (with the aid of a knife with bells on it) from near the organ of hearing, and in this case the left ear is preferred.

We can say, then, that as far as oath-taking is concerned there is a kind of pre-eminence of the right when the hands are involved, and a pre-eminence of the left when it is the ears. (If the Chinese favor the right in the case of the hands [and also the feet], they favor the left in the case of the ears [and also the eyes].) This is because the upper part of the body and the lower part are opposed, for reasons which the doctors will be able to make clear. In China, more than elsewhere, it is necessary for doctors to have (in addition to a technical knowledge, which in itself would not be sufficient) not only a classical education but also a universal science. The principles of their art are based primarily on a knowledge of the macrocosm, and their knowledge of the human body is derived from this.

Now the world does not differ in its structure from the chariot or the house of the Chief. It consists of a roof, which is round (this is the Sky), and a rectangular base (which is the Earth). Between the sky and the earth, connecting them, are one or more columns. A solitary column represents the chief himself; when there are a number of columns (usually four) they stand for ministers, pillars of the state, or for mountains situated at the four corners of space. One of the best known Chinese myths is that of Kong-Kong, an evil minister who rebelled against his sovereign and uprooted a pillar, either the main column or one of the corner columns. He broke Mount Pou-Tcheou, which is a mountain, or a column, situated at the northwest of the Universe. This had grave consequences: Sky and Earth, being no longer Joined by a column to the west, tipped over *in opposite directions*. The Sky leaned towards the west, while the Earth leaned to the east. This explains why it is that the stars move towards the west, and that the rivers of China all flow to the seas to the east. Furthermore, the phenomenon of tipping was complicated by a phenomenon of slipping, so that Sky and Earth are no longer placed one exactly over the other.

This myth has been given learned explanations of an astronomical kind. There is a quite simple explanation, namely, that the myth is intended to account for the fact that the capital, which *ought* to be at the center of the world, is *nevertheless* so situated that at mid-day on the summer solstice the gnomon still throws a shadow. If the world had not been displaced, the gnomon should make no shadow at all at the place which is that of the Chief. But here, now, are the consequences as far as the human body is concerned. Part of the Earth is missing to the west, whereas the Sky is deficient to the east. The Sky is the Above, the Earth is the Below. Now the human body is composed of an above and a below. The head (round) represents the Sky, the feet (rectangular) represent the Earth which they touch. (This is the reason that it has long been forbidden for Chinese sovereigns to display in their court dancers doing the head-stand, for to perform this is precisely to turn the world upside down.) Since the head is the Sky, there is a deficiency in the head, as in the Sky, to the west, whereas close to the Earth, in the lower part of the body, there is a deficiency to the east. *It is enough to know that there is an equivalence between west and right and between east and left*, to see that the right eye must be less good than the left eye, the left ear better than the right ear, and that inversely man must favor the right as far as the feet are concerned—and *also the hands*.

You should not see all this as a simple invention due to scholastic fantasy. The doctors have invented nothing. The idea belongs to ancient folklore and has been translated into the rituals. The practice of cutting off the left ear of prisoners of war is significant; and it may be added that when a bow is drawn at an enemy it is the left eye that is aimed at.

The structure of the microcosm depends exactly, as we have seen, on the structure of the macrocosm. But how is the structure of the latter to be explained? I shall surprise no one here when I say that the macrocosm is explained by the social structure—which is rather complicated. It is from this complication that there arises the alternative preference for the Right or for the Left.

The social structure is ordered by two great principles: (1) by oppositions governed by the category of sex and which are symbolized by the opposition of the Yin and the Yang; (2) by oppositions resulting from the hierarchical organization of the society and corresponding to the opposition between inferior and superior. Here therefore are two couples—Yin and Yang, High and Low—in which the opposition is of a cyclic type and results in an alternation. With these couples is combined the couple Left-Right (not to speak of other couples such as Before and Behind).

Let us start with the opposition of Above and Below, i.e., of superior and inferior. The image of Space, the image of the World, is formed after the representation of the assemblies held by the Chief when he receives his vassals. The Chief holds his reception standing on a dais with his back to the north and his face to the south, i.e., facing the light or the Yang; the vassals prostrate themselves facing north, towards the Yin, and instead of holding their heads up to the Sky they must press their faces against the Earth. Hence a series of equivalences: the Above is equivalent to the Sky, and also to Yang, for when the Chief stands facing the south he receives the full rays of the sun; he thus assimilates the Yang, the luminous principle—it follows also that the *front* of the body is *yang*, that the chest is *yang*. Inversely, the Below is equivalent to Earth, which is equivalent to *Yin*, which is equivalent to Behind, which is equivalent to the back.

All of this evokes an extremely important myth which bears on an essential theme in Chinese mythology, namely, the theme of hierogamy. Sky, Yang, and male are characterized by the fact that they cover, that they embrace and press against their *chest*; Earth, Yin and female, on the contrary, present the *back* and carry on the back. I should mention here that Earth is a mother and Sky is a father, that the mother supplies the *blood* and the father the *breath*. Hence a series of equivalences which are most important in Chinese medicine: the Yang corresponds to the chest, and in the chest is the heart, which is the organ of the *breath*. Thus the doctors know that the heart is a simple vital organ, the Yang corresponding to the uneven. Conversely, the back is Yin: it is associated with the blood, or rather with all the fertile humors, and the vital organs corresponding to the back are the kidneys. The kidneys thus form a double vital organ (even = double) which is associated, moreover, with dancing and with the feet, which are made for touching the Earth (= Yin). . . .

For the rest, Yang and Yin, in their hierogamies, are not reduced to any single position (and this has happily permitted the doctors to adopt a very large number of other equivalences useful for arriving at diagnoses). In one of the most famous books of medicine the back becomes yang, through an inversion in location, and the heart thereby shifts to the back, for it could not remain in the chest and thus become yin (even) without being duplicated.

Here we see a primary opposition, namely, the opposition of Above and Below, Chief and Vassal. The Chief is associated with the south, the Vassal with the north. You will notice that it is he who stands to the north but faces south who is under the influence of South, whereas he who is to the south and faces north is under the influence of North.

The other opposition is that between men and women. Men place themselves to the west, i.e., facing the direction of the rising sun; they are thus equivalent to the east even though they are placed to the west. Women, conversely, who are placed to the east, are equivalent to the direction of the setting sun, the west, a fact which gives rise to a number of ambiguities and complications in etiquette. The Chief, who stands with his face to the south, *has the east to his left*, from which it follows that East equals Left and that West equals Right. These are absolute equivalences which are always valid. The Chief is not limited to holding receptions, standing on his dais. The Chief is an *archer* (which is the title of Chinese lords), and by this title he is mythologically a Sun, especially a rising sun. Consequently, though he faces the south, he is a person of the east. Conversely, the vassals opposite him, while being people who turn their faces to the north, are also people of the west. Hence an essential correlation, viz., the connection of east with south (Yang), and of west with north (Yin). . . .

The structure of the world does not in itself explain the facts concerning the etiquette of Right and Left. The world has a structure, a morphology, which depends on the social structure. It has also a physiology, the essential law of which is a *principle of rotation*, namely, the rhythmic and cyclic alternation of the Yin and the Yang. The principle of etiquette is therefore to render manifest the structural identity of the macrocosm and the microcosms, while taking into account the physiological modifications of the macrocosm, which correspond to different eras, and to changes in the world order, which means in the order of civilization. There are times at which the Yang governs, and there are times which are ruled by the Yin, and at each alternation the principles of etiquette are completely reversed; where the Right previously dominated, the Left comes into the ascendant, and vice versa.

Thus we find in China none of that distrust or hatred of left-handers which is characteristic of other cultures. A left-hander is worth as much as a right-hander. More exactly, there are cultural eras, or physiological phases of the Universe, in which it is fitting to be left-handed, and other phases in which it is appropriate to be right-handed. There are several instructive myths on this point. There are generally six in the family of Suns, and when they give birth it happens that three are born on the left and three on the right. Those who emerge from the right of the maternal body are entirely right-handed, and those who come out of the left are all left-handed. This is to be understood in the most absolute sense. The hero is right-handed, or left-handed, to the point of being hemiplegic. He is alive on only the right, or the left, side of the body. He will be a spirit of the Left or a spirit of the Right. . . .

Such are the ideas which govern mythology and history—for what I have called mythical facts are regarded by a fair number of our contemporaries as historical. But what is valid for mythology or for history is also valid for medicine. Here is the way children are born. The principle of conception corresponds to the point which represents true north, midnight, and the winter solstice. Both male and female come from this and depart from this. The male (Yang) sets off to the left, and the female to the right. Men marry at thirty, women marry at twenty. If we count thirty stations to the left on the chart of the twelve cyclic characters, starting from *tseu* (child, midnight, the first cyclic character), we arrive at the cyclic character *sseu*, and if we count twenty stations on the other side, to the right, we arrive again at the same cyclic character: the female and the male thus meet at the ages of twenty and of thirty respectively, at the character *sseu*. This character represents the embryo; it marks the station appropriate to *real* conceptions. Children, male and female, are therefore born at *sseu*. If the child is a male, it will continue to turn towards the *left*, and as it has to be born at ten months (the Chinese include the first month in their reckoning) the place of its birth will be the cyclic character *yin*: the place of birth of a girl (this time we go to the *right*) will be the cyclic character *chen*. On the chart of the cyclic characters the decimal numbers corresponding to the cyclic numbers *chen* and *yin* are 7 and 8. The entire life of a woman is dominated by the number 7: girls teethe at 7 months, and lose their milk-teeth at 7 years; they are nubile at 14, and the menopause occurs at 49. Boys are dominated by the number 8: their teeth appear at 8 months and are lost at 8 years; they arrive at puberty at 16, and become impotent at 64. This is not a simple invention, either, resulting from the scholastic ingenuity of the doctors, but is implicit in many very ancient rites and a long-lived folklore. You have seen that the one who goes towards the right is the female; the one that goes to the left is the male. When Chinese doctors are asked to discern the sex of a child while it is still in the womb, they can answer without hesitation. All they need to do is to see whether the embryo lies to the left or to the right in the mother's belly. If it is to the left, it is a boy; if it is to the right, it is a girl. When an embryo moves towards the right, this is because it belongs on the right. One that belongs to the left has to move to the left. . . .

This assemblage of facts regarding the mythology of Left and Right brings out the structural correlation which is established in China between the Universe, the human body, and society; all of this, the morphology and physiology of the macrocosm and the microcosms, forms the domain of Etiquette.

Never do we find absolute oppositions: a left-hander is not sinister, and neither is a right-hander. A multitude of rules show the left and the right as predominating alternately. The diversity of times and places imposes, at any point, a very delicate choice between left and right, but *this choice is inspired by a very coherent system of representations.*

Let me illustrate this point with a final example, taken this time from the ritual of serving at table. How should fish be served? According to whether the fish is fresh or dried, matters are entirely different. If it is dried fish, the head must be turned towards the guest. But if it is fresh fish, it is the tail which must be turned towards the guest. Nor is this all: the season has to be taken into account also. If it is summertime, the belly of the fish must be turned to the left; if it is winter, to the right. This is why: winter is the reign of Yin, and Yin, as we have seen, corresponds to the Below; the belly (even though it forms part of the front) is the underneath of the fish; therefore it is *yin.* During winter, in which the Yin reigns, the belly should be the best-nourished part, the fattest and most succulent. The fish is placed with its belly to the right in winter because one has to eat with the right hand, and one begins by eating the good parts. The most succulent morsel must therefore be to the right. In summer, when Yang reigns, everything changes.

We can see, then, how minute are the rules of etiquette. The preeminence of the right or of the left depends always on events, on the occasional circumstances of time and place. I have been able to explain the rule governing service at table because I have found an analysis of it in a competent author of antiquity: it would have been impossible to reconstitute by an act of the imagination the reasons which serve to justify these rules. I shall conclude, therefore, by remarking that when it is a question of etiquette, i.e., of symbolism, any attempt at an ideological interpretation is dangerous. There is only one valid interpretation, namely, that which is given by those who combine a practice of the etiquette with a direct knowledge of the system of symbols by which it is inspired.

TECHNIQUES OF THE BODY

MARCEL MAUSS

CHAPTER ONE: THE NOTION OF TECHNIQUES OF THE BODY

I deliberately say techniques of the body in the plural because it is possible to produce a theory of *the* technique of the body in the singular on the basis of a study, an exposition, a description pure and simple of techniques of the body in the plural. By this expression I mean the ways in which from society to society men know how to use their bodies. In any case, it is essential to move from the concrete to the abstract and not the other way round.

I want to convey to you what I believe is one of the parts of my teaching which is not to be found elsewhere, that I have rehearsed in a course of lectures on descriptive ethnology (the books containing the *Summary instructions* and *Instructions for Ethnographers* are to be published) and have tried out several times in my teaching at the Institut d'Ethnologie of the University of Paris.

When a natural science makes advances, it only ever does so in the direction of the concrete, and always in the direction of the unknown. Now the unknown is found at the frontiers of the sciences, where the professors are at each other's throats, as Goethe put it (though Goethe was not so polite). It is generally in these ill-demarcated domains that the urgent problems lie. Moreover, these uncleared lands are marked. In the natural sciences at present, there is always one obnoxious rubric. There is always a moment when, the science of certain facts not being yet reduced into concepts, the facts not even being organically grouped together, these masses of facts receive that posting of ignorance: "Miscellaneous." This is where we have to penetrate. We can be certain that this is where there are truths to be discovered: first because we know that we are ignorant, and second because we have a lively sense of the quantity of the facts. For many years in my course in descriptive ethnology, I have had to teach in the shadow of the disgrace and opprobrium of the "miscellaneous" in a matter in which in ethnography this rubric "miscellaneous" was truly heteroclite.

I was well aware that walking or swimming, for example, and all sorts of things of the same type, are specific to determinate societies; that the Polynesians do not swim as we do, that my generation did not swim as the present generation does. But what social phenomena did these represent? They were "miscellaneous" social phenomena, and, as this rubric is a horror, I have often thought about this "miscellaneous," at least as often as I have been obliged to discuss it, and often in between times.

Forgive me if, in order to give this notion of techniques of the body shape for you, I tell you about the occasions on which I pursued this general problem and how I managed to pose it clearly. It was a series of steps consciously and unconsciously taken. First, in 1898, I came into contact with someone whose initials I still know, but whose name I can no longer remember. I have been too lazy to look it up. It was the man who wrote an excellent article on "Swimming" for the 1902 edition of the *Encyclopaedia Britannica*, then in preparation. (The articles on "Swimming" in the two later editions are not so good.) He revealed to me the historical and ethno-graphical interest of the question. It was a starting-point, an observational framework. Subsequently—I noticed it myself—we have seen swimming techniques undergo a change, in our generation's life-time. An example will put us in the picture straight away: us, the psychologists, as well as the biologists and sociologists. Previously we were taught to dive after having learned to swim. And when we were learning to dive, we were taught to close our eyes and then to open them under water. Today the technique is the other way round. The whole training begins by getting the children used to keeping their eyes open under water. Thus, even before they can swim, particular care is taken to get the children to control their dangerous but instinctive ocular reflexes, before all else they are familiarized with the water, their fears are suppressed, a certain confidence is created, suspensions and movements are selected. Hence there is a technique of diving and a technique of education in diving which have been discovered in my day. And you can see that it really is a technical education and, as in every technique, there is an apprenticeship in swimming. On the other hand, here our generation has witnessed a complete change in technique; we have seen the breast-stroke with the head out of the water replaced by the different sorts of crawl. Moreover, the habit of swallowing water and spitting it out again has gone. In my day swimmers thought of themselves as a kind of steam-boat. It was stupid, but in fact I still do this: I cannot get rid of my technique. Here then we have a specific technique of the body, a gymnic art perfected in our own day.

But this specificity is characteristic of all techniques. An example: during the War I was able to make many observations on this specificity of techniques.

E.g., the technique of digging. The English troops I was with did not know how to use French spades, which forced us to change 8,000 spades a division when we relieved a French division, and vice versa. This plainly shows that a manual knack can only be learned slowly. Every technique properly so-called has its own form. But the same is true of every attitude of the body. Each society has its own special habits. In the same period I had many opportunities to note the differences between the various armies. An anecdote about marching. You all know that the British infantry marches with a different step from our own: with a different frequency and a different stride. For the moment I am not talking about the English swing or the action of the knees, etc. The Worcester Regiment, having achieved considerable glory alongside French infantry in the Battle of the Aisne, requested Royal permission to have French trumpets and drums, a band of French buglers and drummers. The result was not very encouraging. For nearly six months, in the streets of Bailleul long after the Battle of the Aisne, I often saw the following sight: the regiment had preserved its English march but had set it to a French rhythm. It even had at the head of its band a little French light infantry regimental sergeant major who could blow the bugle and sound the march even better than his men. The unfortunate regiment of tall Englishmen could not march. Their gait was completely at odds. When they tried to march in step, the music would be out of step. With the result that the Worcester Regiment was forced to give up its French buglers. In fact, the bugle-calls adopted army by army earlier, in the Crimean War, were the calls "at ease," "retreat," etc. Thus I saw in a very precise and frequent fashion, not only with the ordinary march, but also at the double and so on, the differences in elementary as well as sporting techniques between the English and the French. Prince Curt Sachs, who is living here in France at present, made the same observation. He has discussed it in several of his lectures. He could recognize the gait of an Englishman and a Frenchman from a long distance.

But these were only approaches to the subject.

A kind of revelation came to me in hospital. I was ill in New York. I wondered where previously I had seen girls walking as my nurses walked. I had the time to think about it. At last I realized that it was at the cinema. Returning to France, I noticed how common this gait was especially in Paris; the girls were French and they too were walking in this way. In fact, American walking fashions had begun to arrive over here, thanks to the cinema. This was an idea I could generalize. The positions of the arms and hands while walking form a social idiosyncrasy, they are not simply a product of some purely individual, almost completely psychical arrangements and mechanisms. For example: I think I can also recognize a girl who has been raised in a convent. In general she will walk with her fists closed. And I can still remember my third-form teacher shouting at me: "Idiot! why do you walk around the whole time with your hands flapping wide open?" Thus there exists an education in walking, too.

Another example: there are polite and impolite positions for the hands at rest. Thus you can be certain that if a child at table keeps his elbows in when he is not eating he is English. A young Frenchman has no idea how to sit up straight; his elbows stick out sideways; he puts them on the table, and so on.

Finally, in running, too, I have seen, you all have seen, the change in technique. Imagine, my gymnastics teacher, one of the top graduates of Joinville around 1860, taught me to run with my fists close to my chest: a movement completely contradictory to all running movements; I had to see the professional runners of 1890 before I realized the necessity of running in a different fashion.

Hence I have had this notion of the social nature of the "habitus" for many years. Please note that I use the Latin word—it should be understood in France —*habitus*. The word translates infinitely better than "*habitude*" (habit or custom), the "*exis*," the "acquired ability" and "faculty" of Aristotle (who was a psychologist). It does not designate those metaphysical *habitudes*, that mysterious "memory," the subjects of volumes or short and famous theses. These "habits" do not just vary with individuals and their limitations, they vary especially between societies, educations, proprieties and fashions, prestiges. In them we should see the techniques and work of collective and individual practical reason rather than, in the ordinary way, merely the soul and its repetitive faculties.

Thus everything moved me towards the position that we in this Society are among those who have adopted, following Comte's example, the position of [Georges] Dumas, for example, who, in the constant relations between the biological and the sociological, leaves but little room for the psychological mediator. And I concluded that it was not possible to have a clear idea of all these facts about running, swimming, etc., unless one introduced a triple consideration instead of a single consideration, be it mechanical and physical, like an anatomical and physiological theory of walking, or on the contrary psychological or sociological. It is the triple viewpoint, that of the "total man," that is needed.

Lastly, another series of facts impressed itself upon me. In all these elements of the art of using the human body, the facts of *education* were dominant. The notion of education could be superimposed on that of imitation. For there are particular children with very strong imitative faculties, others with very weak ones, but all of them go through the same education, such that we can understand the continuity of the concatenations. What takes place is a prestigious imitation. The child, the adult, imitates actions which have succeeded and which he has seen successfully performed by people in whom he has confidence and who have authority over him. The action is imposed from without, from above, even if it is an exclusively biological action, involving his body. The individual borrows the series of movements which constitute it from the action executed in front of him or with him by others. It is precisely this notion of the prestige of the person who performs the ordered, authorized, tested action *vis-à-vis* the imitating individual that contains all the social element. The imitative action which follows contains the psychological element and the biological element.

But the whole, the ensemble, is conditioned by the three elements indissolubly mixed together.

All this is easily linked to a number of other facts. In a book by Elsdon Best that reached here in 1925 there is a remarkable document on the way Maori women (New Zealand) walk. (Do not say that they are primitives, for in some ways I think they are superior to the Celts and Germans.) "Native women adopted a peculiar gait" (the English word is delightful) "that was acquired in youth, a loose-jointed swinging of the hips that looks ungainly to us, but was admired by the Maori. Mothers drilled their daughters in this accomplishment, termed *onioni*, and I have heard a mother say to her girl: '*Ha! Kaore koe e onioni*' (you are not doing the *onioni*) when the young one was neglecting to practice the gait. This was an acquired, not a natural way of walking." To sum up, there is perhaps no "natural way" for the adult. *A fortiori* when other technical facts intervene: to take ourselves, the fact that we wear shoes to walk transforms the positions of our feet: we feel it sure enough when we walk without them.

On the other hand, this same basic question arose for me in a different region, *vis-à-vis* all the notions concerning magical power, beliefs in the not only physical but also moral, magical and ritual effectiveness of certain actions. Here I am perhaps even more on my own terrain than on the adventurous terrain of the psycho-physiology of modes of walking, which is a risky one for me in this company.

Here is a more "primitive" fact, Australian this time: a ritual formula both for hunting and for running. As you will know, the Australian manages to outrun kangaroos, emus, and wild dogs. He manages to catch the possum or phalanger at the top of its tree, even though the animal puts up a remarkable resistance. One of these running rituals, observed a hundred years ago, is that of the hunt for the dingo or wild dog among the tribes near Adelaide. The hunter constantly shouts the following formula:

> Strike (him, i.e. the dingo) with the tuft of eagle feathers (used in initiation, etc.)
> Strike (him) with the girdle, Strike (him) with the string round the head
> Strike (him) with the blood of circumcision
> Strike (him) with the blood of the arm
> Strike (him) with menstrual blood
> Send (him) to sleep, etc.

In another ceremony, that of the possum hunt, the individual carries in his mouth a piece of rock crystal (*kawemukha*), a particularly magical stone, and chants a formula of the same kind, and it is with this support that he is able to dislodge the possum, that he climbs the tree and can stay hanging on to it by his belt, that he can outlast and catch and kill this difficult prey.

The relations between magical procedures and hunting techniques are clear, too universal to need stressing.

The psychological phenomenon I am reporting at this moment is clearly only too easy to know and understand from the normal point of view of the sociologist. But what I want to get at now is the confidence, the psychological *momentum* that can be linked to an action which is primarily a fact of biological resistance, obtained thanks to some words and a magical object.

Technical action, physical action, magico-religious action are confused for the actor. These are the elements I had at my disposal.

All this did not satisfy me. I saw how everything could be described, but not how it could be organized; I did not know what name, what title to give it all.

It was very simple, I just had to refer to the division of traditional actions into techniques and rites, which I believe to be well founded. All these modes of action were techniques, the techniques of the body.

I made, and went on making for several years, the fundamental mistake of thinking that there is technique only when there is an instrument. I had to go back to ancient notions, to the Platonic position on technique, for Plato spoke of a technique of music and in particular of a technique of the dance, and extend these notions. I call technique an action which is *effective* and *traditional* (and you will see that in this it is no different from a magical, religious or symbolic action). It has to be *effective* and *traditional*. There is no technique and no transmission in the absence of tradition. This above all is what distinguishes man from the animals: the transmission of his techniques and very probably their oral transmission.

Allow me, therefore, to assume that you accept my definitions. But what is the difference between the effective traditional action of religion, the symbolic or juridical effective traditional action, the actions of life in common, moral actions on the one hand and the traditional actions of technique on the other? It is that the latter are felt by the author as *actions of a mechanical, physical or physico-chemical order* and that they are pursued with that aim in view.

In this case all that need be said is quite simply that we are dealing with *techniques of the body*. The body is man's first and most natural instrument. Or more accurately, not to speak of instruments, man's first and most natural technical object, and at the same time technical means, is his body. Immediately this whole broad category of what I classified in descriptive sociology as "miscellaneous" disappeared from that rubric and took shape and body: we now know where to range it.

Before instrumental techniques there is the ensemble of techniques of the body. I am not exaggerating the importance of this kind of work, the work of psycho-sociological taxonomy. But it is something: order put into ideas where there was none before. Even inside this grouping of facts, the principle made possible a precise classification. The constant adaptation to a physical, mechanical or chemical aim (e.g. when we drink) is pursued in a series of assembled actions, and assembled for the individual not by himself alone but by all his education, by the whole society to which he belongs, in the place he occupies in it.

Moreover, all these techniques were easily arranged in a system which is common to us, the notion basic to psychologists, particularly [William Halse] Rivers and [Sir Henry] Head, of the symbolic life of the mind; the notion we have of the activity of the consciousness as being above all a system of symbolic assemblages.

I should never stop if I tried to demonstrate to you all the facts that might be listed to make visible this concourse of the body and moral or intellectual symbols. Here let us look for a moment at ourselves. Everything in us all is under command. I am a lecturer for you; you can tell it from my sitting posture and my voice, and you are listening to me seated and in silence. We have a set of permissible or impermissible, natural or unnatural attitudes. Thus we should attribute different values to the act of staring fixedly: a symbol of politeness in the army, and the rudeness of everyday life.

CHAPTER TWO: PRINCIPLES OF THE CLASSIFICATION OF TECHNIQUES OF THE BODY

Two things were immediately apparent given the notion of techniques of the body; they are divided and vary by sex and by age.

I. SEXUAL DIVISION OF TECHNIQUES OF THE BODY (AND NOT JUST SEXUAL DIVISION OF LABOR)

This is a fairly broad subject. The observations of [Robert Mearns] Yerkes and [Wolfgang] Kohler on the position of objects with respect to the body, and especially to the groin, in monkeys provide inspiration for a general disquisition on the different attitudes of the moving body with respect to moving objects in the two sexes. Besides, there are classical observations of man himself on this point. They need to be supplemented. Allow me to suggest this series of investigations to my psychologist friends. I am not very competent in this field and also my time is otherwise engaged. Take the way of closing the fist. A man normally closes his fist with the thumb outside, a woman with her thumb inside; perhaps because she has not been taught to do it, but I am sure that if she were taught, it would prove difficult. Her punching, her delivery of a punch, is weak. And everyone knows that a woman's throwing, of a stone for example, is not just weak, but always different from that of a man: in a vertical instead of a horizontal plane.

Perhaps this is a case of two instructions. For there is a society of men and a society of women. However, I believe that there are also perhaps biological and psychological things involved as well. But there again, the psychologist alone will only be able to give dubious explanations, and he will need the collaboration of two neighboring sciences: physiology, sociology.

2. VARIATIONS OF TECHNIQUES OF THE BODY WITH AGE

The child normally squats. We no longer know how to. I believe that this is an absurdity and an inferiority of our races, civilizations, societies. An example: I lived at the front with Australians (whites). They had one considerable advantage over me. When we made a stop in mud or water, they could sit down on their heels to rest, and the "*flotte*," as it was called, stayed below their heels. I was forced to stay standing up in my boots with my whole foot in the water. The squatting position is, in my opinion, an interesting one that could be preserved in a child. It is a very stupid mistake to take it away from him. All mankind, excepting only our societies, has so preserved it.

It seems besides that in the series of ages of the human race this posture has also changed in importance. You will remember that curvature of the lower limbs was once regarded as a sign of degeneration. A physiological explanation has been given for this racial characteristic. What even [Rudolf Ludwig Karl] Virchow still regarded as an unfortunate degenerate and is in fact simply what is now called Neanderthal man, had curved legs. This is because he normally lived in a squatting position. Hence there are things which we believe to be of a hereditary kind which are in reality physiological, psychological or sociological in kind. A certain form of the tendons and even of the bones is simply the result of certain forms of posture and repose. This is clear enough. By this procedure, it is possible not only to classify techniques, but also to classify their variations by age and sex.

Having established this classification, which cuts across all classes of society, we can now glimpse a third one.

3. CLASSIFICATION OF TECHNIQUES OF THE BODY ACCORDING TO EFFICIENCY

The techniques of the body can be classified according to their efficiency, i.e. according to the results of training. Training, like the assembly of a machine, is the search for, the acquisition of an efficiency. Here it is a human efficiency. These techniques are thus human norms of human training. These procedures that we apply to animals men voluntarily apply to themselves and to their children. The latter are probably the first beings to have been trained in this way, before all the animals which first had to be tamed. As a result I could to a certain extent compare these techniques, them and their transmission, to training systems, and rank them in the order of their effectiveness.

This is the place for the notion of dexterity, so important in psychology, as well as in sociology. But in French we only have the poor term "*habile*," which is a bad translation of the Latin word "*habilis*," far better designating those people with a sense of the adaptation of all their well-coordinated movements to a goal, who are practiced, who "know what they are up to." The English notions of "craft" or "cleverness" (skill, presence of mind and habit combined) imply competence at something. Once again we are clearly in the technical domain.

4. TRANSMISSION OF THE FORM OF THE TECHNIQUES

One last viewpoint: the teaching of techniques being essential, we can classify them according to the nature of this education and training, Here is a new field of studies: masses of details which have not been observed, but should be, constitute the physical education of all ages and both sexes. The child's education is full of so-called details, which are really essential. Take the problem of ambidextrousness for example: our observations of the movements of the right hand and of the left hand are poor and we do not know how much all of them are acquired. A pious Muslim can easily be recognized: even when he has a knife and fork (which is rarely), he will go to any lengths to avoid using anything but his right hand. He must never touch his food with his left hand, or certain parts of his body with his right. To know why he does not make a certain gesture and does make a certain other gesture neither the physiology nor the psychology of motor asymmetry in man is enough, it is also necessary to know the traditions which impose it. Robert Hertz has posed this problem correctly. But reflections of this and other kinds can be applied whenever there is a social choice of the principles of movements. There are grounds for studying all the modes of training, imitation and especially those fundamental fashions that can be called the modes of life, the *model*, the *tonus*, the "matter," the "manners," the "way." Here is the first classification, or rather, four viewpoints.

CHAPTER THREE: A BIOGRAPHICAL LIST OF THE TECHNIQUES OF THE BODY

Another quite different classification is, I would not say more logical, but easier for the observer. It is a simple list. I had thought of presenting to you a series of small tables, of the kind American professors construct. I shall simply follow more or less the ages of man, the normal biography of an individual, as an arrangement of the techniques of the body which concern him or which he is taught.

I. TECHNIQUES OF BIRTH AND OBSTETRICS

The facts are rather little known, and much of the classical information is disputable. Among the best is that of Walter Roth on the Australian tribes of Queensland and on those of British Guiana.

The forms of obstetrics are very variable. The infant Buddha was born with his mother Maya upright and clinging to the branch of a tree. She gave birth standing up. Indian women still in the main give birth in this position. Something we think of as normal, like giving birth lying on one's back, is no more normal than doing so in other positions, e.g., on all fours. There are techniques of giving birth, both on the mother's part and on that of her helpers, of holding the baby, cutting and tying the umbilical cord, caring for the mother, caring for the child. Here are quite a number of questions of some importance. And here are some more: the choice of the child, the exposure of weaklings, the killing of twins are decisive moments in the history of a race. In ancient history and in other civilizations, the recognition of the child is a crucial event.

2. TECHNIQUES OF INFANCY

Rearing and feeding the child. Attitudes of the two inter-related beings; mother and child. Take the child, suckling, etc., carrying, etc. The history of carrying is very important. A child carried next to its mother's skin for two or three years has a quite different attitude to its mother from that of a child not so carried. It has a contact with its mother utterly unlike our children's. It clings to her neck, her shoulder, it sits astride her hip. This remarkable gymnastics is essential throughout its life. And there is another gymnastics for the mother carrying it. It even seems that psychical states arise here which have disappeared from infancy with us. There are sexual contacts, skin contacts, etc.

Weaning. Takes a long time, usually two or three years. The obligation to suckle, sometimes even to suckle animals. It takes a long time for the mother's milk to run dry. Besides this there are relations between weaning and reproduction, suspensions of reproduction during weaning. Mankind can more or less be divided into people with cradles and people without. For there are techniques of the body which presuppose an instrument. Countries with cradles include almost all the peoples of the two Northern hemispheres, those of the Andean region, and also a certain number of Central African populations. In these last two groups, the use of the cradle coincides with a cranial deformation (which perhaps has serious physiological consequences).

The weaned child. It can eat and drink; it is taught to walk; it is trained in vision, hearing, in a sense of rhythm and form and movement, often for dancing and music.

It acquires the notions and practices of physical exercise and breathing. It takes certain postures which are often imposed on it.

3. TECHNIQUES OF ADOLESCENCE

To be observed with men in particular. Less important with girls in those societies to whose study a course in Ethnology is devoted. The big moment in the education of the body is, in fact, the moment of initiation. Because of the way our boys and girls are brought up we imagine that both acquire the same manners and postures and receive the same training everywhere. The idea is already erroneous about ourselves—and it is totally false in so-called primitive countries. Moreover, we describe the facts as if something like our own school, beginning straight away and intended to protect the child and train it for life, had always and everywhere existed. The opposite is the rule. For example: in all black societies the education of the boy intensifies around the age of puberty, while that of women remains traditional, so to speak. There is no school for women. They are at school with their mothers and are formed there continuously, moving directly, with few exceptions, to the married state. The male child enters the society of men where he learns his profession, especially the profession of arms. However, for men as well as women, the decisive moment is that of adolescence. It is at this moment that they learn definitively the techniques of the body that they will retain for the whole of their adult lives.

4. TECHNIQUES OF ADULT LIFE

To list these we can run through the various moments of the day among which coordinated movements and suspensions of movement are distributed. We can distinguish sleep and waking, and in waking, rest and activity.

Technique of sleep. The notion that going to sleep is something natural is totally inaccurate. I can tell you that the War taught me to sleep anywhere, on heaps of stones for example, but that I have never been able to change my bed without a moment of insomnia: only on the second night can I go to sleep quickly.

One thing is very simple: it is possible to distinguish between those societies that have nothing to sleep on except the "floor," and those that have instrumental assistance. The "civilization of latitude 15" discussed by Graebner is characterized among other things by its use of a bench for the neck. This neck-rest is often a totem, sometimes carved with squatting figures of men and totemic animals. There are people with mats and people without (Asia, Oceania, part of America). There are people with pillows and people without. There are populations which lie very close together in a *ring* to sleep, round a fire, or even without a fire. There are primitive ways of getting warm and keeping the feet warm. The Fuegians, who live in a very cold region, cannot warm their feet while they are asleep, having only one blanket of skin (*guanoco*). Finally there is sleep standing up. The Masai can sleep on their feet. I have slept standing up in the mountains. I have often slept on a horse, even sometimes a moving horse: the horse was more intelligent than I was. The old chroniclers of the invasions picture the Huns and Mongols sleeping on horseback. This is still true, and their riders' sleeping does not stop the horses' progress.

There is the use of coverings. People who sleep covered and uncovered. There is the hammock and the way of sleeping hanging up.

Here are a large number of practices which are both techniques of the body and also have profound biological echoes and effects. All this can and must be observed on the ground; hundreds of things still remain to be discovered.

Waking: Techniques of rest. Rest can be perfect rest or a mere suspension of activity: lying down, sitting, squatting, etc. Try squatting. You will realize the torture that a Moroccan meal, for example, eaten according to all the rituals, would cause you. The way of sitting down is fundamental. You can distinguish squatting mankind and sitting mankind. And, in the latter, people with benches and people without benches and daises; people with chairs and people without chairs. Wooden chairs supported by crouching figures are widespread, curiously enough, in all the regions at fifteen degrees of latitude North and along the Equator in both continents. There are people who have tables and people who do not. The table, the Greek "*trapeza*," is far from universal. Normally it is still a carpet, a mat, throughout the East, This is all complicated, for these forms of rest include meals, conversation, etc. Certain societies take their rest in very peculiar positions. Thus, the whole of Nilotic Africa and part of the Chad region, all the way to Tanganyika, is populated by men who rest in the fields like storks. Some manage to rest on one foot without a pole, others lean on a stick. These resting techniques form real characteristics of civilizations, common to a large number of them, to whole families of peoples. Nothing seems more natural to the psychologists; I do not know if they would quite agree with me, but I believe that these postures in the savannah are due to the height of the grasses there and the functions of shepherd or sentry, etc.; they are laboriously acquired by education and preserved.

You have active, generally aesthetic rest; thus even dancing at rest is frequent, etc. I shall return to this.

Techniques of activity, of movement. By definition, rest is the absence of movements, movement the absence of rest. Here is a straightforward list:

Movements of the whole body: climbing; trampling; walking.

—*Walking:* the *habitus* of the body being upright while walking, breathing, rhythm of the walk, swinging the fists, the elbows, progression with the trunk, in advance of the body or by advancing either side of the body alternately (we have got accustomed to moving all the body forward at once). Feet in or out. Extension of the leg. We laugh at the "goose-step." It is the way the German Army can obtain the maximum extension of the leg, given in particular that all

Northerners, high on their legs, like to make steps as long as possible. In the absence of these exercises, we Frenchmen remain more or less knock-kneed. Here is one of those idiosyncrasies which are simultaneously matters of race, of individual mentality and of collective mentality. Techniques such as those of the about-turn are among the most curious. The about-turn "on principle" English-style is so different from our own that it takes considerable study to master it.

—*Running*. Position of the feet, position of the arms, breathing, running magic, endurance. In Washington I saw the chief of the Fire Fraternity of the Hopi Indians who had arrived with four of his men to protest against the prohibition of the use of certain alcoholic liquors in their ceremonies. He was certainly the best runner in the world. He had run 256 miles without stopping. All the Pueblos are accustomed to prodigious physical feats of all kinds. [Henri] Hubert, who had seen them, compared them physically with Japanese athletes. This same Indian was an incomparable dancer.

Finally we reach techniques of active rest which are not simply a matter of aesthetics, but also of bodily games.

—*Dancing*. You have perhaps attended the lectures of M. [Erich Maria] von Hornbostel and M. Curt Sachs. I recommend to you the latter's very fine history of dancing. I accept their division into dances at rest and dances in action. I am less prepared to accept their hypothesis about the distribution of these dances. They are victims to the fundamental error which is the mainstay of a whole section of sociology. There are supposed to be societies with exclusively masculine descent and others with exclusively uterine descent. The uterine ones, being feminized, tend to dance on the spot; the others, with descent by the male, take their pleasure in moving about.

Curt Sachs has better classified these dances into extravert and introvert dances. We are plunged straight into psychoanalysis, which is probably quite well-founded here. In fact the sociologist has to see things in a more complex way. Thus, the Polynesians and in particular the Maori, shake very greatly, even on the spot, or move about very much when they have the space to do so.

Men's dancing and women's dancing should be distinguished, for they are often opposed.

Lastly we should realize that dancing in a partner's arms is a product of modern European civilization. Which shows you that things we find natural are historical. Moreover, they horrify everyone in the world but ourselves.

I move on to the techniques of the body which are also a function of vocations and part of vocations or more complex techniques.

—*Jumping*. We have witnessed a transformation of jumping techniques. We all jumped from a spring-board and, once again, full-face. I am glad to say that this has stopped. Now people jump, fortunately, from one side. Jumping lengthways, sideways, up and down. Standing jump, pole-jump. Here was return to the objects of the reflections of our friends [Wolfgang] Kohler, [Paul] Guillaume and [Ignace] Meyerson: the comparative psychology of man and animals. I won't say anything more about it. These techniques are infinitely variable.

—*Climbing*. I can tell you that I'm very bad at climbing trees, though reasonable on mountains and rocks. A difference of education and hence of method.

A method of getting up trees with a belt encircling the tree and the body is crucial among all so-called primitives. But we do not have the use of this belt. We see telegraph workers climbing with crampons, but no belt. This procedure should be taught them.

The history of mountaineering methods is very noteworthy. It has made fabulous progress in my lifetime.

—*Descent*. Nothing makes me so dizzy as watching a Kabyle going down-stairs in Turkish slippers (*babouches*). How can he keep his feet without the slippers coming off? I have tried to see, to do it, but I can't understand.

Nor can I understand how women can walk in high heels. Thus there is a lot even to be observed, let alone compared.

—*Swimming*. I have told you what I think. Diving, swimming; use of supplementary means; air-floats, planks, etc. We are on the way to the invention of navigation. I was one of those who criticized the de Rouges' book on Australia, demonstrated their plagiarisms, believed they were grossly inaccurate. Along with so many others I held their story for a fable: they had seen the Niol-Niol (N.W. Australia) riding cavalcades of great sea-turtles. But now we have excellent photographs in which these people can be seen riding turtles. In the same way [Robert Sutherland] Rattray noted the story of pieces of wood on which people swim among the Ashanti. Moreover, it has been confirmed for the natives of almost all the lagoons of Guinea, Porto-Novo in our own colonies.

—*Forceful movements*. Pushing, pulling, lifting. Everyone knows what a back-heave is. It is an acquired technique, not just a series of movements. Throwing, up or along the ground, etc., the way of holding the object to be thrown between the fingers is noteworthy and undergoes great variation.

—*Holding*. Holding between the teeth. Use of the toes. the arm-pit, etc.

....

This study of mechanical movements has got off to a good start. It is the formation of mechanical "pairs of elements" with the body. You will recall [Franz] Reuleaux's great theory about the formation of these pairs of elements. And here the great name of [Louis-Hubert] Farabeuf will not be forgotten. As soon as I use my fist, and *a fortiori*, when a man had a "Chellean hand-axe" in his hand, these "pairs of elements" are formed.

This is the place for conjuring tricks, sleight of hand, athletics, acrobatics, etc. I must tell you that I had and still have a great admiration for jugglers and gymnasts.

Techniques of care for the body. Rubbing, washing, soaping. This dossier is hardly a day old. The inventors of soap were not the Ancients, they did not use it. It was the Gauls. And on the other hand, independently, in the whole of Central and North East of South America they soaped themselves with *quillaia* bark or "brazil," hence the name of the empire.

—*Care of the mouth.* Coughing and spitting technique. Here is a personal observation. A little girl did not know how to spit and this made every cold she had much worse. I made inquiries. In her father's village and in her father's family in particular, in Berry, people do not know how to spit. I taught her to spit. I gave her four *sous* per spit. As she was saving up for a bicycle she learned to spit. She is the first person in her family who knows how to spit.

—*Hygiene in the needs of nature.*
Here I could list innumerable facts for you.

Consumption techniques.
—*Eating.* You will remember the story [Harald] Hoffding repeats about the Shah of Persia. The Shah was the guest of Napoleon III and insisted on eating with his fingers. The Emperor urged him to use a golden fork. "You don't know what a pleasure you are missing," the Shah replied. Absence and use of knives. An enormous factual error is made by [W. J.] McGee, who believed he had observed that the Seri (Indians of the Madeleine Peninsula, California), having no notion of knives, were the most primitive human beings. They did not have knives for eating, that is all.

—*Drinking.* It would be very useful to teach children to drink straight from the source, the fountain, etc., or from puddles of water, etc., to pour their drinks straight down their throats, etc.

Techniques of Reproduction.
Nothing is more technical than sexual positions. Very few writers have had the courage to discuss this Question. We should be grateful to M. [Friedrich

Saloman] Krauss for having published his great collection of *Anthropophyteia*. Consider for example the technique of the sexual position consisting of this: the woman's legs hang by the knees from the man's elbows. It is a technique *specific* to the whole Pacific, from Australia to lower Peru, via the Behring Straits—very rare, so to speak, elsewhere.

There are all the techniques of normal and abnormal sexual acts. Contact of the sexual organs, mingling of breath, kisses, etc. Here sexual techniques and sexual morals are closely related.

Lastly there are the *techniques of the care of the abnormal:* massages, etc. But let us move on.

CHAPTER FOUR: GENERAL CONSIDERATIONS

General questions may perhaps be of more interest to you than these lists of techniques that I have paraded before you at rather too great a length.

What emerges very clearly from them is the fact that we are everywhere faced with physio-psycho-sociological assemblages of series of actions. These actions are more or less habitual and more or less ancient in the life of the individual and the history of the society.

Let us go further: one of the reasons why these series may more easily be assembled in the individual is precisely because they are assembled by and for social authority. As a corporal this is how I taught the reason for exercise in close order, marching four abreast and in step. I ordered the soldiers not to march in step drawn up in ranks and in two files four abreast, and I obliged the squad to pass between two of the trees in the courtyard. They marched on top of one another. They realized that what they were being made to do was not so stupid. In group life as a whole there is a kind of education of movements in close order.

In every society, everyone knows and has to know and learn what he has to do in all conditions. Naturally, social life is not exempt from stupidity and abnormalities. Error may be a principle. The French Navy only recently began to teach its sailors to swim. But example and order, that is the principle. Hence there is a strong sociological causality in all these facts. I hope you will accept that I am right.

On the other hand, since these are movements of the body, this all presupposes an enormous biological and physiological apparatus. What is the breadth of the linking psychological cogwheel? I deliberately say cogwheel. A Comtian would say that there is no gap between the social and the biological. What I can tell you is that here I see psychological facts as connecting cogs and not as causes, except in moments of creation or reform. Cases of invention, of laying down principles, are rare. Cases of adaptation are an individual psychological matter. But in general they are governed by education, and at least by the circumstances of life in common, of contact.

On the other hand there are two big questions on the agenda for psychology: the question of individual capacities, of technical orientation, and the question of salient features, of bio-typology, which may concur with the brief investigations I have just made. The great advances of psychology in the last few years have not, in my opinion, been made *vis-à-vis* each of the so-called faculties of psychology, but in psychotechnics, and in the analysis of psychological "wholes."

Here the ethnologist comes up against the big questions of the psychical possibilities of such a race and such a biology of such a people. These are fundamental questions. I believe that here, too, whatever the appearances, we are dealing with biologico-sociological phenomena. I think that the basic education in all these techniques consists of an adaptation of the body to their use. For example, the great tests of stoicism, etc., which constitute initiation for the majority of mankind, have as their aim to teach composure, resistance, seriousness, presence of mind, dignity, etc. The main utility I see in my erstwhile mountaineering was this education of my composure, which enabled me to sleep upright on the narrowest ledge overlooking an abyss.

I believe that this whole notion of the education of races that are selected on the basis of a determinate efficiency is one of the fundamental moments of history itself: education of the vision, education in walking—ascending, descending, running. It consists especially of education in composure. And the latter is above all a retarding mechanism, a mechanism inhibiting disorderly movements; this retardation subsequently allows a coordinated response of coordinated movements setting off in the direction of a chosen goal. This resistance to emotional seizure is something fundamental in social and mental life. It separates out, it even classifies the so-called primitive societies: according to whether they display more brutal, unreflected, unconscious reactions or on the contrary more isolated, precise actions governed by a clear consciousness. It is thanks to society that there is an intervention of consciousness.

It is not thanks to unconsciousness that there is an intervention of society. It is thanks to society that there is the certainty of pre-prepared movements, domination of the conscious over emotion and unconsciousness. It is right that the French Navy is now to make it obligatory for its sailors to learn to swim, From here we easily move on to much more philosophical problems. I don't know whether you have paid attention to what our friend [Marcel]

Granet has already pointed out in his great investigations into the techniques of Taoism, its techniques of the body, breathing techniques in particular. I have studied the Sanskrit texts of Yoga enough to know that the same things occur in India. I believe precisely that at the bottom of all our mystical states there are techniques of the body which we have not studied, but which were perfectly studied by China and India, even in very remote periods. This socio-psycho-biological study should be made. I think that there are necessarily biological means of entering into "communication with God." Although in the end breath technique, etc., is only the basic aspect in India and China, I believe this technique is much more widespread. At any rate, on this point we have the methods to understand a great many facts which we have not understood hitherto. I even believe that all the recent discoveries in reflex therapy deserve our attention, ours, the sociologists,' as well as that of biologists and psychologists . . . much more competent than ourselves.

SYMBOLS IN NDEMBU RITUAL

VICTOR TURNER

Among the Ndembu of Zambia (formerly Northern Rhodesia), the importance of ritual in the lives of the villagers in 1952 was striking. Hardly a week passed in a small neighborhood, without a ritual drum being heard in one or another of its villages. By "ritual" I mean prescribed formal behavior for occasions not given over to technological routine, having reference to beliefs in mystical beings or powers. The symbol is the smallest unit of ritual which still retains the specific properties of ritual behavior; it is the ultimate unit of specific structure in a ritual context. Since this essay is in the main a description and analysis of the structure and properties of symbols, it will be enough to state here, following the *Concise Oxford Dictionary*, that a "symbol" is a thing regarded by general consent as naturally typifying or representing or recalling something by possession of analogous qualities or by association in fact or thought. The symbols I observed in the field were, empirically, objects, activities, relationships, events, gestures, and spatial units in a ritual situation.

Following the advice and example of Professor Monica Wilson, I asked Ndembu specialists as well as laymen to interpret the symbols of their ritual. As a result, I obtained much exegetic material. I felt that it was methodologically important to keep observational and interpretative materials distinct from one another. The reason for this will soon become apparent.

I found that I could not analyze ritual symbols without studying them in time series in relation to other "events," for symbols are essentially involved in social process. I came to see performances of ritual as distinct phases in the social processes whereby groups became adjusted to internal changes and adapted to their external environment. From this standpoint the ritual symbol becomes a factor in social action, a positive force in an activity field. The symbol becomes associated with human interests, purposes, ends, and means, whether these are explicitly formulated or have to be inferred from the observed behavior. The structure and properties of a symbol become those of a dynamic entity, at least within its appropriate context of action.

The structure and properties of ritual symbols may be inferred from three classes of data: (1) external form and observable characteristics; (2) interpretations offered by specialists and by laymen; (3) significant contexts largely worked out by the anthropologist.

Here is an example. At *Nkang'a*, the girl's puberty ritual, a novice is wrapped in a blanket and laid at the foot of a *mudyi* sapling. The *mudyi* tree *Diplorhyncus condylocarpon* is conspicuous for its white latex, which exudes in milky beads if the thin bark is scratched. For Ndembu this is its most important observable characteristic, and therefore I propose to call it "the milk tree" henceforward. Most Ndembu women can attribute several meanings to this tree. In the first place, they say that the milk tree is the "senior" (*mukulumpi*) tree of the ritual. Each kind of ritual has this "senior" or, as I will call it, "dominant" symbol. Such symbols fall into a special class which I will discuss more fully later. Here it is enough to state that dominant symbols are regarded not merely as means to the fulfillment of the avowed purposes of a given ritual, but also and more importantly refer to values that are regarded as ends in themselves, that is, to axiomatic values. Secondly, the women say with reference to its observable characteristics that the milk tree stands for human breast milk and also for the breasts that supply it. They relate this meaning to the fact that *Nkang'a* is performed when a girl's breasts begin to ripen, not after her first menstruation, which is the subject of another and less elaborate ritual. The main theme of *Nkang'a* is indeed the tie of nurturing between mother and child, not the bond of birth. This theme of nurturing is expressed at *Nkang'a* in a number of supplementary symbols indicative of the act of feeding and of foodstuff. In the third place, the women describe the milk tree as "the tree of a mother and her child." Here the reference has shifted from description of a biological act, breast feeding, to a social tie of profound significance both in domestic relations and in the structure of the widest Ndembu community. This latter meaning is brought out most clearly in a text I recorded from a male ritual specialist. I translate literally.

> The milk tree is the place of all mothers of the lineage (*ivumu*, literally "womb" or "stomach"). It represents the ancestress of women and men. The milk tree is where our ancestress slept when she was initiated. "To initiate" here means the dancing of women round and round the milk tree where the novice sleeps. One ancestress after another slept there down to our grandmother and our mother and ourselves the children. That is the

place of our tribal custom (*muchidi*), where we began, even men just the same, for men are circumcised under a milk tree.

This text brings out clearly those meanings of the milk tree which refer to principles and values of social organization. At one level of abstraction the milk tree stands for matriliny, the principle on which the continuity of Ndembu society depends. Matriliny governs succession to office and inheritance of property, and it vests dominant rights of residence in local units. More than any other principle of social organization it confers order and structure on Ndembu social life. Beyond this, however, "*mudyi*" means more than matriliny, both according to this text and according to many other statements I have collected. It stands for tribal custom (*muchidi wetu*) itself. The principle of matriliny, the backbone of Ndembu social organization, as an element in the semantic structure of the milk tree, itself symbolizes the total system of interrelations between groups and persons that makes up Ndembu society. Some of the meanings of important symbols may themselves be symbols, each with its own system of meanings. At its highest level of abstraction, therefore, the milk tree stands for the unity and continuity of Ndembu society. Both men and women are components of that spatiotemporal continuum. Perhaps that is why one educated Ndembu, trying to cross the gap between our cultures, explained to me that the milk tree was like the flag above the administrative headquarters. "*Mudyi* is our flag," he said.

When discussing the milk tree symbolism in the context of the girls' puberty ritual, informants tend to stress the harmonizing, cohesive aspects of the milk tree symbolism. They also stress the aspect of dependence. The child depends on its mother for nutriment; similarly, say the Ndembu, the tribesman drinks from the breasts of tribal custom. Thus nourishment and learning are equated in the meaning content of the milk tree. I have often heard the milk tree compared to "going to school"; the child is said to swallow instruction as a baby swallows milk and *kapudyi*, the thin cassava gruel Ndembu liken to milk. Do we not ourselves speak of "a thirst for knowledge?" Here the milk tree is a shorthand for the process of instruction in tribal matters that follows the critical episode in both boys' and girls' initiation—circumcision in the case of the boys and the long trial of lying motionless in that of the girls. The mother's role is the archetype of protector, nourisher, and teacher. For example, a chief is often referred to as the "mother of his people," while the hunter-doctor who initiates a novice into a hunting cult is called "the mother of huntsmanship (*mama dawnuyang'a*)." An apprentice circumciser is referred to as "child of the circumcision medicine" and his instructor as "mother of the circumcision medicine." In all the senses hitherto described, the milk tree represents harmonious, benevolent aspects of domestic and tribal life.

However, when the third mode of interpretation, contextual analysis, is applied, the interpretations of informants are contradicted by the way people actually behave with reference to the milk tree. It becomes clear that the milk tree represents aspects of social differentiation and even opposition between the components of a society which ideally it is supposed to symbolize as a harmonious whole. The first relevant context we shall examine is the role of the milk tree in a series of action situations within the framework of the girls' puberty ritual. Symbols, as I have said, produce action, and dominant symbols tend to become focuses in interaction. Groups mobilize around them, worship before them, perform other symbolic activities near them, and add other symbolic objects to them, often to make composite shrines. Usually these groups of participants themselves stand for important components of the secular social system, whether these components consist of corporate groups, such as families and lineages, or of mere categories of persons possessing similar characteristics, such as old men, women, children, hunters, or widows. In each kind of Ndembu ritual a different group or category becomes the focal social element. In *Nkang'a* this focal element is the unity of Ndembu women. It is the women who dance around the milk tree and initiate the recumbent novice by making her the hub of their whirling circle. Not only is the milk tree the "flag of the Ndembu"; more specifically, in the early phases of *Nkang'a*, it is the "flag" of Ndembu women. In this situation it does more than focus the exclusiveness of women; it mobilizes them in opposition to the men. For the women sing songs taunting the men and for a time will not let men dance in their circle. Therefore, if we are to take account of the operational aspect of the milk tree symbol, including not only what Ndembu say about it but also what they do with it in its "meaning," we must allow that it distinguishes women as a social category and indicates their solidarity.

The milk tree makes further discriminations. For example, in certain action contexts it stands for the novice herself. One such context is the initial sacralization of a specific milk tree sapling. Here the natural property of the tree's immaturity is significant. Informants say that a young tree is chosen because the novice is young. A girl's particular tree symbolizes her new social personality as a mature woman. In the past and occasionally today, the girl's puberty ritual was part of her marriage ritual, and marriage marked her transition from girlhood to womanhood. Much of the training and most of the symbolism of *Nkang'a* are concerned with making the girl a sexually accomplished spouse, a fruitful woman, and a mother able to produce a generous supply of milk. For each girl this is a unique process. She is initiated alone and is the center of public attention and care. From her point of view it is her *Nkang'a*, the most thrilling and self-gratifying phase of her life. Society recognizes and encourages these sentiments, even though it also prescribes certain trials and hardships for the novice, who must suffer before she is glorified on the last day of the ritual. The milk tree, then, celebrates the coming-of-age of a new social personality, and distinguishes her from all other women at this one moment in her life. In terms of its action context, the milk tree here also expresses the conflict between the girl and the moral community of adult women she is entering. Not without reason is the milk tree site known as "the place of death" or "the place of suffering," terms also applied to the site where boys are circumcised, for the girl novice must not move a muscle throughout a whole hot and clamant day.

In other contexts, the milk tree site is the scene of opposition between the novice's own mother and the group of adult women. The mother is debarred from attending the ring of dancers. She is losing her child, although later she recovers her as an adult co-member of her lineage. Here we see the conflict between the matricentric family and the wider society which, as I have said, is dominantly articulated by the principle of matriliny. The relationship between mother and daughter persists throughout the ritual, but its content is changed. It is worth pointing out that, at one phase in *Nkang'a*, mother and daughter interchange portions of clothing. This may perhaps be related to the Ndembu custom whereby mourners wear small portions of a dead relative's clothing. Whatever the interchange of clothing may mean to a psychoanalyst—and here we arrive at one of the limits of our present anthropological competence—it seems not unlikely that Ndembu intend to symbolize the termination for both mother and daughter of an important aspect of their relationship. This is one of the symbolic actions—one of very few—about which I found it impossible to elicit any interpretation in the puberty ritual. Hence it is legitimate to infer, in my opinion, that powerful unconscious wishes, of a kind considered illicit by Ndembu, are expressed in it.

Opposition between the tribeswomen and the novice's mother is mimetically represented at the milk tree towards the end of the first day of the puberty ritual. The girl's mother cooks a huge meal of cassava and beans—both kinds of food are symbols in *Nkang'a*, with many meanings—for the women visitors, who eat in village groups and not at random. Before eating, the women return to the milk tree from their eating place a few yards away and circle the tree in procession. The mother brings up the rear holding up a large spoon full of cassava and beans. Suddenly she shouts: "Who wants the cassava of *chipwamp-*

wilu?" All the women rush to be first to seize the spoon and eat from it. "*Chipwampwilu*" appears to be an archaic word and no one knows its meaning. Informants say that the spoon represents the novice herself in her role of married woman, while the food stands both for her reproductive power (*lusemu*) and her role as cultivator and cook. One woman told my wife: "It is lucky if the person snatching the spoon comes from the novice's own village. Otherwise, the mother believes that her child will go far away from her to a distant village and die there. The mother wants her child to stay near her." Implicit in this statement is a deeper conflict than that between the matricentric family and mature female society. It refers to another dominant articulating principle of Ndembu society, namely virilocal marriage according to which women live at their husbands' villages after marriage. Its effect is sometimes to separate mothers from daughters by considerable distances. In the episode described, the women symbolize the matrilineal cores of villages. Each village wishes to gain control through marriage over the novice's capacity to work. Its members also hope that her children will be raised in it, thus adding to its size and prestige. Later in *Nkang'a* there is a symbolic struggle between the novice's matrilineal kin and those of her bridegroom, which makes explicit the conflict between virilocality and matriliny.

Lastly, in the context of action situation, the milk tree is sometimes described by informants as representing the novice's own matrilineage. Indeed, it has this significance in the competition for the spoon just discussed, for women of her own village try to snatch the spoon before members of other villages. Even if such women do not belong to her matrilineage but are married to its male members, they are thought to be acting on its behalf. Thus, the milk tree in one of its action aspects represents the unity and exclusiveness of a single matrilineage with a local focus in a village against other such corporate groups. The conflict between yet another subsystem and the total system is given dramatic and symbolic form.

By this time, it will have become clear that considerable discrepancy exists between the interpretations of the milk tree offered by informants and the behavior exhibited by Ndembu in situations dominated by the milk tree symbolism. Thus, we are told that the milk tree represents the close tie between mother and daughter. Yet the milk tree separates a daughter from her mother. We are also told that the milk tree stands for the unity of Ndembu society. Yet we find that in practice it separates women from men, and some categories and groups of women from others. How are these contradictions between principle and practice to be explained?

I am convinced that my informants genuinely believed that the milk tree represented only the linking and unifying aspects of Ndembu social organization. I am equally convinced that the role of the milk tree in action situations, where it represents a focus of specified groups in opposition to other groups, forms an equally important component of its total meaning. Here the important question must be asked, "meaning for whom?" For if Ndembu do not recognize the discrepancy between their interpretation of the milk tree symbolism and their behavior in connection with it, does this mean that the discrepancy has no relevance for the social anthropologist? Indeed, some anthropologists claim, with Nadel (1954:108), that "uncomprehended symbols have no part in social enquiry; their social effectiveness lies in their capacity to indicate, and if they indicate nothing to the actors, they are, from our point of view, irrelevant, and indeed no longer symbols (whatever their significance for the psychologist or psychoanalyst)." Professor Monica Wilson (1957:6) holds a similar point of view. She writes that she stresses "Nyakyusa interpretations of their own rituals, for anthropological literature is bespattered with symbolic guessing, the ethnographer's interpretations of the rituals of other people." Indeed, she goes so far as to base her whole analysis of Nyakyusa ritual on "the Nyakyusa translation or interpretation of the symbolism." In my view, these investigators go beyond the limits of salutary caution and impose serious, and even arbitrary, limitations on themselves. To some extent, their difficulties derive from their failure to distinguish the concept of symbol from that of a mere sign. Although I am in complete disagreement with his fundamental postulate that the collective unconscious is the main formative principle in ritual symbolism, I consider that Carl Jung (1949:601) has cleared the way for further investigation by making just this distinction. "A sign," he says, "is an analogous or abbreviated expression of a known thing. But a symbol is always the best possible expression of a relatively *unknown* fact, a fact, however, which is none the less recognized or postulated as existing." Nadel and Wilson, in treating most ritual symbols as signs, must ignore or regard as irrelevant some of the crucial properties of such symbols.

THREE PROPERTIES OF RITUAL SYMBOLS

Before we can interpret, we must further classify our descriptive data, collected by the methods described above. Such a classification will enable us to state some of the properties of ritual symbols. The simplest property is that of condensation. Many things and actions are represented in a single formation. Secondly, a dominant symbol is a *unification of disparate significata*. The disparate *significata* are inter-connected by virtue of their common possession of analogous qualities or by association in fact or thought. Such qualities or links of association may in themselves be quite trivial or random or widely distributed over a range of phenomena. Their very generality enables them to bracket together the most diverse ideas and phenomena. Thus, as we have seen, the milk tree stands for, *inter alia*, women's breasts, motherhood, a novice at *Nkang'a*, the principle of matriliny, a specific matrilineage, learning, and the unity and persistence of Ndembu society. The themes of nourishment and dependence run through all these diverse *significata*.

The third important property of dominant ritual symbols is *polarization of meaning*. Not only the milk tree but all other dominant Ndembu symbols possess two clearly distinguishable poles of meaning. At one pole is found a cluster of *significata* that refer to components of the moral and social orders of Ndembu society, to principles of social organization, to kinds of corporate grouping, and to the norms and values inherent in structural relationships. At the other pole, the *significata* are usually natural and physiological phenomena and processes. Let us call the first of these the "ideological pole," and the second the "sensory pole." At the sensory pole, the meaning content is closely related to the outward form of the symbol. Thus one meaning of the milk tree—breast milk—is closely related to the exudation of milky latex from the tree. One sensory meaning of another dominant symbol, the *mukula* tree, is blood; this tree secretes a dusky red gum.

At the sensory pole are concentrated those *significata* that may be expected to arouse desires and feelings; at the ideological pole one finds an arrangement of norms and values that guide and control persons as members of social groups and categories. The sensory, emotional *significata* tend to be "gross" in a double sense. In the first place, they are gross in a general way, taking no account of detail or the precise qualities of emotion. It cannot be sufficiently stressed that such symbols are social facts, "collective representations," even though their appeal is to the lowest common denominator of human feeling. The second sense of "gross" is "frankly, even flagrantly, physiological." Thus, the milk tree has the gross meanings of breast milk, breasts, and the process of breast feeding. These are also gross in the sense that they represent items of universal Ndembu experience. Other Ndembu symbols, at their sensory poles of meaning, represent such themes as blood, male and female genitalia, semen, urine, and feces. The same symbols, at their ideological poles of meaning, represent the unity and continuity of social groups, primary and associational, domestic, and political.

THE SITUATIONAL SUPPRESSION OF CONFLICT FROM INTERPRETATION

Emotion and praxis, indeed, give life and coloring to the values and norms, but the connection between the behavioral expression of conflict and the normative components of each kind of ritual, and of its dominant symbols, is seldom explicitly formulated by believing actors. Only if one were to personify a society, regarding it as some kind of supra-individual entity, could one speak of "unconsciousness" here. Each individual participant in the *Nkang'a* ritual is well aware that kin quarrel most bitterly over rights and obligations conferred by the principle of matriliny, but that awareness is situationally held back from verbal expression: the participants must behave as if conflicts generated by matriliny were irrelevant.

This does not mean, as Nadel (1954) considers, that what is not verbalized is in fact irrelevant either to the participants or to the anthropologist. On the contrary, in so far as the anthropologist considers problems of social action to fall within his purview, the suppression from speech of what might be termed "the behavioral meaning" of certain dominant symbols is highly relevant. The fact is that any kind of coherent, organized social life would be impossible without the assumption that certain values and norms, imperatives and prohibitions, are axiomatic in character, ultimately binding on everyone. However, for many reasons, the axiomatic quality of these norms is difficult to maintain in practice, since in the endless variety of real situations, norms considered equally valid in abstraction are frequently found to be inconsistent with one another, and even mutually to conflict.

Furthermore, social norms, by their very nature, impose unnatural constraints on those whose biopsychical dispositions impel them to supranormal or abnormal behavior, either fitfully or regularly. Social life in all organized groups appears to exhibit a cycle or oscillation between periods when one set of axiomatic norms is observed and periods dominated by another set. Thus, since different norms govern different aspects or sectors of social behavior, and, more importantly, since the sectors overlap and interpenetrate in reality, causing norm-conflict, the validity of several major norms has to be reaffirmed in isolation from others and outside the contexts in which struggles and conflicts arise in connection with them. This is why one so often finds in ritual that dogmatic and symbolic emphasis is laid on a single norm or on a cluster of closely, and on the whole harmoniously, interrelated norms in a single kind of ritual.

Yet, since at major gatherings of this sort, people assemble not as aggregates of individuals but as social personalities arrayed and organized by many principles and norms of grouping, it is by no means a simple matter to assert the clear situational paramountcy of the norms to be commemorated and extolled. Thus, in the Ndembu boys' circumcision ritual, relationships between social categories, such as men and women, old men and young men, circumcised and uncircumcised, and the norms governing such relationships, are given formal representation, but the members of the ritual assembly come as members of corporate groups, such as villages and lineages, which in secular life are in rivalry with one another. That this rivalry is not mysteriously and wonderfully dispelled by the circumcision ritual becomes abundantly clear from the number of quarrels and fights that can be observed during public dances and beer drinking in the intervals between phases of the ritual proper. Here people quarrel as members of groupings that are not recognized in the formal structure of the ritual.

It may be said that any major ritual that stresses the importance of a single principle of social organization only does so by blocking the expression of other important principles. Sometime the submerged principles, and the norms and customs through which they become effective, are given veiled and disguised representation in the symbolic pattern of the ritual; sometimes, as in the boys' circumcision ritual, they break through to expression in the spatial and temporal interstices of the procedure. In this essay we are concerned principally with the effects of the suppression on the meaning-structure of dominant symbols.

For example, in the frequently performed *Nkula* ritual, the dominant symbols are a cluster of red objects, notably red clay (*mukundu*) and the *mukula* tree mentioned previously. In the context of *Nkula*, both of these are said to represent menstrual blood and the "blood of birth," which is the blood that accompanies the birth of a child. The ostensible goal of the ritual is to coagulate the patient's menstrual blood, which has been flowing away in menorrhagia, around the fetus in order to nourish it. A series of symbolic acts are performed to attain this end. For example, a young *mukula* tree is cut down by male doctors and part of it is carved into the shape of a baby, which is then inserted into a round calabash medicated with the blood of a sacrificed cock, with red clay, and with a number of other red ingredients. The red medicines here, say the Ndembu, represent desired coagulation of the patient's menstrual blood, and the calabash is a symbolic womb. At the ideological pole of meaning, the *mukula* tree and the medicated calabash both represent (as the milk tree does) the patient's matrilineage and, at a higher level of abstraction, the principle of matriliny itself. This is also consistent with the fact that *ivumu*, the term for "womb," also means "matrilineage." In this symbolism the procreative, rather than the nutritive, aspect of motherhood is stressed. However, Ndembu red symbolism, unlike the white symbolism of which the milk tree symbolism is a species, nearly always has explicit reference to violence, to killing, and, at its most general level of meaning, to breach, both in the social and natural orders. Although informants, when discussing this *Nkula* ritual specifically, tend to stress the positive, feminine aspects of parturition and reproduction, other meanings of the red symbols, stated explicitly in other ritual contexts, can be shown to make their influence felt in *Nkula*. For example, both red clay and the *mukula* tree are dominant symbols in the hunter's cult, where they mean the blood of animals, the red meat of game, the inheritance through either parent of hunting prowess, and the unity of all initiated hunters. It also stands for the hunter's power to kill. The same red symbols, in the context of the *Wubanji* ritual performed to purify a man who has killed a kinsman or a lion or leopard (animals believed to be reincarnated hunter kin of the living), represent the blood of homicide. Again, in the boys' circumcision ritual, these symbols stand for the blood of circumcised boys. More seriously still, in divination and in antiwitchcraft rituals, they stand for the blood of witches' victims, which is exposed in necrophagous feasts.

Most of these meanings are implicit in *Nkula*. For example, the female patient, dressed in skins like a male hunter and carrying a bow and arrow, at one phase of the ritual performs a special hunter's dance. Moreover, while she does this, she wears in her hair, just above the brow, the red feather of a lourie bird. Only shedders of blood, such as hunters, man-slayers, and circumcisers, are customarily entitled to wear this feather. Again, after the patient has been given the baby figurine in its symbolic womb, she dances with it in a style of dancing peculiar to circumcisers when they brandish aloft the great *nfunda* medicine of the circumcision lodge. Why then is the woman patient identified with male bloodspillers? The field context of these symbolic objects and items of behavior suggests that the Ndembu feel that the woman, in wasting her menstrual blood and in failing to bear children, is actively renouncing her expected role as a mature married female. She is behaving like a male killer, not like a female nourisher. The situation is analogous, though modified by matriliny, to the following pronouncement in the ancient Jewish Code of Qaro: "Every man is bound to marry a wife in order to beget children, and he who fails of this duty is as one who sheds blood."

One does not need to be a psychoanalyst, one only needs sound sociological training, acquaintance with the total Ndembu symbolic system, plus ordinary common sense, to see that one of the aims of the ritual is to make the woman accept her lot in life as a childbearer and rearer of children for her lineage. The symbolism suggests that the patient is unconsciously rejecting her female role, that indeed she is guilty; indeed, "*mbayi*," one term for menstrual blood, is etymologically connected with "*ku-baya*" (to be guilty). I have not time here to present further evidence of symbols and interpretations, both in *Nkula* and in cognate rituals, which reinforce this explanation. In the situation of *Nkula*, the dominant principles celebrated and reanimated are those of matriliny, the mother-child bond, and tribal continuity through matriliny. The norms in which these are expressed are those governing the behavior of mature women, which ascribe to them the role appropriate to their sex. The suppressed or submerged principles and norms, in this situation, concern and control the personal and corporate behavior deemed appropriate for man.

The analysis of *Nkula* symbolism throws into relief another major function of ritual. Ritual adapts and periodically readapts the biopsychical individual to the basic conditions and axiomatic values of human social life. In redressive rituals, the category to which *Nkula* belongs, the eternally rebellious individual is converted for a while into a loyal citizen. In the case of *Nkula*, a female individual whose behavior is felt to demonstrate her rebellion against, or at least her reluctance to comply with, the biological and social life patterns of her sex, is both induced and coerced by means of precept and symbol to accept her culturally prescribed destiny.

I have suggested in this essay that different aspects of ritual symbolism can be analyzed within the framework of structuralist theory and of cultural anthropology respectively. As I have said, this would be to treat ritual symbols as timeless entities. Many useful conclusions can be arrived at by these methods, but the essential nature, both of dominant symbols and of constellations of instrumental symbols, is that they are dynamic factors. Static analysis would here presuppose a corpse, and, as Jung (1949) says, "a symbol is alive." It is alive only in so far as it is "pregnant with meaning" for men and women, who interact by observing, transgressing, and manipulating for private ends the norms and values that the symbol expresses. If the ritual symbol is conceptualized as a force in a field of social action, its critical properties of condensation, polarization, and unification of disparities become intelligible and explicable. On the other hand, conceptualizing the symbol as if it were an object and neglecting its role in action often lead to a stress on only those aspects of symbolism which can be logically and consistently related to one another to form an abstract unitary system. In a field situation, the unity of a symbol or a symbolic configuration appears as the resultant of many tendencies converging towards one another from different areas of biophysical and social existence. The symbol is an independent force which is itself a product of many opposed forces.

CONCLUSION: THE ANALYSIS OF SYMBOLS IN SOCIAL PROCESSES

Let me outline briefly the way in which I think ritual symbols may fruitfully be analyzed. Performances of ritual are phases in broad social processes, the span and complexity of which are roughly proportional to the size and degree of differentiation of the groups in which they occur. One class of ritual is situated near the apex of a whole hierarchy of redressive and regulative institutions that correct deflections and deviations from customarily prescribed behavior. Another class anticipates deviations and conflicts. This class includes periodic rituals and life-crisis rituals. Each kind of ritual is a patterned process in time, the units of which are symbolic objects and serialized items of symbolic behavior.

The symbolic constituents may themselves be classed into structural elements, or "dominant symbols," which tend to be ends in themselves, and variable elements, or "instrumental symbols," which serve as means to the explicit or implicit goals of the given ritual. In order to give an adequate explanation of the meaning of a particular symbol, it is necessary first to examine the widest action-field context, that, namely, in which the ritual itself is simply a phase. Here one must consider what kinds of circumstances give rise to a performance of ritual, whether these are concerned with natural phenomena, economic and technological processes, human life-crises, or with the breach of crucial social relationships. The circumstances will probably determine what sort of ritual is performed. The goals of the ritual will have overt and implicit reference to the antecedent circumstances and will in turn help to determine the meaning of the symbols. Symbols must now be examined within the context of the specific ritual. It is here that we enlist the aid of indigenous informants. It is here also that we may be able to speak legitimately of "levels" of interpretation, for laymen will give the investigator simple and exoteric meanings, while specialists will give him esoteric explanations and more elaborate texts. Next, behavior directed towards each symbol should be noted, for such behavior is an important component of its total meaning.

We are now in a position to exhibit the ritual as a system of meanings, but this system acquires additional richness and depth if it is regarded as itself constituting a sector of the Ndembu ritual system, as interpreted by informants and as observed in action. It is in comparison with other sectors of the total system, and by reference to the dominant articulating principles of the total system, that we often become aware that the overt and ostensible aims and purposes of a given ritual conceal unavowed, and even "unconscious," wishes and goals. We also become aware that a complex relationship exists between the overt and the submerged, and the manifest and latent patterns of meaning. As social anthropologists we are potentially capable of analyzing the social aspect of this relationship. We can examine, for example, the relations of dependence and independence between the total society and its parts, and the relations between different kinds of parts, and between different parts of the same kind. We can see how the same dominant symbol, which in one kind of ritual stands for one kind of social group or for one principle of organization, in another kind of ritual stands for another kind of group or principle, and in its aggregate of meanings stands for unity and continuity of the widest Ndembu society, embracing its contradictions.

REFERENCES

Jung, Carl G. 1949. *Psychological Types*. London: Routledge and Kegan Paul.
Nadel, S. F. 1954. *Nupe Religion*. London: Routledge and Kegan Paul.
Wilson, Monica. 1957. *Rituals of Kinship among the Nyakyusa*. London: Oxford University Press, for the International African Institute.

THE SOCIAL SKIN

TERENCE S. TURNER

Man is born naked but is everywhere in clothes (or their symbolic equivalents). We cannot tell how this came to be, but we can say something about
why it should be so and what it means.

Decorating, covering, uncovering or otherwise altering the human form in
accordance with social notions of everyday propriety or sacred dress, beauty or
solemnity, status or changes in status, or on occasion of the violation and
inversion of such notions, seems to have been a concern of every human
society of which we have knowledge. This objectively universal fact is associated with another of a more subjective nature—that the surface of the body
seems everywhere to be treated, not only as the boundary of the individual as a
biological and psychological entity but as the frontier of the social self as well.
As these two entities are quite different, and as cultures differ widely in the
ways they define both, the relation between them is highly problematic. The
problems involved, however, are ones that all societies must solve in one way or
another, because upon the solution must rest a society's ways of "socializing"
individuals, that is, of integrating them into the societies to which they belong,
not only *as* children but throughout their lives.

The surface of the body, as the common frontier of society, the social self,
and the psycho-biological individual, becomes the symbolic stage upon which
the drama of socialization is enacted, and bodily adornment (in all its culturally
multifarious forms, from body-painting to clothing and from feather head-
dresses to cosmetics) becomes the language through which it is expressed. The
adornment and public presentation of the body, however inconsequential or
even frivolous a business it may appear to individuals, is for cultures a serious
matter: *de la vie serieuse*, as Durkheim said of religion. Wilde observed that the
feeling of being in harmony with the fashion gives a man a measure of security
he rarely derives from his religion. The seriousness with which we take questions of dress and appearance is betrayed by the way we regard not taking them
seriously as an index, either of a "serious" disposition or of serious psychological problems. As Lord Chesterfield remarked:

Dress is a very foolish thing; and yet it is a very foolish thing for a man not to be well dressed, according to his rank and way of life; and it is so far from being a disparagement to any man's understanding, that it is rather a proof of it, to be as well dressed as those whom he lives with: the difference in this case, between a man of sense and a fop, is, that the fop values himself upon his dress; and the man of sense laughs at it, at the same time that he knows that he must not neglect it (cited in Bell 1949:13).

The most significant point of this passage is not the explicit assertion that a man of sense should regard dress with a mixture of contempt and attentiveness, but the implicit claim that by doing so, and thus maintaining his appearance in a way compatible with "those he lives with," he defines himself as a man of sense. The uneasy ambivalence of the man of sense, whose "sense" consists in conforming to a practice he laughs at, is the consciousness of a truth that seems as scandalous today as it did in the eighteenth century. This is that culture, which we neither understand nor control, is not only the necessary medium through which we communicate our social status, attitudes, desires, beliefs and ideals (in short, our identities) to others, but also to a large extent constitutes these identities, in ways with which we are compelled to conform regardless of our self-consciousness or even our contempt. Dress and bodily adornment constitute one such cultural medium, perhaps the one most specialized in the shaping and communication of personal and social identity.

The Kayapo are a native tribe of the southern borders of the Amazon forest. They live in widely scattered villages which may attain populations of several hundred. The economy is a mixture of forest horticulture, and hunting and gathering. The social organization of the villages is based on a relatively complex system of institutions, which are clearly defined and uniform for the population as a whole. The basic social unit is the extended family household, in which residence is based on the principle that men must leave their maternal households as boys and go to live in the households of their wives upon marriage. In between they live as bachelors in a "men's house," generally built in the center of the circular village plaza, round the edges of which are ranged the "women's houses" (as the extended family households are called). Women, on the other hand, remain from birth to death in the households into which they are born.

The Kayapo possess a quite elaborate code of what could be called "dress," a fact which might escape notice by a casual Western observer because it does not involve the use of clothing. A well turned out adult Kayapo male, with his large lower-lip plug (a saucer-like disc some six centimeters across), penis sheath (a small cone made of palm leaves covering the *glans penis*), large holes pierced through the ear lobes from which hang small strings of beads, overall body paint in red and black patterns, plucked eyebrows, eyelashes and facial hair, and head shaved to a point at the crown with the hair left long at the sides and back, could on the other hand hardly leave the most insensitive traveler with the impression that bodily adornment is a neglected art among the Kayapo. There are, however, very few Western observers, including anthropologists, who have ever taken the trouble to go beyond the superficial recording of such exotic paraphernalia to inquire into the system of meanings and values which it evokes for its wearers. A closer look at Kayapo bodily adornment discloses that the apparently naked savage is as fully covered in a fabric of cultural meaning as the most elaborately draped Victorian lady or gentleman.

The first point that should be made about Kayapo notions of propriety in bodily appearance is the importance of cleanliness. All Kayapo bathe at least once a day. To be dirty, and especially to allow traces of meat, blood or other animal substances or food to remain on the skin, is considered not merely slovenly or dirty but actively anti-social. It is, moreover, dangerous to the health of the unwashed person. "Health" is conceived as a state of full and proper integration into the social world, while illness is conceived in terms of the encroachment of natural, and particularly animal forces upon the domain of social relations. Cleanliness, as the removal of all "natural" excrescence from the surface of the body, is thus the essential first step in "socializing" the interface between self and society, embodied in concrete terms by the skin. The removal of facial and bodily hair carries out this same fundamental principle of transforming the skin from a mere "natural" envelope of the physical body into a sort of social filter, able to contain within a social form the biological forces and libidinal energies that lie beneath.

The mention of bodily hair leads on to a consideration of the treatment of the hair of the head. The principles that govern coiffure are consistent with the general notions of cleanliness, hygiene, and sociality, but are considerably more developed, and accord with those features of the head-hair which the Kayapo emphasize as setting it apart from bodily hair (it is even called by a different name).

Hair, like skin, is a "natural" part of the surface of the body, but unlike skin it continually grows outwards, erupting from the body into the social space beyond it. Inside the body, beneath the skin, it is alive and growing; outside, beyond the skin, it is dead and without sensation, although its growth manifests the unsocialized biological forces within. The hair of the head thus focuses the dynamic and unstable quality of the frontier between the "natural," bio-libidinous forces of the inner body and the external sphere of social relations. In this context, hair offers itself as a symbol of the libidinal energies of the self and of the never-ending struggle to constrain within acceptable forms their eruption into social space.

So important is this symbolic function of hair as a focus of the socializing function, not only among the Kayapo but among Central Brazilian tribes in general, that variations in coiffure have become the principal visible means of distinguishing one tribe from another. Each people has its own distinctive hair-style, which stands as the emblem of its own culture and social community (and as such, in its own eyes, for the highest level of sociality to have been attained by humanity). The Kayapo tribal coiffure, used by both men and women, consists of shaving the hair above the forehead upwards to a point at the crown, leaving the hair long at the back and sides of the head (unless the individual belongs to one of the special categories of people who wear their hair cut short, as described below). Men may tease up a little widow's peak at the point of the triangular shaved area. The sides of this area are often painted in black with bands of geometrical patterns.

Certain categories of people in Kayapo society are *privileged* to wear their hair long. Others must keep it cut short. Nursing infants, women who have borne children, and men who have received their penis sheaths and have been through initiation (that is, those who have been socially certified as able to carry on sexual relations) wear their hair long. Children and adolescents of both sexes (girls from weaning to childbirth, boys from weaning to initiation) and those mourning the death of a member of their immediate family (for example, a spouse, sibling or child) have their hair cut short.

To understand this social distribution of long and short hair it is necessary to comprehend Kayapo notions about the nature of family relations. Parents are thought to be connected to their children, and siblings to one another, by a tie that goes deeper than a mere social or emotional bond. This tie is imagined as a sort of spiritual continuation of the common physical substance that they share through conception and the womb. This relation of biological participation lasts throughout life but is broken by death. The death of a person's child or sibling thus directly diminishes his or her own biological being and energies. Although spouses lack the intrinsic biological link of blood relations, their sexual relationship constitutes a "natural" procreative, libidinal community that is its counterpart. Inasmuch as both sorts of biological relationship are cut off by death, cutting off the hair, conceived as the extension of the biological energy of the self into social space, is the symbolically appropriate response to the death of a spouse as well as a child.

The same concrete logic accounts for the treatment of children's hair. While a child is still nursing, it is still, as it were, an extension of the biological being and energies of its parents, and above all, at this stage, the mother. In these terms nursing constitutes a kind of external and attenuated final stage of pregnancy. Weaning is the decisive moment of the "birth" of the child as a separate biological and social being. Thus nursing infants' hair is never cut, and is left to grow as long as that of sexually active adults: infants at this stage *are* still the extensions of the biological and sexual being of their long-haired parents. Cutting the infant's hair at the onset of weaning aptly symbolizes the severance of this bio-sexual continuity (or, as we would say, its repression). Henceforth, the child's hair remains short as a sign of its biological separation from its parents, on the one hand, and the undeveloped state of its own bio-sexual powers on the other. When these become strong enough to be socially extended, through sexual intercourse and procreation, as the basis of a new family, the hair is once again allowed to grow to full length. For men this point is considered to arrive at puberty, and specifically with the bestowal of a penis sheath, which is ideally soon followed by initiation (a symbolic "marriage" which signals marriageability, or "bachelorhood," rather than being a binding union in and of itself).

The discrepancy in the timing of the return to long hair for the two sexes reflects a fundamental difference in Kayapo notions of their respective social roles. "Society" is epitomized for the Kayapo by the system of communal societies and age-sets centred on the men's house. These collective organizations are primarily a male domain, as their association with the men's house suggests, although women have certain societies of their own. The communal societies are defined in terms of the criteria for recruitment, and this is always defined as a corollary of some important transformation in family or household structure (such as a boy's moving out of his maternal family household to the men's house, marriage, the birth of children, etc.). These transformations in family relations are themselves associated with key points in the process of growth and sexual development.

The structure of communal groups, then, constitutes a sort of sociological mechanism for reproducing, not only itself but the structure of the extended family households that form the lower level or personal sphere of Kayapo social organization. This communal institutional structure, on the other hand, is itself defined in terms of the various stages of the bio-sexual development of men (and to a much lesser extent, women). All this comes down to the proposition that men reproduce society through the transformation of their "natural" biological and libidinal powers into collective social form. This conception can be found elaborated in Kayapo mythology.

Women, by contrast, reproduce the natural biological individual, and, as a corollary, the elementary family, which the Kayapo conceive as a "natural" or infra-social set of essentially physical relations. Inasmuch as the whole Kayapo system works on the principle of the co-optation of "natural" forces and their channeling into social form, it follows that women's biological forces of reproduction should be exercised only within the framework of the structure of social relations reproduced by men. The effective social extension of a woman's biological reproductive powers therefore occurs at the moment of the first childbirth within the context of marriage, husband and household. This is, accordingly, the moment at which a woman begins to let her hair grow long again. For men, as we have seen, the decisive social co-option of libidinal energy or reproductive power comes earlier, at the point at which those powers are publicly appropriated for purposes of the reproduction of the collective social order. This is the moment symbolically marked by the bestowal of the penis sheath at puberty.

The penis sheath, then, symbolizes the collective appropriation of male powers of sexual reproduction for the purposes of social reproduction. To the Kayapo, the appropriation of "natural" or biological powers for social purposes implies the suppression of their "natural" or socially unrestrained forms of expression. The penis sheath works as a symbol of the channeling of male libidinal energies into social form by effectively restraining the spontaneous, "natural" expression of male sexuality: in a word, erection. The sheath, the small cone of woven palm leaf, is open at both the wide and narrow ends. The wide end fits over the tip of the penis, while the narrow end has an aperture just wide enough to enable the foreskin to be drawn through it. Once pulled through, it bunches up in a way that holds the sheath down on the *glans penis*, and pushes the penis as a whole back into the body. This obviously renders erection impossible. A public erection, or even the publicly visible protrusion of the *glans penis* through the foreskin without erection, is as embarrassing for a Kayapo male as walking naked through one's town or workplace would be for a Westerner. It is the action of the sheath in preventing such an eventuality that is the basis of its symbolic meaning.

Just as the cutting or growing of hair becomes a code for defining and expressing a whole system of ideas about the nature of the individual and society and the relations between the two, so other types of bodily adornment are used to express other modalities of the same basic relationships.

Pierced ears, ear-plugs, and lip-plugs comprise a similar distinct complex of social meanings. Here the emphasis is on the socialization, not of sexual powers, but of the faculties of understanding and active self-expression. The

Kayapo distinguish between passive and active modes of knowing. Passive understanding is associated with hearing, active knowledge of how to make and do things with seeing. The most important aspect of the socialization of the passive faculty of understanding is the development of the ability to "hear" language. To be able to hear and understand speech is spoken of in terms of "having a hole in one's ear"; to be deaf is "to have the hole in one's ear closed off." The ear lobes of infants of both sexes are pierced, and large cigar-shaped ear-plugs, painted red, are inserted to stretch the holes to a diameter of two or three centimeters (I shall return to the significance of the red color). At weaning (by which time the child has learned to speak and understand language) the ear-plugs are removed, and little strings of beads like earrings are tied through the holes to keep them open. Kayapo continue to wear these bead earrings, or simply leave their ear-lobe-holes empty throughout adult life. I suggest that the piercing and stretching of these secondary, social "holes-in-the-ear" through the early use of the ear-plugs for infants is a metaphor for the socialization of the understanding, the opening of the ears to language and all that implies, which takes place during the first years of infancy.

The lip-plug, which reaches such a large size among older men, is incontestably the most striking piece of Kayapo finery. Only males have their lips pierced. This happens soon after birth, but at first only a string of beads with a bit of shell is placed in the hole to keep it open. After initiation, young bachelors begin to put progressively larger wooden pins through the hole to enlarge it. This gradual process continues through the early years of adult manhood, but accelerates when a man graduates to the senior male grade of "fathers-of-many-children." These are men of an age to have become heads of their wives' households, with married daughters and thus sons-in-law living under their roofs as quasi-dependents. Such men have considerable social authority, but they wield it, not within the household itself (which is considered a woman's domain) but rather in the public arena of the communal men's house, in the form of political oratory. Public speaking, in an ornate and blustering style, is the most characteristic attribute of senior manhood, and is the essential medium of political power. An even more specialized form of speaking, a kind of metrical chanting known as *ben* is the distinctive prerogative of chiefs, who are called "chanters" in reference to the activity that most embodies their authority.

Public speaking, and chanting as its more rarified and potent form, is the supreme expression of the values of Kayapo society considered as a politically ordered hierarchy. Senior men, and, among them, chiefs, are the dominant figures in this hierarchy, and it can therefore be said that oratory and chanting as public activities express this dominance as a value implicit in the Kayapo social order. The lip-plug of the senior male, as a physical expression of the oral assertiveness and pre-eminence of the orator, embodies the social dominance and expressivencss of the senior males of whom it is the distinctive badge.

The senior male lip-plug is in these terms the complement of the pierced ears of both sexes and the infantile ear-plugs from which they derive. The former is associated with the active expression and political construction of the social order, while the latter betoken the receptiveness to such expressions as the attribute of all socialized persons. Speaking and "hearing" (that is, understanding and conforming) are the complementary and interdependent functions that constitute the Kayapo polity. Through the symbolic medium of bodily adornment, the body of every Kayapo becomes a microcosm of the Kayapo body politic.

As a man grows old he retires from active political life. He speaks in public less often, and on the occasions when he does it is to assume an elder statesman's role of appealing to common values and interests rather than to take sides. The transformation from the politically active role of the senior man to the more honorific if less dynamic role of elder statesman is once again signaled by a change in the style and shape of the lip-plug. The simplest form this can take is a diminution in the size of the familiar wooden disc. It may, however, take the form of the most precious and prestigious object in the entire Kayapo wardrobe—the cylindrical lip-plug of ground and polished rock crystal worn only by elder males. These neolithic valuables, which may reach six inches in length and one inch in diameter, with two small flanges at the upper end to keep them from sliding through the hole in the lip, require immense amounts of time to make and are passed down as heirlooms within families. They are generally clear to milky white in color. White is associated with old age and with ghosts, and thus in general terms with the transcendence of the social divisions and transformations whose qualities are evoked by the two main Kayapo colors, black and red. This quality of transcendence of social conflict, and of direct involvement in the processes of suppression and appropriation of libidinal energies and their transformation into social form which constitute Kayapo public life in its political and ritual aspects, is characteristic of the content of the oratory of old men, and is what lends it its great if relatively innocuous prestige. Once again, then, we find that the symbolic qualities of the lip-plug match the social qualities of the speech of its wearer.

Before the advent of Western clothes, Kayapo of both sexes and all ages constantly went about with their bodies painted (many still do, especially in the more remote villages). The Kayapo have raised body painting to an art, and the variety and elaborateness of the designs are apt to seem overwhelming upon first acquaintance. Analysis, however, reveals that a few simple principles run through the variation of forms and styles and lend coherence to the whole. These principles, in turn, can be seen to add a further dimension to the total system of meanings conveyed by Kayapo bodily adornment.

There are two main aspects to the Kayapo art of body painting, one concerning the association of the two main colors used (red and black) on distinct zones of the body, the other concerning the two basic styles employed in painting that part of the body for which black is used.

To begin with the first aspect, the use of the two colors, black and red, and their association with different regions of the body reveal yet another dimension of Kayapo ideas about the makeup of the person as biological being and social actor. Black is applied to the trunk of the body, the upper arms and thighs. Black designs or stripes are also painted on the cheeks, forehead, and occasionally across the eyes or mouth. Red is applied to the calves and feet, forearms and hands, and face, especially around the eyes. Sometimes it is smeared over black designs already painted on the face, to render the whole face red.

Black is associated with the idea of transformation between society and unsocialized nature. The word for black is applied to the zone just outside the village that one passes through to enter the "wild" forest (the domain of nature). It is also the word for death (that is, the first phase of death, while the body is still decomposing and the soul has not yet forsaken its social ties to become a ghost: ghosts are white). In both of these usages, the term for black applies to a spatial or temporal zone of transition between the social world and the world of natural or infra-social forces that is closed off from society proper and lies beyond its borders. It is therefore appropriate that black is applied to the surface of those parts of the body conceived to be the seat of its "natural" powers and energies (the trunk, internal and reproductive organs, major muscles, etc.) that are in themselves beyond the reach of socialization (an analogy might be drawn here to the Freudian notion of the id). The black skin becomes the repressive boundary between the natural powers of the individual and the external domain of social relations.

Red, by contrast, is associated with notions of vitality, energy and intensification. It is applied to the peripheral points of the body that come directly into contact with the outside world (the hands and feet, and the face with its sensory organs, especially the eyes). The principle here seems to be the intensification of the individual's powers of relating to the external (that is, primarily, the social) world. Notice that the opposition between *red* (intensification, vitalization) and *black* (repression) coincides with that between the *peripheral* and *central* parts of the body, which is itself treated as a form of the relationship between the *surface* and *inside* of the body respectively. The contrasting use of the two colors thus establishes a binary classification of the human body and its powers and relates that classification back to the conceptual oppositions, *inside: surface: outside,* that underlies the system of bodily adornment as a whole.

Turning now to the second major aspect of the system of body painting, that is, the two main styles of painting in black, the best place to begin is with the observation that one style is used primarily for children and one primarily for adults. The children's style is by far the more elaborate. It consists of intricate geometrical designs traced in black with a narrow stylus made from the central rib of a leaf. A child's entire body from the neck to below the knees, and down the arms to below the elbows, is covered. To do the job properly requires a couple of hours. Mothers (occasionally doting aunts or grandmothers) spend much time in this way keeping their children "well dressed."

The style involves building up a coherent overall pattern out of many individually insignificant lines, dots, etc. The final result is unique, as a snowflake is unique. The idiosyncratic nature of the design reflects the relationship between the painter and the child being decorated. Only one child is painted at a time, in his or her own house, by his or her own mother or another relation. All of this reflects the social position of the young child and the nature of the process of socialization it is undergoing. The child is the object of a prolonged and intensive process of creating a socially acceptable form out of a myriad of individually unordered elements. It must lie still and submit to this process, which requires a certain amount of discipline. The finished product is the unique expression of the child's relationship to its own mother and household. It is not a collectively stereotyped pattern establishing a common identity with children from other families. This again conforms with the social situation of the child, which is not integrated into communal society above the level of its particular family.

Boys cease to be painted in this style, except for rare ceremonial occasions, when they leave home to live in the men's house. Older girls and women, however, continue to paint one another in this way as an occasional pastime. This use of the infantile style by women reflects the extent to which they remain identified with their individual families and households, in contrast to men's identification with collective groups at the communal level.

The second style, which can be used for children when a mother lacks the time or inclination for a full-scale job in the first style, is primarily associated with adults. It consists of standardized designs, many of which have names (generally names of the animals they are supposed to resemble). These designs are simple, consisting of broad strokes that can be applied quickly with the hand, rather than by the time-consuming stylus method. Their social context of application is typically collective: men's age sets gathered in the men's house, or women's societies, which meet fortnightly in the village plaza for the purpose of painting one another. On such occasions, a uniform style is generally used for the whole group (different styles may be used to distinguish structurally distinct groups, such as bachelors and mature men).

The second style is thus typically used by fully socialized adults, acting in a collective capacity (that is, at a level defined by common participation in the structure of the community as a whole rather than at the individual family level). Collective action (typically, though not necessarily, of a ritual character) is "socializing" in the higher sense of directly constituting and reproducing the structure of society as a whole: those painted in the adult style are thus acting, not in the capacity of *objects* of socialization, but as its *agents*. The "animal" quality of the designs is evocative of this role; the Kayapo conceive of collective society-constituting activities, like their communal ceremonies, as the transformation of "natural" or animal qualities into social form by means of collective social replication. The adult style, with its "animal" designs applied collectively to social groups as an accompaniment to collective activity, epitomizes these meanings and ideas. The contrasts between the children's and adults' styles of body painting thus model key contrasts in the social attributes of children and adults, specifically, their relative levels of social integration or, which comes to the same thing, their degree of "socialization."

The greater part of Kayapo communal activity consists of the celebration of long and complex ceremonies, which generally take the form of collective dances by all the men and boys or all the women and girls of the village, and occasionally of both. These sacred events are always distinguished by collective body-painting and renewed coiffures in the tribal pattern, as well as by numerous special items of ritual regalia, such as feather head-dresses, elaborate bracelets, ear- and lip-plugs of special design, belts and leg bands hung with noise-making objects like tapir or peccary hooves etc. The more important ceremonies are rites either of "baptism," that is, the bestowal of ceremonially prestigious names, or initiation. Certain items of regalia distinguish those actually receiving names or being initiated from the mass of celebrants. In a more fundamental sense, the entire repertoire of ceremonial costume marks ceremonial activity in contrast to everyday activities and relations. In the ceremonial context itself, the contrast is preserved between the celebrants of the ceremony and the non-participating spectators, who wear no special costume. An important group in the latter category are the parents of the children being named or initiated in the ceremony, who may not take a direct part in the dancing but must work hard to supply the many dancers with food. In this role they are treated with great rudeness and disrespect by the actual celebrants, who shout at them to bring food and then complain loudly about its quantity and quality. . . .

The names bestowed in the great naming ceremonies belong to a special class of "beautiful" names which are passed down from certain categories of kinsmen (mother's brother, and both maternal and paternal grandfathers for boys, father's sister and both grandmothers for girls). In keeping with the ritual prestige of the names being transmitted to them, the children being honored in the ceremony are adorned with special regalia, notably elaborate bracelets with bead and feather pendants. Initiands are similarly distinguished by bracelets, although they are so huge as to cover the whole forearm, and are exceptionally heavy and bulky in construction. The initiation ceremony itself is named after these bracelets; it is known as "the black bracelets," or literally "black bone marrow." The name at first strikes one as odd, since the bracelets are painted bright red. It may be suggested that the symbolic appropriateness of bracelets as the badges of initiands and baptisands derives from the same set of ideas about the connotations of different parts of the body and the associated color symbolism. If the extremities of the body represent the extension of the psycho-biological level of the self into social space, and if the hands are in a sense the prototypical extremities in this regard, elaborate bracelets are an apt symbol for the imposition of social form upon this extension. This is, of course, what is happening in the ceremonies in question. In the case of initiation into manhood, which involves a first, symbolic marriage, both the repression of childish, merely individualistic libido and the accentuation of sexuality and procreativity in the service of social reproduction are involved. Black and red, as we have seen, are the symbols of repression and sensory accentuation, respectively, and the accentuation of sexuality and procreativity in the service of social reproduction is involved. That what is "blackened" or repressed is the inner substance of the bones aptly conveys the idea of the suppression of the pre-social, biological basis of social relatedness, while the actual redness of the so-called "black" bracelets through which this is achieved simultaneously expresses the activation of this basis in the social form represented by the bracelets themselves. The initiation bracelets thus condense within themselves a number of the fundamental principles of the whole system of bodily adornment and the social concepts it expresses.

Among the ordinary ritual celebrants, there is considerable variation within the standard categories of ritual wear, such as feather head-dresses, ear-plugs, necklaces, bracelets and belts or leg bands already described. Many of these variations (for example, the use of feathers from a particular sort of bird for a head-dress, or distinctive materials such as wound cotton string or perhaps fresh-water mussel shells for ear-plugs, or the breast plumage of the red macaw for bracelets, etc.) are passed down like names themselves from uncle or grandfather to nephew or grandson (or the corresponding female categories). They thus denote an aspect of the social identity of the wearer that he or she owes to his or her relationship with a particular kinsman. These distinctive items of ritual dress make up the "paraphernalia" mentioned above that is bestowed, in parallel to, but separately from, names, in the ritual setting. The Kayapo call such accessories by their general (and only) term for "valuables," "wealth," or "riches."

Ask a Kayapo why he is wearing a certain sort of head-dress for a ceremonial dance, and he will be likely to answer, "It is my wealth." Ask him why he dances, or indeed why the ceremony is being performed, and he will almost certainly answer, "To be beautiful" ("For the sake of beauty" would be an equally accurate Englishing of the usual Kayapo expression). Wealth and beauty are closely connected notions among the Kayapo, and both refer to aspects of the person coded by items of prestigious ritual dress. Certain "beautiful" names are, in fact, associated with specific forms of adornment (that is, with certain types of "wealth") such as ear-plugs. "Beautiful people" (those who have received "beautiful" names in ceremonies) generally possess more "wealth" than "common people" who have not gone through a ritual baptism, and thus possess only "common" names.

The connection of "beauty" with "wealth" in the form of bodily adornment is strikingly expressed in the lyrics of the choral hymns sung by the massed ritual celebrants as they dance, with uniform steps which vary with each song, the successive rites that constitute a "great" or "beautiful" naming ceremony. These songs are almost invariably those of animals, especially birds, the muses of Kayapo lyric and dance, who have communicated them to humans in various ways. Two verses from the vast Kayapo repertoire may serve as examples. A bird proudly boasts to his human listener,

Can we [birds] not reach up to the sky?
Why, we can snatch the very clouds,
Wind them round our legs as bracelets,
And sit thus, regaled by their thunder.

Another calls out a stirring summons to earthbound humanity:

> I fly among the branches [rays] of the sun;
> I fly among the branches of the sun.
> I perch on its branches and
> Sit gazing over the whole world.

> Throw yourselves into the sky beside me!
> Throw yourselves into the sky beside me!
> Cover yourselves with the blood and feathers of birds
> And follow after me!

The admonition about blood and feathers refers to the technique of covering the body of a dancer with his or her own blood, which is then used as an adhesive to which the breast plumage or down of macaws, vultures or eagles is fixed. This is perhaps the most prestigious ("beautiful") form of sacred costume. These verses may help one to grasp the connotations of the fact that dancing (and by extension, the celebration of any ritual) is called "flying" in Kayapo, and of the term for the most common item of ceremonial adornment, the feather head-dress, which is the ritual form of the word for "bird."

Birds fly, and "can scan" the *whole* world. They are not confined by its divisions, but transcend them in a way that to a Kayapo seems the supreme natural metaphor for the direct experience of totality, the integration of the self through the perception of the wholeness of the world. This principle of wholeness, the transcendental integration of what ordinary human (that is, social) life separates and puts at odds, is the essence of the Kayapo notion of "beauty."

Two aspects of this notion, as embodied in items of ritual costume and the sacred activities in which they are worn, seem incongruous and even contradictory in the context of what has already been said about everyday Kayapo "dress" and its underlying assumptions. First, whereas everyday bodily adornment stresses the imposition of social form upon the "natural" energies and powers of the individual, ritual costume (such as feather head-dresses, body painting with "animal" designs or the covering of the body with blood and feathers) seems to represent the opposite idea: that is, the imposition of natural form upon social actors engaged in what are the most important social activities of all, the great sacred performances that periodically reconstitute the fabric of society itself. Secondly, sacred costume, together with the notion that the ritual songs and dances themselves originate among wild animals and birds, seems to reverse the meaning of everyday bodily adornment. The latter is implicitly based upon a relation between a "natural" core (the human body, or on the sociological level, the elementary family) and a "social" periphery (the space outside the body in which social interaction takes place, or in the structure of kinship, the more distant blood relations outside the immediate family who bestow names and ritual paraphernalia upon the child). Ritual space, in contrast, seems based on the relationship between a "natural" periphery (the jungle beyond the village boundary, as the abode of the birds and animals that are the sources both of ritual costume and of the rituals themselves) and a "social" center (the central plaza of the village, where the sacred dances are actually performed).

This inversion of space and the fundamental relationship of nature to society encoded in sacred costume turns out upon closer examination to parallel two other inversions in the organization of everyday social relations which form the very basis of the sacred ceremonial system. The first of these involves the types of kinship relations involved in the key ritual relations of name giving and the bestowing of ritual "wealth" (such as special paraphernalia), as contrasted with those involved in the transformations of family relations marked by the everyday complex of bodily adornment (penis sheath, ear- and lip-plugs, etc.). The latter typically involve immediate family relations like parent–child, husband–wife, or the key extended-family relationship of son-in-law to father-in-law. These relations have in common that they directly link status within the family, or two families by marriage; they may therefore be thought of as *direct* relations. Ritual relations, on the other hand, connect grandparents and grandchildren, uncles and nephews, aunts and nieces: they skip over the connecting relatives (the parents of the children receiving the ritual names or paraphernalia, who are themselves the children or siblings of the name-bestowing relatives) and may thus be described as *indirect* relations.

The structure of *direct* relations functions as a sort of ladder or series of steps up which the developing individual moves from status to status within the structure of the families, extended families and households to which he or she belongs in the course of his or her life. The first step in this process is the "natural" domain of immediate family relations, within which the individual is at first defined as merely an extension of his or her parents' "natural" powers of reproduction. The course of the life cycle from there on is a series of steps by which the individual is detached from this primal "natural" unity and integrated into the social life of the community. As a corollary of this process, his or her own "natural" powers develop until they can in turn become the basis of a new family. The highest step in the socialization process, however, comes when this second "natural" family unit is dispersed, and the individual becomes a parent-in-law, thus moving into the prestigious role of extended family household head at home and, in the case of men, political leader in the community at large. Each major step in this process is marked, as we have seen, by modifications of bodily adornment of the "everyday" sort. The overall form of the process is that of a progressive evacuation of "natural" energy and powers from the "central" sources of body and elementary family and its extrapolation into social forms and powers standing in a "peripheral" relation to these sources in physical and social space. The result is that "social" structure is created at the expense of the evacuation and dispersal of the "natural" powers and relational units (elementary families) that comprise its foundation. The "natural" at any given moment is the socially unintegrated: embodied at the beginning of the process by the newborn infant or new family, as yet not completely absorbed into the wider community of social relations, by the end of the process it is represented by the scattered members of dispersed families, whose younger members have gone on to form new families. The integration of society made possible by this transformation is achieved at the cost of the disintegration of the primal natural community of the immediate family and the externalisation and social appropriation of the "natural" powers of the individual.

Seen in this perspective the ritual system represents a balancing of accounts. The dispersed *direct*, "natural" relations of the parents (their parents and siblings from the family they have left behind) now become the key *indirect* relations whose identification with the children of these same parents becomes the point of the ritual. I use the term "identification" deliberately, for this is what is implied by the sharing of personal names and idiosyncratic items of ritual costume. The point of this identification is, of course, that it reasserts a connection between, or in other words reintegrates, the dispersed, disintegrated or "natural" relations of the parents' previous families with the not-yet-socially-integrated relations of the parents' present family (their children). This integration, however, is achieved, not on the basis of *direct* relations as is characteristic of "natural" groups like the elementary family, but on a new, *indirect* footing with no natural basis; in short, on a purely "social" basis. The new integrated "whole" that is established through ritual action is thus defined simultaneously as reintegrating and transcending the "natural" basis of social relations. It therefore becomes the quintessential prototype of "social" relationship, and as such, the appropriate vehicle of the basic components of individual social identity, personal names, distinctive ritual dress and other personal "wealth."

In terms of social space, what has happened from the point of view of the central name-receiving individual and her or his family is that "socialization" has been achieved through the transference of attributes of the identity of "natural" relations located on the *periphery* of the family to the actor located at its *center*. In terms of the equilibrium of "natural" and "social" forces and qualities, the prospective evacuation of natural powers from the individual and family has been offset (in advance) by an infusion of social attributes, which are themselves the products of the reintegration, through the social mechanism of ritual action, of elements of the "natural" infrastructure dispersed as the price of social integration.

The symbolic integration of "natural" attributes from the periphery of social space as aspects of the social identity of actors located at its center, can now be understood as a metaphorical embodiment of the integrative ("beautiful") structural properties of the social relations evoked by the rituals. The regalia, of course, does more than simply encode this process. It is the concrete medium (along with names) through which the identity of the ritual celebrants is simultaneously redefined, "socialized," and infused with "beauty."

The foregoing analysis should help to clarify the full meaning of the Kayapo notion of beauty as wholeness, integration or completion. In its primary context, that of sacred ritual, it is the value associated with the creation of a *social* whole based on *indirect* (mediate) relations, capable of reintegrating the dismembered elements of simpler *natural* wholes (elementary families) constituted *by direct* (immediate) relations. "Beauty" is an ideal expression of society itself in its holistic capacity. It is, as such, one of the primary values of Kayapo social life.

Just as the value of beauty is associated with the complex of social relations and cultural notions involved in ritual action, so the complex of relations and categories that constitutes the social structure of everyday life is focused upon another general social value. This value is in a sense the opposite of beauty, since it pertains to relations of separation, opposition, and inequality. We may call it "dominance," meaning by that the combination of prestige, authority, individual autonomy and ability to control others that accrues in increasing measure to individuals as they move through the stages of social development, passing from lower to higher status in the structure of the extended family and community. It is doubtless significant that the Kayapo themselves do not name this intrinsically divisive value with any term in their own language, whereas they continually employ the adjective "beautiful" in connection with the most varied activities, including those of a divisive (and thus "ugly") character which they wish to put in a better light.

The lack of a term notwithstanding, there can be no doubt of the existence of the value in question and its importance as the organizing focus of Kayapo social and political life outside the ritual sphere. It reaches its highest and most concentrated expression in the public activities of senior men, for example their characteristic activity of aggressively flamboyant oratory. The lip-plug, and particularly the senior man's lip-plug as the largest and most spectacularly obtrusive in the entire age-graded lip-plug series, is directly associated with this value as a quality of male, and particularly senior male, social identity. "Dominance" is, however, to be understood as a symbolic attribute, a culturally imputed quality expressing a person's place in the hierarchy of extended family and community structure, rather than as a relation of naked power or forcible oppression. It is, as such, an expression of the whole edifice of age, sex, family and communal status-categories marked by the whole system of everyday bodily accoutrements described earlier. Younger men and women can acquire this quality in some measure within their own proper spheres, making due allowance for the fact that in the context of community-wide relations they are subordinate to the dominant senior males.

These two values, "dominance" and "beauty," inform the social activities and goals of every Kayapo, and constitute the most general purposes of social action and the most important qualities of personal identity. The identity of social actors is constituted as much by the goals towards which they direct their activities, as by their classification according to status, sex, age, degree of socialization, etc. I have tried to show that bodily adornment encodes these values as well as the other sorts of categories; it may thus be said to define the total social identity of the individual, meaning his or her subjective identity as a social actor, as well as objective identity conceived in terms of a set of social categories. It does this by mediating between individuals, considered both in their objective and subjective capacities, and society, also considered both in its objective capacity as a structure of relations and its subjective capacity as a system of values. I have attempted to demonstrate that the symbolic mediation effected by the code of bodily adornment in both these respects is, in the terms of Kayapo culture, systematically and finely attuned to the nuances of Kayapo social relations. The structure of Kayapo society, including its highest values and its most fundamental conceptual presuppositions (such as the nature-culture relationship, the modes of expression and understanding, the character of "socialization," etc.) could be read from the paint and ornaments of a representative collection of Kayapo of all ages, sexes, and secular and ritual roles.

Bodily adornment, considered as a symbolic medium, is not unique in these respects: every society has a number of such media or languages, the most important among which is of course language itself. The distinctive place of the adornment of the body among these is that it is the medium most directly and concretely concerned with the construction of the individual as social actor or cultural "subject." This is a fundamental concern of all societies and social groups, and this is why the imposition of a standardized symbolic form upon the body, as a symbol or "objective correlative" of the social self, invariably becomes a serious business for all societies, regardless of whether their members as individuals consciously take the matter seriously or not.

It may be suggested that the "construction of the subject," is a process which is broadly similar in all human societies, and the study of systems of bodily adornment is one of the best ways of comprehending what it involves. As the Kayapo example serves to illustrate, it is essentially a question of the conflation of certain basic types of social notions and categories, among which can be listed categories of time and space, modes of activity (for example, individual or collective, secular or sacred), types of social status (sex, age, family roles, political positions, etc.), personal qualities (degree of "socialization," relative passivity or activity as a social actor, etc.) and modes of social value, for example, "dominance" or "beauty." In any given society, of course, these basic categories will be combined in culturally idiosyncratic ways to constitute the symbolic medium of bodily adornment, and these synthetic patterns reveal much about the basic notions of value, social action, and person- or self-hood of the culture in question.

In the case of the Kayapo, three broad synthetic clusters of meanings and values of this type emerge from analysis. One is concerned with the Kayapo notion of socialization, conceived as the transformation of "natural" powers and attributes into social forms. The basic symbolic vehicle for this notion, after the general concern for cleanliness, is the form of body painting by which the trunk is contrasted with the extremities as black and red zones, respectively. This fundamental mapping of the body's "natural" and "social" areas is inflected, at a higher level of articulation, by hair style. The contrast between long and short hair is used to mark the successive phases of the development and social extension of the individual's libidinous and reproductive powers. Finally, the penis sheath (correlated with the shift to long hair for men) serves to mark the decisive point in the social appropriation of male reproductive powers and, perhaps more important, the collective nature of this appropriation. A second major complex concerns the distinction and relationship between the passive and active qualities of social agency. The basic indicator here is again body painting, in this case the distinction between the infantile and adult styles. This basic distinction is once again inflected by the set composed of pierced ears and ear-plugs, on the one hand, and pierced lips and lip-plugs on the other. This set adds the specific meanings associated with the notions of hearing and speaking as passive knowledge and the active expression of decisions and programs of action, respectively. Finally, both these clusters are cross-cut by a broad distinction between modes of activity. The most strongly marked distinction here is between secular and sacred (ritual) action, with the latter distinguished from the former by a rich variety of regalia. This distinction, however, may be considered a heightened inflection of the more basic distinction between individual or family-level activities and communal activities, not all of which are of a sacred character. Secular men's house gatherings or meetings of women's societies, for example, may be accompanied by collective painting and perhaps the wearing of simple head-dresses of palm leaves, even though there is no ceremony.

An important structural principle emerges from this analysis of the Kayapo system—the hierarchical or iterative structure of the symbolic code. Each major cluster of symbolic meanings is seen to be arranged in a series of increasingly specific modulations or inflections of the general notions expressed by the most basic symbol in the cluster. A second structural principle is the multiplicative character of the system as a whole. By this I mean that the three basic clusters are necessarily simultaneously present or conflated in the "dress" of any individual Kayapo at any time. One cannot paint an infant or adult in the appropriate style without at the same time observing the concentric distinction between trunk and extremities common to both styles.

The conflation first of the levels of meaning within each cluster, secondly of each cluster with the others, and finally of the more basic categories of meaning and value listed above that are combined in different ways to form each cluster, is what I mean by "the construction of the cultural subject" or actor. (It is sometimes necessary to speak in terms of *collective* subjects, such as the class of young men, or of workers, but for the sake of simplicity I shall leave this issue aside here.) It is, by the same token, the construction of the social universe within which he or she acts (that is, an aspect of that construction). As the Kayapo example suggests, this is a dynamic process that proceeds as it were in opposite directions at the same time, towards equilibrium or equilibrated growth at both the individual and social levels (it goes without saying that in speaking of equilibrium I am referring to cultural ideals rather than concrete realities, either social or individual).

In the Kayapo case, the externalization of the internal biological and libidinous ("natural") powers of individuals as the basis of social reproduction, and the socialization of "external" natural powers as the basis of social structure and the social identity of actors otherwise defined only as biological extensions of their parents, are clearly metaphorical inversions of one another.

Each complements the other, just as the social values respectively associated with the two aspects of the process, "dominance" and "beauty," complement each other; a balance between the two processes and their associated values is the ideal state of Kayapo society as a dynamic equilibrium. It is also, and equally, the basis of the unity and balance of the personality of the socialized individual, likewise conceived as a dynamic equilibrium.

The point I have sought to demonstrate is that this balance between opposing yet complementary forces, which is the most fundamental structural principle of Kayapo society, is systematically articulated and, as it were, played out on the bodies of every member of Kayapo society through the medium of bodily adornment. This finding supports the general hypothesis with which we began, namely that the surface of the body becomes, in any human society, a boundary of a peculiarly complex kind, which simultaneously separates domains lying on either side of it and conflates different levels of social, individual and intra-psychic meaning. The skin (and hair) are the concrete boundary between the self and the other, the individual and society. It is, however, a truism to which our investigation has also attested, that the "self" is a composite product of social and "natural" (libidinous) components.

REFERENCE

Bell, Q. 1949. *On Human Finery*. New York: A. A. Wyn.

PART II

Philosophical Studies, or

Learning How to Think Embodiment

Philosophical Studies, or

Learning How to Think Embodiment

INTRODUCTION

Post-Enlightenment European and North American philosophers have attended more to the life of the mind than to the nature of bodily life. As a result
it would be difficult to assemble a continuous tradition of reflection on bodies
within philosophy proper. Yet philosophies that take material existence seriously are not new. One need only consult the early writings of Marx to find a
complex engagement with the "sensuous existence" of "real active men, as they
are conditioned by a definite development of their productive forces." Thus it
is appropriate to begin reading the philosophy of embodiment with Marx and
Engels's "The German Ideology," written in the 1840s, which is extracted in this
part. From this seminal work of materialist philosophy and critique, we turn to
more recent works, not all by philosophers, that have engaged in systematic
thinking about bodies and material life. The essays and excerpts included in
part II are works that have had a significant impact in anthropology, or they
are lesser known works by authors who are directly engaging embodiment in a
philosophical mode. They have in common a commitment to some form of
materialism and a striving to get beyond any simplistic mind-body divide.

"The German Ideology" is one of the greatest examples of Marx's characteristic sarcasm. The essay begins with an extended polemical critique of German Idealist philosophy. Attacking the Young Hegelians (e.g., David Strauss,
Max Stirner, Bruno Bauer, and Arnold Rupe) for their abstraction from the
"real existing world" and the hubris of their claims to be resolving universal
problems, Marx and Engels propose instead a "real, positive science, . . . the
representation of the practical activity, of the practical process of development
of men." Thus they suggest that a materialist knowledge would ultimately
replace philosophy as an independent branch of scholarly endeavor.

Marx's and Engels's essay is notable both for its clear statement of the
centrality of human productive life to a Marxist materialism and for its deep
seriousness about language, ideology, and consciousness. In making their famous argument that consciousness is "the direct efflux" of material behavior,

Marx and Engels turn definitively to the body: "In the [materialist] method, which conforms to real life, [the starting point] is the real living individuals themselves, and consciousness is considered solely as *their* consciousness." Yet this is a body for which sensuous experience and the life of language are an irreducible aspect of its existence. See, for example, the discussion of language and consciousness under the heading "History: Fundamental Conditions." Clearly, concrete human life is conscious and articulate and thus fundamentally cultural—going far beyond the brute existence of a biological organism. But the basic materiality of human life, figured as productive activity, must form the starting point of positive knowledge.

Turning from Marx and Engels to the mid-twentieth-century Marxist critic Walter Benjamin, one finds him making an anthropological gesture in the essay "On the Mimetic Faculty" (1978). Here he grounds certain fundamental elements of consciousness in mimesis, a presumably universal faculty of human nature. This capacity for mimicry is interesting precisely because it goes beyond a "monkey see, monkey do" kind of imitation. Rather. Benjamin is interested in the human elaboration of "nonsensuous similarity." Humans forge relations of similarity on the basis of many criteria, not all of them immediately obvious to the senses. Thus mimesis can be active in the formation of abstract categories as well as in the modeling of a behavior or a form after something visible or palpable. But nonsensuous similarities are still far from being purely ideal or cognitive. Benjamin's greatest interest in this essay, in fact, is to assert a certain materiality to language and thought. He suggests that in earlier epochs the mimetic faculty functioned via resemblance to link microcosm with macrocosm and to infuse the social with ritual significance. (Michel Foucault explores this theme of resemblance in premodern Europe at length in *The Order of Things*.) But even in the modern world, with all its pointed disenchantments, language retains some of its material magic. The semiotic richness of words bears flashes of a more archaic similarity, which "flits past" as we use language. Thus Benjamin suggests that the creativity and depth of significance of language both originate in a fundamental human instinct, the mimetic faculty, and remain an organic and largely unconscious part of human life. As in so much of his other writing, he refuses any simple classification of experience into material and ideal, body and mind, or economic base and ideological superstructure.

Maurice Merleau-Ponty also strives to overcome the classic dualisms that have afflicted philosophy. His *The Phenomenology of Perception* (1962) is among the foundational texts for the anthropology of embodiment. As he says in the preface to his book, he sets about "re-achieving a direct and primitive contact

with the world" through an encount[er]
tions provided by scientists, histor[ians]
project involved a reflective engage[ment]
body, an attempt to "return to things[ ]
"precedes knowledge." This too is a[ ]
Merleau-Ponty's rejection of a philo[sophy]
suspicion of various forms of empiri[cism]
that turns almost exclusively to the i[ ]
earlier phenomenologists, Merleau-[Ponty]
world-making activity and to see the[ ]
experiences of the embodied, intenti[onal]
Kantian transcendental ego that coul[d]
an abstracted and disembodied persp[ective]

The philosopher Ian Hacking, d[rawing on the]
phenomenological tradition, develops in "Making Up People" a "dynamic
nominalism" that he has drawn from Foucault. Thus he is able to take a
somewhat different tack on the hoary dualism of subject and object, mind and
body. Noting that our classifications and our classes "conspire to emerge hand
in hand," he describes how certain types of people (the schizophrenic, the
homosexual, even the suicide) have literally come into existence as experts
developed language to speak of them. As names are crafted, real types appear
to fit those names. Even where these names are utterly mundane, he suggests,
as in the case of the Parisian *garçons de café*, certain styles or embodied roles are
elaborated to suit the demands of the classification. Avoiding any simplistic
functionalism that might suggest that discourse determines the real, he nevertheless turns philosophical attention from abstract universals to history and
to "the intricacies of real life." If bodies are to be "known" they cannot be
reduced to some universal organic level prior to culture and history. Bodies are
always already infused with ideology.

This Foucauldian presumption applies very rigorously to questions of sex
and gender, as Judith Butler has extensively argued. Rejecting the commitment
of some feminists to the naturalness of sex (as opposed to the social constructedness of gender), Butler, in *Bodies That Matter*, interrogates the tendency of feminists and other theorists to resort to materiality as an unquestioned foundation of both their ideas and their politics. Asking "how and why
'materiality' has become a sign of irreducibility" (and, an anthropologist might
add, naturalness, universality, human nature, and so on), she explores a way of
thinking that refuses to presume an opposition between constructedness and a
materiality that still centers the body proper. She opposes those who fear that

people who think the world is too "DD"? because his socialized than? is socialized than?

attention to the processes through which our lived worlds have been built will make stubborn social and economic constraints look as mutable as mere ideas. Arguing that a refusal to examine processes of construction has its own dangers, she suggests that naturalization of the material world reproduces an all-too-convenient conflation of the material with the necessary, and the mental with the optional. To destabilize these assumptions, she notes the etymological polysemy of the term *matter*, which refers both to materializing and to meaning, or "mattering"; she then turns to a discussion of matter in Aristotle and Foucault. Butler elucidates Foucault's materialism when she points out that, for him, "'materiality' designates a certain effect of power, or, rather, *is* power in its formative and constituting effects." Most of Butler's extended discussion of Plato and Luce Irigaray has been deleted here, but enough remains to raise the question of what is excluded when absolute priority, or pure otherness vis-à-vis consciousness, is relegated to the domain of the material. Finally, Butler addresses some of the ethical and political questions that afflict these questions of the body and the material. She invites us to evaluate "not the materiality of sex but the sex of materiality" as a way to open new and more courageous forms of feminist analysis.

In the final extract of this section Bruno Latour issues a polemical challenge to all previous philosophies of materiality and consciousness, reality and fiction. Crediting Descartes with having bequeathed to modernity the bodiless observer, a "brain-in-a-vat," he diagnoses the modern epistemological point of view as increasingly separated from a "withering" body, helplessly contemplating a reality forced ever more radically to the exteriors of human experience. In a rapid and breezy survey of the entire modern tradition of philosophy and the human sciences, he denounces the narrow range of dualisms deployed in empiricism and rationalism, phenomenology, positivism, neurophilosophy, and the social sciences *tout court*. In one stroke, it would seem, the many subtle discriminations and insights of Enlightenment civilization must be banished if we are to restore reality to its proper and natural importance.

As usual, Latour parodies himself as he denounces others, reminding us that what is on offer in these pages are merely "unsavory pellets of textbook philosophy." Yet his most central arguments here are fundamental to the process of moving beyond the body proper. Most recent scholarship on the body, for instance, has found it essential to relativize and denaturalize the body, demonstrating always the deep level at which lived bodies are socially constructed. But this commitment to contingency does not mean that the bodies we describe are not perfectly real. Much of Latour's mad dash through the philosophical underpinnings of the human sciences is motivated by his desire to show how it has come to be that constructed bodies are seen, strangely, as not actually real. Thus the non-dualist ontology he seeks to revive moves beyond the entrenched habits of critique to open up new fields and networks for empirical appreciation.

Inhabiting these new fields and networks are diverse arrays of humans and non-humans enjoying multiple forms of agency. In this essay Latour characterizes non-humans in a distinctly carnal idiom: facts, machines, and theories have "roots, blood vessels, networks, rhizomes, and tendrils"; non-humans have "history, flexibility, culture, blood." These new "actants" remind us of Donna Haraway's "disturbingly lively" information machines; they make it ever more difficult to think of our cyborg selves as simple flesh and blood, body and soul. The collective social world envisioned in Latour's polemic is loaded with bodies, all of them thoroughly material and carnal, but nowhere do we find the a priori "individual" seen as the privileged partner of "society."

The readings gathered in this section all concern themselves with material life and the philosophical challenges of thinking about it. All of them brilliantly resist any simple relations of determination or priority between mind and body, subject and object, matter and consciousness. They push beyond the commonsense boundaries of the body proper and transgress the stubborn classification of experience which would separate the mental from the biological, the subjective from the objective, and the ideal from the material. Though some philosophers have done much to overcome these invidious classifications, perhaps no writing in modern Western languages can entirely escape the persistent dualism of body and mind. It is this dyad that serious thinking about collective, material human life—that is, anthropology in it broadest sense—must work to overcome.

ADDITIONAL READINGS

Bachelard, Gaston. 1964. *The Poetics of Space.* New York: Orion Press.

Bergson, Henri. 1929 (1911). *Matter and Memory.* New York: Macmillan.

Biagioli, Mario, ed. 1999. *The Science Studies Reader.* New York: Routledge.

Haraway, Donna J. 1991. *Simians, Cyborgs, and Women: The Reinvention of Nature.* New York: Routledge.

James, William. 1985. *The Varieties of Religious Experience.* Cambridge: Harvard University Press.

Martin, Dale B. 1995. *The Corinthian Body.* New Haven: Yale University Press.

Massumi, Brian. 2002. *Parables for the Virtual: Movement, Affect, Sensation.* Durham: Duke University Press.

KARL MARX AND FRIEDRICH ENGELS

IDEALISM AND MATERIALISM

THE ILLUSIONS OF GERMAN IDEOLOGY

As we hear from German ideologists, Germany has in the last few years gone through an unparalleled revolution. The decomposition of the Hegelian philosophy, which began with [David] Strauss, has developed into a universal ferment into which all the "powers of the past" are swept. In the general chaos mighty empires have arisen only to meet with immediate doom, heroes have emerged momentarily only to be buried back into obscurity by bolder and stronger rivals. It was a revolution beside which the French Revolution was child's play, a world struggle beside which the struggles of the Diadochi [successors of Alexander the Great] appear insignificant. Principles ousted one another, heroes of the mind overthrew each other with unheard-of rapidity, and in the three years 1842–45 more of the past was swept away in Germany than at other times in three centuries.

[margin handwritten note: not talk of literal Death but still ↑]

All this is supposed to have taken place in the realm of pure thought.

Certainly it is an interesting event we are dealing with: the putrescence of the absolute spirit. When the last spark of its life had failed, the various components of this *caput mortuum* began to decompose, entered into new combinations and formed new substances. The industrialists of philosophy, who till then had lived on the exploitation of the absolute spirit, now seized upon the new combinations. Each with all possible zeal set about retailing his apportioned share. This naturally gave rise to competition, which, to start with, was carried on in moderately staid bourgeois fashion. Later when the German market was glutted, and the commodity in spite of all efforts found no response in the world market, the business was spoiled in the usual German manner by fabricated and fictitious production, deterioration in quality, adulteration of the raw materials, falsification of labels, fictitious purchases, bill-jobbing and a credit system devoid of any real basis. The competition turned into a bitter struggle, which is now being extolled and interpreted to us as a revolution of world significance, the begetter of the most prodigious results and achievements.

If we wish to rate at its true value this philosophic charlatanry, which awakens even in the breast of the honest German citizen a glow of national pride, if we wish to bring out clearly the pettiness, the parochial narrowness of this whole Young-Hegelian movement and in particular the tragicomic contrast between the illusions of these heroes about their achievements and the actual achievements themselves, we must look at the whole spectacle from a standpoint beyond the frontiers of Germany.

German criticism has, right up to its latest efforts, never quitted the realm of philosophy. Far from examining its general philosophic premises, the whole body of its inquiries has actually sprung from the soil of a definite philosophical system, that of Hegel. Not only in their answers but in their very questions there was a mystification. This dependence on Hegel is the reason why not one of these modern critics has even attempted a comprehensive criticism of the Hegelian system, however much each professes to have advanced beyond Hegel. Their polemics against Hegel and against one another are confined to this—each extracts one side of the Hegelian system and turns this against the whole system as well as against the sides extracted by the others. To begin with they extracted pure, unfalsified Hegelian categories such as "substance" and "self-consciousness," later they desecrated these categories with more secular names such as "species," "the Unique," "Man," etc.

The entire body of German philosophical criticism from Strauss to [Max] Stirner is confined to criticism of *religious* conceptions. The critics started from real religion and actual theology. What religious consciousness and a religious conception really meant was determined variously as they went along. Their advance consisted in subsuming the allegedly dominant metaphysical, political, juridical, moral and other conceptions under the class of religious or theological conceptions; and similarly in pronouncing political, juridical, moral consciousness as religious or theological, and the political, juridical, moral man—"*man*" in the last resort—as religious. The dominance of religion was taken for granted. Gradually every dominant relationship was pronounced a religious relationship and transformed into a cult, a cult of law, a cult of the State, etc. On all sides it was only a question of dogmas and belief in dogmas. The world sanctified to an ever increasing extent till at last our venerable Saint Max [Stirner] was able to canonize it *en bloc* and thus dispose of it once for all.

The Old Hegelians had *comprehended* everything as soon as it was reduced to an Hegelian logical category. The Young Hegelians *criticized* everything by attributing to it religious conceptions or by pronouncing it a theological matter. The Young Hegelians are in agreement with the Old Hegelians in their belief in the rule of religion, of concepts, of a universal principle in the existing world. Only, the one party attacks this dominion as usurpation, while the other extols it as legitimate.

Since the Young Hegelians consider conceptions, thoughts, ideas, in fact all the products of consciousness, to which they attribute an independent existence, as the real chains of men (just as the Old Hegelians declared them the true bonds of human society) it is evident that the Young Hegelians have to fight only against these illusions of consciousness. Since, according to their fantasy, the relationships of men, all their doings, their chains and their limitations are products of their consciousness, the Young Hegelians logically put to men the moral postulate of exchanging their present consciousness for human, critical or egoistic consciousness, and thus of removing their limitations. This demand to change consciousness amounts to a demand to interpret reality in another way, i.e., to recognize it by means of another interpretation. The Young-Hegelian ideologists, in spite of their allegedly "world-shattering" statements, are the staunchest conservatives. The most recent of them have found the correct expression for their activity when they declare they are only fighting against "*phrases.*" They forget, however, that to these phrases they themselves are only opposing other phrases, and that they are in no way combating the real existing world when they are merely combating the phrases of this world. The only results which this philosophic criticism could achieve were a few (and at that thoroughly one-sided) elucidations of Christianity from the point of view of religious history; all the rest of their assertions are only further embellishments of their claim to have furnished, in these unimportant elucidations, discoveries of universal importance.

It has not occurred to any one of these philosophers to inquire into the connection of German philosophy with German reality, the relation of their criticism to their own material surroundings. The premises from which we begin are not arbitrary ones, not dogmas, but real premises from which abstraction can only be made in the imagination. They are the real individuals, their activity and the material conditions under which they live, both those which they find already existing and those produced by their activity. These premises can thus be verified in a purely empirical way.

The first premise of all human history is, of course, the existence of living human individuals. Thus the first fact to be established is the physical organization of these individuals and their consequent relation to the rest of nature. Of course, we cannot here go either into the actual physical nature of man, or into the natural conditions in which man finds himself—geological, orohydro-graphical, climatic and so on. The writing of history must always set out from these natural bases and their modification in the course of history through the action of men.

Men can be distinguished from animals by consciousness, by religion or anything else you like. They themselves begin to distinguish themselves from animals as soon as they begin to *produce* their means of subsistence, a step which is conditioned by their physical organization. By producing their means of subsistence men are indirectly producing their actual material life.

The way in which men produce their means of subsistence depends first of all on the nature of the actual means of subsistence they find in existence and have to reproduce. This mode of production must not be considered simply as being the production of the physical existence of the individuals. Rather it is a definite form of activity of these individuals, a definite form of expressing their life, a definite *mode of life* on their part. As individuals express their life, so they are. What they are, therefore, coincides with their production, both with *what* they produce and with *how* they produce. The nature of individuals thus depends on the material conditions determining their production.

This production only makes its appearance with the *increase of population.* In its turn this presupposes the *intercourse* of individuals with one another. The form of this intercourse is, again, determined by production.

The relations of different nations among themselves depend upon the extent to which each has developed its productive forces, the division of labor and internal intercourse. This statement is generally recognized. But not only the relation of one nation to others, but also the whole internal structure of the nation itself depends on the stage of development reached by its production and its internal and external intercourse. How far the productive forces of a nation are developed is shown most manifestly by the degree to which the division of labor has been carried. Each new productive force, insofar as it is not merely a quantitative extension of productive forces already known (for instance, the bringing into cultivation of fresh land), causes a further development of the division of labor.

The division of labor inside a nation leads at first to the separation of industrial and commercial from agricultural labor, and hence to the separation of *town* and *country* and to the conflict of their interests. Its further development leads to the separation of commerical from industrial labor. At the same time through the division of labor inside these various branches there develop various divisions among the individuals cooperating in definite kinds of labor. The relative position of these individual groups is determined by the methods employed in agriculture, industry and commerce (patriarchalism, slavery, estates, classes). These same conditions are to be seen (given a more developed intercourse) in the relations of different nations to one another.

The various stages of development in the division of labor are just so many different forms of ownership, i.e., the existing stage in the division of labor determines also the relations of individuals to one another with reference to the material, instrument, and product of labor.

. . . The fact is, therefore, that definite individuals who are productively active in a definite way enter into these definite social and political relations. Empirical observation must in each separate instance bring out empirically, and without any mystification and speculation, the connection of the social and political structure with production. The social structure and the State are continually evolving out of the life-process of definite individuals, but of individuals, not as they may appear in their own or other people's imagination, but as they *really* are: i.e., as they operate, produce materially and hence as they work under definite material limits, presuppositions and conditions independent of their will.

The production of ideas, of conceptions, of consciousness, is at first directly interwoven with the material activity and the material intercourse of men, the language of real life. Conceiving, thinking, the mental intercourse of men, appear at this stage as the direct efflux of their material behavior. The same applies to mental production as expressed in the language of politics, laws, morality, religion, metaphysics, etc. of a people. Men are the producers of their conceptions, ideas, etc.—real, active men, as they are conditioned by a definite development of their productive forces and of the intercourse corresponding to these, up to its furthest forms. Consciousness can never be anything else than conscious existence, and the existence of men is their actual life-process. If in all ideology men and their circumstances appear upside-down as in a *camera obscura*, this phenomenon arises just as much from their historical life-process as the inversion of objects on the retina does from their physical life-process.

In direct contrast to German philosophy which descends from heaven to earth, here we ascend from earth to heaven. That is to say, we do not set out from what men say, imagine, conceive, nor from men as narrated, thought of, imagined, conceived, in order to arrive at men in the flesh. We set out from real, active men, and on the basis of their real life-process we demonstrate the development of the ideological reflexes and echoes of this life-process. The phantoms formed in the human brain are also, necessarily, sublimates of their material life-process, which is empirically verifiable and bound to material premises. Morality, religion, metaphysics, all the rest of ideology and their corresponding forms of consciousness, thus no longer retain the semblance of independence. They have no history, no development; but men, developing their material production and their material intercourse, alter, along with this their real existence, their thinking and the products of their thinking. Life is not determined by consciousness, but consciousness by life. In the first method of approach the starting point is consciousness taken as the living individual; in the second method, which conforms to real life, it is the real living individuals themselves, and consciousness is considered solely as *their* consciousness.

This method of approach is not devoid of premises. It starts out from the real premises and does not abandon them for a moment. Its premises are men, not in any fantastic isolation and rigidity, but in their actual, empirically perceptible process of development under definite conditions. As soon as this active life-process is described, history ceases to be a collection of dead facts as it is with the empiricists (themselves still abstract), or an imagined activity of imagined subjects, as with the idealists.

Where speculation ends—in real life—there real, positive science begins: the representation of the practical activity, of the practical process of development of men. Empty talk about consciousness ceases, and real knowledge has to take its place. When reality is depicted, philosophy as an independent branch of knowledge loses its medium of existence. At the best its place can only be taken by a summing-up of the most general results, abstractions which arise from the observation of the historical development of men. Viewed apart from real history, these abstractions have in themselves no value whatsoever. They can only serve to facilitate the arrangement of historical material, to indicate the sequence of its separate strata. But they by no means afford a recipe or schema, as does philosophy, for neatly trimming the epochs of history. On the contrary, our difficulties begin only when we set about the observation and the arrangement—the real depiction—of our historical material, whether of a past epoch or of the present. The removal of these difficulties is governed by premises which it is quite impossible to state here, but which only the study of the actual life-process and the activity of the individuals of each epoch will make evident. We shall select here some of these abstractions, which we use in contradistinction to the ideologists, and shall illustrate them by historical examples.

HISTORY: FUNDAMENTAL CONDITIONS

Since we are dealing with the Germans, who are devoid of premises, we must begin by stating the first premise of all human existence and, therefore, of all history, the premise, namely, that men must be in a position to live in order to be able to "make history." But life involves before everything else eating and drinking, a habitation, clothing and many other things. The first historical act is thus the production of the means to satisfy these needs, the production of material life itself. And indeed this is an historical act, a fundamental condition of all history, which today, as thousands of years ago, must daily and hourly be fulfilled merely in order to sustain human life. Even when the sensuous world is reduced to a minimum, to a stick as with Saint Bruno [Bauer], it presupposes the action of producing the stick. Therefore in any interpretation of history one has first of all to observe this fundamental fact in all its significance and all its implications and to accord it its due importance. It is well known that the Germans have never done this, and they have never, therefore, had an *earthly* basis for history and consequently never an historian. The French and the English, even if they have conceived the relation of this fact with so-called history only in an extremely one-sided fashion, particularly as long as they remained in the toils of political ideology, have nevertheless made the first attempts to give the writing of history a materialistic basis by being the first to write histories of civil society, of commerce and industry.

The second point is that the satisfaction of the first need (the action of satisfying, and the instrument of satisfaction which has been acquired) leads to new needs; and this production of new needs is the first historical act. Here we recognize immediately the spiritual ancestry of the great historical wisdom of the Germans, who, when they run out of positive material and when they can serve up neither theological nor political nor literary rubbish, assert that this is not history at all, but the "prehistoric era." They do not, however, enlighten us as to how we proceed from this nonsensical "prehistory" to history proper; although, on the other hand, in their historical speculation they seize upon this "prehistory" with especial eagerness because they imagine themselves safe there from interference on the part of "crude facts," and, at the same time, because there they can give full rein to their speculative impulse and set up and knock down hypotheses by the thousand.

The third circumstance which, from the very outset, enters into historical development, is that men, who daily remake their own life, begin to make other men, to propagate their kind: the relation between man and woman, parents and children, the *family*. The family, which to begin with is the only social relationship, becomes later, when increased needs create new social relations and the increased population new needs, a subordinate one (except in Germany), and must then be treated and analyzed according to the existing empirical data, not according to "the concept of the family," as is the custom in Germany. These three aspects of social activity are not of course to be taken as three different stages, but just as three aspects or, to make it clear to the Germans, three "moments," which have existed simultaneously since the dawn of history and the first men, and which still assert themselves in history today.

The production of life, both of one's own in labor and of fresh life in procreation, now appears as a double relationship: on the one hand as a natural, on the other as a social relationship. By social we understand the co-operation of several individuals, no matter under what conditions, in what manner and to what end. It follows from this that a certain mode of production, or industrial stage, is always combined with a certain mode of co-operation, or social stage, and this mode of co-operation is itself a "productive force." Further, that the multitude of productive forces accessible to men determines the nature of society, hence, that the "history of humanity" must always be studied and treated in relation to the history of industry and exchange. But it is also clear how in Germany it is impossible to write this sort of history, because the Germans lack not only the necessary power of comprehension and the material but also the "evidence of their senses," for across the Rhine you cannot have any experience of these things since history has stopped happening. Thus it is quite obvious from the start that there exists a materialistic connection of men with one another, which is determined by their needs and their mode of production, and which is as old as men themselves. This connection is ever taking on new forms, and thus presents a "history" independently of the existence of any political or religious nonsense which in addition may hold men together.

Only now, after having considered four moments, four aspects of the primary historical relationships, do we find that man also possesses "consciousness," but, even so, not inherent, not "pure" consciousness. From the start the "spirit" is afflicted with the curse of being "burdened" with matter, which here makes its appearance in the form of agitated layers of air, sounds, in short, of language. Language is as old as consciousness; language *is* practical consciousness that exists also for other men, and for that reason alone it really exists for me personally as well; language, like consciousness, only arises from the need, the necessity, of intercourse with other men. Where there exists a relationship, it exists for me: the animal does not enter into "*relations*" with anything, it does not enter into any relation at all. For the animal, its relation to others does not exist as a relation. Consciousness is, therefore, from the very beginning a social product, and remains so as long as men exist at all. Consciousness is at first, of course, merely consciousness concerning the *immediate* sensuous environment and consciousness of the limited connection with other persons and things outside the individual who is growing self-conscious. At the same time it is consciousness of nature, which first appears to men as a completely alien, all-powerful and unassailable force, with which men's relations are purely animal and by which they are overawed like beasts; it is thus a purely animal consciousness of nature (natural religion) just because nature is as yet hardly modified historically. (We see here immediately: this natural religion or this particular relation of men to nature is determined by the form of society and vice versa. Here, as everywhere, the identity of nature and man appears in such a way that the restricted relation of men to nature determines their restricted relation to one another, and their restricted relation to one another determines men's restricted relation to nature.) On the other hand, man's consciousness of the necessity of associating with the individuals around him is the beginning of the consciousness that he is living in society at all. This beginning is as animal as social life itself at this stage. It is mere herd-consciousness, and at this point man is distinguished from sheep only by the fact that with him consciousness takes the place of instinct or that his instinct is a conscious one. This sheep-like or tribal consciousness receives its further development and extension through increased productivity, the increase of needs, and, what is fundamental to both of these, the increase of population. With these there develops the division of labor, which was originally nothing but the division of labor in the sexual act, then that division of labor which develops spontaneously or "naturally" by virtue of natural predisposition (e.g., physical strength), needs, accidents, etc. etc. Division of labor only becomes truly such from the moment when a division of material and mental labor appears. (The first form of ideologists, *priests*, is concurrent.) From this moment onwards consciousness *can* really flatter itself that it is something other than consciousness of existing practice, that it *really* represents something without representing something real; from now on consciousness is in a position to emancipate itself from the world and to proceed to the formation of "pure" theory, theology, philosophy, ethics, etc. But even if this theory, theology, philosophy, ethics, etc. comes into contradiction with the existing relations, this can only occur because existing social relations have come into contradiction with existing forces of production; this, moreover, can also occur in a particular national sphere of relations through the appearance of the contradiction, not within the national orbit, but between this national consciousness and the practice of other nations, i.e., between the national and the general consciousness of a nation (as we see it now in Germany).

Moreover, it is quite immaterial what consciousness starts to do on its own: out of all such muck we get only the one inference that these three moments, the forces of production, the state of society, and consciousness, can and must come into contradiction with one another, because the *division of labor* implies the possibility, nay the fact that intellectual and material activity—enjoyment and labor, production and consumption—devolve on different individuals, and that the only possibility of their not coming into contradiction lies in the negation in its turn of the division of labor. It is self-evident, moreover, that "spectres," "bonds," "the higher being," "concept," "scruple," are merely the idealistic, spiritual expression, the conception apparently of the isolated individual, the image of very empirical fetters and limitations, within which the mode of production of life and the form of intercourse coupled with it move.

PRIVATE PROPERTY AND COMMUNISM

With the division of labor, in which all these contradictions are implicit, and which in its turn is based on the natural division of labor in the family and the separation of society into individual families opposed to one another, is given simultaneously the *distribution*, and indeed the *unequal* distribution, both quantitative and qualitative, of labor and its products, hence property: the nucleus; the first form of which lies in the family, where wife and children are the slaves of the husband. This latent slavery in the family, though still very crude, is the first property, but even at this early stage it corresponds perfectly to the definition of modern economists who call it the power of disposing of the labor-power of others. Division of labor and private property are, moreover, identical expressions: in the one the same thing is affirmed with reference to activity as is affirmed in the other with reference to the product of the activity.

Further, the division of labor implies the contradiction between the interest of the separate individual or the individual family and the communal interest of all individuals who have intercourse with one another. And indeed, this communal interest does not exist merely in the imagination, as the "general interest," but first of all in reality, as the mutual interdependence of the individuals among whom the labor is divided. And finally, the division of labor offers us the first example of how, as long as man remains in natural society, that is, as long as a cleavage exists between the particular and the common interest, as long, therefore, as activity is not voluntarily, but naturally, divided, man's own deed becomes an alien power opposed to him, which enslaves him instead of being controlled by him. For as soon as the distribution of labor comes into being, each man has a particular, exclusive sphere of activity, which is forced upon him and from which he cannot escape. He is a hunter, a fisherman, a shepherd, or a critical critic, and must remain so if he does not want to lose his means of livelihood; while in communist society, where nobody has one exclusive sphere of activity but each can become accomplished in any branch he wishes, society regulates the general production and thus makes it possible for me to do one thing today and another tomorrow, to hunt in the morning, fish in the afternoon, rear cattle in the evening, criticize after dinner, just as I have a mind, without ever becoming hunter, fisherman, shepherd or critic. This fixation of social activity, this consolidation of what we ourselves produce into an objective power above us, growing out of our control, thwarting our expectations, bringing to naught our calculations, is one of the chief factors in historical development up till now.

And out of this very contradiction between the interest of the individual and that of the community the latter takes an independent form as the *State*, divorced from the real interests of individual and community, and at the same time as an illusory communal life, always based, however, on the real ties existing in every family and tribal conglomeration—such as flesh and blood, language, division of labor on a larger scale, and other interests—and especially, as we shall enlarge upon later, on the classes, already determined by the division of labor, which in every such mass of men separate out, and of which one dominates all the others. It follows from this that all struggles within the State, the struggle between democracy, aristocracy, and monarchy, the struggle for the franchise, etc. etc. are merely the illusory forms in which the real struggles of the different classes are fought out among one another. Of this the German theoreticians have not the faintest inkling, although they have received a sufficient introduction to the subject in the *Deutsch-Französische Jahrbücher* and *Die heilige Familie*. Further, it follows that every class which is struggling for mastery, even when its domination, as is the case with the proletariat, postulates the abolition of the old form of society in its entirety and of domination itself, must first conquer for itself political power in order to represent its interest in turn as the general interest, which immediately it is forced to do. Just because individuals seek *only* their particular interest, which for them does not coincide with their communal interest, the latter will be imposed on them as an interest "alien" to them, and "independent" of them, as in its turn a particular, peculiar "general" interest; or they themselves must remain within this discord, as in democracy. On the other hand, too, the *practical* struggle of these particular interests, which constantly *really* run counter to the communal and illusory communal interests, makes *practical* intervention and control necessary through the illusory "general" interest in the form of the State.

The social power, i.e., the multiplied productive force, which arises through the co-operation of different individuals as it is determined by the division of labor, appears to these individuals, since their co-operation is not voluntary but has come about naturally, not as their own united power, but as an alien force existing outside them, of the origin and goal of which they are ignorant, which they thus cannot control, which on the contrary passes through a peculiar series of phases and stages independent of the will and the action of man, nay even being the prime governor of these.

How otherwise could for instance property have had a history at all, have taken on different forms, and landed property, for example, according to the different premises given, have proceeded in France from parcellation to centralization in the hands of a few, in England from centralization in the hands of a few to parcellation, as is actually the case today? Or how does it happen that trade, which after all is nothing more than the exchange of products of various individuals and countries, rules the whole world through the relation of supply and demand—a relation which, as an English economist says, hovers over the earth like the fate of the ancients, and with invisible hand allots fortune and misfortune to men, sets up empires and overthrows empires, causes nations to rise and to disappear—while with the abolition of the basis of private property, with the communistic regulation of production (and, implicit in this, the destruction of the alien relation between men and what they themselves produce), the power of the relation of supply and demand is dissolved into nothing, and men get exchange, production, the mode of their mutual relation, under their own control again?

In history up to the present it is certainly an empirical fact that separate individuals have, with the broadening of their activity into world-historical activity, become more and more enslaved under a power alien to them (a pressure which they have conceived of as a dirty trick on the part of the so-called universal spirit, etc.), a power which has become more and more enormous and, in the last instance, turns out to be the *world market*. But it is just as empirically established that, by the overthrow of the existing state of society by the communist revolution (of which more below) and the abolition of private property which is identical with it, this power, which so baffles the German theoreticians, will be dissolved; and that then the liberation of each single individual will be accomplished in the measure in which history becomes transformed into world history. From the above it is clear that the real intellectual wealth of the individual depends entirely on the wealth of his real connections. Only then will the separate individuals be liberated from the various national and local barriers, be brought into practical connection with the material and intellectual production of the whole world and be put in a position to acquire the capacity to enjoy this all-sided production of the whole earth (the creations of man). *All-round* dependence, this natural form of the *world-historical* co-operation of individuals, will be transformed by this communist revolution into the control and conscious mastery of these powers, which, born of the action of men on one another, have till now overawed and governed men as powers completely alien to them. Now this view can be expressed again in speculative-idealistic, i.e., fantastic, terms as "self-generation of the species" ("society as the subject"), and thereby the consecutive series of interrelated individuals connected with each other can be conceived as a single individual, which accomplishes the mystery of generating itself. It is clear here that individuals certainly make *one another*, physically and mentally, but do not make themselves.

This "alienation" (to use a term which will be comprehensible to the philosophers) can, of course, only be abolished given two *practical* premises. For it to become an "intolerable" power, i.e., a power against which men make a revolution, it must necessarily have rendered the great mass of humanity "propertyless," and produced, at the same time, the contradiction of an existing world of wealth and culture, both of which conditions presuppose a great increase in productive power, a high degree of its development. And, on the other hand, this development of productive forces (which itself implies the actual empirical existence of men in their *world-historical,* instead of local, being) is an absolutely necessary practical premise because without it *want* is merely made general, and with *destitution* the struggle for necessities and all the old filthy business would necessarily be reproduced, and furthermore, because only with this universal development of productive forces is a *universal* intercourse between men established, which produces in all nations simultaneously the phenomenon of the "propertyless" mass (universal competition), makes each nation dependent on the revolutions of the others, and finally has put *world-historical,* empirically universal individuals in place of local ones. Without this, (I) communism could only exist as a local event; (2) the *forces* of intercourse themselves could not have developed as *universal,* hence intolerable powers: they would have remained homo-bred conditions surrounded by superstition; and (3) each extension of intercourse would abolish local communism. Empirically, communism is only possible as the act of the dominant peoples "all at once" and simultaneously, which presupposes the universal development of productive forces and the world intercourse bound up with communism. Moreover, the mass of *propertyless* workers—the utterly precarious position of labor-power on a mass scale cut off from capital or from even a limited satisfaction and, therefore, no longer merely temporarily deprived of work itself as a secure source of life—presupposes the *world market* through competition. The proletariat can thus only exist *world-historically*, just as communism, its activity, can only have a "world-historical" existence. World-historical existence of individuals means, existence of individuals which is directly linked up with world history.

Communism is for us not a *state of affairs* which is to be established, an *ideal* to which reality [will] have to adjust itself. We call communism the *real* movement which abolishes the present state of things. The conditions of this movement result from the premises now in existence.

THE ILLUSION OF THE EPOCH

CIVIL SOCIETY AND THE CONCEPTION OF HISTORY

The form of intercourse determined by the existing productive forces at all previous historical stages, and in its turn determining these, is *civil society*. The latter, as is clear from what we have said above, has as its premises and basis the simple family and the multiple, the so-called tribe; the more precise determinants of this society are enumerated in our remarks above. Already here we see how this civil society is the true source and theater of all history, and how absurd is the conception of history held hitherto, which neglects the real relationships and confines itself to high-sounding dramas of princes and states.

Civil society embraces the whole material intercourse of individuals within a definite stage of the development of productive forces. It embraces the whole commercial and industrial life of a given stage and, insofar, transcends the State and the nation, though, on the other hand again, it must assert itself in its foreign relations as nationality, and inwardly must organize itself as State. The word "civil" society [*bürgerliche Gesellschaft*] emerged in the eighteenth century, when property relationships had already extricated themselves from the ancient and medieval communal society. Civil society as such only develops with the bourgeoisie; the social organization evolving directly out of production and commerce, which in all ages forms the basis of the State and of the rest of the idealistic superstructure, has, however, always been designated by the same name.

History is nothing but the succession of the separate generations, each of which exploits the materials, the capital funds, the productive forces handed down to it by all preceding generations, and thus, on the one hand, continues the traditional activity in completely changed circumstances and, on the other, modifies the old circumstances with a completely changed activity. This can be speculatively distorted so that later history is made the goal of earlier history, e.g., the goal ascribed to the discovery of America is to further the eruption of the French Revolution. Thereby history receives its own special aims and becomes "a person ranking with other persons" (to wit: "Self-Consciousness, Criticism, the Unique," etc.), while what is designated with the words "destiny," "goal," "germ," or "idea" of earlier history is nothing more than an abstraction formed from later history, from the active influence which earlier history exercises on later history.

The further the separate spheres, which interact on one another, extend in the course of this development, the more the original isolation of the separate nationalities is destroyed by the developed mode of production and intercourse and the division of labor between various nations naturally brought forth by these, the more history becomes world history. Thus, for instance, if in England a machine is invented, which deprives countless workers of bread in India and China, and overturns the whole form of existence of these empires, this invention becomes a world-historical fact. Or again, take the case of sugar and coffee, which have proved their world-historical importance in the nineteenth century by the fact that the lack of these products, occasioned by the Napoleonic Continental System, caused the Germans to rise against Napoleon, and thus became the real basis of the glorious Wars of Liberation of 1813. From this it follows that this transformation of history into world history is not indeed a mere abstract act on the part of the "self-consciousness," the world spirit, or of any other metaphysical spectre, but a quite material, empirically verifiable act, an act the proof of which every individual furnishes as he comes and goes, eats, drinks and clothes himself.

This conception of history depends on our ability to expound the real process of production, starting out from the material production of life itself, and to comprehend the form of intercourse connected with this and created by this mode of production (i.e., civil society in its various stages), as the basis of all history; and to show it in its action as State, to explain all the different theoretical products and forms of consciousness, religion, philosophy, ethics, etc. etc. and trace their origins and growth from that basis; by which means, of course, the whole thing can be depicted in its totality (and therefore, too, the reciprocal action of these various sides on one another). It has not, like the idealistic view of history, in every period to look for a category, but remains constantly on the real *ground* of history; it does not explain practice from the idea but explains the formation of ideas from material practice; and accordingly it comes to the conclusion that all forms and products of consciousness cannot be dissolved by mental criticism, by resolution into "self-consciousness" or transformation into "apparitions," "spectres," "fancies," etc.

but only by the practical overthrow of the actual social relations which gave rise to this idealistic humbug; that not criticism but revolution is the driving force of history, also of religion, of philosophy and all other types of theory. It shows that history does not end by being resolved into "self-consciousness" as "spirit of the spirit," but that in it at each stage there is found a material result: a sum of productive forces, an historically created relation of individuals to nature and to one another, which is handed down to each generation from its predecessor; a mass of productive forces, capital funds and conditions, which, on the one hand, is indeed modified by the new generation, but also on the other prescribes for it its conditions of life and gives it a definite development, a special character. It shows that circumstances make men just as much as men make circumstances.

This sum of productive forces, capital funds and social forms of inter- course, which every individual and generation finds in existence as something given, is the real basis of what the philosophers have conceived as "substance" and "essence of man," and what they have deified and attacked; a real basis which is not in the least disturbed, in its effect and influence on the develop- ment of men, by the fact that these philosophers revolt against it as "self- consciousness" and the "Unique." These conditions of life, which different generations find in existence, decide also whether or not the periodically re- curring revolutionary convulsion will be strong enough to overthrow the basis of the entire existing system. And if these material elements of a complete revolution are not present (namely, on the one hand the existing productive forces, on the other the formation of a revolutionary mass, which revolts not only against separate conditions of society up till then, but against the very "production of life" till then, the "total activity" on which, it was based), then, as far as practical development is concerned, it is absolutely immaterial whether the *idea* of this revolution has been expressed a hundred times already, as the history of communism proves.

In the whole conception of history up to the present this real basis of history has either been totally neglected or else considered as a minor matter quite irrelevant to the course of history. History must, therefore, always be written according to an extraneous standard; the real production of life seems to be primeval history, while the truly historical appears to be separated from ordinary life, something extra-superterrestrial. With this the relation of man to nature is excluded from history and hence the antithesis of nature and history is created. The exponents of this conception of history have conse- quently only been able to see in history the political actions of princes and States, religious and all sorts of theoretical struggles, and in particular in each historical epoch have had to *share the illusion of that epoch*. For instance, if an epoch imagines itself to be actuated by purely "political" or "religious" motives, although "religion" and "politics" are only forms of its true motives, the historian accepts this opinion. The "idea," the "conception" of the people in question about their real practice, is transformed into the sole determining, active force, which controls and determines their practice. When the crude form in which the division of labor appears with the Indians and Egyptians calls forth the caste-system in their State and religion, the historian believes that the caste-system is the power which has produced this crude social form. While the French and the English at least hold by the political illusion, which is moderately close to reality, the Germans move in the realm of the "pure spirit," and make religious illusion the driving force of history. The Hegelian philosophy of history is the last consequence, reduced to its "finest expression," of all this German historiography, for which it is not a question of real, nor even of political, interests, but of pure thoughts, which consequently must appear to Saint Bruno as a series of "thoughts" that devour one another and are finally swallowed up in "self-consciousness."

play on identity or experience?

ON THE MIMETIC FACULTY

WALTER BENJAMIN

Nature creates similarities. One need only think of mimicry. The highest capacity for producing similarities, however, is man's. His gift of seeing resemblances is nothing other than rudiment of the powerful compulsion in former times to become and behave like something else. Perhaps there is none of his higher functions in which his mimetic faculty does not play a decisive role.

This faculty has a history, however, in both the phylogenetic and the ontogenetic sense. As regards the latter, play is to many its school. Children's play is everywhere permeated by mimetic modes of behavior, and its realm is by no means limited to what one person can imitate in another. The child plays at being not only a shopkeeper or teacher but also a windmill and a train. Of what use to him is this schooling to his mimetic faculty?

The answer presupposes an understanding of the phylogenetic significance of the mimetic faculty. Here it is not enough to think of what we understand today by the concept of similarity. As is known, the sphere of life that formerly seemed to be governed by the law of similarity was comprehensive; it ruled both microcosm and macrocosm. But these natural correspondences are given their true importance only if seen as stimulating and awakening the mimetic faculty in man. It must be borne in mind that neither mimetic powers nor mimetic objects remain the same in the course of thousands of years. Rather, we must suppose that the gift of producing similarities—for example, in dances, whose oldest function this was—and therefore also the gift of recognizing them have changed with historical development.

The direction of this change seems definable as the increasing decay of the mimetic faculty. For clearly the observable world of modern man contains only minimal residues of the magical correspondences and analogies that were familiar to ancient peoples. The question is whether we are concerned with the decay of this faculty or with its transformation. Of the direction in which the latter might lie some indications may be derived, even if indirectly, from astrology.

We must assume in principle that in the remote past the processes considered imitable included those in the sky. In dance, on other cultic occasions, such imitation could be produced, such similarity manipulated. But if the mimetic genius was really a life-determining force for the ancients, it is not difficult to imagine that the newborn child was thought to be in full possession of this gift, and in particular to be perfectly molded on the structure of cosmic being.

Allusion to the astrological sphere may supply a first reference point for an understanding of the concept of nonsensuous similarity. True, our existence no longer includes what once made it possible to speak of this kind of similarity: above all the ability to produce it. Nevertheless we, too, possess a canon according to which the meaning of nonsensuous similarity can be at least partly clarified. And this canon is language.

From time immemorial the mimetic faculty has been conceded some influence on language. Yet this was done without foundation: without consideration of a further meaning, still less a history, of the mimetic faculty. But above all such notions remained closely tied to the commonplace, sensuous area of similarity. All the same, imitative behavior in language formation was acknowledged under the name of onomatopoeia. Now if language, as is evident, is not an agreed system of signs, we shall be constantly obliged to have recourse to the kind of thoughts that appear in their most primitive form as the onomatopoeic mode of explanation. The question is whether this can be developed and adapted to improved understanding.

"Every word—and the whole of language," it has been asserted, is onomatopoeic. It is difficult to conceive in any detail the program that might be implied by this proposition. However, the concept of nonsensuous similarity is of some relevance. For if words meaning the same thing in different languages are arranged about that thing as their center we have to inquire how they all— while often possessing not the slightest similarity to one another—are similar to what they signify at their center. Yet this kind of similarity may be explained not only by the relationships between words meaning the same thing in different languages, just as, in general, our reflections cannot be restricted to the spoken word. They are equally concerned with the written word. And here it is noteworthy that the latter—in some cases perhaps more vividly than the spoken word—illuminates, by the relation of its written form to what it signifies, the nature of nonsensuous similarity. In brief it is nonsensuous similarity that establishes the ties not only between the spoken and the signified but also between the written and the signified, and equally between the spoken and the written.

Graphology has taught us to recognize in handwriting images that the unconscious of the writer conceals in it. It may be supposed that the mimetic process that expresses itself in this way in the activity of the writer was, in the very distant times in which script originated, of utmost importance for writing. Script has thus become, like language, an archive of nonsensuous similarities, of nonsensuous correspondences.

This aspect of language as script, however, does not develop in isolation from its other, semiotic aspect. Rather, the mimetic element in language can, like a flame, manifest itself only through a kind of bearer. This bearer is the semiotic element. Thus the coherence of words or sentences is the bearer through which, like a flash, similarity appears. For its production by man—like its perception by him—is in many cases, and particularly the most important, limited to flashes. It flits past. It is not improbable that the rapidity of writing and reading heightens the fusion of the semiotic and the mimetic in the sphere of language.

"To read what was never written." Such reading is the most ancient: reading before all languages, from the entrails, the stars, or dances. Later the mediating link of a new kind of reading, of runes and hieroglyphs, came into use. It seems fair to suppose that these were the stages by which the mimetic gift, which was once the foundation of occult practices, gained admittance to writing and language. In this way language may be seen as the highest level of mimetic behavior and the most complete archive of nonsensuous similarity: a medium into which the earlier powers of mimetic production and comprehension have passed without residue, to the point where they have liquidated those of magic.

What is phenomenology? It may seem strange that this question has still to be asked half a century after the first works of Husserl. The fact remains that it has by no means been answered. Phenomenology is the study of essences; and according to it, all problems amount to finding definitions of essences: the essence of perception, or the essence of consciousness, for example. But phenomenology is also a philosophy which puts essences back into existence, and does not expect to arrive at an understanding of man and the world from any starting point other than that of their "facticity." It is a transcendental philosophy which places in abeyance the assertions arising out of the natural attitude, the better to understand them; but it is also a philosophy for which the world is always "already there" before reflection begins—as an inalienable presence; and all its efforts are concentrated upon re-achieving a direct and primitive contact with the world, and endowing that contact with a philosophical status. It is the search for a philosophy which shall be a "rigorous science," but it also offers an account of space, time and the world as we live them. It tries to give a direct description of our experience as it is, without taking account of its psychological origin and the causal explanations which the scientist, the historian or the sociologist may be able to provide. Yet Husserl in his last works mentions a "genetic phenomenology," and even a "constructive phenomenology." One may try to do away with these contradictions by making a distinction between Husserl's and Heidegger's phenomenologies; yet the whole of *Sein und Zeit* springs from an indication given by Husserl and amounts to no more than an explicit account of the "naturlicher Weltbegriff" or the "Lebenswelt" which Husserl, towards the end of his life, identified as the central theme of phenomenology, with the result that the contradiction reappears in Husserl's own philosophy. The reader pressed for time will be inclined to give up the idea of covering a doctrine which says everything, and will wonder whether a philosophy which cannot define its scope deserves all the discussion which has gone on around it, and whether he is not faced rather by a myth or a fashion.

Even if this were the case, there would still be a need to understand the prestige of the myth and the origin of the fashion, and the opinion of the responsible philosopher must be that *phenomenology can be practiced and identified as a manner or style of thinking, that it existed as a movement before arriving at complete awareness of itself as a philosophy*. It has been long on the way, and its adherents have discovered it in every quarter, certainly in Hegel and Kierkegaard, but equally in Marx, Nietzsche and Freud. A purely linguistic examination of the texts in question would yield no proof; we find in texts only what we put into them, and if ever any kind of history has suggested the interpretations which should be put on it, it is the history of philosophy. We shall find in ourselves, and nowhere else, the unity and true meaning of phenomenology. It is less a question of counting up quotations than of determining and expressing in concrete form this *phenomenology for ourselves* which has given a number of present-day readers the impression, on reading Husserl or Heidegger, not so much of encountering a new philosophy as of recognizing what they had been waiting for. Phenomenology is accessible only through a phenomenological method. Let us, therefore, try systematically to bring together the celebrated phenomenological themes as they have grown spontaneously together in life. Perhaps we shall then understand why phenomenology has for so long remained, at an initial stage, as a problem to be solved and a hope to be realized.

It is a matter of describing, not of explaining or analyzing. Husserl's first directive to phenomenology, in its early stages, to be a "descriptive psychology," or to return to the "things themselves," is from the start a foreswearing of science. I am not the outcome or the meeting-point of numerous causal agencies which determine my bodily or psychological make-up. I cannot conceive myself as nothing but a bit of the world, a mere object of biological, psychological or sociological investigation. I cannot shut myself up within the realm of science. All my knowledge of the world, even my scientific knowledge, is gained from my own particular point of view, or from some experience of the world without which the symbols of science would be meaningless. The whole universe of science is built upon the world as directly experienced, and if we want to subject science itself to rigorous scrutiny and arrive at a precise assessment of its meaning and scope, we must begin by reawakening the basic experience of the world of which science is the second-order expression. Science has not and never will have, by its nature, the same significance *qua* form of being as the world which we perceive, for the simple reason that it is a rationale or explanation of that world. I am not a living creature nor even a "man," nor again even "a consciousness" endowed with all the characteristics which zoology, social anatomy or inductive psychology recognize in these various products of the natural or historical process—I am the absolute source, my existence does not stem from my antecedents, from my physical and social environment; instead it moves out towards them and sustains them, for I alone bring into being for myself (and therefore into being in the only sense that the word can have for me) the tradition which I elect to carry on, or the horizon whose distance from me would be abolished—since that distance is not one of its properties—if I were not there to scan it with my gaze. Scientific points of view, according to which my existence is a moment of the world's, are always both naive and at the same time dishonest, because they take for granted, without explicitly mentioning it, the other point of view, namely, that of consciousness, through which from the outset a world forms itself round me and begins to exist for me. To return to things themselves is to return to that world which precedes knowledge, of which knowledge always *speaks*, and in relation to which every scientific schematization is an abstract and derivative sign-language, as is geography in relation to the countryside in which we have learned beforehand what a forest, a prairie or a river is.

This move is absolutely distinct from the idealist return to consciousness, and the demand for a pure description excludes equally the procedure of analytical reflection on the one hand, and that of scientific explanation on the other. Descartes and particularly Kant *detached* the subject, or consciousness, by showing that I could not possibly apprehend anything as existing unless I first of all experienced myself as existing in the act of apprehending it. They presented consciousness, the absolute certainty of my existence for myself, as the condition of there being anything at all; and the act of relating as the basis of relatedness. It is true that the act of relating is nothing if divorced from the spectacle of the world in which relations are found; the unity of consciousness in Kant is achieved simultaneously with that of the world. And in Descartes methodical doubt does not deprive us of anything, since the whole world, at least in so far as we experience it, is reinstated in the *Cogito*, enjoying equal certainty, and simply labelled "thought of. . . ." But the relations between subject and world are not strictly bilateral: if they were, the certainty of the world would, in Descartes, be immediately given with that of the *Cogito*, and Kant would not have talked about his "Copernican revolution." Analytical reflection starts from our experience of the world and goes back to the subject as to a condition of possibility distinct from that experience revealing the all-embracing synthesis as that without which there would be no world. To this extent it ceases to remain part of our experience and offers, in place of an account, a reconstruction. It is understandable, in view of this, that Husserl, having accused Kant of adopting a "faculty psychologism," should have urged, in place of a noetic analysis which bases the world on the synthesizing activity of the subject, his own "*noematic reflection*" which remains within the object and, instead of begetting it, brings to light its fundamental unity.

The world is there before any possible analysis of mine, and it would be artificial to make it the outcome of a series of syntheses which link, in the first place sensations, then aspects of the object corresponding to different perspectives, when both are nothing but products of analysis, with no sort of prior reality. Analytical reflection believes that it can trace back the course followed by a prior constituting act and arrive, in the "inner man"—to use Saint Augustine's expression—at a constituting power which has always been identical with that inner self. Thus reflection is carried off by itself and installs itself in an impregnable subjectivity, as yet untouched by being and time. But this is very ingenuous, or at least it is an incomplete form of reflection which loses sight of its own beginning. When I begin to reflect my reflection bears upon an unreflective experience; moreover my reflection cannot be unaware of itself as an event, and so it appears to itself in the light of a truly creative act, of a changed structure of consciousness, and yet it has to recognize, as having priority over its own operations, the world which is given to the subject because the subject is given to himself. The real has to be described, not constructed or formed. Which means that I cannot put perception into the same category as the syntheses represented by judgments, acts or predications. My field of perception is constantly filled with a play of colors, noises and fleeting tactile sensations which I cannot relate precisely to the context of my clearly perceived world, yet which I nevertheless immediately "place" in the world, without ever confusing them with my daydreams. Equally constantly I weave dreams round things. I imagine people and things whose presence is not incompatible with the context, yet who are not in fact involved in it: they are ahead of reality, in the realm of the imaginary. If the reality of my perception were based solely on the intrinsic coherence of "representations," it ought to be forever hesitant and, being wrapped up in my conjectures on probabilities, I ought to be ceaselessly taking apart misleading syntheses, and reinstating in reality stray phenomena which I had excluded in the first place. But this does not happen. The real is a closely woven fabric. It does not await our judgment before incorporating the most surprising phenomena, or before rejecting the most plausible figments of our imagination. Perception is not a science of the world, it is not even an act, a deliberate taking up of a position; it is the background from which all acts stand out, and is presupposed by them. The world is not an object such that I have in my possession the law of its making; it is the natural setting of, and field for, all my thoughts and all my explicit perceptions. Truth does not "inhabit" only "the inner man," or more accurately, there is no inner man, man is in the world, and only in the world does he know himself. When I return to myself from an excursion into the realm of dogmatic common sense or of science, I find, not a source of intrinsic truth, but a subject destined to the world.

Probably the chief gain from phenomenology is to have united extreme subjectivism and extreme objectivism in its notion of the world or of rationality. Rationality is precisely proportioned to the experiences in which it is disclosed. To say that there exists rationality is to say that perspectives blend, perceptions confirm each other, a meaning emerges. But it should not be set in a realm apart, transposed into absolute Spirit, or into a world in the realist sense. The phenomenological world is not pure being, but the sense which is revealed where the paths of my various experiences intersect, and also where my own and other people's intersect and engage each other like gears. It is thus inseparable from subjectivity and intersubjectivity, which find their unity when I either take up my past experiences in those of the present, or other people's in my own. For the first time the philosopher's thinking is sufficiently conscious not to anticipate itself and endow its own results with reified form in the world. The philosopher tries to conceive the world, others and himself and their interrelations. But the meditating Ego, the "impartial spectator" [*uninteressierter Zuschauer*] do not rediscover an already given rationality, they "establish themselves," and establish it, by an act of initiative which has no guarantee in being, its justification resting entirely on the effective power which it confers on us of taking our own history upon ourselves.

The phenomenological world is not the bringing to explicit expression of a pre-existing being, but the laying down of being. Philosophy is not the reflection of a pre-existing truth, but, like art, the act of bringing truth into being. One may well ask how this creation is *possible*, and if it does not recapture in things a pre-existing Reason. The answer is that the only pre-existent Logos is the world itself, and that the philosophy which brings it into visible existence does not begin by being *possible*, it is actual or real like the world of which it is a part, and no explanatory hypothesis is clearer than the act whereby we take up this unfinished world in an effort to complete and conceive it. Rationality is not a *problem*. There is behind it no unknown quantity which has to be determined by deduction, or, beginning with it, demonstrated inductively. We witness every minute the miracle of related experiences, and yet nobody knows better than we do how this miracle is worked, for we are ourselves this network of relationships. The world and reason are not problematical. We may say, if we wish, that they are mysterious, but their mystery defines them: there can be no question of dispelling it by some "solution," it is on the hither side of all solutions. True philosophy consists in re-learning to look at the world, and in this sense a historical account can give meaning to the world quite as "deeply" as a philosophical treatise. We take our fate in our hands, we become responsible for our history through reflection, but equally by a decision on which we stake our life, and in both cases what is involved is a violent act which is validated by being performed.

Phenomenology, as a disclosure of the world, rests on itself, or rather provides its own foundation. All cognitions are sustained by a "ground" of postulates and finally by our communication with the world as primary embodiment of rationality. Philosophy, as radical reflection, dispenses in principle with this resource. As, however, it too is in history, it too exploits the world and constituted reason. It must therefore put to itself the question which it puts to all branches of knowledge, and so duplicate itself infinitely, being, as Husserl says, a dialogue or infinite meditation, and, in so far as it remains faithful to its intention, never knowing where it is going. The unfinished nature of phenomenology and the inchoative atmosphere which has surrounded it are not to be taken as a sign of failure, they were inevitable because phenomenology's task was to reveal the mystery of the world and of reason. If phenomenology was a movement before becoming a doctrine or a philosophical system, this was attributable neither to accident, nor to fraudulent intent. It is as painstaking as the works of Balzac, Proust, Valéry or Cézanne—by reason of the same kind of attentiveness and wonder, the same demand for awareness, the same will to seize the meaning of the world or of history as that meaning comes into being. In this way it merges into the general effort of modern thought . . .

THE PHENOMENAL FIELD

It will now be seen in what direction the following chapters will carry their inquiry. "Sense experience" has become once more a question for us. Empiricism had emptied it of all mystery by bringing it down to the possession of a quality. This had been possible only at the price of moving far from the ordinary acceptation of the word. Between sense experience and knowing, common experience establishes a difference which is not that between the quality and the concept. This rich notion of sense experience is still to be found in Romantic usage, for example, in Herder. It points to an experience in which we are given not "dead" qualities, but active ones. A wooden wheel placed on the ground is not, *for sight*, the same thing as a wheel bearing a load. A body at rest because no force is being exerted upon it is again for sight not the same thing as a body in which opposing forces are in equilibrium. The light of a candle changes its appearance for a child when, after a burn, it stops attracting the child's hand and becomes literally repulsive. Vision is already inhabited by a meaning (*sens*) which gives it a function in the spectacle of the world and in our existence. The pure *quale* would be given to us only if the world were a spectacle and one's own body a mechanism with which some impartial mind made itself acquainted. Sense experience, on the other hand, invests the quality with vital value, grasping it first in its meaning for us, for that heavy mass which is our body, whence it comes about that it always involves a reference to the body. The problem is to understand these strange relationships which are woven between the parts of the landscape, or between it and me as incarnate subject, and through which an object perceived can concentrate in itself a whole scene or become the *imago* of a whole segment of life. Sense experience is that vital communication with the world which makes it present as a familiar setting of our life. It is to it that the perceived object and the perceiving subject owe their thickness. It is the intentional tissue which the effort to know will try to take apart. With the problem of sense experience, we rediscover that of association and passivity. They have ceased to be problematical because the classical philosophies put themselves either below or above them, giving them everything or nothing: sometimes association was understood as a mere *de facto* co-existence, sometimes derived from an intellectual construction; sometimes, passivity was imported from things into the mind, and sometimes analytical reflection would find in it an activity of understanding. Whereas these notions take on their full meaning if sense experience is distinguished from quality: then association or rather "affinity," in the Kantian sense, is the central phenomenon of perceptual life, since it is the constitution, without any ideal model, of a significant grouping. The distinction between the perceptual life and the concept, between passivity and spontaneity is no longer abolished by analytical reflection, since we are no longer forced by the atomism of sensation to look to some connecting activity for our principle of all co-ordination. Finally, after sense experience, understanding also needs to be redefined, since the general connective function ultimately attributed to it by Kantianism is now spread over the whole intentional life and no longer suffices to distinguish it. We shall try to bring out in relation to perception, both the instinctive substructure and the superstructures erected upon it by the exercise of intelligence. As Cassirer puts it, by mutilating perception from above, empiricism mutilated it from below too: the impression is as devoid of instinctive and affective meaning as of ideal significance. One might add that mutilating perception from below, treating it immediately as knowledge and forgetting its existential content, amounts to mutilating it from above, since it involves taking for granted and passing over in silence the decisive moment in perception: the upsurge of a *true* and *exact* world. Reflection will be sure of having precisely located the center of the phenomenon if it is equally capable of bringing to light its vital inherence and its rational intention.

So, "sensation" and "judgment" have together lost their apparent clearness: we have observed that they were clear only as long as the prejudice in favor of the world was maintained. As soon as one tried by means of them, to picture consciousness in the process of perceiving, to revive the forgotten perceptual experience, and to relate them to it, they were found to be inconceivable. By dint of making these difficulties more explicit, we were drawn implicitly into a new kind of analysis, into a new dimension in which they were destined to disappear. The criticism of the constancy hypothesis and more generally the reduction of the idea of "the world" opened up a *phenomenal field* which now has to be more accurately circumscribed, and suggested the rediscovery of a direct experience which must be, at least provisionally, assigned its place in relation to scientific knowledge, and to psychological and philosophical reflection.

Science and philosophy have for centuries been sustained by unquestioning faith in perception. Perception opens a window on to things. This means that it is directed, quasi-teleologically, towards a *truth in itself* in which the reason underlying all appearances is to be found. The tacit thesis of perception is that at every instant experience can be coordinated with that of the previous instant and that of the following, and my perspective with that of other consciousnesses—that all contradictions can be removed, that monadic and intersubjective experience is one unbroken text—that what is now indeterminate for me could become determinate for a more complete knowledge, which is as it were realized in advance in the thing, or rather which is the thing itself. Science has first been merely the sequel or amplification of the process which constitutes perceived things. Just as the thing is the invariant of all sensory fields and of all individual perceptual fields, so the scientific concept is the means of fixing and objectifying phenomena. Science defined a theoretical state of bodies not subject to the action of any force, and *ipso facto* defined force, reconstituting with the aid of these ideal components the processes actually observed. It established statistically the chemical properties of pure bodies, deducing from these those of empirical bodies, and seeming thus to hold the plan of creation or in any case to have found a reason immanent in the world. The notion of geometrical space, indifferent to its contents, that of pure movement which does not by itself affect the properties of the object, provided phenomena with a setting of inert existence in which each event could be related to physical conditions responsible for the changes occurring, and therefore contributed to this freezing of being which appeared to be the task of physics. In thus developing the concept of the thing, scientific knowledge was not aware that it was working on a presupposition. Precisely because perception, in its vital implications and prior to any theoretical thought, is presented as perception of a being, it was not considered necessary for reflection to undertake a genealogy of being, and it was therefore confined to seeking the conditions which make being possible. Even if one took account of the transformations of determinant consciousness, even if it were conceded that the constitution of the object is never completed, there was nothing to add to what science said of it; the natural object remained an ideal unity for us and, in the famous words of Lachelier, a network of general properties. It was no use denying any ontological value to the principles of science and leaving them with only a methodical value, for this reservation made no essential change as far as philosophy was concerned, since the sole conceivable being remained defined by scientific method. The living body, under these circumstances, could not escape the determinations which alone made the object into an object and without which it would have had no place in the system of experience. The value predicates which the reflecting judgement confers upon it had to be sustained, in being, by a foundation of physico-chemical properties. In ordinary experience we find a fittingness and a meaningful relationship between the gesture, the smile and the tone of a speaker. But this reciprocal relationship of expression which presents the human body as the outward manifestation of a certain manner of being-in-the-world, had, for mechanistic physiology, to be resolved into a series of causal relations.

It was necessary to link to centripetal conditions the centrifugal phenomenon of expression, reduce to third person processes that particular way of dealing with the world which we know as behavior, bring experience down to the level of physical nature and convert the living body into an interiorless thing. The emotional and practical attitudes of the living subject in relation to the world were, then, incorporated into a psycho-physiological mechanism. Every evaluation had to be the outcome of a transfer whereby complex situations became capable of awakening elementary impressions of pleasure and pain, impressions bound up, in turn, with nervous processes. The impelling intentions of the living creature were converted into objective movements: to the will only an instantaneous fiat was allowed, the execution of the act being entirely given over to a nervous mechanism. Sense experience, thus detached from the affective and motor functions, became the mere reception of a quality, and physiologists thought they could follow, from the point of reception to the nervous centers, the projection of the external world in the living body. The latter, thus transformed, ceased to be my body, the visible expression of a concrete Ego, and became one object among all others. Conversely, the body of another person could not appear to me as encasing another Ego. It was merely a machine, and the perception of the other could not really be *of the other*, since it resulted from an inference and therefore placed behind the automaton no more than a consciousness in general, a transcendent cause and not an inhabitant of his movements. So we no longer had a grouping of factors constituting the self co-existing in a world. The whole concrete content of "psychic states" resulting, according to the laws of psychophysiology and psychology, from a universal determinism, was integrated into the *in-itself*. There was no longer any real *for-itself* other than the thought of the scientist which perceives the system and which alone ceases to occupy any place in it. Thus, while the living body became an exterior without interior, subjectivity became an interior without exterior, an impartial spectator. The naturalism of science and the spiritualism of the universal constituting subject, to which reflection on science led, had this in common, that they leveled out experience: in face of the constituting I, the empirical selves are objects. The empirical Self is a hybrid notion, a mixture of in-itself and for-itself, to which reflective philosophy could give no status. In so far as it has a concrete content it is inserted in the system of experience and is therefore not a subject; in so far as it *is* a subject, it is empty and resolves itself into the transcendental subject. The ideality of the object, the objectification of the living body, the placing of spirit in an axiological dimension having no common measure with nature, such is the transparent philosophy arrived at by pushing further along the route of knowledge opened up by perception. It could be held that perception is an incipient science, science a methodical and complete perception, since science was merely following uncritically the ideal of knowledge set up by the perceived thing.

Now this philosophy is collapsing before our eyes. The natural object was the first to disappear and physics has itself recognized the limits of its categories by demanding a recasting and blending of the pure concepts which it had adopted. For its part the organism presents physico-chemical analysis not with the practical difficulties of a complex object, but with the theoretical difficulty of a meaningful being. In more general terms the idea of a universe of thought or a universe of values, in which all thinking lives come into contact and are reconciled, is called into question. Nature is *not* in itself geometrical, and it appears so only to a careful observer who contents himself with macrocosmic data. Human society is *not* a community of reasonable minds, and only in fortunate countries where a biological and economic balance has locally and temporarily been struck has such a conception of it been possible. The experience of Chaos, both on the speculative and the other level, prompts us to see rationalism in a historical perspective which it set itself on principle to avoid, to seek a philosophy which explains the upsurge of reason in a world not of its making and to prepare the substructure of living experience without which reason and liberty are emptied of their content and wither away. We shall no longer hold that perception is incipient science, but conversely that classical science is a form of perception which loses sight of its origins and believes itself complete. The first philosophical act would appear to be to return to the world of actual experience which is prior to the objective world, since it is in it that we shall be able to grasp the theoretical basis no less than the limits of that objective world, restore to things their concrete physiognomy, to organisms their individual ways of dealing with the world, and to subjectivity its inherence in history. Our task will be, moreover, to rediscover phenomena, the layer of living experience through which other people and things are first given to us, the system "Self-others-things" as it comes into being to reawaken perception and foil its trick of allowing us to forget it as a fact and as perception in the interest of the object which it presents to us and of the rational tradition to which it gives rise.

This phenomenal field is not an "inner world," the "phenomenon" is not a "state of consciousness," or a "mental fact," and the experience of phenomena is not an act of introspection or an intuition, in Bergson's sense. It has long been the practice to define the object of psychology by saying that it was "without extension" and "accessible to one person only," with the result that this peculiar object could be grasped only by means of a special kind of act, "internal perception" or introspection, in which subject and object were mingled and knowledge achieved by an act of coinciding. The return to the "immediate data of consciousness" became therefore a hopeless enterprise since the philosophical scrutiny was trying to *be* what it could not, in principle, *see*. The difficulty was not only to destroy the prejudice of the exterior, as nil philosophies urge the beginner to do, or to describe the mind in a language made for representing things. It was much more fundamental, since interiority, defined by the impression, by its nature evaded every attempt to express it. It was not only the imparting of philosophical intuitions to others which became difficult—or rather reduced itself to a sort of incantation designed to induce in them experiences comparable to the philosopher's—but the philosopher himself could not be clearly aware of *what* he saw in the instant, since he would have had to think it, that is, fix and distort it. The immediate was therefore a lonely, blind and mute life. The return to the phenomenal presents none of these peculiarities. The sensible configuration of an object or a gesture, which the criticism of the constancy hypothesis brings before our eyes, is not grasped in some inexpressible coincidence, it "is understood" through a sort of act of appropriation which we all experience when we say that we have "found" the rabbit in the foliage of a puzzle, or that we have "caught" a slight gesture. Once the prejudice of sensation has been banished, a face, a signature, a form of behavior cease to be mere "visual data" whose psychological meaning is to be sought in our inner experience, and the mental life of others becomes an immediate object, a whole charged with immanent meaning. More generally it is the very notion of the immediate which is transformed: henceforth the immediate is no longer the impression, the object which is one with the subject, but the meaning, the structure, the spontaneous arrangement of parts. My own "mental life" is given to me in precisely the same way, since the criticism of the constancy hypothesis teaches me to recognize the articulation and melodic unity of my behavior as original data of inner experience, and since introspection, when brought down to its positive content, consists equally in making the immanent meaning of any behavior explicit. Thus what we discover by going beyond the prejudice of the objective world is not an occult inner world. . . .

Psychological reflection, once begun, then, outruns itself through its own momentum. Having recognized the originality of phenomena in relation to the objective world, since it is through them that the objective world is known to us, it is led to integrate with them every possible object and to try to find out how that object is constituted through them. At the same time the phenomenal field becomes a transcendental field. Since it is now the universal focus of knowledge, consciousness definitely ceases to be a particular region of being, a certain collection of "mental" contents; it no longer resides or is no longer confined within the domain of "forms" which psychological reflection had first recognized, but the forms, like all things, exist for it. It can no longer be a question of describing the world of living experience which it carries within itself like some opaque datum, it has to be constituted. The process of making explicit, which had laid bare the "lived-through" world which is prior to the objective one, is put into operation upon the "lived-through" world itself, thus revealing, prior to the phenomenal field, the transcendental field. The system "Self-others-world" is in its turn taken as an object of analysis and it is now a matter of awakening the thoughts which constitute other people, myself as individual subject, and the world as a pole of my perception. This new "reduction" would then recognize only one true subject, the thinking Ego. This move from *naturata* to *naturans*, from constituted to constituting, would complete the thematizing begun by psychology and would leave nothing implicit or tacitly accepted in my knowledge. It would enable me to take complete possession of my experience, thus equating thinking and thought. Such is the ordinary perspective of a transcendental philosophy, and also, to all appearances at least, the program of a transcendental phenomenology. Now the phenomenal field as we have revealed it in this chapter, places a fundamental difficulty in the way of any attempt to make experience directly and totally explicit. It is true that psychologism has been left behind, that the meaning and structure of the percept are for us no longer the mere outcome of psycho-physiological events, that rationality is no longer a fortunate accident bringing together dispersed sensations, and that the Gestalt is recognized as primary. But although the Gestalt may be expressible in terms of some internal law, this law must not be considered as a model on which the phenomena of structure are built up. Their appearance is not the external unfolding of a pre-existing reason. It is not *because* the "form" produces a certain state of equilibrium, solving a problem of maximum coherence and, in the Kantian sense, making a world possible, that it enjoys a privileged place in our perception; it is the very appearance of the world and not the condition of its possibility; it is the birth of a norm and is not realized according to a norm; it is the identity of the external and the internal and not the projection of the internal in the external. Although, then, it is not the outcome of some circulation of mental states in themselves, neither is it an idea. The Gestalt of a circle is not its mathematical law but its physiognomy. The recognition of phenomena as an original order is a condemnation of empiricism as an *explanation* of order and reason in terms of a coming together of facts and of natural accidents, but it leaves reason and order themselves with the character of facticity. If a universal constituting consciousness were possible, the opacity of the fact would disappear. If then we want reflection to maintain, in the object on which it bears, its descriptive characteristics, and thoroughly to understand that object, we must not consider it as a mere return to a universal reason and see it as anticipated in unreflective experience, we must regard it as a creative operation which itself participates in the facticity of that experience. That is why phenomenology, alone of all philosophies, talks about a transcendental *field*. This word indicates that reflection never holds, arrayed and objectified before its gaze, the whole world and the plurality of monads, and that its view is never other than partial and of limited power. It is also why phenomenology is phenomenology, that is, a study of the *advent* of being to consciousness, instead of presuming its possibility as given in advance. It is striking how transcendental philosophies of the classical type never question the possibility of effecting the complete disclosure which they always assume *done somewhere*. It is enough for them that it should be necessary, and in this way they judge what is by what ought to be, by what the idea of knowledge requires. In fact, the thinking Ego can never abolish its inherence in an individual subject, which knows all things in a particular perspective. Reflection can never make me stop seeing the sun two hundred yards away on a misty day, or seeing it "rise" and "set," or thinking with the cultural apparatus with which my education, my previous efforts, my personal history, have provided me. I never actually collect together, or call up simultaneously, all the primary thoughts which contribute to my perception or to my present conviction. A critical philosophy attaches in the last analysis no importance to this resistance offered by passivity, as if it were not necessary to become the transcendental subject in order to have the right to affirm it. It tacitly assumes, consequently, that the philosopher's thinking is not conditioned by any situation. Starting from the spectacle of the world, which is that of a nature open to a plurality of thinking subjects, it looks for the conditions which make possible this unique world presented to a number of empirical selves, and finds it in a transcendental ego in which they participate without dividing it up, because it is not a Being, but a Unity or a Value. This is why the problem of the knowledge of other people is never posed in Kantian philosophy: the transcendental ego which it discusses is just as much other people's as mine, analysis is from the start located outside me, and has nothing to do but to determine the general conditions which make possible a world for an ego—myself or others equally—and so it never comes up against the question: *who is thinking?* If on the other hand contemporary philosophy takes this as its main theme, and if other people become a problem for it, it is because it is trying to achieve a more radical self-discovery. Reflection cannot be thorough-going, or bring a complete elucidation of its object, if it does not arrive at awareness of itself as well as of its results. We must not only adopt a reflective attitude, in an impregnable *Cogito*, but furthermore reflect on this reflection, understand the natural situation which it is conscious of succeeding and which is therefore part of its definition; not merely practice philosophy, but realize the transformation which it brings with it in the spectacle of the world and in our existence. Only on this condition can philosophical knowledge become absolute knowledge, and cease to be a speciality or a technique. So there will be no assertion of an absolute Unity, all the less doubtful for not having had to "come into Being."

The core of philosophy is no longer an autonomous transcendental subjectivity, to be found everywhere and nowhere: it lies in the perpetual beginning of reflection, at the point where an individual life begins to reflect on itself. Reflection is truly reflection only if it is not carried outside itself, only if it knows itself as reflection-on-an-unreflective-experience, and consequently as a change in structure of our existence. We earlier attacked Bergsonian intuitionism and introspection for seeking to know by coinciding. But at the opposite extremity of philosophy, in the notion of a universal constituting consciousness, we encounter an exactly corresponding mistake. Bergson's mistake consists in believing that the thinking subject can become fused with the object thought about, and that knowledge can swell and be incorporated into being. The mistake of reflective philosophies is to believe that the thinking subject can absorb into its thinking or appropriate without remainder the object of its thought, that our being can be brought down to our knowledge. As thinking subject we are never the unreflective subject that we seek to know; but neither can we become wholly consciousness, or make ourselves into the transcendental consciousness. If we were consciousness, we would have to have before us the world, our history and perceived objects in their uniqueness as systems of transparent relationships. Now even when we are not dealing with psychology, when we try to comprehend, in direct reflection and without the help of the varied associations of inductive thought, what a perceived movement, or a circle, are, we can elucidate this singular fact only by varying it somewhat through the agency of imagination, and then fastening our thought upon the invariable element of this mental experience. We can get through to the individual only by the hybrid procedure of finding an *example*, that is, by stripping it of its facticity. Thus it is questionable whether thought can ever quite cease to be inductive, and whether it can assimilate any experience to the point of taking up and appropriating its whole texture. A philosophy becomes transcendental, or radical, not by taking its place in absolute consciousness without mentioning the ways by which this is reached, but by considering itself as a problem; not by postulating a knowledge rendered totally explicit, but by recognizing as the fundamental philosophic problem *this presumption* on reason's part.

EXPERIENCE AND OBJECTIVE THOUGHT: THE PROBLEM OF THE BODY

Our perception ends in objects, and the object once constituted, appears as the reason for all the experiences of it which we have had or could have. For example, I see the next-door house from a certain angle, but it would be seen differently from the right bank of the Seine, or from the inside, or again from an aeroplane: the house *itself* is none of these appearances: it is, as Leibniz said, the geometrized projection of these perspectives and of all possible perspectives, that is, the perspectiveless position from which all can be derived, the house seen from nowhere. But what do these words mean? Is not to see always to see from somewhere? To say that the house itself is seen from nowhere is surely to say that it is invisible! Yet when I say that I see the house with my own eyes, I am saying something that cannot be challenged; I do not mean that my retina and crystalline lens, my eyes as material organs, go into action and cause me to see it; with only myself to consult, I can know nothing about this. I am trying to express in this way a certain manner of approaching the object, the "gaze" in short, which is as indubitable as my own thought, as directly known by me. We must try to understand how vision can be brought into being from somewhere without being enclosed in its perspective. . . .

Taken in itself—and as an object it demands to be taken thus—the object has nothing cryptic about it; it is completely displayed and its parts co-exist while our gaze runs from one to another, its present does not cancel its past, nor will its future cancel its present. The positing of the objects therefore makes us go beyond the limits of our actual experience which is brought up against and halted by an alien being, with the result that finally experience believes that it extracts all its own teaching from the object. It is this *ek-stase* of experience which causes all perception to be perception of something.

Obsessed with being, and forgetful of the perspectivism of my experience, I henceforth treat it as an object and deduce it from a relationship between objects. I regard my body, which is my point of view upon the world, as one of the objects of that world. My recent awareness of my gaze as a means of knowledge I now repress, and treat my eyes as bits of matter. They then take their place in the same objective space in which I am trying to situate the external object and I believe that I am producing the perceived perspective by the projection of the objects on my retina. In the same way I treat my own perceptual history as a result of my relationships with the objective world; my present, which is my point of view on time, becomes one moment of time among all the others, my duration a reflection or abstract aspect of universal time, as my body is a mode of objective space. In the same way, finally, if the objects which surround the house or which are found in it remained what they are in perceptual experience, that is, acts of seeing conditioned by a certain perspective, the house would not be posited as an autonomous being. Thus the positing of one single object, in the full sense, demands the compositive bringing into being of all these experiences in one act of manifold creation. Therein it exceeds perceptual experience and the synthesis of horizons—as the notion of a *universe*, that is to say, a completed and explicit totality, in which the relationships are those of reciprocal determination, exceeds that of a *world*, or an open and indefinite multiplicity of relationships which are of reciprocal implication. I detach myself from my experience and pass to the *idea*. Like the object, the idea purports to be the same for everybody, valid in all times and places, and the individuation of an object in an objective point of time and space finally appears as the expression of a universal positing power. I am no longer concerned with my body, nor with time, nor with the world, as I experience them in ante-predicative knowledge, in the inner communion that I have with them. I now refer to my body only as an idea, to the universe as idea, to the idea of space and the idea of time. Thus "objective" thought (in Kierkegaard's sense) is formed—being that of common sense and of science— which finally causes us to lose contact with perceptual experience, of which it is nevertheless the outcome and the natural sequel. The whole life of consciousness is characterized by the tendency to posit objects since it is consciousness, that is to say, self-knowledge, only in so far as it takes hold of itself and draws itself together in an identifiable object. And yet the absolute positing of a single object is the death of consciousness, since it congeals the whole of existence, as a crystal placed in a solution suddenly crystallizes it.

We cannot remain in this dilemma of having to fail to understand either the subject or the object. We must discover the origin of the object at the very center of our experience: we must describe the emergence of being and we must understand how, paradoxically, there is *for us* an *in-itself*. In order not to prejudge the issue, we shall take objective thought on its own terms and not ask it any questions which it does not ask itself. If we are led to rediscover experience behind it, this shift of ground will be attributable only to the difficulties which objective thought itself raises. Let us consider it then at work in the constitution of our body *as* object, since this is a crucial moment in the genesis of the objective world. It will be seen that one's own body evades, even within science itself, the treatment to which it is intended to subject it. And since the genesis of the objective body is only a moment in the constitution of the object, the body, by withdrawing from the objective world, will carry with it the intentional threads linking it to its surrounding and finally reveal to us the perceiving subject as the perceived world.

MAKING UP PEOPLE

IAN HACKING

Were there any perverts before the latter part of the nineteenth century? According to Arnold Davidson, "The answer is NO . . .—Perversion was not a disease that lurked about in nature, waiting for a psychiatrist with especially acute powers of observation to discover it hiding everywhere. It was a disease created by a new (functional) understanding of disease." Davidson is not denying that there have been odd people at all times. He is asserting that perversion, as a disease, and the pervert, as a diseased person, were created in the late nineteenth century. Davidson's claim, one of many now in circulation, illustrates what I call making up people.

I have three aims: I want a better understanding of claims as curious as Davidson's; I would like to know if there could be a general theory of making up people, or whether each example is so peculiar that it demands its own nongeneralizable story; and I want to know how this idea "making up people" affects our very idea of what it is to be an individual. I should warn that my concern is philosophical and abstract; I look more at what people might be than at what we are. I imagine a philosophical notion I call dynamic nominalism, and reflect too little on the ordinary dynamics of human interaction.

First we need more examples. I study the dullest of subjects, the official statistics of the nineteenth century. They range, of course, over agriculture, education, trade, births, and military might, but there is one especially striking feature of the avalanche of numbers that begins around 1820. It is obsessed with *analyse morale*, namely, the statistics of deviance. It is the numerical analysis of suicide, prostitution, drunkenness, vagrancy, madness, crime, *les miserables*. Counting generated its own subdivisions and rearrangements. We find classifications of over 4,000 different crisscrossing motives for murder and requests that the police classify each individual suicide in 21 different ways. I do not believe that motives of these sorts or suicides of these kinds existed until the practice of counting them came into being.

New slots were created in which to fit and enumerate people. Even national and provincial censuses amazingly show that the categories into which people fall change every ten years. Social change creates new categories of people, but the counting is no mere report of developments. It elaborately, often philanthropically, creates new ways for people to be.

People spontaneously come to fit their categories. When factory inspectors in England and Wales went to the mills, they found various kinds of people there, loosely sorted according to tasks and wages. But when they had finished their reports, millhands had precise ways in which to work, and the owner had a clear set of concepts about how to employ workers according to the ways in which he was obliged to classify them.

I am more familiar with the creation of kinds among the masses than with interventions that act upon individuals, though I did look into one rare kind of insanity. I claim that multiple personality as an idea and as a clinical phenomenon was invented around 1875: only one or two possible cases per generation had been recorded before that time, but a whole flock of them came after. I also found that the clinical history of split personality parodies itself—the one clear case of classic symptoms was long recorded as two, quite distinct, human beings, each of which was multiple. There was "the lady of MacNish," so called after a report in *The Philosophy of Sleep*, written by the Edinburgh physician Robert MacNish in 1832, and there was one Mary R. The two would be reported in successive paragraphs as two different cases, although in fact Mary Reynolds was the very split-personality lady reported by MacNish.

Mary Reynolds died long before 1875, but she was not taken up as a case of multiple personality until then. Not she but one Felida X got the split-personality industry under way. As the great French psychiatrist Pierre Janet remarked at Harvard in 1906, Felida's history "was the great argument of which the positivist psychologists made use at the time of the heroic struggles against the dogmatism of Cousin's school. But for Felida, it is not certain that there would be a professorship of psychology at the Collège de France." Janet held precisely that chair. The "heroic struggles" were important for our passing conceptions of the self, and for individuality, because the split Felida was held to refute the dogmatic transcendental unity of apperception that made the self prior to all knowledge.

After Felida came a rush of multiples. The syndrome bloomed in France and later flourished in America, which is still its home. Do I mean that there were no multiples before Felida? Yes. Except for a very few earlier examples, which after 1875 were reinterpreted as classic multiples, there was no such syndrome for a disturbed person to display or to adopt.

I do not deny that there are other behaviors in other cultures that resemble multiple personality. Possession is our most familiar example—a common form of Renaissance behavior that died long ago, though it was curiously hardy in isolated German villages even late in the nineteenth century. Possession was not split personality, but if you balk at my implication that a few people (in committee with their medical or moral advisers) almost choose to become splits, recall that tormented souls in the past have often been said to have in some way chosen to be possessed, to have been seeking attention, exorcism, and tranquility.

I should give one all-too-tidy example of how a new person can be made up. Once again I quote from Janet, whom I find the most open and honorable of the psychiatrists. He is speaking to Lucie, who had the once-fashionable but now-forgotten habit of automatic writing. Lucie replies to Janet in writing without her normal self's awareness:

Janet. Do you understand me?
Lucie (*writes*). No.
J. But to reply you must understand me!
L. Oh yes, absolutely.
J. Then what are you doing?
L. Don't know.
J. It is certain that someone is understanding me.
L. Yes.
J. Who is that?
L. Somebody besides Lucie.
J. Aha! Another person. Would you like to give her a name?
L. No.
J. Yes. It would be far easier that way.
L. Oh well. If you want: Adrienne.
J. Then, Adrienne, do you understand me?
L. Yes.

If you think this is what people used to do in the bad old days, consider poor Charles, who was given a whole page of *Time* magazine on October 25, 1982 (p. 70). He was picked up wandering aimlessly and was placed in the care of Dr. Malcolm Graham of Daytona Beach, who in turn consulted with Dr. William Rothstein, a notable student of multiple personality at the University Hospital in Columbia, South Carolina. Here is what is said to have happened:

After listening to a tape recording made in June of the character Mark, Graham became convinced he was dealing with a multiple personality.

Graham began consulting with Rothstein, who recommended hypnosis. Under the spell, Eric began calling his characters. Most of the personalities have been purged, although there are three or four being treated, officials say. It was the real personality that signed a consent form that allowed Graham to comment on the case.

Hypnosis elicited Charles, Eric, Mark, and some 24 other personalities. When I read of such present-day manipulations of character, I pine a little for Mollie Fancher, who gloried in the personalities of Sunbeam, Idol, Rosebud, Pearl, and Ruby. She became somewhat split after being dragged a mile by a horse car. She was not regarded as especially deranged, nor in much need of "cure." She was much loved by her friends, who memorialized her in 1894 in a book with the title *Mollie Fancher, The Brooklyn Enigma: An Authentic Statement of Facts in the Life of Mollie J. Fancher, The Psychological Marvel of the Nineteenth Century.*

The idea of making up people has, I said, become quite widespread. *The Making of the Modern Homosexual* is a good example; "Making" in this title is close to my "making up." The contributors by and large accept that the homosexual and the heterosexual as kinds of persons (as ways to be persons, or as conditions of personhood), came into being only toward the end of the nineteenth century. There has been plenty of same-sex activity in all ages, but not, *Making* argues, same-sex people and different-sex people. I do not wish to enter the complexities of that idea, but will quote a typical passage from this anthology to show what is intended: "One difficulty in transcending the theory of gender inversion as the basis of the specialized homosexual identity was the rather late historical development of more precise conceptions of components of sexual identity. [fn:] It is not suggested that these components are 'real' entities, which awaited scientific 'discovery.' However once the distinctions were made, new realities effectively came into being."

Note how the language here resembles my opening quotation: "not a disease . . . in nature, waiting for . . . observation to discover it" versus "not . . . 'real' entities, which awaited scientific 'discovery.'" Moreover, this author too suggests that "once the distinctions were made, new realities effectively came into being."

This theme, the homosexual as a kind of person, is often traced to a paper by Mary Macintosh, "The Homosexual Role," which she published in 1968 in *Social Problems.* That journal was much devoted to "labeling theory," which asserts that social reality is conditioned, stabilized, or even created by the labels we apply to people, actions, and communities. Already in 1963 "A Note on the

Uses of Official Statistics" in the same journal anticipated my own inferences about counting.

But there is a currently more fashionable source of the idea of making up people, namely, Michel Foucault, to whom both Davidson and I are indebted. A quotation from Foucault provides the epigraph—following one from Nietzsche—for *The Making of the Modern Homosexual*; and although its authors cite some 450 sources, they refer to Foucault more than anyone else. Since I shall be primarily concerned with labeling, let me state at once that for all his famous fascination with discourse, naming is only one element in what Foucault calls the "constitution of subjects" (in context a pun, but in one sense the making up of the subject): "We should try to discover how it is that subjects are gradually, progressively, really and materially constituted through a multiplicity of organisms, forces, energies, materials, desires, thoughts, etc."

Since so many of us have been influenced by Foucault, our choice of topic and time may be biased. My examples dwell in the nineteenth century and are obsessed with deviation and control. Thus among the questions on a complete agenda, we should include these two: Is making up people intimately linked to control? Is making up people itself of recent origin? The answer to both questions might conceivably be yes. We may be observing a particular medico-forensic-political language of individual and social control. Likewise, the sheer proliferation of labels in that domain during the nineteenth century may have engendered vastly more kinds of people than the world had ever known before.

Partly in order to distance myself for a moment from issues of repression, and partly for intrinsic interest, I would like to abstract from my examples. If there were some truth in the descriptions I and others have furnished, then making up people would bear on one of the great traditional questions of philosophy, namely, the debate between nominalists and realists. The author I quoted who rejects the idea that the components of the homosexual identity are real entities, has taken a time-worn nominalist suggestion and made it interesting by the thought that "once the distinctions were made, new realities effectively came into being."

You will recall that a traditional nominalist says that stars (or algae, or justice) have nothing in common except our names ("stars," "algae," "justice"). The traditional realist in contrast finds it amazing that the world could so kindly sort itself into our categories. He protests that there are definite sorts of objects in it, at least stars and algae, which we have painstakingly come to recognize and classify correctly. The robust realist does not have to argue very hard that people also come sorted. Some are thick, some thin, some dead, some alive. It may be a fact about human beings that we notice who is fat and who is dead, but the fact itself that some of our fellows are fat and others are dead has nothing to do with our schemes of classification.

The realist continues: consumption was not only a sickness but also a moral failing, caused by defects of character. That is an important nineteenth-century social fact about TB. We discovered in due course, however, that the disease is transmitted by bacilli that divide very slowly and that we can kill. It is a fact about us that we were first moralistic and later made this discovery, but it is a brute fact about tuberculosis that it is a specific disease transmitted by microbes. The nominalist is left rather weakly contending that even though a particular kind of person, the consumptive, may have been an artifact of the nineteenth century, the disease itself is an entity in its own right, independently of how we classify.

It would be foolhardy, at this conference, to have an opinion about one of the more stable human dichotomies, male and female. But very roughly, the robust realist will agree that there may be what really are physiological borderline cases, once called "hermaphrodites." The existence of vague boundaries is normal: most of us are neither tall nor short, fat nor thin. Sexual physiology is unusually abrupt in its divisions. The realist will take the occasional compulsive fascination with transvestitism, or horror about hermaphrodites (so well described by Stephen Greenblatt in this volume), as human (nominalist) resistance to nature's putative aberrations. Likewise the realist will assert that even though our attitudes to gender are almost entirely nonobjective and culturally ordained, gender itself is a real distinction.

I do not know if there were thoroughgoing, consistent, hard-line nominalists who held that every classification is of our own making. I might pick that great British nominalist Hobbes out of context: "How can any man imagine that the names of things were imposed by their natures?" Or I might pick Nelson Goodman.

Let me take even the vibrant Hobbes, Goodman, and their scholastic predecessors as pale reflections of a perhaps nonexistent static nominalist, who thinks that all categories, classes, and taxonomies are given by human beings rather than by nature and that these categories are essentially fixed throughout the several eras of humankind. I believe that static nominalism is doubly wrong: I think that many categories come from nature, not from the human mind, and I think our categories are not static. A different kind of nominalism —1 call it dynamic nominalism—attracts my realist self, spurred on by theories about the making of the homosexual and the heterosexual as kinds of persons or by my observations about official statistics. The claim of dynamic nominalism is not that there was a kind of person who came increasingly to be recognized by bureaucrats or by students of human nature but rather that a kind of person came into being at the same time as the kind itself was being invented. In some cases, that is, our classifications and our classes conspire to emerge hand in hand, each egging the other on.

Take four categories: horse, planet, glove, and multiple personality. It would be preposterous to suggest that the only thing horses have in common is that we call them horses. We may draw the boundaries to admit or to exclude Shetland ponies, but the similarities and differences are real enough. The planets furnish one of T. S. Kuhn's examples of conceptual change. Arguably the heavens looked different after we grouped Earth with the other planets and excluded Moon and Sun, but I am sure that acute thinkers had discovered a real difference. I hold (most of the time) that strict nominalism is unintelligible for horses and the planets. How could horses and planets be so obedient to our minds? Gloves are something else: we manufacture them. I know not which came first, the thought or the mitten, but they have evolved hand in hand. That the concept "glove" fits gloves so well is no surprise; we made them that way. My claim about making up people is that in a few interesting respects multiple personalities (and much else) are more like gloves than like horses. The category and the people in it emerged hand in hand.

How might a dynamic nominalism affect the concept of the individual person? One answer has to do with possibility. Who we are is not only what we did, do, and will do but also what we might have done and may do. Making up people changes the space of possibilities for personhood. Even the dead are more than their deeds, for we make sense of a finished life only within its sphere of former possibilities. But our possibilities, although inexhaustible, are also bounded. If the nominalist thesis about sexuality were correct, it simply wasn't possible to be a heterosexual kind of person before the nineteenth century, for that kind of person was not there to choose. What could that mean? What could it mean in general to say that possible ways to be a person can from time to time come into being or disappear? Such queries force us to be careful about the idea of possibility itself.

We have a folk picture of the gradations of possibility. Some things, for example, are easy to do, some hard, and some plain impossible. What is impossible for one person is possible for another. At the limit we have the statement: "With men it is impossible, but not with God: for with God, all things are possible" (Mark 10:27). (Christ had been saying that it is easier for a camel to pass through the eye of a needle than for a rich man to enter the kingdom of heaven.) Degrees of possibility are degrees in the ability of some agent to do or make something. The more ability, the more possibility, and omnipotence makes anything possible. At that point, logicians have stumbled, worrying about what were once called "the eternal truths" and are now called "logical necessities." Even God cannot make a five-sided square, or so mathematicians say, except for a few such eminent dissenters as Descartes. Often this limitation on omnipotence is explained linguistically, being said to reflect our unwillingness to call anything a five-sided square.

There is something more interesting that God can't do. Suppose that Arnold Davidson, in my opening quotation about perversion, is literally correct. Then it was not possible for God to make George Washington a pervert. God could have delayed Washington's birth by over a century, but would that have been the same man? God could have moved the medical discourse back 100-odd years. But God could not have simply made him a pervert, the way He could have made him freckled or had him captured and hung for treachery. This may seem all the more surprising since Washington was but eight years older than the Marquis de Sade—and Krafft-Ebing has sadomasochism among the four chief categories of perversion. But it follows from Davidson's doctrine that de Sade was not afflicted by the disease of perversion, nor even the disease of sadomasochism either.

Such strange claims are more trivial than they seem; they result from a contrast between people and things. Except when we interfere, what things are doing, and indeed what camels are doing, does not depend on how we describe them. But some of the things that we ourselves do are intimately connected to our descriptions. Many philosophers follow Elizabeth Anscombe and say that intentional human actions must be "actions under a description." This is not mere lingualism, for descriptions are embedded in our practices and lives. But if a description is not there, then intentional actions under that description cannot be there either: that, apparently, is a fact of logic.

Elaborating on this difference between people and things: what camels, mountains, and microbes are doing does not depend on our words. What happens to tuberculosis bacilli depends on whether or not we poison them with BCG vaccine, but it does not depend upon how we describe them. Of course we poison them with a certain vaccine in part because we describe them in certain ways, but it is the vaccine that kills, not our words. Human action is more closely linked to human description than bacterial action is. A century ago I would have said that consumption is caused by bad air and sent the patient to the Alps. Today, I may say that TB is caused by microbes and prescribe a two-year course of injections. But what is happening to the microbes and the patient is entirely independent of my correct or incorrect description, even though it is not independent of the medication prescribed. The microbes' possibilities are delimited by nature, not by words. What is curious about human action is that by and large what I am deliberately doing depends on the possibilities of description. To repeat, this is a tautological inference from what is now a philosopher's commonplace, that all intentional acts are acts under a description. Hence if new modes of description come into being, new possibilities for action come into being in consequence.

Let us now add an example to our repertoire; let it have nothing to do with deviancy, let it be rich in connotations of human practices, and let it help furnish the end of a spectrum of making up people opposite from the multiple personality. I take it from Jean-Paul Sartre, partly for the well-deserved fame of his description, partly for its excellence as description, partly because Sartre is our premium philosopher of choice, and partly because recalling Sartre will recall an example that returns me to my origin. Let us first look at Sartre's magnificent humdrum example. Many among us might have chosen to be a waiter or waitress and several have been one for a time. A few men might have chosen to be something more specific, a Parisian *garçon de café*, about whom Sartre writes in his immortal discussion of bad faith: "His movement is quick and forward, a little too precise, a little too rapid. He comes toward the patrons with a step a little too quick. He bends forward a little too eagerly, his eyes express an interest too solicitous for the order of the customer." Psychiatrists and medical people in general try to be extremely specific in describing, but no description of the several classical kinds of split personality is as precise (or as recognizable) as this. Imagine for a moment that we are reading not the words of a philosopher who writes his books in cafes but those of a doctor who writes them in a clinic. Has the *garçon de café* a chance of escaping treatment by experts? Was Sartre knowing or merely anticipating when he concluded this very paragraph with the words: "There are indeed many precautions to imprison a man in what he is, as if we lived in perpetual fear that he might escape from it, that he might break away and suddenly elude his condition." That is a good reminder of Sartre's teaching: possibility, project, and prison are one of a piece.

Sartre's antihero chose to be a waiter. Evidently that was not a possible choice in other places, other times. There are servile people in most societies, and servants in many, but a waiter is something specific, and *garçon de café* more specific. Sartre remarks that the waiter is doing something different when he pretends to play at being a sailor or a diplomat than when he plays at being a waiter in order to be a waiter. I think that in most parts of, let us say, Saskatchewan (or in a McDonald's anywhere), a waiter playing at being a *garçon de café* would miss the mark as surely as if he were playing at being a diplomat while passing over the french fries. As with almost every way in which it is possible to be a person, it is possible to be a *garçon de café* only at a certain time, in a certain place, in a certain social setting. The feudal serf putting food on my lady's table can no more choose to be a *garçon de café* than he can choose to be lord of the manor. But the impossibility is evidently different in kind.

It is not a technical impossibility. Serfs may once have dreamed of travel to the moon; certainly their lettered betters wrote or read adventures of moon travel. But moon travel was impossible for them, whereas it is not quite impossible for today's young waiter. One young waiter will, in a few years, be serving steaks in a satellite. Sartre is at pains to say that even technical limitations do not mean that you have fewer possibilities. For every person, in every era, the world is a plenitude of possibilities. "Of course," Sartre writes, "a contemporary of Duns Scotus is ignorant of the use of the automobile or the aeroplane. . . . For the one who has no relation of any kind to these objects and the techniques that refer to them, there is a kind of absolute, unthinkable and undecipherable nothingness. Such a nothing can in no way limit the For-itself that is choosing itself; it cannot be apprehended as a lack, no matter how we consider it." Passing to a different example, he continues, "The feudal world offered to the vassal lord of Raymond VI infinite possibilities of choice; we do not possess more."

"Absolute, unthinkable and undecipherable nothingness" is a great phrase. That is exactly what being a multiple personality, or being a *garçon de café*, was to Raymond's vassal. Many of you could, in truth, be neither a Parisian waiter nor a split, but both are thinkable, decipherable somethingnesses. It would be possible for God to have made you one or the other or both, leaving the rest of the world more or less intact. That means, to me, that the outer reaches of your space as an individual are essentially different from what they would have been had these possibilities not come into being.

Thus the idea of making up people is enriched; it applies not to the unfortunate elect but to all of us. It is not just the making up of people of a kind that did not exist before: not only are the split and the waiter made up, but each of us is made up. We are not only what we are but what we might have been, and the possibilities for what we might have been are transformed.

Hence anyone who thinks about the individual, the person, must reflect on this strange idea, of making up people. Do my stories tell a uniform tale? Manifestly not. The multiple personality, the homosexual or heterosexual person, and the waiter form one spectrum among many that may color our perception here.

Suppose there is some truth in the labeling theory of the modern homosexual. It cannot be the whole truth, and this for several reasons, including one that is future-directed and one that is past-directed. The future-directed fact is that after the institutionalization of the homosexual person in law and official morality, the people involved had a life of their own, individually and collectively. As gay liberation has amply proved, that life was no simple product of the labeling.

The past-directed fact is that the labeling did not occur in a social vacuum, in which those identified as homosexual people passively accepted the format. There was a complex social life that is only now revealing itself in the annals of academic social history. It is quite clear that the internal life of innumerable clubs and associations interacted with the medico-forensic-journalistic labeling. At the risk of giving offense, I suggest that the quickest way to see the contrast between making up homosexuals and making up multiple personalities is to try to imagine split-personality bars. Splits, insofar as they are declared, are under care, and the syndrome, the form of behavior, is orchestrated by a team of experts. Whatever the medico-forensic experts tried to do with their categories, the homosexual person became autonomous of the labeling, but the split is not.

The *garçon de café* is at the opposite extreme. There is of course a social history of waiters in Paris. Some of this will be as anecdotal as the fact that croissants originated in the cafes of Vienna after the Turkish siege was lifted in 1683: the pastries in the shape of a crescent were a mockery of Islam. Other parts of the story will be structurally connected with numerous French institutions. But the class of waiters is autonomous of any act of labeling, At most the name *garçon de café* can continue to ensure both the inferior position of the waiter and the fact that he is male. Sartre's precise description does not fit the *fille de salle,* that is a different role.

I do not believe there is a general story to be told about making up people. Each category has its own history. If we wish to present a partial framework in which to describe such events, we might think of two vectors. One is the vector of labeling from above, from a community of experts who create a "reality" that some people make their own. Different from this is the vector of the autonomous behavior of the person so labeled, which presses from below, creating a reality every expert must face. The second vector is negligible for the split but powerful for the homosexual person. People who write about the history of homosexuality seem to disagree about the relative importance of the two vectors. My scheme at best highlights what the dispute is about. It provides no answers.

The scheme is also too narrow. I began by mentioning my own dusty studies in official statistics and asserted that these also, in a less melodramatic way, contribute to making up people. There is a story to tell here, even about Parisian waiters, who surface in the official statistics of Paris surprisingly late, in 1881. However, I shall conclude with yet another way of making up people and human acts, one of notorious interest to the existentialist culture of a couple of generations past. I mean suicide, the option that Sartre always left open to the For-itself. Suicide sounds like a timeless option. It is not. Indeed it might be better described as a French obsession.

There have been cultures, including some in recent European history, that knew no suicide. It is said that there were no suicides in Venice when it was the noblest city of Europe. But can I seriously propose that suicide is a concept that has been made up? Oddly, that is exactly what is said by the deeply influential Esquirol in his 1823 medical-encyclopedia article on suicide. He mistakenly asserts that the very word was devised by his predecessor Sauvages. What is true is this: suicide was made the property of medics only at the beginning of the nineteenth century, and a major fight it was too. It was generally allowed that there was the noble suicide, the suicide of honor or of state, but all the rest had to be regarded as part of the new medicine of insanity. By mid-century it would be contended that there was no case of suicide that was not preceded by symptoms of insanity.

This literature concerns the doctors and their patients. It exactly parallels a statistical story. Foucault suggests we think in terms of "two poles of development linked together by a whole cluster of intermediary relations." One pole centers on the individual as a speaking, working, procreating entity he calls an "anatomo-politics of the human body." The second pole, "focused on the species body," serves as the "basis of the biological processes: propagation, births, and mortality, the level of health, life expectancy and longevity." He calls this polarity a "biopolitics of the population." Suicide aptly illustrates patterns of connection between both poles. The medical men comment on the bodies and their past, which led to self-destruction; the statisticians count and classify the bodies. Every fact about the suicide becomes fascinating. The statisticians compose forms to be completed by doctors and police, recording everything from the time of death to the objects found in the pockets of the corpse. The various ways of killing oneself are abruptly characterized and become symbols of national character. The French favor carbon monoxide and drowning; the English hang or shoot themselves.

By the end of the nineteenth century there was so much information about French suicides that Durkheim could use suicide to measure social pathology. Earlier, a rapid increase in the rate of suicide in all European countries had caused great concern. More recently authors have suggested that the growth may have been largely apparent, a consequence of improved systems of reporting. It was thought that there were more suicides because more care was taken to report them. But such a remark is unwittingly ambiguous: reporting brought about more suicides. I do not refer to suicide epidemics that follow a sensational case, like that of von Kleist, who shot his lover and then himself on the Wannsee in 1811—an event vigorously reported in every European capital. I mean instead that the systems of reporting positively created an entire ethos of suicide, right down to the suicide note, an art form that previously was virtually unknown apart from the rare noble suicide of state. Suicide has of course attracted attention in all times and has invited such distinguished essayists as Cicero and Hume. But the distinctively European and American pattern of suicide is a historical artifact. Even the unmaking of people has been made up.

Naturally my kinds of making up people are far from exhaustive. Individuals serve as role models and sometimes thereby create new roles. We have only to think of James Clifford's contribution to this volume, "On Ethnographic Self-Fashioning: Conrad and Malinowski." Malinowski's book largely created the participant-observer cultural-relativist ethnographer, even if Malinowski himself did not truly conform to that role in the field. He did something more important—he made up a kind of scholar. The advertising industry relies on our susceptibilities to role models and is largely engaged in trying to make up people. But here nominalism, even of a dynamic kind, is not the key. Often we have no name for the very role a model entices us to adopt.

Dynamic nominalism remains an intriguing doctrine, arguing that numerous kinds of human beings and human acts come into being hand in hand with our invention of the categories labeling them. It is for me the only intelligible species of nominalism, the only one that can even gesture at an account of how common names and the named could so tidily fit together. It is of more human interest than the arid and scholastic forms of nominalism because it contends that our spheres of possibility, and hence our selves, are to some extent made up by our naming and what that entails. But let us not be overly optimistic about the future of dynamic nominalism. It has the merit of bypassing abstract hand-wringing and inviting us to do serious philosophy, namely, to examine the intricate origin of our ideas of multiple personality or of suicide. It is, we might say, putting some flesh on that wizened figure, John Locke, who wrote about the origin of ideas while introspecting at his desk. But just because it invites us to examine the intricacies of real life, it has little chance of being a general philosophical theory. Although we may find it useful to arrange influences according to Foucault's poles and my vectors, such metaphors are mere suggestions of what to look for next. I see no reason to suppose that we shall ever tell two identical stories of two different instances of making up people.

FROM *BODIES THAT MATTER*

JUDITH BUTLER

"the body"
p. 164 ~~Toward~~ Irreductability?
Unstructure?

Within some quarters of femi
to retrieve body from what is
poststructuralism. In another
that poststructuralism, under
matter as a contemporary cate
must now be reformulated in

project of greater ethical and political value. The terms of these debates are
difficult and unstable ones, for it is difficult to know in either case who or what
is designated by the term "poststructuralism," and perhaps even more difficult
to know what to retrieve under the sign of "the body." And yet these two
signifiers have for some feminists and critical theorists seemed fundamentally
antagonistic. One hears warnings like the following: If everything is discourse,
what happens to the body? If everything is a text, what about violence and
bodily injury? Does anything *matter* for poststructuralism?

It has seemed to many, I think, that in order for feminism to proceed as a
critical practice, it must ground itself in the sexed specificity of the female
body. Even as the category of sex is always reinscribed as gender, that sex must
still be presumed as the irreducible point of departure for the various cultural
constructions it has come to bear. And this presumption of the material irreducibility of sex has seemed to ground and to authorize feminist epistemologies and ethics, as well as gendered analyses of various kinds. In an effort to
displace the terms of this debate, I want to ask how and why "materiality" has
become a sign of irreducibility, that is, how is it that the materiality of sex is
understood as that which only bears cultural constructions and, therefore,
cannot be a construction? What is the status of this exclusion? Is materiality
a site or surface that is excluded from the process of construction, as that
through which and on which construction works? Is this perhaps an enabling
or constitutive exclusion, one without which construction cannot operate?
What occupies this site of unconstructed materiality? And what kinds of con-

structions are foreclosed through the figuring of this site as outside or beneath construction itself? → mapping; innerworkings?.

In what follows, what is at stake is less a theory of cultural construction than a consideration of the scenography and topography of construction. This scenography is orchestrated by and as a matrix of power that remains disarticulated if we presume constructedness and materiality as necessarily oppositional notions. In the place of materiality, one might inquire into other foundationalist premises that operate as political "irreducibles." Instead of rehearsing the theoretical difficulties that emerge by presuming the notion of the subject as a foundational premise or by trying to maintain a stable distinction, recourse to matter and to the materiality of sex is necessary in order to establish that irreducible specificity that is said to ground feminist practice. And here the question is not whether or not there ought to be reference to matter, just as the question never has been whether or not there ought to be speaking about women. This speaking will occur, and for feminist reasons, it must; the category of women does not become useless through deconstruction, but becomes one whose uses are no longer reified as "referents," and which stand a chance of being opened up, indeed, of coming to signify in ways that none of us can predict in advance. Surely, it must be possible both to use the term, to use it tactically even as one is, as it were, used and positioned by it, and also to subject the term to a critique which interrogates the exclusionary operations and differential power-relations that construct and delimit feminist invocations of "women." This is, to paraphrase . . . from [Gayatri] Spivak, the critique of something useful, the critique of something we cannot do without. Indeed, I would argue that it is a critique without which feminism loses its democratizing potential through refusing to engage—take stock of, and become transformed by—the exclusions which put it into play.

Something similar is at work with the concept of materiality, which may well be "something without which we cannot do anything." What does it mean to have recourse to materiality, since it is clear from the start that matter has a history (indeed, more than one) and that the history of matter is in part determined by the negotiation of sexual difference. We may seek to return to matter as prior to discourse to ground our claims about sexual difference only to discover that matter is fully sedimented with discourses on sex and sexuality that prefigure and constrain the uses to which that term can be put. Moreover, we may seek recourse to matter in order to ground or to verify a set of injuries or violations only to find that *matter itself is founded through a set of violations,* ones which are unwittingly repeated in the contemporary invocation.

posit by way of language a materiality outside of language is still to posit that materiality, and the materiality so posited will retain that positing as its constitutive condition. Derrida negotiates the question of matter's radical alterity with the following remark: "I am not even sure that there can be a 'concept' of an absolute exterior."[1] To have the concept of matter is to lose the exteriority that the concept is supposed to secure. Can language simply refer to materiality, or is language also the very condition under which materiality may be said to appear?

If matter ceases to be matter once it becomes a concept, and if a concept of matter's exteriority to language is always something less than absolute, what is the status of this "outside"? Is it produced by philosophical discourse in order to effect the appearance of its own exhaustive and coherent systematicity? What is cast out from philosophical propriety in order to sustain and secure the borders of philosophy? And how might this repudiation return?

MATTERS OF FEMININITY

The classical association of femininity with materiality can be traced to a set of etymologies which link matter with *mater* and *matrix* (or the womb) and, hence, with a problematic of reproduction. The classical configuration of matter as a site *of generation* or *origination* becomes especially significant when the account of what an object is and means requires recourse to its originating principle. When not explicitly associated with reproduction, matter is generalized as a principle of origination and causality. In Greek, *hyle* is the wood or timber out of which various cultural constructions are made, but also a principle of origin, development, and teleology which is at once causal and explanatory. This link between matter, origin, and significance suggests the indissolubility of classical Greek notions of materiality and signification. That which matters about an object is its matter.

In both the Latin and the Greek, matter (*materia* and *hyle*) is neither a simple, brute positivity or referent nor a blank surface or slate awaiting an external signification, but is always in some sense temporalized. This is true for Marx as well, when "matter" is understood as a principle *of transformation*, presuming and inducing a future. The matrix is an originating and formative principle which inaugurates and informs a development of some organism or object. Hence, for Aristotle, "matter is potentiality [*dynameos*], form actuality." In reproduction, women are said to contribute the matter; men, the form. The Greek *hyle* is the wood that already has been cut from trees, instrumentalized and instrumentalizable, artifactual, on the way to being put to use.

Materia in Latin denotes the stuff out of which things are made, not only the timber for houses and ships but whatever serves as nourishment for infants: nutrients that act as extensions of the mother's body. Insofar as matter appears in these cases to be invested with a certain capacity to originate and to compose that for which it also supplies the principle of intelligibility, then matter is clearly defined by a certain power of creation and rationality that is for the most part divested from the more modern empirical deployments of the term. To speak within these classical contexts *of bodies that matter* is not an idle pun, for to be material means to materialize, where the principle of that materialization is precisely what "matters" about that body, its very intelligibility. In this sense, to know the significance of something is to know how and why it matters, where "to matter" means at once "to materialize" and "to mean."

Obviously, no feminist would encourage a simple return to Aristotle's natural teleologies in order to rethink the "materiality" of bodies. I want to consider, however, Aristotle's distinction between body and soul to effect a brief comparison between Aristotle and Foucault in order to suggest a possible contemporary redeployment of Aristotelian terminology. At the end of this brief comparison, I will offer a limited criticism of Foucault, which will then lead to a longer discussion of [Luce] Irigaray's deconstruction of materiality in Plato's *Timaeus*. It is in the context of this second analysis that I hope to make clear how a gendered matrix is at work in the constitution of materiality (although it is obviously present in Aristotle as well), and why feminists ought to be interested, not in taking materiality as an irreducible, but in conducting a critical genealogy of its formulation.

ARISTOTLE/FOUCAULT

For Aristotle the soul designates the actualization of matter, where matter is understood as fully potential and unactualized. As a result, he maintains in *de Anima* that the soul is "the first grade of actuality of a naturally organized body."[2] He continues, "That is why we can wholly dismiss as unnecessary the question whether the soul and the body are one: it is as meaningless to ask whether the wax and the shape given to it by the stamp are one, or generally the matter [*hyle*] of a thing and that of which it is the matter [*hyle*]."[3] In the Greek, there is no reference to "stamps," but the phrase, "the shape given by the stamp" is contained in the single term, "*schema*." *Schema* means form, shape, figure, appearance, dress, gesture, figure of a syllogism, and grammatical form. If matter never appears without its *schema*, that means that it only appears under a certain grammatical form and that the principle of its recognizability, its characteristic gesture or usual dress, is indissoluble from what constitutes its matter.

In Aristotle, we find no clear phenomenal distinction between materiality and intelligibility, and yet for other reasons Aristotle does not supply us with the kind of "body" that feminism seeks to retrieve. To install the principle of intelligibility in the very development of a body is precisely the strategy of a natural teleology that accounts for female development through the rationale of biology. On this basis, it has been argued that women ought to perform certain social functions and not others, indeed, that women ought to be fully restricted to the reproductive domain.

We might historicize the Aristotelian notion of the *schema* in terms of culturally variable principles of formativity and intelligibility. To understand the *schema* of bodies as a historically contingent nexus of power/discourse is to arrive at something similar to what Foucault describes in *Discipline and Punish* as the "materialization" of the prisoner's body. This process of materialization is at stake as well in the final chapter of the first volume of *The History of Sexuality* when Foucault calls for a "history of bodies" that would inquire into "the manner in which what is most material and vital in them has been invested."[4]

At times it appears that for Foucault the body has a materiality that is ontologically distinct from the power relations that take that body as a site of investments. And yet, in *Discipline and Punish*, we have a different configuration of the relation between materiality and investment. There the soul is taken as an instrument of power through which the body is cultivated and formed. In a sense, it acts as a power-laden schema that produces and actualizes the body itself.

We can understand Foucault's references to the "soul" as an implicit reworking of the Aristotelian formulation. Foucault argues in *Discipline and Punish* that the "soul" becomes a normative and normalizing ideal according to which the body is trained, shaped, cultivated, and invested; it is an historically specific imaginary ideal (*ideal speculatif*) under which the body is effectively materialized. Considering the science of prison reform, Foucault writes, "The man described for us, whom we are invited to free, is already in himself the effect of a subjection [*assujettissement*] much more profound than himself. A 'soul' inhabits him and brings him to existence, which is itself a factor in the mastery that power exercises over the body. The soul is the effect and instrument of a political anatomy; the soul is the prison of the body."[5]

This "subjection," or *assujettissement*, is not only a subordination but a securing and maintaining, a putting into place of a subject, a subjectivation. The "soul brings [the prisoner] to existence"; and not fully unlike Aristotle, the soul described by Foucault as an instrument of power, forms and frames the body, stamps it, and in stamping it, brings it into being. Here "being" belongs in quotation marks, for ontological weight is not presumed, but always conferred. For Foucault, this conferral can take place only within and by an operation of power. This operation produces the subjects that it subjects; that is, it subjects them in and through the compulsory power relations effective as their formative principle. But power is that which forms, maintains, sustains, and regulates bodies at once, so that, strictly speaking, power is not a subject who acts on bodies as its distinct objects. The grammar which compels us to speak that way enforces a metaphysics of external relations, whereby power acts on bodies but is not understood to form them. This is a view of power as an external relation that Foucault himself calls into question.

Power operates for Foucault in the *constitution* of the very materiality of the subject, in the principle which simultaneously forms and regulates the "subject" of subjectivation. Foucault refers not only to the materiality of the body of the prisoner but to the materiality of the body of the prison. The materiality of the prison, he writes, is established to the extent that [*dans la mesure ou*] it is a vector and instrument of power. Hence, the prison is *materialized* to the extent that it is *invested with power*, or, to be grammatically accurate, there is no prison prior to its materialization. Its materialization is coextensive with its investiture with power relations, and materiality is the effect and gauge of this investment. The prison comes to be only within the field of power relations, but more specifically, only to the extent that it is invested or saturated with such relations, that such a saturation is itself formative of its very being. Here the body is not an independent materiality that is invested by power relations external to it, but it is that for which materialization and investiture are coextensive.

"Materiality" designates a certain effect of power or, rather, *is* power in its formative or constituting effects. Insofar as power operates successfully by constituting an object domain, a field of intelligibility, as a taken-for-granted ontology, its material effects are taken as material data or primary givens. These material positivities appear *outside* discourse and power, as its incontestable referents, its transcendental signifieds. But this appearance is precisely the moment in which the power/discourse regime is most fully dissimulated and most insidiously effective. When this material effect is taken as an epistemological point of departure, a *sine qua non* of some political argumentation, this is a move of empiricist foundationalism that, in accepting this constituted effect as a primary given, successfully buries and masks the genealogy of power relations by which it is constituted.

getting out of mind/body duality *(chicken or egg but either matter?)*

Insofar as Foucault traces the process of materialization as an investiture of discourse and power, he focuses on that dimension of power that is productive and formative. But we need to ask what constrains the domain of what is materializable, and whether there are *modalities* of materialization—as Aristotle suggests, and Althusser is quick to cite. To what extent is materialization governed by principles of intelligibility that require and institute a domain of radical *unintelligibility* that resists materialization altogether or that remains radically dematerialized? Does Foucault's effort to work the notions of discourse and materiality through one another fail to account for not only what is *excluded from* the economies of discursive intelligibility that he describes, but what *has to be excluded* for those economies to function as self-sustaining systems?

This is the question implicitly raised by Luce Irigaray's analysis of the form/matter distinction in Plato. This argument is perhaps best known from the essay "Plato's Hystera," in *Speculum of the Other Woman*, but is trenchantly articulated as well in the less well-known essay, "Une Mere de Glace," also in *Speculum*.

Irigaray's task is to reconcile neither the form/matter distinction nor the distinctions between bodies and souls or matter and meaning. Rather, her effort is to show that those binary oppositions are formulated through the exclusion of a field of disruptive possibilities. Her speculative thesis is that those binaries, even in their reconciled mode, are part of a phallogocentric economy that produces the "feminine" as its constitutive outside. Irigaray's intervention in the history of the form/matter distinction underscores "matter" as the site at which the feminine is excluded from philosophical binaries. Inasmuch as certain phantasmatic notions of the feminine are traditionally associated with materiality, these are specular effects which confirm a phallogocentric project of autogenesis. And when those specular (and spectral) feminine figures are taken to be the feminine, the feminine is, she argues, fully erased by its very representation. The economy that claims to include the feminine as the subordinate term in a binary opposition of masculine/feminine excludes the feminine, produces the feminine as that which must be excluded for that economy to operate. . . .

There is for Irigaray, always, a matter that exceeds matter, where the latter is disavowed for the autogenetic form/matter coupling to thrive. Matter occurs in two modalities: First, as a metaphysical concept that serves a phallogocentrism; second, as an ungrounded figure, worrisomely speculative and catachrestic, that marks for her the possible linguistic site of a critical mime.

excluding some bodies out of lived histories

To play with mimesis is thus, for a woman, to try to recover the place of her exploitation by discourse, without allowing herself to be simply reduced to it. It means to resubmit herself—inasmuch as she is on the side of the "perceptible," of "matter"—to "ideas," in particular to ideas about herself, that are elaborated in/by a masculine logic, but so as to make "visible," by an effect of playful repetition, what was supposed to remain invisible: the cover up of a possible operation of the feminine in language.[6]

So perhaps here is the return of essentialism, in the notion of a "feminine in language"? And yet, she continues by suggesting that *miming* is that very operation of the feminine in language. To mime means to participate in precisely that which is mimed, and if the language mime is the language of phallogocentrism, then this is only a specifically feminine language to the extent that the feminine is radically implicated in the very terms of a phallogocentrism it seeks to rework. The quotation continues, "[to play with mimesis means] 'to unveil' the fact that, if women are such good mimics, it is because they are not simply resorbed in this function. *They also remain elsewhere*, another case of the persistence of 'matter.' "[7]

They mime phallogocentrism, but they also expose what is covered over by the mimetic self-replication of that discourse. For Irigaray, what is broken with and covered over is the linguistic operation of metonymy, a closeness and proximity which appears to be the linguistic residue of the initial proximity of mother and infant. It is this metonymic excess in every mime, indeed, in every metaphorical substitution, that is understood to disrupt the seamless repetition of the phallogocentric norm.

To claim, though, as Irigaray does, that the logic of identity is potentially disruptible by the insurgence of metonymy, and then to identify this metonymy with the repressed and insurgent feminine is to consolidate the place of the feminine in and as the irruptive chora, that which cannot be figured, but which is necessary for any figuration. That is, of course, to figure that chora nevertheless, and in such a way that the feminine is "always" the outside, and the outside is "always" the feminine. This is a move that at once positions the feminine as the unthematizable, the non-figurable, but which, in identifying the feminine with that position, thematizes and figures, and so makes use of the phallogocentric exercise to produce this identity which "is" the non-identical.

There are good reasons, however, to reject the notion that the feminine monopolizes the sphere of the excluded here. Indeed, to enforce such a monopoly redoubles the effect of foreclosure performed by the phallogocentric discourse itself, one which "mimes" its founding violence in a way that works against the explicit claim to have found a linguistic site in metonymy that works as disruption. After all, Plato's scenography of intelligibility depends on the exclusion of women, slaves, children, and animals, where slaves are characterized as those who do not speak his language, and who, in not speaking his language, are considered diminished in their capacity for reason. This xenophobic exclusion operates through the production of racialized Others, and those whose "natures" are considered less rational by virtue of their appointed task in the process of laboring to reproduce the conditions of private life. This domain of the less than rational human bounds the figure of human reason, producing that "man" as one who is without a childhood; is not a primate and so is relieved of the necessity of eating, defecating, living and dying; one who is not a slave, but always a property holder; one whose language remains originary and untranslatable. This is a figure of disembodiment, but one which is nevertheless a figure of a body, a bodying forth of a masculinized rationality, the figure of a male body which is not a body, a figure in crisis, a figure that enacts a crisis it cannot fully control. This figuration of masculine reason as disembodied body is one whose imaginary morphology is crafted through the exclusion of other possible bodies. This is a materialization of reason which operates through the dematerialization of other bodies, For the Feminine, strictly speaking, has no morphe, no morphology, no contour. For it is that which contributes to the contouring of things, but is itself undifferentiated, without boundary. The body that is reason dematerializes the bodies that may not properly stand for reason or its replicas, and yet this is a figure in crisis, for this body of reason is itself the phantasmatic dematerialization of masculinity, one which requires that women and slaves, children and animals be the body, perform the bodily functions, that it will not perform. . . .

Irigaray does not always help matters here, for she fails to follow through the metonymic link between women and these other Others, idealizing and appropriating the "elsewhere" as the Feminine. But what is the "elsewhere" of Irigaray's "elsewhere"? If the Feminine is not the only or primary kind of being that is excluded from the economy of masculinist reason, what and who is excluded in the course of Irigaray's analysis?

IMPROPER ENTRY: PROTOCOLS OF SEXUAL DIFFERENCE

The above analysis has considered not the materiality of sex, but the sex of materiality. In other words, it has traced materiality as the site at which a certain drama of sexual difference plays itself out. The point of such an exposition is not only to warn against an easy return to the *materiality* of the body or the materiality of sex, but to show that to invoke matter is to invoke a sedimented history of sexual hierarchy and sexual erasures which should surely be an *object* of feminist inquiry, but which would be quite problematic as a *ground* of feminist theory. To return to matter requires that we return to matter as a *sign* which in its redoublings and contradictions enacts an inchoate drama of sexual difference.

If there is an occupation and reversal of the master's discourse, it will come from many quarters, and those resignifying practices will converge in ways that scramble the self-replicating presumptions of reason's mastery. For if the copies speak, or if what is merely material begins to signify, the scenography of reason is rocked by the crisis on which it was always built. And there will be no way finally to delimit the elsewhere of Irigaray's elsewhere, for every oppositional discourse will produce its outside, an outside that risks becoming installed as its non-signifying inscriptional space. And whereas this can appear as the necessary and founding violence of any truth-regime, it is important to resist that theoretical gesture of pathos in which exclusions are simply affirmed as sad necessities of signification. The task is to refigure this necessary "outside" as a future horizon, one in which the violence of exclusion is perpetually in the process of being overcome. But of equal importance is the preservation of the outside, the site where discourse meets its limits, where the opacity of what is not included in a given regime of truth acts as a disruptive site of linguistic impropriety and unrepresentability, illuminating the violent and contingent boundaries of that normative regime precisely through the inability of that regime to represent that which might pose a fundamental threat to its continuity. In this sense, radical and inclusive representability is not precisely the goal: to include, to speak as, to bring in every marginal and excluded position within a given discourse is to claim that a singular discourse meets its limits nowhere, that it can and will domesticate all signs of difference. If there is a violence necessary to the language of politics, then the risk of that violation might well be followed by another in which we begin, without ending, without mastering, to own—and yet never fully to own—the exclusions by which we proceed.

NOTES

1. Jacques Derrida, *Positions*, ed. Alan Bass (Chicago: University of Chicago Press, 1978), 64.

2. Aristotle, "De Anima," in *The Basic Works of Aristotle*, trans. Richard McKeon (New York: Random House, 1941), bk.2, ch.1, 412a10, p. 555

3. Ibid., bk.2, ch.1, 412b7–8.

4. Foucault, *The History of Sexuality*, vol. 1 (Paris: Gallimard, 1978), 200.

5. Foucault, *Discipline and Punish: The Birth of the Prison* (New York: Pantheon, 1977), 30.

6. Luce Irigaray, "The Power of Discourse" in *This Sex Which Is Not One* (Ithaca: Cornell University Press, 1985), 76.

7. Ibid.

sensory engagement, we actively make it through our experience - Butler is talking about the lenses that are not considered by yr own lens?

Sex/gender duality
gender importance than sex

• how subjects are made 16/17

DO YOU BELIEVE IN REALITY?

BRUNO LATOUR

"I have a question for you," he said, taking out of his pocket a crumpled piece of paper on which he had scribbled a few key words. He took a breath: "Do you believe in reality?"

"But of course!" I laughed. "What a question! Is reality something we have to believe in?

He had asked me to meet him for a private discussion in a place I found as bizarre as the question: by the lake near the chalet, in this strange imitation of a Swiss resort located in the tropical mountains of Teresopolis in Brazil. Has reality truly become something people have to believe in, I wondered, the answer to a serious question asked in a hushed and embarrassed tone? Is reality something like God, the topic of a confession reached after a long and intimate discussion? Are there people on earth who *don't* believe in reality?

When I noticed that he was relieved by my quick and laughing answer, I was even more baffled, since his relief proved clearly enough that he had antici-pated a *negative* reply, something like, "Of course not! Do you think I am that naïve?" This was not a joke, then: he really was concerned, and his query had been in earnest.

"I have two more questions," he added, sounding more relaxed. "Do we know more than we used to?"

"But of course! A thousand times more!"

"But is science cumulative?" he continued with some anxiety, as if he did not want to be won over too fast.

"I guess so," I replied, "although I am less positive on this one, since the sciences also forget so much, so much of their past and so much of their bygone research programs—but, on the whole, let's say yes. Why are you asking me these questions? Who do you think I am?"

I had to switch interpretations fast enough to comprehend both the mon-ster he was seeing me as when he raised these questions and his touching openness of mind in daring to address such a monster privately. It must have taken courage for him to meet with one of these creatures that threatened, in his view, the whole establishment of science, one of these people from a mysterious field called "science studies," of which he had never before met a flesh-and-blood representative but which—at least so he had been told—was another threat to science in a country, America, where scientific inquiry had never had a completely secure foothold.

He was a highly respected psychologist, and we had both been invited by the Wenner-Gren Foundation to a gathering made up of two-thirds scientists and one-third "science students." This division itself, announced by the organizers, baffled me. How could we be pitted *against* the scientists? That we are studying a subject matter does not mean that we are attacking it. Are biologists anti-life, astronomers anti-stars, immunologists anti-antibodies? Besides, I had taught for twenty years in scientific schools, I wrote regularly in scientific journals, I and my colleagues lived on contract research carried out on behalf of many groups of scientists in industry and in the academy. Was I not part of the French scientific establishment? I was a bit vexed to be excluded so casually. Of course I am just a philosopher, but what would my friends in science studies say? Most of them have been trained in the sciences, and several of them, at least, pride themselves on *extending* the scientific outlook to science itself. They could be labeled as members of another discipline or another subfield, but certainly not as "anti-scientists" meeting halfway with scientists, as if the two groups were opposing armies conferring under a flag of truce before returning to the battlefield!

I could not get over the strangeness of the question posed by this man I considered a colleague, yes, a colleague (and who has since become a good friend). If science studies has achieved anything, I thought, surely it has *added* reality to science, not withdrawn any from it. Instead of the stuffed scientists hanging on the walls of the armchair philosophers of science of the past, we have portrayed lively characters, immersed in their laboratories, full of passion, loaded with instruments, steeped in know-how, closely connected to a larger and more vibrant milieu. Instead of the pale and bloodless objectivity of science, we have all shown, it seemed to me, that the many nonhumans mixed into our collective life through laboratory practice have a history, flexibility, culture, blood—in short, all the characteristics that were denied to them by the humanists on the other side of the campus. Indeed, I have naively thought, if scientists have a faithful ally, it is we, the "science students" who have managed over the years to interest scores of literary folk in science and technology, readers who were convinced, until science studies came along, that "science does not think" as Heidegger, one of their masters, had said.

The psychologist's suspicion struck me as deeply unfair, since he did not seem to understand that in this guerrilla warfare being conducted in the no-man's-land between the "two cultures," *we were the ones* being attacked by militants, activists, sociologists, philosophers, and technophobes of all hues, precisely because of our interest in the inner workings of scientific facts. Who loves the sciences, I asked myself, more than this tiny scientific tribe that has learned to open facts, machines, and theories with all their roots, blood vessels, networks, rhizomes, and tendrils? Who believes more in the objectivity of science than those who claim that it can be turned into an object of inquiry?

Then I realized that I was wrong. What I would call "adding realism to science" was actually seen, by the scientists at this gathering, as a threat to the calling of science, as a way of decreasing its stake in truth and their claims to certainty. How has this misunderstanding come about? How could I have lived long enough to be asked in all seriousness the incredible question: "Do you believe in reality?" The distance between what I thought we had achieved in science studies and what was implied by this question was so vast that I needed to retrace my steps a bit. . . .

THE STRANGE INVENTION OF AN "OUTSIDE" WORLD

There is no natural situation on earth in which someone could be asked this strangest of all questions: "Do you believe in reality?" To ask such a question one has to become so *distant* from reality that the fear of *losing* it entirely becomes plausible—and this fear itself has an intellectual history that should at least be sketched. Without this detour we would never be able to fathom the extent of the misunderstanding between my colleague and me, or to measure the extraordinary form of radical realism that science students have been uncovering.

I remember that my colleague's question was not so new. My compatriot Descartes had raised it against himself when asking how an isolated mind could be *absolutely* as opposed to relatively sure of anything about the outside world. Of course, he framed his question in a way that made it impossible to give the only reasonable answer, which we in science studies have slowly re-discovered three centuries later: that we are *relatively* sure of many of the things with which we are daily engaged through the practice of our laborato-ries. By Descartes's time this sturdy relativism, based on the number of *rela-tions* established with the world, was already with the past, a once-passable path now lost in a thicket of brambles. Descartes was asking for absolute certainty from a brain-in-a-vat, a certainty that was not needed when the brain (or the mind) was firmly attached to its body and the body thoroughly involved in its normal ecology. As in Curt Siodmak's novel *Donovan's Brain*, absolute certainty is the sort of neurotic fantasy that only a surgically removed mind would look for after it had lost everything else. Like a heart taken out of a young woman who has just died in an accident and soon to be transplanted into someone else's thorax thousands of miles away, Descartes's mind requires artificial life-support to keep it viable. Only a mind put in the strangest position, looking at a world *from the inside out* and linked to the outside by nothing but the tenuous connection of the *gaze*, will throb in the constant fear of losing reality; only such a bodiless observer will desperately look for some absolute life-supporting survival kit.

For Descartes the only route by which his mind-in-a-vat could re-establish some reasonably pure connection with the outside world was through God. My friend the psychologist was thus right to phrase his query using the same formula I had learned in Sunday school: "Do you believe in reality?"—"Credo in unum Deum," or rather, "Credo in unam realitam," as my friend Donna Haraway kept chanting in Teresopolis! After Descartes, however, many people thought that going through God to reach the world was a bit expensive and far-fetched. They looked for a shortcut. They wondered whether the world could *directly* send us enough information to produce a stable image of itself in our minds.

But in asking this question the empiricists kept going along the same path. They did not retrace their steps. They never plugged the wriggling and squiggling brain back into its withering body. They were still dealing with a mind looking through the gaze at a lost outside world. They simply tried to train it to recognize patterns. God was out, to be sure, but the *tabula rasa* of the empiricists was as disconnected as the mind in Descartes's times. The brain-in-a-vat simply exchanged one survival kit for another. Bombarded by a world reduced to meaningless stimuli, it was supposed to extract from these stimuli everything it needed to recompose the world's shapes and stories. The result was like a badly connected TV set, and no amount of tuning made this precursor of neural nets produce more than a fuzzy set of blurry lines, with white points falling like snow. No shape was recognizable. Absolute certainty was lost, so precarious were the connections of the senses to a world that was pushed even further outside. There was too much static to get any clear picture.

The solution came, but in the form of a catastrophe from which we are only now beginning to extricate ourselves. Instead of retracing their steps and taking the other path at the forgotten fork in the road, philosophers abandoned even the claim to absolute certainty, and settled instead on a makeshift solution that preserved at least some access to an outside reality. Since the empiricists' associative neural net was unable to offer clear pictures of the lost world, this must prove, they said, that the mind (still in a vat) extracts *from itself* everything it needs to form shapes and stories. Everything, that is, except the reality itself. Instead of the fuzzy lines of the poorly tuned TV set, we got the fixed tuning grid, molding the confused static, dots, and lines of the empiricist channel into a steady picture held in place by the mindset's predesigned categories. Kant's *a priori* started this extravagant form of constructivism, which neither Descartes, with his detour through God, nor Hume, with his shortcut to associated stimuli, would ever have dreamed of.

Now, with the Konigsberg broadcast, everything was ruled by the mind itself and reality came in simply to say that it was there, indeed, and not imaginary! For the banquet of reality, the mind provided the food, and the inaccessible things-in-themselves to which the world had been reduced simply dropped by to say "We are here, what you eat is not dust," but otherwise remained mute and stoic guests. If we abandon absolute certainty, Kant said, we can at least retrieve universality as long as we remain inside the restricted sphere of science, to which the world outside contributes decisively but minimally. The rest of the quest for the absolute is to be found in morality, another *a priori* certainty that the mind-in-a-vat extracts from its own wiring. Under the name of a "Copernican Revolution" Kant invented this science-fiction nightmare: the outside world now turns around the mind-in-a-vat, which dictates most of that world's laws, laws it has extracted from itself without help from anyone else. A crippled despot now ruled the world of reality. This philosophy was thought, strangely enough, to be the deepest of all, because it had at once managed to abandon the quest for absolute certainty and to retain it under the banner of "universal *a prioris*," a clever sleight of hand that hid the lost path even deeper in the thickets.

Do we really have to swallow these unsavory pellets of textbook philosophy to understand the psychologist's question? I am afraid so, because otherwise the innovations of science studies will remain invisible. The worst is yet to come. Kant had invented a form of constructivism in which the mind-in-a-vat built everything by itself but not entirely without constraints: what it learned from itself had to be universal and could be elicited only by some experimental contact with a reality out there, a reality reduced to its barest minimum, but there nonetheless. For Kant there was still something that revolved around the crippled despot, a green planet around this pathetic sun. It would not be long before people realized that this "transcendental Ego," as Kant named it, was a fiction, a line in the sand, a negotiating position in a complicated settlement to avoid the complete loss of the world or the complete abandonment of the quest for absolute certainty. It was soon replaced by a more reasonable candidate, *society*. Instead of a mythical Mind giving shape to reality, carving it, cutting it, ordering it, it was now the prejudices, categories, and paradigms of a group of people living together that determined the representations of every one of those people. This new definition, however, in spite of the use of the word "social," had only a superficial resemblance to the realism in which we science students have become attached, and which I will outline over the course of this book.

First, the replacement of the despotic Ego with the sacred "society" did not retrace the philosopher's steps but went even *further* in distancing the individual's vision, now a "view of the world," from the definitely lost outside world. Between the two, society interposed its filters; its paraphernalia of biases, theories, cultures, traditions, and standpoints became an opaque window. Nothing of the world could pass through so many intermediaries and reach the individual mind. People were now locked not only into the prison of their own categories but into that of their social groups as well. Second, this "society" itself was just a series of minds-in-a-vat, many minds and many vats to be sure, but each of them still composed of that strangest of beasts: a detached mind gazing at an outside world. Some improvement! If prisoners were no longer in isolated cells, they were now confined to the same dormitory, the same collective mentality. Third, the next shift, from one Ego to multiple cultures, jeopardized the only good thing about Kant, that is, the universality of the *a priori* categories, the only bit of ersatz absolute certainty he had been able to retain. Everyone was not locked in the same prison any more; now there were *many* prisons, incommensurable, unconnected. Not only was the mind disconnected from the world, but each collective mind, each culture was disconnected from the others. More and more progress in a philosophy dreamed up, it seems, by prison wardens.

But there was a fourth reason, even more dramatic, even sadder, that made this shift to "society" a catastrophe following fast on the heels of the Kantian revolution. The claims to knowledge of all these poor minds, prisoners in their long row of vats, were now made part of an even more bizarre history, were now associated with an even more ancient threat, *the fear of mob rule*. If my friend's voice quivered as he asked me "Do you believe in reality?" it was not only because he feared that all connection with the outside world might be lost, but above all because he worried that I might answer, "Reality depends on whatever the mob thinks is right at any given time." It is the resonance of these two fears, the *loss* of any certain access to reality and the *invasion* by the mob, that makes his question at once so unfair and so serious.

But before we disentangle this second threat, let me finish with the first one. The sad story, unfortunately, does not end here. However incredible it seems, it is possible to go even further along the wrong path, always thinking that a more radical solution will solve the problems accumulated from the past decision. One solution, or more exactly another clever sleight of hand, is to become so very pleased with the loss of absolute certainty and universal *a prioris* that one rejoices in abandoning them. Every defect of the former position is now taken to be its best quality. Yes, we have lost the world. Yes, we are forever prisoners of language. No, we will never regain certainty. No, we will never get beyond our biases. Yes, we will forever be stuck within our own selfish standpoint. Bravo! Encore! The prisoners are now gagging even those who ask them to look out their cell windows; they will "deconstruct," as they say—which means destroy in slow motion—anyone who reminds them that there was a time when they were free and when their language bore a connection with the world.

Who can avoid hearing the cry of despair that echoes deep down, carefully repressed, meticulously denied, in these paradoxical claims for a joyous, jubilant, free construction of narratives and stories by people forever in chains? But even if there *were* people who could say such things with a blissful and light heart (their existence is as uncertain to me as that of the Loch Ness monster, or, for that matter, as uncertain as that of the real world would be to these mythical creatures), how could we avoid noticing that we have not moved an inch since Descartes? That the mind is still in its vat, excised from the rest, disconnected, and contemplating (now with a blind gaze) the world (now lost in darkness) from the very same bubbling glassware? Such people may be able to smile smugly instead of trembling with fear, but they are still descending further and further along the spiraling curves of the same hell. At the end of this chapter we will meet these gloating prisoners again.

In our century, though, a second solution has been proposed, one that has occupied many bright minds. This solution consists of taking only a *part* of the mind out of the vat and then doing the obvious thing, that is, offering it a body again and putting the reassembled aggregate back into relation with a world that is no longer a spectacle at which we gaze but a lived, self-evident, and unreflexive extension of ourselves. In appearance, the progress is immense, and the descent into damnation suspended, since we no longer have a mind dealing with an outside world, but a lived world to which a semi-conscious and intentional body is now attached.

Unfortunately, however, in order to succeed, this emergency operation must chop the mind into even smaller pieces. The real world, the one known by science, is left entirely to itself. Phenomenology deals only with the world-for-a-human-consciousness. It will teach us a lot about how we never distance ourselves from what we see, how we never gaze at a distant spectacle, how we are always immersed in the world's rich and lived texture, but, alas, this knowledge will be of no use in accounting for how things really are, since we will never be able to escape from the narrow focus of human intentionality. Instead of exploring the ways we can shift from standpoint to standpoint, we will always be fixed in the human one. We will hear much talk about the real, fleshy, pre-reflexive lived world, but this will not be enough to cover the noise of the second ring of prison doors slamming even more tightly shut behind us. For all its claims to overcoming the distance between subject and object—as if this distinction were something that could be overcome!—phenomenology leaves us with the most dramatic split in this whole sad story: a world of science left entirely to itself, entirely cold, absolutely inhuman; and a rich lived world of intentional stances entirely limited to humans, absolutely divorced from what things are in and for themselves. A slight pause on the way down before sliding even further in the same direction.

Why not choose the opposite solution and forget the mind-in-a-vat altogether? Why not let the "outside world" invade the scene, break the glassware, spill the bubbling liquid, and turn the mind into a brain, into a neuronal machine sitting inside a Darwinian animal struggling for its life? Would that not solve all the problems and reverse the fatal downward spiral? Instead of the complex "life-world" of the phenomenologists, why not study the adaptation of humans, as naturalists have studied all other aspects of "life"? If science can invade everything, it surely can put an end to Descartes's long-lasting fallacy and make the mind a wriggling and squiggling part of nature. This would certainly please my friend the psychologist—or would it? No, because the ingredients that make up this "nature," this hegemonic and all-encompassing nature, which would now include the human species, are the *very same ones* that have constituted the spectacle of a world viewed from inside by a brain-in-a-vat. Inhuman, reductionist, causal, law-like, certain, objective, cold, unanimous, absolute—all these expressions do not pertain to nature *as such*, but to nature viewed through the deforming prism of the glass vessel!

If there is something unattainable, it is the dream of treating nature as a homogeneous unity in order to unify the different views the sciences have of it! This would require us to ignore too many controversies, too much history, too much unfinished business, too many loose ends. If phenomenology abandoned science to its destiny by limiting it to human intention, the opposite move, studying humans as "natural phenomena," would be even worse: it would abandon the rich and controversial human history of science—and for what? The averaged-out orthodoxy of a few neurophilosophers? A blind Darwinian process that would limit the mind's activity to a struggle for survival to "fit" with a reality whose true nature would escape us forever? No, no, we can surely do better, we can surely stop the downward slide and retrace our steps, retaining both the history of humans' involvement in the making of scientific facts and the sciences' involvement in the making of human history.

PART III

Fundamental Processes, or

Denaturalizing the Given

Fundamental Processes, or

Denaturalizing the Given

INTRODUCTION

The basic facts of embodiment seem to provide a foundation for humanness itself. Processes like respiration, bipedal locomotion, alimentation, mammalian sexual reproduction, and elimination are, we often feel, a lowest common denominator for the species, with variations that should be interesting only to the biologist. Differences introduced into these preeminently organic domains by cultural activity such as the culinary or military or erotic arts appear to be superimposed on a natural base that need not be interrogated by the historian or social anthropologist. These scholars can confine their attentions, it has seemed, to the meanings and values attached to the life of the body, without searching for cultural variation in material being.

Yet anthropology and history have revealed many instances of body processes that challenge Euro-western common sense about the natural foundations of our humanity. Some "unnatural" processes were once the bread and butter of a rather sensationalist anthropology of the body, one that reveled in presenting images of stretched lips and earlobes, self-flagellation, tattooing and scarification, and people hanging themselves from meat hooks. This was denaturalization for the thrill of it, with little effort made to present the social and historical contexts within which these body practices evolved and had a social role. It was easy to dismiss the modified bodies thus presented as bizarre exceptions to all the natural rules governing our own commonsense embodiment. Our own nature, as residents of a bourgeois metropolitan world, could go unchallenged.

In part III, and elsewhere in this book, there are examples of anthropological and historical research that make any such dismissal much more difficult. Kristofer Schipper's short and puzzling guide to breathing the Taoist way, for example, suggests that even the most foundational physiological processes can be subject to deeply formative training. Caroline Walker Bynum's study of eucharistic devotions among thirteenth-century women mystics demands, partly by the sheer multiplicity of her examples, that social and historical sense be made of a practice that at first seems self-destructive and illusory. And Henry Abelove's ironic reading of historical demographics suggests that "sexual intercourse" is perhaps not simply "doing what comes naturally."

These examples of the situated life of the "unnatural" body serve to challenge assumptions about both human nature and bourgeois common sense. They should lead us to interrogate not just the cultural meanings and values attached to bodily life, but the very material form of human embodiment and its many ways of being natural and cultural at once. Very often, this process of bracketing our own nature, seeing it as exotic and contingent on variable historical conditions, does not require that we go as far afield as the exotic locales associated with an earlier cross-cultural comparison. In the age of the cyborg (Haraway 1991), and as Lock's article in this part demonstrates, the most unnatural bodily situations are part of the fabric of everyday life in some sectors of contemporary society.

Part III begins, however, with a classic example of cross-cultural comparison. Near the center of his magisterial ethnography of the Nuer (based on fieldwork in the 1930s), E. E. Evans-Pritchard turns to the practical life of the body by describing Nuer experience of time and space. Though he presents this material as concepts, his account always tends toward a phenomenological appreciation of how time and space were not just thought but lived among the Nuer. He often writes as if time and space actually took the form that Nuer took for granted. Thus, for example, the seasons of *Tot* and *Mai* are divided more according to actors' "direction of attention" (planning for impending residence in the village, for example) than according to "objective" markers of environmental change. Evans-Pritchard emphasizes the way in which temporally mandated *activity* produces a sense of time and a foundation for decision making, arguing that "the calendar is a relation between a cycle of activities and a conceptual cycle and the two cannot fall apart, since the conceptual cycle is dependent on the cycle of activities from which it derives its meaning and function." Because activities and concepts "cannot fall apart," Nuer did not reify time as if it were something that could be spent, passed, wasted, and so on. In a typically understated snippet of implicit comparison, Evans-Pritchard remarks, "Nuer are fortunate." In our clock-driven lives, can we even imagine what embodiment must have been like in the East African villages Evans-Pritchard visited?

Caroline Walker Bynum offers another example of a challenge to our ways of imagining fundamental processes, this one drawn from the past of Western civilization. Bynum is well known for her scholarly exploration of the body practices of medieval mystics. Drawing on hagiographic and other ecclesiastical writings, she describes medieval religious experience, gender assumptions, and the logical relation between them. Though her chapter has been shortened here, one is unfailingly impressed in all of Bynum's writing with the sheer number of examples she is able to adduce of people pushing bodily experience to its limits in the name of the spiritual. Thanks to her research, we can no longer see "holy anorexia," religious ecstasy in the Eucharist, or techniques of mortification of the flesh as bizarre or deviant in their original context. In the extract presented here, Bynum examines the special interest thirteenth-century religious women showed in the Eucharist, the eating and drinking of Christ's body and blood. Linking this sacrament to other forms of self-mortification and asceticism, thirteenth-century religious women experienced Eucharist as a participation in Christ's humanity, the carnal aspects of which were especially associated with the female state. Similarly, they saw illness, hunger, and pain as imitation of Christ, "an effort to plumb the depths of Christ's humanity at the moment of his most insistent and terrifying humanness—the moment of his dying." She concludes that "to soar toward Christ as lover and bride, to sink into the stench and torment of the Crucifixion, to eat God, was for the woman only to give religious significance to what she already was." Perhaps this article reminds us indirectly why in our own era feminist thought and the anthropology of the body have had a linked history.

Kristofer Schipper, the author of *The Taoist Body*, drew upon extensive personal training in Taoist ritual techniques he received in southern Taiwan during the 1970s to emphasize the carnal concreteness of this monastic system. In this extract describing some disciplined breathing techniques of Taoist adepts, proper respiration is seen as being possible only after deliberate training and lengthy practice. A form of breathing attuned to cosmic temporality leads to a healthier and more complete existence, producing a body that can eventually come into harmony with the Way. These techniques are only part of the fundamentally corporeal religious training Schipper reports. Though his book was based on a very personal engagement with only one of many strains of Taoist practice, it demonstrated very clearly the role of deliberate embodiment in East Asian religious life. Far from being a matter of beliefs or even worldviews, Taoist practice encourages an embodiment suffused with agency and discipline, one that offers a better way of living to those who can learn it. This Way is more natural only in a very special sense, since it requires rigorous training, extending to an activity as basic as breathing, to achieve.

Henry Abelove denaturalizes in a more ironic mode, taking a skeptical stance on conventional demographic history in his short piece on England's eighteenth-century population explosion. Abelove's insistence on using quotation marks around some standard historical usages such as "long eighteenth century" (conventionally understood as the 1680s to the 1830s) and even the very ordinary term "sexual intercourse" signals his skepticism about the ways in which the past has been carved up into standard periods and self-evident categories of behavior. Here he suggests that a marked birthrate increase in late eighteenth-century England is not well accounted for by demographic historians' usual explanations: more marriages, an earlier age of marriage, and more illegitimate births. Put another way, it seems that population historians are unwilling to state the most obvious thing about their data, which Abelove does state: "Sexual intercourse, so-called, became importantly more popular" in late eighteenth-century England. Something changed, he suggests, and he speculates that it was a shift in "conceptions of what sex is, what it is for." Prior to the shift into high fertility, sexual activity outside marriage may have been quite diverse and—crucially—often nonreproductive. Abelove suggests that higher birthrates were most closely linked with the new discursive and phenomenological centrality accorded to production as the Industrial Revolution got under way. Is it possible that, as productive activity began to take center stage in early modern Europe and the disciplinary regimes documented by historians such as E. P. Thompson took hold (see part VIII), nonproductive sexual activities came to be less popular or stigmatized? Abelove suggests that something changes in the tradition of very diverse cross-sex sexual behaviors in Europe: "They are reorganized and reconstructed in the late eighteenth century as foreplay." In other words, the "proper" or "natural" stages of sexual activity—foreplay leading to "sexual intercourse as such"—are neither proper nor natural, but rather may have come into existence in a historical development. Abelove hypothesizes that "the invention of foreplay is an aspect of the history of capitalism," coming to be through the same process in which sexual intercourse—the "(re)productive" act—becomes especially privileged. Abelove's speculations are very suggestive about new avenues for historical research; even as a starting point, they force us to consider the genuine possibility that sexual activity may have no natural form, trajectory, or outcome.

But what of birth and death? Surely these are the unavoidable limit conditions that ground the naturalness of everyone's embodiment. The ethnographic studies that conclude this part suggest otherwise, however. In her book *Twice Dead: Organ Transplants and the Reinvention of Death*, Margaret Lock discusses that bodily state in which patients exhibit biological signs of both life and death. Lock's focus is on the invention of the diagnosis of brain death, a condition that has been recognized legally for decades in many countries as the end of human life. Biomedical definitions of brain death are applied when the patient's entire brain is irreversibly damaged; patients in this condition cannot exist for more than a very brief time without life support. Lock examines the pressures that were brought to bear more than thirty years ago, largely from the medical profession, so that brain-dead hybrids came to be recognized as no longer persons. Therefore, with consent, organs could be legally procured from the brain-dead for transplant. However, medical and ethical disputes continue to this day about whether brain-dead patients are ultimately alive or dead. Lock derives many of her insights from ethnographic work in clinical settings, not only in North America but also in Japan, where for complex political and cultural reasons brain death was not legally recognized until a few years ago. The unresolved dilemmas of hospital (non)death force us to recognize that biological, social, and individual death are not of the same order and that to understand either death or life merely as biological events or states is entirely inadequate.

Included here is a short excerpt from an essay by Lock in which individuals who are recipients of an organ transplant give voice to their feelings about the experience. Although some people insist that their lives are essentially unchanged, a more common response is to report a sense of being radically transformed—of being embodied differently; one's mode of being in the world is irrevocably altered because of, above all, the knowledge that the body part was a "gift" from another human being. New human relationships often result from such radically remade bodies, although they are all too often ties of fantasy only, owing to the medical requirement of anonymity if the donor is not a living relation. As the excerpt shows, however, the cementing of actual, long-lasting attachments can and does take place, bringing about a profound change not only in the body itself and in subjectivity, but at times in communal and even in international relations.

Part III concludes with a reflection on birth taken from Anna Lowenhaupt Tsing's ethnography of the Meratus Dayak inhabitants of mountainous areas of southeastern Kalimantan, Borneo. In this complex study of an "out-of-the-way place" Tsing considers both the uniqueness of Meratus culture and practice and its many linkages to global capitalist processes and to the Indonesian state. One of the strengths of her treatment of these issues is her presentation of narratives by shamans, which she frames in an interrogative mode, asking how these very different—and yet profoundly mimetic—texts might be understood. One of the stars of her story is Uma Adang, a visionary who makes claims to both political and cosmological leadership through her idiosyncratic version of the world. Uma Adang's orations, made up of fragments and parodies of traditional texts, official propaganda, and more standard histories, have persuaded not a few residents of the Meratus mountains to adhere to her system of history and practice. Tsing finds much to inspire us as theorists in these puzzling and "messy" narratives.

The extract included here takes up the question of how the fetus develops in the womb. Most interesting and challenging to objectivist biological embryology is the notion that the fetus develops its fate, *rajaki* or "life's wisdom," in full in the womb before experiencing the trauma of birth. On being born, *rajaki* is scattered, and the life lived in this world is spent gathering one's own wisdom and experience up again. Seeing life as a painstaking recovery and reassembly of something once known and possessed in the womb—the desires, the strengths, and the ultimate end of an individual life—contradicts our assumptions about what is naturally possible. At the same time, though, it seems to explain some of the preternatural air of wisdom we feel we can see in the all-knowing gazes of newborns. Tsing suggests that Uma Adang's narrative challenges some prevailing representations—especially those retailed by anti-abortion interests—of fetuses as transcendently innocent children. And she invites us to reflect on the implications for our politics and our experiences of the image of the "once-wise fetus."

ADDITIONAL READINGS

Crary, Jonathan, Michel Feher, Hal Foster, and Sanford Kwinter, eds. 1989. *Fragments for a History of the Human Body*, Parts 1–3. 3 vols. New York: Zone Books.

Csordas, Thomas J. 1994. *The Sacred Self: A Cultural Phenomenology of Charismatic Healing*. Berkeley: University of California Press.

Stafford, Barbara. 1991. *Body Criticism: Imaging the Unseen in Enlightenment Art and Medicine*. New York: Zone Books.

TIME AND SPACE

E. E. EVANS-PRITCHARD

In describing Nuer concepts of time we may distinguish between those that are mainly reflections of their relations to environment, which we call oecological time, and those that are reflections of their relations to one another in the social structure, which we call structural time. Both refer to successions of events which are of sufficient interest to the community for them to be noted and related to each other conceptually. The larger periods of time are almost entirely structural, because the events they relate are changes in the relation-ship of social groups. Moreover, time-reckoning based on changes in nature and man's response to them is limited to an annual cycle and therefore cannot be used to differentiate longer periods than seasons. Both, also, have limited and fixed notations. Seasonal and lunar changes repeat themselves year after year, so that a Nuer standing at any point of time has conceptual knowledge of what lies before him and can predict and organize his life accordingly. A man's structural future is likewise already fixed and ordered into different periods, so that the total changes in status a boy will undergo in his ordained passage through the social system, if he lives long enough, can be foreseen. Structural time appears to an individual passing through the social system to be entirely progressive, but, as we shall see, in a sense this is an illusion. Oecological time appears to be, and is, cyclical.

The ecological cycle is a year. Its distinctive rhythm is the backwards and forwards movement from villages to camps, which is the Nuer's response to the climatic dichotomy of rains and drought. The year (*ruon*) has two main seasons, *tot* and *mai. Tot*, from about the middle of March to the middle of September, roughly corresponds to the rise in the curve of rainfall, though it does not cover the whole period of the rains. Rain may fall heavily at the end of September and in early October, and the country is still flooded in these months which belong, nevertheless, to the *mat* half of the year, for it com-mences at the decline of the rains—not at their cessation—and roughly covers the trough of the curve, from about the middle of September to the middle of March. The two seasons therefore only approximate to our division into rains and drought, and the Nuer classification aptly summarizes their way of looking at the movement of time, the direction of attention in marginal months being as significant as the actual climatic conditions. In the middle of September Nuer turn, as it were, towards the life of fishing and cattle camps and feel that village residence and horticulture lie behind them. They begin to speak of camps as though they were already in being, and long to be on the move. This restlessness is even more marked towards the end of the drought when, noting cloudy skies, people turn towards the life of villages and make preparations for striking camp. Marginal months may therefore be classed as *tot* or *mai*, since they belong to one set of activities but presage the other set, for the concept of seasons is derived from social activities rather than from the climatic changes which determine them, and a year is to Nuer a period of village residence (*cieng*) and period of camp residence (*wec*).

I have already noted the significant physical changes associated with rains and drought. . . . I have also described . . . the oecological movement that follows these physical changes where it affects human life to any degree. Seasonal variations in social activities, on which Nuer concepts of time are primarily based, have also been indicated and, on the economic side, recorded at some length. . . .

The movements of the heavenly bodies other than the sun and the moon, the direction and variation of winds, and the migration of some species of birds are observed by the Nuer, but they do not regulate their activities in relation to them nor use them as points of reference in seasonal time-reckoning. The characters by which seasons are most clearly defined are those which control the movements of the people: water, vegetation, movements of fish, etc.; it being the needs of the cattle and variations in food-supply which chiefly translate oecological rhythm into the social rhythm of the year, and the contrast between modes of life at the height of the rains and at the height of the drought which provides the conceptual poles in time-reckoning.

Besides these two main seasons of *tot* and *mai* Nuer recognize two subsidiary seasons included in them, being transitional periods between them. The four seasons are not sharp divisions but overlap. Just as we reckon summer and winter as the halves of our year and speak also of spring and autumn, so Nuer reckon *tot* and *mai* as halves of their year and speak also of the seaons of *rwil* and *jiom*. *Rwil* is the time of moving from camp to village and of clearing and planting, from about the middle of March to the middle of June, before the rains have reached their peak. It counts as part of the *tot* half of the year, though it is contrasted with *tot* proper, the period of full village life and horticulture, from about the middle of June to the middle of September. *Jiom*, meaning 'wind,' is the period in which the persistent north wind begins to blow and people harvest, fish from dams, fire the bush, and form early camps, from about the middle of September to the middle of December. It counts as part of the *mai* half of the year, though it is contrasted with *mai* proper, from about the middle of December to the middle of March, when the main camps are formed. Roughly speaking, therefore, there are two major seasons of six months and four minor seasons of three months, but these divisions must not be regarded too rigidly since they are not so much exact units of time as rather vague conceptualizations of changes in oecological relations and social activities which pass imperceptibly from one state to another. . . .

Nuer would soon be in difficulties over their lunar calendar if they consistently counted the succession of moons, but there are certain activities associated with each month, the association sometimes being indicated by the name of the month. The calendar is a relation between a cycle of activities and a conceptual cycle and the two cannot fall apart, since the conceptual cycle is dependent on the cycle of activities from which it derives its meaning and function. Thus a twelve-month system does not incommode Nuer, for the calendar is anchored to the cycle of oecological changes. In the month of *kur* one makes the first fishing dams and forms the first cattle camps, and since one is doing these things it must be *kur* or thereabouts. Likewise in *dwat* one breaks camp and returns to the villages, and since people are on the move it must be *dwat* or thereabouts. Consequently the calendar remains fairly stable and in any section of Nuerland there is general agreement about the name of the current month.

In my experience Nuer do not to any great extent use the names of the months to indicate the time of an event, but generally refer instead to some outstanding activity in process at the time of its occurrence, e.g., at the time of early camps, at the time of weeding, at the time of harvesting, etc., and it is easily understandable that they do so, since time is to them a relation between activities. During the rains the stages in the growth of millet and the steps taken in its culture are often used as points of reference. Pastoral activities, being largely undifferentiated throughout the months and seasons, do not provide suitable points.

There are no units of time between the month and day and night. People indicate the occurrence of an event more than a day or two ago by reference to some other event which took place at the same time or by counting the number of intervening "sleeps" or, less commonly, "suns." There are terms for to-day, to-morrow, yesterday, etc., but there is no precision about them. When Nuer wish to define the occurrence of an event several days in advance, such as a dance or wedding, they do so by reference to the phases of the moon: new moon, its waxing, full moon, its waning, and the brightness of its second quarter. When they wish to be precise they state on which night of the waxing or waning an event will take place, reckoning fifteen nights to each and thirty to the month. They say that only cattle and the Anuak can see the moon in its invisible period. The only terms applied to the nightly succession of lunar phases are those that describe its appearance just before, and in, fullness.

The course of the sun determines many points of reference, and a common way of indicating the time of events is by pointing to that part of the heavens the sun will then have reached in its course. There are also a number of expressions, varying in the degree of their precision, which describe positions of the sun in the heavens, though, in my experience, the only ones commonly employed are those that refer to its more conspicuously differentiated movements: the first stroke of dawn, sunrise, noon, and sunset. It is, perhaps, significant that there are almost as many points of reference between 4 and 6 A.M. as there are for the rest of the day. This may be chiefly due to striking contrasts caused by changes in relations of earth to sun during these two hours, but it may be noted, also, that the points of reference between them are more used in directing activities, such as starting on journeys, rising from sleep, tethering cattle in kraals, gazelle hunting, etc., than points of reference during most of the rest of the day, especially in the slack time between 1 and 3 P.M. There are also a number of terms to describe the time of night. They are to a very limited extent determined by the course of the stars. Here again, there is a richer terminology for the transition period between day and night than during the rest of the night and the same reasons may be suggested to explain this fact. There are also expressions for distinguishing night from day, forenoon from afternoon, and that part of the day which is spent from that part which lies ahead.

Except for the commonest of the terms for divisions of the day they are little used in comparison with expressions which describe routine diurnal activities. The daily timepiece is the cattle clock, the round of pastoral tasks, and the time of day and the passage of time through a day are to a Nuer primarily the succession of these tasks and their relations to one another. The better demarcated points are taking of the cattle from byre to kraal, milking, driving of the adult herd to pasture, milking of the goats and sheep, driving of the flocks and calves to pasture, cleaning of byre and kraal, bringing home of the flocks and calves, the return of the adult herd, the evening milking, and the enclosure of the beasts in byres. Nuer generally use such points of activity, rather than concrete points in the movement of the sun across the heavens, to co-ordinate events. Thus a man says, "I shall return at milking," "I shall start off when the calves come home," and so forth.

Oecological time-reckoning is ultimately, of course, entirely determined by the movement of the heavenly bodies, but only some of its units and notations are directly based on these movements, e.g., month, day, night, and some parts of the day and night, and such points of reference are paid attention to and selected as points only because they are significant for social activities. It is the activities themselves, chiefly of an economic kind, which are basic to the system and furnish most of its units and notations, and the passage of time is perceived in the relation of activities to one another. Since activities are dependent on the movement of the heavenly bodies and since the movement of the heavenly bodies is significant only in relation to the activities one may often refer to either in indication of the time of an event. Thus one may say, "In the *jiom* season" or "At early camps," "The month of *Dwaf*" or "The return to villages," "When the sun is warming up" or "At milking." The movements of the heavenly bodies permit Nuer to select natural points that are significant in relation to activities. Hence in linguistic usage nights, or rather "sleeps," are more clearly defined units of time than days, or "suns," because they are undifferentiated units of social activity, and months, or rather "moons," though they are clearly differentiated units of natural time, are little employed as points of reference because they are not clearly differentiated units of activity, whereas the day, the year, and its main seasons are complete occupational units.

Certain conclusions may be drawn from this quality of time among the Nuer. Time has not the same value throughout the year. Thus in dry season camps, although daily pastoral tasks follow one another in the same order as in the rains, they do not take place at the same time, are more a precise routine owing to the severity of seasonal conditions, especially with regard to water and pasturage, and require greater co-ordination and co-operative action. On the other hand, life in the dry season is generally uneventful, outside routine tasks, and oecological and social relations are more monotonous from month to month than in the rains when there are frequent feasts, dances, and ceremonies. When time is considered as relations between activities it will be understood that it has a different connotation in rains and drought. In the drought the daily time-reckoning is more uniform and precise while lunar reckoning receives less attention, as appears from the lesser use of names of months, less confidence in stating their order, and the common East African trait of two dry-season months with the same name (*tiop in dit* and *tiop in tot*), the order of which is often interchanged. The pace of time may vary accordingly, since perception of time is a function of systems of time-reckoning, but we can make no definite statement on this question.

Though I have spoken of time and units of time the Nuer have no expression equivalent to "time" in our language, and they cannot, therefore, as we can, speak of time as though it were something actual, which passes, can be wasted, can be saved, and so forth. I do not think that they ever experience the same feeling of fighting against time or of having to co-ordinate activities with an abstract passage of time, because their points of reference are mainly the activities themselves, which are generally of a leisurely character. Events follow a logical order, but they are not controlled by an abstract system, there being no autonomous points of reference to which activities have to conform with precision. Nuer are fortunate.

Also they have very limited means of reckoning the relative duration of periods of time intervening between events, since they have few, and not well-defined or systematized, units of time. Having no hours or other small units of time they cannot measure the periods which intervene between positions of the sun or daily activities. It is true that the year is divided into twelve lunar units, but Nuer do not reckon in them as fractions of a unit. They may be able to state in what month an event occurred, but it is with great difficulty that they reckon the relation between events in abstract numerical symbols. They think much more easily in terms of activities and of successions of activities and in terms of social structure and of structural differences than in pure units of time.

We may conclude that the Nuer system of time-reckoning within the annual cycle and parts of the cycle is a series of conceptualizations of natural changes, and that the selection of points of reference is determined by the significance which these natural changes have for human activities.

....

In a sense all time is structural since it is a conceptualization of collateral, coordinated, or co-operative activities: the movements of a group. Otherwise time concepts of this kind could not exist, for they must have a like meaning for every one within a group. Milking-time and meal-times are approximately the same for all people who normally come into contact with one another, and the movement from villages to camps has approximately the same connotation everywhere in Nuerland, though it may have a special connotation for a particular group of persons. There is, however, a point at which we can say that time concepts cease to be determined by oecological factors and become more determined by structural interrelations, being no longer a reflection of man's dependence on nature, but a reflection of the interaction of social groups.

The year is the largest unit of oecological time. Nuer have words for the year before last, last year, this year, next year, and the year after next. Events which took place in the last few years are then the points of reference in time-reckoning, and these are different according to the group of persons who make use of them: joint family, village, tribal section, tribe, etc. One of the commonest ways of stating the year of an event is to mention where the people of the village made their dry season camps, or to refer to some evil that befell their cattle. A joint family may reckon time in the birth of calves of their herds. Weddings and other ceremonies, fights, and raids may likewise give points of time, though in the absence of numerical dating no one can say without lengthy calculations how many years ago an event took place. Moreover, since time is to Nuer an order of events of outstanding significance to a group, each group has its own points of reference and time is consequently relative to structural space, locally considered. This is obvious when we examine the names given to years by different tribes, or sometimes by adjacent tribes, for these are floods, pestilences, famines, wars, etc., experienced by the tribe. In course of time the names of years are forgotten and all events beyond the limits of this crude historical reckoning fade into the dim vista of long long ago. Historical time, in this sense of a sequence of outstanding events of significance to a tribe, goes back much farther than the historical time of smaller groups, but fifty years is probably its limit, and the farther back from the present day the sparser and vaguer become its points of reference.

However, Nuer have another way of stating roughly when events took place; not in numbers of years, but by reference to the age-set system. Distance between events ceases to be reckoned in time concepts as we understand them and is reckoned in terms of structural distance, being the relation between groups of persons. It is therefore entirely relative to the social structure. Thus a Nuer may say that an event took place after the *Thut* age-set was born or in the initiation period of the *Boiloc* age-set, but no one can say how many years ago it happened. Time is here reckoned in sets. If a man of the *Dangunga* set tells one that an event occurred in the initiation period of the *Thut* set he is saying that it happened three sets before his set, or six sets ago. . . . We cannot accurately translate a reckoning in sets into a reckoning in years, but . . . we can roughly estimate a ten-year interval between the commencement of successive sets. There are six sets in existence, the names of the sets are not cyclic, and the order of extinct sets, all but the last, are soon forgotten, so that an age-set reckoning has seven units covering a period of rather under a century. . . .

We have restricted our discussion to Nuer systems of time-reckoning and have not considered the way in which an individual perceives time. The subject bristles with difficulties. Thus an individual may reckon the passage of time by reference to the physical appearance and status of other individuals and to changes in his own life-history, but such a method of reckoning time has no wide collective validity. We confess, however, that our observations on the matter have been slight and that a fuller analysis is beyond our powers. We have merely indicated those aspects of the problem which are directly related to the description of modes of livelihood which has gone before and to the description of political institutions which follows.

We have remarked that the movement of structural time is, in a sense, an illusion, for the structure remains fairly constant and the perception of time is no more than the movement of persons, often as groups, through the structure. Thus age-sets succeed one another forever, but there are never more than six in existence and the relative positions occupied by these six sets at any time are fixed structural points through which actual sets of persons pass in endless succession. Similarly . . . the Nuer system of lineages may be considered a fixed system, there being a constant number of steps between living persons and the founder of their clan and the lineages having a constant position relative to one another. However many generations succeed one another, the depth and range of lineages does not increase unless there has been structural change.

Beyond the limits of historical time we enter a plane of tradition in which a certain element of historical fact may be supposed to be incorporated in a complex of myth. Here the points of reference are the structural ones we have indicated. At one end this plane merges into history; at the other end into myth. Time perspective is here not a true impression of actual distances like that created by our dating technique, but a reflection of relations between lineages, so that the traditional events recorded have to be placed at the points where the lineages concerned in them converge in their lines of ascent. The events have therefore a position in structure, but no exact position in historical time as we understand it. Beyond tradition lies the horizon of pure myth, which is always seen in the same time perspective. One mythological event did not precede another, for myths explain customs of general social significance rather than the interrelations of particular segments and are, therefore, not structurally stratified. Explanations of any qualities of nature or of culture are drawn from this intellectual ambient which imposes limits on the Nuer world and makes it self-contained and entirely intelligible to Nuer in the relation of its parts. The world, peoples, and cultures all existed together from the same remote past.

It will have been noted that the Nuer time dimension is shallow. Valid history ends a century ago, and tradition, generously measured, takes us back only ten to twelve generations in lineage structure, and if we are right in supposing that lineage structure never grows, it follows that the distance between the beginning of the world and the present day remains unalterable. Time is thus not a continuum, but is a constant structural relationship between two points, the first and last persons in a line of agnatic descent. How shallow is Nuer time may be judged from the fact that the tree under which mankind came into being was still standing in Western Nuerland a few years ago!

Beyond the annual cycle, time-reckoning is a conceptualization of the social structure, and the points of reference are a projection into the past of actual relations between groups of persons. It is less a means of co-ordinating events than of co-ordinating relationships, and is therefore mainly a looking-backwards, since relationships must be explained in terms of the past.

WOMEN MYSTICS AND EUCHARISTIC DEVOTION

IN THE THIRTEENTH CENTURY

CAROLINE WALKER BYNUM

Early in the thirteenth century, the hagiographer James of Vitry described thus the eucharistic piety of the beguine, Mary of Oignies:

> The holy bread strengthened her heart; the holy wine inebriated her, rejoicing her mind; the holy body fattened her; the vitalizing blood purified her by washing. And she could not bear to abstain from such solace for long. For it was the same to her to live as to eat the body of Christ, and this it was to die, to be separated from the sacrament by having for a long time to abstain. . . . Indeed she felt all delectation and all savor of sweetness in receiving it, not just within her soul but even in her mouth. . . . Sometimes she happily accepted her Lord under the appearance of a child, sometimes as a taste of honey, sometimes as a sweet smell, and sometimes in the pure and gorgeously embellished marriage bed of the heart. And when she was not able to bear any longer her thirst for the vivifying blood, sometimes after the mass was over, she would remain for a long time contemplating the empty chalice on the altar. . . .

In such stories and quotations, which I could multiply by the dozens, we see reflected the most prominent, characteristically female concern in thirteenth-century religiosity: devotion to the eucharist.

The attention which thirteenth-century women paid to the eucharist has been noticed before. But scholars have tended to correlate eucharistic concern either with order (particularly the Cistercians or the Dominicans), or with region (particularly the Low Countries or southern Germany), or with type of religious life (particularly nuns or recluses). If one reads widely in thirteenth-century saints' lives and spiritual treatises, however, it is glaringly obvious that laywomen, recluses, tertiaries, beguines, nuns of all orders and those women (especially common in the early thirteenth century) who wandered from one type of life to another were inspired, compelled, comforted and troubled by the eucharist to an extent found in only a few male writers of the period. In this essay I want not only to illustrate the importance of women in the development of this aspect of thirteenth-century piety, but also to explain why women's religiosity expressed itself in eucharistic devotion.

DEVOTION TO THE EUCHARIST: A FEMALE CONCERN

The centrality of the eucharist to women and of women in the propagation of eucharistic devotion is easy to demonstrate. Women were especially prominent in the creation and spread of special devotions, such as the feast of Corpus Christi (revealed to Juliana of Cornillon and established as a result of other efforts and those of Eva of St. Martin) or the devotion to the Sacred Heart (found especially in the Flemish saint, Lutgard of Aywières, and the many visions of the nuns of the Saxon monastery of Helfta). Some thirteenth-century women (e.g., Ida of Nivelles and the Viennese beguine Agnes Blannbekin) made vocational decisions or changed orders out of desire to receive the eucharist more frequently. Stories of people racing from church to church to attend as many masses as possible were usually told of women—for example, Hedwig of Silesia. Moreover, female visions and miracles make up such a large proportion of the total number of eucharistic miracles known from the thirteenth century that the eucharistic miracle almost seems a female genre (and this despite the fact that women were less than a quarter of thirteenth-century saints). . . .

EUCHARIST AND ECSTASY

To thirteenth-century women, the mass and the reception or adoration of the eucharist were closely connected with mystical union or ecstasy, which was frequently accompanied by paramystical phenomena. To some extent, reception of Christ's body and blood was a substitute for ecstasy—a union that anyone, properly prepared by confession or contrition, could achieve. To receive was to become Christ—by eating, by devouring and being devoured. No "special effects" were necessary. In the eucharist Christ was available to the beginner as well as to the spiritually trained. This is what the German beguine Mechtild of Magdeburg means when she says:

> Yet I, least of all souls,
> Take Him in my hand,
> Eat Him and drink Him,
> And do with Him what I will!

Why then should I trouble myself
As to what the angels experience?

Simply to eat Christ is enough; it *is* to achieve union.

For many holy women, however, the eating and the ecstasy fused; para-mystical phenomena were expected. The biographer of Douceline of Marseilles, for example, tells us that the saint went into ecstasy every time she communicated. So automatic was the experience that the countess of Provence, who wanted to observe a real ecstatic, tried to manipulate Douceline into receiving communion alongside her so she could see the accompanying trance; and Douceline had to sneak off to church on the eve of feast days in order to escape the crowds that went so far as to scale the grillwork to observe her. By the early fourteenth century, Agnes Blannbekin thought everyone would experience the taste of the honeycomb upon receiving the host, and Christina Ebner was puzzled by the presence at Engelthal of a nun who did *not* have visions and ecstasies. . . .

EUCHARISTIC DEVOTION AND THE HUMANITY OF CHRIST

The eucharist was, however, more than an occasion for ecstasy. It was also a moment of encounter with that *humanitas Christi* which was such a prominent theme of women's spirituality. For thirteenth-century women this humanity was, above all, Christ's physicality, his corporality, his being-in-the-body-ness; Christ's humanity was Christ's body and blood.

Both in a eucharistic context and outside it, the humanity of Christ was often described as "being eaten." To popular twelfth-century metaphors for union (often ultimately of Neoplatonic origin)—metaphors of light, of darkness, of wine diffused in water—women added the insistent image and experience of flesh taken into flesh. Lutgard of Aywières rejected an earthly suitor, calling him "thou food of death"; she later nursed from the breast of Christ so that afterward her own saliva was sweet to the taste, and she received Christ as a lamb who stood on her shoulders and sucked from her mouth. Angela of Foligno nursed from Christ and saw him place the heads of other "sons" (the friars) into the wound in his side. Anna Vorchtlin of Engelthal exclaimed, upon receiving a vision of the baby Jesus: "If I had you, I would eat you up, I love you so much!" Mechtild of Magdeburg spoke of mystical union as "eating God." And Ida of Louvain was able to eat Christ almost at will by reciting John 1.14. For, whenever she spoke the words *Verbum caro factum est*—which she inserted into the Hours whenever possible—she tasted the Word on her tongue and felt flesh in her mouth; when she chewed it, it was like honey, "not [her biographer tells us] a phantasm but like any other kind of food." This example makes clear to the modern reader how insistently Christ's humanity was thought of as flesh, as food (*corpus, caro, carnis*), eaten in the eucharist, a substitute for the meat many women denied themselves in long fasts. Moreover, the incorporation of self into Christ or of Christ into self was so much a matter of flesh swallowing flesh that women who were not able to eat still received and digested Christ's physicality. For example, when the host was placed on the breast of the dying Juliana Falconieri it simply sank into her body and disappeared.

The humanity of Christ with which women joined in the eucharist was the physical Jesus of the manger and of Calvary. Women from all walks of life saw in the host and the chalice Christ the baby, Christ the bridegroom, Christ the tortured body on the cross. The nuns of Helfta, Töss and Engelthal, the Franciscan tertiary Angela of Foligno, the Carthusian Beatrice of Ornacieux and the beguine Mary of Oignies saw Christ as a baby in the host. Agnes of Montepulciano and Margaret of Faenza became so intoxicated with the pleasure of holding the baby that they refused to give him up. Ida of Louvain bathed him and played with him. Tiedala of Nivelles could not rest until her spiritual friend, a monk of Villers, also received the baby at his breast. Lukardis of Oberweimar, Lutgard of Aywieres, Margaret of Ypres, Ida of Louvain and Marguerite of Oingt received Christ as a beautiful young man; Angela of Foligno and Adelheid Langmann married him in the eucharist. (To Adelheid, he gave the host as pledge instead of a wedding ring!) Most prominent, however, was the Christ of the cross. No religious woman failed to experience Christ as wounded, bleeding and dying. Women's efforts to imitate this Christ involved *becoming* the crucified, not just patterning themselves after or expanding their compassion toward, but *fusing with*, the body on the cross. Both in fact and in imagery the *imitatio*, the fusion, was achieved in two ways: through asceticism and through eroticism.

Thirteenth-century women joined with the crucifix through physical suffering, both involuntary and voluntary—that is, through illness and through self-mortification. Ernst Benz has pointed out the special prominence of illness as a theme in women's visions and, if anything, he has underestimated its significance. But the most important thing to note is that there is no sharp line between illness induced by the eucharist and illness cured by the eucharist, nor between illness and self-torture or mutilation. We see this particularly in the case of stigmata, where it is sometimes not only impossible to tell whether the wounds are inner or outer, but also impossible to tell how far the appearance is miraculous and how far it is self-induced. Horrible pain, twisting of the body, bleeding—whether inflicted by God or by oneself—were not an effort to destroy the body, *not* a punishment of physicality, not primarily an effort to shear away a source of lust, not even primarily an identification with the martyrs (although this was a subsidiary theme). Illness and asceticism were rather *imitatio Christi*, an effort to plumb the depths of Christ's humanity at the moment of his most insistent and terrifying humanness—the moment of his dying.

Mary of Oignies, in a frenzied vision of the crucifix, cut off pieces of her own flesh and buried them in the ground to keep the secret of what she had done. Lukardis of Oberweimar drove the middle finger of each hand, hard as a nail, through the palm of the opposite hand, until the room rang with the sound of the hammering; and stigmata "miraculously" (says her biographer) appeared. Beatrice of Omacieux thrust a nail completely through her hands and only clear water flowed from the wound. Mary of Oignies refused prayers for relief of illness; Gertrude of Helfta embraced her illness as a source of grace; Beatrice of Nazareth, who desired the torments of illness, was healed almost against her wishes; Margaret of Ypres so desired to join with Christ's suffering that she prayed for her illness to last beyond the grave. Illness and asceticism are common themes in the *Nonnenbücher*, where, for example, the nuns expose themselves to bitter cold or pray to be afflicted with leprosy. Even religious women who were not ill desired to identify with Christ through loneliness and persecution (for example, Mechtild of Magdeburg and Hadewijch). Angela of Foligno, whose asceticism was less intense than that of some of the northern nuns, drank the scabs from lepers' wounds and found them "as sweet as communion." Common ascetic practices included thrusting nettles into one's breasts, wearing hair shirts, binding one's flesh tightly with twisted ropes, enduring extreme sleep and food deprivation, performing thousands of genuflections and praying barefoot in winter. Among the more bizarre manifestations were rolling in broken glass, jumping into ovens, hanging oneself from a gibbet, and praying while standing on one's head (skirts clinging, miraculously and modestly, around one's ankles). The author of the nuns' book of Unterlinden, in the Alsace, wrote:

> In Advent and Lent, all the sisters, coming into the chapter house after Matins, or in some other suitable place, hack at themselves cruelly, hostilely lacerating their bodies until the blood flows, with all kinds of whips, so that the sound reverberates all over the monastery and rises to the ears of the Lord of hosts sweeter than all melody. . . .

And she called the results of such discipline *stigmata*.

Not only do illness and self-mortification fuse with each other as each fuses with the experience of the cross; suffering and ecstasy also fuse. In one of the most touching of all thirteenth-century lives, an anonymous biographer describes Alice of Schaerbeke:

> And a little after this, . . . as she surpassed in virtues what could be expected from the number of her years, God wished to purge her within . . . because she was his spouse. . . . And so that she would be free to rest with God alone and dally in the cubicle of her mind as in a bridal chamber and be inebriated with the sweetness of his odor . . . , he gave her an incurable disease, leprosy. And the first night when she was sequestered from the convent because of the fear of contagion, she was afflicted with such sadness her heart was wounded. . . . [So she cried and prayed at God's feet]. . . . And when she had learned from what she experienced to take refuge in the most secure harbor of God, she ran to Christ's breasts and wounds, in every tribulation or anguish, every depression or dryness, like a little child drinking from its mother's breasts, and by that liquid she felt her members restored.

Thus, the most feared of all diseases becomes a bridal bed, wounds are the source of a mother's milk; the physicality into which the woman sinks is unspeakable suffering and unspeakable joy.

Physical union with Christ is thus described not only in images of disease and torment but also in images of marriage and sexual consummation; it sometimes culminates in what appears to be orgasm—as in Hadewijch's beautiful vision quoted above. Although scholars have, of course, suggested that such reactions are sublimated sexual desire, it seems inappropriate to speak of "sublimation." In the eucharist and in ecstasy, a male Christ was handled and loved; sexual feelings were, as certain contemporary commentators (like David of Augsburg) clearly realized, not so much translated into another medium as simply set free.

Asceticism and eroticism sometimes fused so completely that it is hard to know under which category to place a mystic like Ida of Louvain, who went mad from desire for the eucharist and had to be put in chains, or Beatrice of Nazareth, who consulted a spiritual adviser to find out whether God would sanction her effort to drive herself literally "crazy" as a way of "following" him. We misunderstand the power of the erotic, nuptial mysticism of Low Country figures like Hadewijch and Beatrice of Nazareth if we project onto their image of lover seeking Lover stereotyped notions of brides as passive and submissive. Their search for Christ took them through a frenzy they called insanity (*orewoet* in Flemish, *aestus* or *insania amoris* in Latin).

Erotic imagery is unimportant in some women's writing. And nuptial language is often most elaborate in *male* biographers, who may have had their own reasons for describing women they admired and loved in erotic metaphors. But the image of bride or lover was clearly a central metaphor for the woman mystic's union with Christ's humanity. In the twelfth century, Hildegard of Bingen actually dressed her nuns as brides when they went forward to receive communion. And Hadewijch and Mechtild of Magdeburg, women given voice by the emergence of the vernaculars, found in secular love poetry the vocabulary and the pulsating rhythms to speak of the highest of all loves.

ECSTASY AND THE POWER OF THE RECIPIENT

Women's devotion to the body and blood of Christ was thus an affirmation of the religious significance of physicality and emotionality. The eucharist was, to medieval women, a moment at which they were released into ecstatic union; it was also a moment at which the God with whom they joined was supremely human because supremely vulnerable and fleshly. But why were these aspects of the eucharist so central to female piety? Why did ecstasy and *humanitas Christi* matter so much to women?

Part of the answer seems to be that women's ecstasy or possession served as an alternative to the authority of priestly office. From the time of the Gregorian reform of the late eleventh century, theology and spirituality increasingly stressed priesthood and preaching as the way of imitating Christ. By the thirteenth century the priest, momentarily divinized by the Christ whom he held in his hands at the consecration, and the friar, imitating the life of Jesus in poverty, begging, penance and preaching, were the admired male roles. Increasingly prohibited from even minor "clerical" tasks (such as veiling nuns or touching altar vessels) and never permitted full evangelical poverty or wandering, women emerged—in their own eyes and in the eyes of their male advisers—as what I have elsewhere called a "prophetic" or "charismatic" alternative. Thus, the eucharist and the paramystical phenomena that often accompanied it were substitutes for priesthood in two complementary senses. First, eucharistic ecstasy was a means by which women claimed "clerical" power for themselves or bypassed the power of males, or criticized male abuse of priestly authority. Second, ecstasy was a means of endowing women's nonclerical status—their status as lay recipients—with special spiritual significance.

In the visions that women received at mass, they sometimes acquired metaphorical priesthood. I have discussed elsewhere the fact that women's visions sometimes gave them general authorization to prophesy and teach and hear confessions. What should be noted here is that eucharistic visions occasionally projected women, in metaphor and vision, into access to the altar, even into the role of celebrant—things strictly forbidden to them. For example, a woman who had loved Juliana of Cornillon very much in her life saw her after death at mass, assisting the priest. Angela of Foligno, feeling that the celebrant was unworthy, had a vision of Christ bleeding on the cross and angels said to her: "Oh you who are pleasing to God, behold, he has been administered to you . . . in order that you may administer and present him to others." Benevenuta of Bojano saw the Virgin administering the chalice in a vision. And, in addition to the infrequent visions in which women actually see themselves or other women as priests or acolytes, there are hundreds of visions in which Christ himself gives the chalice or the host to a nun or beguine or lay-woman who is unable to receive, either because of illness or because the clergy prevent it. . . .

But women's ecstatic eucharistic devotion was not merely a bypassing or criticizing of priests, nor was it primarily a claim to charismatic authority that competed with or substituted for priesthood. Women's devotion to the mass and the eucharistic elements was also an endowing of their role as "nonpriests" with a new spiritual importance. In a recent study of canonization, André Vauchez has shown that women were 50 percent of the laity canonized in the thirteenth century and 71.4 percent after 1305. By the later Middle Ages, the male saint was usually a cleric or a friar; the ideal layperson was female. As representative laity, women were quintessentially recipients. It is thus not surprising that almost all thirteenth-century eucharistic miracles of recipients are female miracles. And it is significant that the occasional male who receives directly from Christ or angels or doves—that is, whose act of *receiving* is given special recognition—is not a priest, but a layman. Both men and women seem to have seen the religious roles of men and women as different. Women's eucharistic devotion was the devotion of those who receive *rather than* consecrate, those who are lay *rather than* clergy, those whose closeness to God and whose authorization to serve others come through intimacy and direct inspiration *rather than* through office or worldly power. . . .

WOMEN AND PHYSICALITY IN THE THEOLOGICAL TRADITION

Concern with the literal following of Jesus, with the problem and the opportunity of physicality, was thus a basic theme in thirteenth-century religiosity. But it was reflected and espoused especially intensely in women's lives and in women's writing. For this, there are specific theological reasons. To put it simply, the weight of the Western tradition had long told women that physicality was particularly their problem.

Some modern commentators have made much of the fact that certain patristic figures argued that woman *qua* woman was not created in God's image, although woman *qua* human being was. This is a complex point—and certainly in thirteenth-century legal and theological writing it was often interpreted as referring to woman's social role (namely, her subordination to man in the family) more than to her anatomical nature or biological role. But in any case it was not absorbed by late medieval women (even married women) as a prohibition of their approach to God, their imitation of Christ. Their writing is full of references to being created in God's image and likeness.

Women were also told that, allegorically speaking, woman was to man what matter is to spirit—that is, that they symbolized the physical, lustful, material, appetitive part of human nature, whereas man symbolized the spiritual or mental. The roots of this idea were multiple, scientific as well as theological; and it did unquestionably influence women writers. The first great woman theologian, Hildegard of Bingen, knew the tradition, and indeed argued against some of its implications. Women do not, however, seem to have drawn from such teaching a debilitating sense of female incapacity. Most of the references to womanly weakness in thirteenth-century spiritual writing come from the pens of male biographers. These biographers occasionally compliment women saints on their "virility." But women writers by and large either ignore their own sex, using mixed-gender imagery for the self (as did Gertrude the Great and Hadewijch), or embrace their femaleness as a sign of closeness to Christ. Where they do refer to female weakness, the reference is often, as Peter Dronke has argued, an ironic comment on male failure to achieve even the level of virtue of "weak women." If anything, women drew from the traditional notion of the female as physical a special emphasis on their own redemption by a Christ who was supremely physical because supremely human. They sometimes even extrapolated from this to the notion that, in Christ, divinity is to humanity as male is to female.

As I have explained elsewhere, such an idea did not imply that the human Christ was a body without a soul (a clearly heretical Christological position), nor did it deny Christ's divinity. But as spirituality in general came increasingly to stress Christ's humanity as manifested in his physicality, women—who were the special symbol of the physical—suggested that that physical, tangible humanity might be symbolized or understood as female. And the doctrine of the Virgin Birth contributed to this. One could argue that all of Christ's humanity had to come from Mary because Christ had no human father. So in some sense Mary could be seen as adding humanity to the Logos. This is in fact exactly what Hildegard of Bingen and Mechtild of Magdeburg do argue. Hildegard describes that which is redeemed by Christ—the physicality that comes from Mary—as feminine; and this is enhanced by her sense that Christ's body is also *ecclesia*, which is equally feminine. In a eucharistic vision, Hildegard saw woman (*muliebris imago*) receiving from Christ, hanging on the cross, a dowry of his blood, while a voice said: "Eat and drink the body and blood of my Son to abolish the prevarication of Eve and receive your true inheritance." Although the priesthood is, to Hildegard, both revered and essential, the priest enters this eucharistic vision only after Holy Church; and the image of both sinful *and saved* humanity is the image of woman. A century later, Mechtild of Magdeburg argued that the Incarnation joined the Logos (the preexistent Son of God) with a pure humanity, created along with Adam but preserved as pure in Mary after the fall. Thus, Mary became a kind of preexistent humanity of Christ. Such a notion is reinforced even in iconography, where we find that Mary has a place of honor on eucharistic tabernacles. For Mary, the source of Christ's physicality and his humanity, is in some sense the reliquary or chest that houses Christ's body. To Hildegard of Bingen as to Marguerite of Oingt, she was explicitly the *tunica humanitatis*, the clothing of humanity which God put on in the Incarnation.

Mary was, of course, important in women's spirituality. Particularly in southern European saints' lives, the theme of *imitatio Mariae Virginis* is strong. The biographer of Douceline of Marseilles, for example, actually sees Douceline as imitating the poverty of Mary, whereas her beloved Francis imitated the poverty of Christ directly. But the reverence for Mary that we find in thirteenth-century women mystics is less a reverence for a "representative woman" than a reverence for the bearer and conduit of the Incarnation. The ultimate identification was with Christ as human. Some women saints swoon with Mary before the cross; *all* women saints swoon on the cross with Christ himself.

Thus, women theologians took from the theological and scientific tradition a notion that male is to female as soul is to matter and elaborated it in their own way as an identification with the human Christ in his physicality. Modern claims that medieval women were alienated from a male Christ (i.e., a God not of their gender) quite miss the point; these women saw themselves less in terms of gender than in terms of matter. Modern claims that women were deprived of a sense of self-worth or forced into denial of their sexuality by the traditional association of woman with the physical also miss the point: these women found physicality, as they understood it, redeemed and expressed by a human God. Contrary to what some recent interpretations have asserted, thirteenth-century women seem to have concluded from their physicality an intense conviction of their *ability* to imitate Christ without role or gender inversion. To soar toward Christ as lover and bride, to sink into the stench and torment of the Crucifixion, to eat God, was for the woman only to give religious significance to what she already was. So, female devotion to the eucharist—and to the dying or the infant or the bridegroom Christ—expresses a special confidence in the Incarnation. If the Incarnation meant that the whole human person was capable of redemption, then what woman was seen as being—even in the most misogynist form of the Christian tradition—was caught up into God in Christ. And if the agony of the Crucifixion was less sacrifice or victory than the redemption of that which is human (matter joined to form), then the Crucifixion could be imaged as death or as eating or as orgasm (all especially human—bodily—experiences). Women mystics seem to have felt that they *qua* women were not only *also* but even *especially* saved in the Incarnation.

ON BREATH

KRISTOFER M. SCHIPPER

Spiritual exercise is closely tied to the breath; calling on the main gods must be done in connection with breathing exercises.

> Grind your teeth twenty-four times, swallow your breath twelve times, and repeat the following formula: "Five viscera, six receptacles, your true gods all go up to the great assembly in the Scarlet Palace."

Grinding the teeth resounds inside the body as does beating the drum to summon the gods to assemble and participate at the great parade. Swallowing the breath is done in the following way: bend back the tongue so that the tip touches the palate, fill the mouth with saliva, inhale through the nose, hold your breath and swallow the saliva, exhale through the mouth. This breathing exercise knows all kinds of variations. Here is an easy one:

> The Old Lord says: "The Gate of the Obscure Female is the root of Heaven and Earth; like a slender thread that unwinds forever, we use it without ever exhausting it."
>
> This means that the mouth and the nose are the gates of Heaven and Earth through which we inhale and exhale the energies (*ch'i*) *of yin* and *yang*, of life and death.
>
> Each morning, stand up and face the South. Place both hands on your knees, then pressing softly on the two joints, exhale the impure *ch'i* and inhale the pure one. This is called "exhaling the old and inhaling the new." Hold your breath for a long time, then very softly let it out. As a rule, rub yourself during this exercise with your hands on the left and the right, from top to bottom, in front and behind. As you absorb the breath, think of the Original *Ch'i* of the Great Harmony which descends to the genitalia and spreads from there to the five viscera and the four limbs, which all receive its beneficial action, like the mountain that absorbs the clouds, like the earth that drinks up the rain. When the breath has been properly diffused through the body, you will feel your belly stir lightly. When this happens ten

times, the body is filled with a feeling of well-being, the complexion is fresh, hearing and vision are clear, appetite is good, and health complaints go away.

This method should be practiced between midnight and noon, when the energies are alive. During the other half of the day—from noon to midnight —the *ch'i* are dead; one should not do these exercises.

This easy exercise can become the starting point for more advanced ones. Breath-holding can be rhythmical: taking the time necessary for normal breathing as the basic unit, hold the breath during a certain number of units—three, five, seven, or nine. These numbers correspond to the symbolic values of the viscera: heart, kidneys, lungs, and liver. Each time the breath is held in this way, one should guide it *mentally* to the particular organ. Indeed, this mental function is that of the central organ, the spleen, which receives that symbolic number, one. This kind of exercise is called *tao-yin*, "guiding the energies (*ch'i*)." They are usually performed while lying on the back with the legs slightly apart and may be accompanied by massages that enhance circulation.

The *ch'i* which are thus guided towards the different organs may also be vocalized. While inhaling, clench the teeth and make the sound *shhii*; while exhaling, the sounds are:

—For the spleen: *hou*!
—For the heart: *kha*!
—For the lungs: *si*!
—For the kidneys: *tse*!
—For the liver: *su*!

The exact sounds are of course impossible to reproduce in writing. They must be learned from a master. This exercise of the Six Breaths (*liu-ch'i*, one inhaling and five exhaling) is meant to heal illnesses of the viscera through the vocalized expulsion of air. More than a meditative practice, it is a rhythmical expression of the body's functions.

Guiding the energies (*tao-yin*) is not alone stimulated through concentration and massage but through physical exercises as well. One of the earliest forms of moving *tao-yin* was the Dance of the Five Animals in which certain stereotyped gestures of the tiger, the bear, the deer, the monkey and the owl are imitated. Among the manuscripts of the second century B.C. discovered at the site of Ma-wang tui there is a richly illustrated text on different exercises of this and other types. Intended to develop suppleness and relaxation and to improve circulation, these exercises were the origin of today's Taoist gymnastics, of which *T'ai-chi ch'üan* ("boxing of the Highest Ultimate") is an example. This wonderful method of harmony and well-being is a martial art for the defense of the inner world. The slow, supple dance of *T'ai-chi ch'üan*, performed with no apparent effort, is for everybody an excellent initiation into the very essentials of Taoism. It requires no special equipment, very little space, and no prior training, yet it is so efficient that even thinking through the movements provides some benefit. In the same way as reciting the *Book of the Yellow Court*, this form of "boxing" is a rhythmical expression which guides the breathing and which, through daily practice, conditions one for the Keeping of the One.

These preparatory exercises are not simple repetitive movements that can be done whenever one feels like it. They require precise timing. The day is divided into two halves: that of the living *ch'i* and that of the dead *ch'i*. There is, moreover, a system of correspondence between the hours of the day and the four directions and the center, and thus the absorption of the *ch'i* of the five viscera should be performed at certain times (morning for the liver, noon for the heart, et cetera). The sixty-day cycle is part of the spatiotemporal order, as are the twenty-four calendar-nodes. There is even a method for regulating the daily breathing exercises according to the sixty-four hexagrams of the *I-ching*.

The daily exercises can and should be made to coincide with the time cycle. To every period of the time cycle corresponds a part of the body, to each hour a god with whom one can communicate through a kind of perpetual prayer. At every moment, the body changes; it floats through time, and the regulation of our life can never exist without the cyclical time structure. The daily preparatory exercises already constitute an entrance into the cosmic rhythm, a way of participating in the spontaneous evolution of nature. As soon as the practitioner enters into this universal movement, he becomes one with the great mutation of all beings.

This harmony with time, moreover, is necessary in order to realize one's female nature. Phases, periods, and critical days do belong much more to the existential experience of a woman than to that of a man. The *Tao-te ching* stresses this connection between the practice of Keeping the One, cyclical time, and female nature:

> Can you keep the turbulent *p'o*, prevent them from leaving; embrace the One without its leaving you?
> Can you control your breathing and make it as soft as a child's?
> Can you purify your vision of the mystery so that it loses all distortion?
> Can you, by Non-Action, watch over the people and rule the [inner] land?

Can you, by opening and closing (at given times) your natural gates, realize your female nature?

Can you, by Non-Knowledge, let the white light penetrate all the regions of the [inner] space?

The cycles of fertility and gestation lead us to understand that the human body is a time machine. Chinese medicine does take this very much into account, in diagnosis as well as for therapy. The ordering of the inner world demands that one submit to the rules of time and continuously prepares oneself for the Work of the Tao, and this entails, so to speak, a "cosmologization" of the individual. The initial and preliminary stage of Keeping the One concerns in the first place breathing exercises which allow one to restore the equilibrium and the harmony of the body's energies. This, as we have seen, is the Work of the *ch'i* (*ch'i-kung*). Practiced with regularity, these preliminaries are a way of setting the body to music. After a while, the exercises and the discipline become spontaneous and the integration into the cosmic rhythm is achieved with less and less effort, while the individual's spiritual strength increases with the natural completion of the greater and lesser cycles. "No need for divine aid to become immortal," says the *Book of the Yellow Court*. "The continual accumulation of energies, year after year, is enough."

SOME SPECULATIONS ON THE HISTORY OF

"SEXUAL INTERCOURSE" DURING THE

"LONG EIGHTEENTH CENTURY" IN ENGLAND

HENRY ABELOVE

My purpose in this paper is simple. I intend to try to deduce, from the mathematically impressive recent scholarship on English demography, some conclusions about the development of English sexual behavior. Once I have explained these conclusions about sexual behavior, I shall also speculate just a little on their significance.

It has been known for many years that the population of England increased mightily during the period which is now called the "long eighteenth century" —the period stretching from the 1680s to the 1830s. In 1681 the population stood at about 4.93 million; in 1831, it was 13.28 million. But what hasn't been known with any certainty, at least until recently, was whether this extraordinary increase of people was due to a decline in mortality, to an increase in fertility, or to a combination of both. Historians, of course, have argued about the matter, some favoring one answer, some favoring another; but since their arguments have been based on scanty or broken data, nothing of what they have said could be really established and command general assent.

This arguing was stilled in 1981, when the demographers Wrigley and Schofield, both associated with the Cambridge Group for the History of Population and Social Structure, published their seven-hundred-some-page *magnum opus* on English population history. Their book virtually settled the question.[1] At last sufficient data had been gathered (the data came from more than four hundred parish registers), and a statistical technique sufficiently sophisticated for the job of interpreting them (the technique is called "back-projection") had been utilized. Two years later, in 1983, Wrigley repeated the findings of the book in short form in an article for the journal *Past and Present*. He called the article "The Growth of Population in Eighteenth-Century England: A Conundrum Resolved," and to that decisive-sounding title he may actually have been entitled.[2]

What the work of Wrigley and Schofield shows is that the mighty increase in English population during the long eighteenth century was due to a combination of a decline in mortality and a rise in fertility, but that of these two factors, a rise in fertility was much the more important. It isn't that the decline in mortality was a negligible matter. In the 1680s the average life expectancy, as Wrigley and Schofield establish it, was 32.4 years. By the 1820s it was 38.7 years. So life had lengthened and mortality declined by a bit more than six years on average from the start of the period until the end. But Wrigley and Schofield can demonstrate mathematically that a rise in fertility, which occurred during the same period, contributed two and a half times as much to the outcome of population increase as did the decline in mortality, substantial as that was.

The chief cause, then, of the increase in population in England during the long eighteenth century was a rise in fertility; and this rise was realized, as Wrigley and Schofield show, in several different ways. First of all, more women got married. At the start of the period, about 15 percent of all women who survived through the years of their fertility never married. By the end of the period, no more than half that percentage of all women who survived through the years of their fertility never married. This drop in the percentage of single women seems to have occurred mostly in the latter part of the eighteenth century. Second, the average age of the first marriage for men as well as women fell by about three years, from twenty-six to twenty-three. This drop in the age of first marriage seems also to have occurred mostly during the latter part of the eighteenth century. Finally, there was a marked increase in the rate of illegitimate births. At the start of the period only about 2 percent of all births were illegitimate. At the end of the period it was about 8 percent. This rise in the illegitimacy rate is more important than the figures may suggest. Another way of expressing the same rise, a way which may make its importance plainer, is to say that at the start of the period fewer than one-tenth of all first births were illegitimate but that at the end of the period about a quarter were. I should add that an additional quarter were legitimate but prenuptially conceived, and that this figure, too, represented a marked increase. Like the drop in average age of first marriage, like the increase in the percentage of women marrying, the rise in illegitimacy and in prenuptial pregnancy seems to have occurred mostly during the latter part of the century.

To sum up, a rise in fertility was realized in these ways: more women got married than had done before, women and men married earlier in their lives than they had done before, and women had more illegitimate children and prenuptial pregnancies than they had done before. Chiefly because of the rise in fertility so realized, and only very secondarily because of a concurrent decline in mortality, England's population grew from 4.93 million to 13.28 million in the course of about 150 years, but with special acceleration during the latter part of the eighteenth century.

What all this means is that there was a remarkable increase in the *incidence* of cross-sex genital intercourse (penis in vagina, vagina around penis, with seminal emission uninterrupted) during the late eighteenth century in England. I mean that the particular kind of sexual expression which we moderns often name tendentiously "sexual intercourse" became importantly more popular at that time in England, and so much more popular that by means of that enhanced popularity alone, without any assistance from a decline in mortality, England's population could have doubled in a relatively short span. With the assistance of a decline in mortality, the population did actually more than double.

That is my deduction from the demographic data—that sexual intercourse so-called became importantly more popular in late eighteenth-century England—and I believe that the deduction is irresistible. It is, however, a deduction that the demographers do not make. They do not say it; they do not seem to see it. I should guess that for all of us, whether or not we are demographers, seeing, saying, deducing such propositions on the history of sexual behavior may be peculiarly difficult. It isn't primarily a matter of embarrassment, of a fear of indecorum, though embarrassment and fear may of course play a part in restraining some of us. It's more a matter of a very strong feeling we're likely to have that such deductions and observations are, first, too bizarre to be cogent, yet, second, too obvious to be worth seeing and saying. We can easily feel both ways simultaneously. On some other occasion I should like to talk further about that discomfiting dual feeling. For now, I want to remark that in my opinion the feeling is ideologically determined and that in the measure we give way to it and allow it to govern us we reinforce that essentialism which so disempowers us both as historians and as political beings.

If we take seriously the deduction I've put to you, then many lines of inquiry open before us. We may, for instance, want to ask *why* sexual intercourse so-called should have become so much more popular in late eighteenth-century England. Nor need we be deterred from asking that question by our realization that we don't in the least know how to go about answering it, that we don't even know what would constitute an adequate answer to it. By permitting ourselves to ask it, we may eventually learn how to answer it. We may also want to ask whether or not this change in sexual practice is related to any change in late eighteenth-century English conceptions of what sex is, what it is for.

Returning for the moment to the demographers, we may note that although they don't make the deduction I've put to you nor ask the questions

I've just asked, they do try to account for their data by asking another question. Their question is: Why did people marry earlier in life, and why did more women marry?[3] One worrisome disadvantage to this question is that it doesn't save the appearances, that it doesn't fully respond to the data on fertility that they themselves present. Their data show not only a fall in the average age of marriage partners at first marriage, with consequences for population, not only a higher percentage of women marrying, with consequences for population, but also a rise in the rate of illegitimacy and prenuptial conception, with consequences for population—that is to say, their data show more sexual intercourse so-called both inside and outside of marriage. On the other hand, their question has also a certain advantage. Even if the answering of the question couldn't explain what needs explaining, the question is at least comfortable. The demographers imagine that they know, and maybe most of us would imagine that we knew, how to go about trying to provide an answer for a question such as theirs. When the point at issue is understood simply as earlier marrying and more marrying, then surely we would look almost automatically for a rise in the wage-rate as the likely cause. The demographers do look for a rise in the wage-rate, but to their obvious disappointment they cannot find the causal connection they expect. Wages go up, but they go up a good thirty years in advance of the fertility rate; and thirty years is, as the demographers concede, probably too long an interval to fit into a causal argument of the sort they want to make. Still, their failure to answer the question they ask is less remarkable than their evading their own data with that question. What they do is defensively transform something that ought to be a problem in the history of sexual behavior into a problem in the history of nuptiality so that they can proceed comfortably, if unsuccessfully.

I'd like to suggest that the conclusion I've deduced from the demographers' data, that sexual intercourse so-called came to be importantly more popular in late eighteenth-century England, can perhaps be related to another contemporary experience. The new popularity of intercourse so-called doesn't follow from, or depend on, a rise in wages; and it is only when we misconceive the data to be understood as simply about marrying that we are tempted even to look to wage-rates for an explanation. But the new popularity of intercourse so-called does correlate rather well with a dramatic rise in virtually all indices of production, a rise which the textbooks call the onset of the Industrial Revolution and which as we know distinguished late eighteenth-century England. I don't mean to imply that this rise in production, which was probably the biggest since the invention of agriculture in prehistoric times, caused the rise in intercourse so-called, nor do I mean to imply that the rise in intercourse so-called caused the rise in production. Neither of these causal arguments would seem to me to be sound, and both would of course depend on a too-easy and conventional distinction between the sexual and material realms. What does seem to me at least conceivable, though I am just speculating in saying so, is that the rise in production (the privileging of production) and the rise in the popularity of the sexual act which uniquely makes for reproduction (the privileging of intercourse so-called) may be aspects of the same phenomenon. Viewed from different perspectives, this phenomenon could be called either capitalism or the discourse of capitalism or modern heterosexuality or the discourse of modern heterosexuality.

It is of course true that sexual intercourse so-called had been valued in some measure or another, on some grounds or another, before the late eighteenth century and in every previous European society about which we know anything. But it is also true that production had been valued before the late eighteenth century and in every previous European society about which we know anything. What happens to production in the late eighteenth-century England is nevertheless new. While production increases importantly, it also becomes discursively and phenomenologically central in ways that it had never been before. Behaviors, customs, usages which are judged to be nonproductive, like the traditional plebeian conception of time, according to which Mondays and maybe Tuesdays and Wednesdays as well are free days, play days, rest days ("St. Monday" was the plebeian phrase), come under extraordinary and ever-intensifying negative pressure.[4] If I should be right in speculating that the rise in popularity of sexual intercourse so-called in late eighteenth-century England is an aspect of the same phenomenon that includes the rise in production, then we should expect to find that sexual intercourse so-called becomes at this time and in this place discursively and phenomenologically central in ways that it had never been before; that nonreproductive sexual behaviors come under extraordinary negative pressure; and finally that both developments happen in ways that testify to their relatedness, even to their unity.

I cannot say that I have such findings to present to you. As I mentioned before, I am just at a point of speculative beginning. But I can say something about where my attention is currently directed. The earlier part of the long eighteenth century, before the big rise in production and in the incidence of sexual intercourse so-called, was an era of relatively late marriage, low illegitimacy and prenuptial pregnancy rates, and a relatively high rate of non-marrying for women. According to some students of plebeian sexuality, like Flandrin, Quaife, and Bray, this was also an era of very diverse sexual practice. If outside of marriage plebeians typically avoided sexual intercourse so-called (penis in vagina, vagina around penis, with seminal emission uninterrupted), they were nevertheless typically sexually active. They practiced mutual mastur-bation, oral sex, anal sex, display and watching (or to use the more common and pejorative terms, "exhibitionism" and "voyeurism"), and much else be-sides, on a cross-sex basis and in some now uncertain measure on a same-sex basis as well.[5]

What happens to the tradition of same-sex sexual behaviors in the late eighteenth century is something that I shall put aside for now. It is an intricate problem and demands extended and separate treatment. As for what happens to the tradition of very diverse cross-sex sexual behaviors, my hypothesis is: They are reorganized and reconstructed in the late eighteenth century as fore-play. They don't disappear, they aren't ruled out, as the incidence of inter-course so-called increases, but they are relegated and largely confined to the position of the preliminary. From the late eighteenth century on, they are construed as what precedes that sexual behavior which alone is privileged, intercourse so-called. On this hypothesis, the invention of foreplay—an im-portant passage in the making of modern heterosexuality—is to be understood as homologous with the crowding of St. Monday, Tuesday, and Wednesday into Sunday, the first day of the work week—an important passage in the making of capitalism. Rest doesn't disappear, isn't ruled out, as production rises; but rest is relegated and largely confined to the position of the preliminary. To put it differently, I hypothesize that the invention of foreplay is an aspect of the history of capitalism, that the invention of industrial work-discipline is an aspect of the history of heterosexuality, and that both developments are in an important sense the same.

NOTES

1. E. A. Wrigley and R. S. Schofield, *The Population History of England, 1541–1871: A Reconstruction* (Cambridge, Mass: Harvard University Press, 1981). Of course there has been *some* criticism. See, for instance, Peter Lindert, "English Living Standards, Popula-tion Growth, and Wrigley-Schofield," *Explorations in Economic History* 20 (April 1983), 131–55, and Louis Henry, "La Population de l'Angleterre de 1541 à 1871," *Population* 38 (July-October 1983), 781–826.

2. E. A. Wrigley, "The Growth of Population in Eighteenth-Century England: A Conundrum Resolved," *Past and Present* 98 (February 1983), 121–50. Wrigley has re-cently reprinted this essay with little revision in E. A. Wrigley, *People, Cities, and Wealth: The Transformation of Traditional Society* (Oxford and New York: Basil Blackwell, 1987), 215–41.

3. Wrigley, "Conundrum," 134.

4. See, for instance, E. P. Thompson, "Time, Work-Discipline, and Industrial Capitalism," *Past and Present* 38 (December 1967), 56–97.

5. Jean Louis Flandrin, *Families in Former Times: Kinship, Household, and Sexuality*, trans. R. Southern (Cambridge: Cambridge University Press, 1979); G. R. Quaife, *Wanton Wenches and Wayward Wives: Peasants and Illicit Sex in Early Seventeenth Century England* (New Brunswick, N.J.: Rutgers University Press, 1979); Alan Bray, *Homosexuality in Renaissance England* (London: Gay Men's Press, 1982).

HUMAN BODY PARTS AS THERAPEUTIC TOOLS:
CONTRADICTORY DISCOURSES AND
TRANSFORMED SUBJECTIVITIES
MARGARET LOCK

Over the past half-century, the development and refinement of the technology of tissue and organ transplantation have enabled us to make routine use of human bodies as therapeutic tools. Appropriation of human cadavers and body parts for medical purposes has a long history that commenced in classical Greece. This history is not a savory one—very often, vivisection of criminals or marginalized people was involved. As late as the mid–nineteenth century, in both Europe and North America, bodies obtained for medical dissection were frequently procured through foul means, and a disproportionate number were bodies of the poor or minority peoples.

It was not until the first part of the twentieth century that medical knowledge advanced sufficiently that blood could be transfused, and then, later, solid organs were transplanted, bringing about a confusion of body boundaries and mingling of body parts never before possible. Some rather crude experimentation with organ transplants in the early years revealed that body parts cannot be grafted at random and biological rules of blood and tissue typing must be adhered to faithfully. Solid organs, more often than not, are never fully accepted by recipient bodies, so that lifelong use of immunosuppressants is necessary. Despite this difficulty, organ transplants have been routinized with apparent ease and become part of the health care systems of virtually all countries in the world able to support the necessary technology. This suggests that the majority of health care professionals and policy makers assume that making use of organs obtained from willing donors, whether living or dead at the time of procurement, is a rational, worthwhile, and relatively unproblematic endeavor. . . .

Before the removal of organs from donors and their preparation for use as therapeutic tools can come about, the necessary technology must be in place and, furthermore, human organs have to be understood as fungible. Moreover, donors must be designated as dead prior to organ removal. I point out what seems obvious today as a reminder that it is only over the past 40 years that we have gradually come to accept organ procurement as commonplace; during this time, for the most part, a utilitarian drive to maximize available organs has dominated any deeper examination of the issues involved.

In addition to assuring that death has indeed taken place, a tacit agreement must also exist that the body will not be violated through organ removal, and, to this end, conceptualization of organs by the medical profession as mere objects is enabling. However, organs for transplant are, by definition, alive; although objectified, they cannot be reduced to mere things, even in the minds of involved physicians, and they retain, therefore, a hybrid-like status.

Mixed metaphors associated with human organs encourage confusion about their worth. The language of medicine insists that human body parts are material entities, devoid entirely of identity whether located in donors or in recipients. However, to promote donation, organs are animated with a life force that, it is argued, can be gifted, and donor families are not discouraged from understanding donation as permitting their relatives to "live on" in the bodies of recipients. Organ donation is very often understood as creating meaning out of a senseless, accidental death through the use of a technologically mediated path to transcendence, although the enforced anonymity of donor families ensures that no earthly ties of solidarity between recipients and donor families are formed except on rare occasions.

Despite the enforced cloak of anonymity associated with donors, it has been shown on many occasions that large numbers of recipients experience a frustrated sense of obligation about the need to repay the family of the donor for the extraordinary act of benevolence that has brought them back from the brink of death. The "tyranny of the gift" has been well documented in the transplant world, but it is not merely a desire to try to settle accounts that is at work when people want to know more about the donor. It is abundantly clear that donated organs very often represent much more than mere biological body parts; the life with which they are animated is experienced by recipients as personified, an agency that manifests itself in some surprising ways and profoundly influences subjectivity.

A conversation I had a few years ago with a heart transplant surgeon was most revealing in this respect. This surgeon was responding to stories that have been circulating for some time now about a debate taking place in several of the American states as to whether prisoners on death row should have the option of donating their organs for transplant before they are put to death. The argument is that prisoners should be given the choice of making a "gift" to society just before their lives are extinguished. Perhaps those among the prisoners who are believers will even go straight to heaven.

This surgeon was uncomfortable about the idea of organ donations made by Death Row prisoners, not so much because he was concerned about the highly questionable ethics (Can one make an "informed choice" in such circumstances?) but about receiving a heart that had been taken out of the body of a murderer. He said to me, with some embarrassment, "I wouldn't like to have a murderer's heart put into my body," then added hastily, trying to make a joke out of the situation, "I might find myself starting to change."

A good number of organ recipients worry about the gender, ethnicity, skin color, personality, and social status of their donors, and many believe that their mode of being-in-the-world is radically changed after a transplant, thanks to the power and vitality diffusing from the organ they have received. This situation leads to contradictions and confusion, even among health care professionals, it seems. Organ donation is promoted making use of the metaphor of "the gift of life," so that organs are indirectly attributed with a transcendent life force by many people involved with the transplant world. Once transplanted, however, if the recipient attributes the "life-saving" organ with animistic qualities for more than a few weeks, then he or she is severely reprimanded, even thought of as exhibiting pathology.

Interviews that I carried out in 1996 with 30 transplant recipients living in Montreal reveal that just under half are very matter-of-fact about the organs they have received. These people insist that after an interim period of a few months, they ceased to be concerned about the source of the new organ encased in their bodies and resumed their lives as best they could, unchanged in any profound way except for a daily regime of massive doses of medication. The responses of the remaining recipients were different: They produced emotionally charged accounts about their donors (about whom, in reality, they knew very little), the particular organ they had received, and often about their transformed subjectivity.

Forty-one-year-old Stefan Rivet falls into the first group. He is a kidney recipient, doing well when interviewed 5 years after the transplant. He says,

> *Rivet:* I heard about the donor, even though I wasn't supposed to. It was a woman between 20 and 25. She was in a car accident. You know, don't you, that you can't meet the family because the doctors think it would be too emotional? But I wrote a letter to them, it must have been a terrible time for them, and I wanted to thank them.
>
> *Lock:* Did you find it hard to write that letter?
>
> *Rivet:* No, no, it wasn't hard for me. Like saying "thank you" to someone if they do something for you, that's just the way it was.

Lock: Did you feel at all strange because it was a woman's kidney?

Rivet: No. At first you wonder how could a female kidney work in a man. You think about it. But once the doctor tells you that it works exactly the same in men and women you don't question things any more. It doesn't bug me. I have my kidney, and I can live, that's all you really worry about.

When I first woke up in hospital I was worried. Of course, I didn't know whose kidney it was then, all you know is that there's a strange organ in there and you hope that it works; you don't want anything to go wrong. After a while though, you adapt and you stop thinking about it, except that it's really important to take the pills. I just say now that it's my second life.

In contrast to recipients such as Stefan Rivet, many others undergo a rather dramatic transformative experience. One such was Katherine White, who first received a kidney transplant in 1982, and then, in 1994, after that kidney failed and her own liver was also in jeopardy, she received a double transplant of liver and kidney. Six months after the second surgery, she had this to say:

White: I have no idea who the donor was, all I know is that both the kidney and liver came from one person because you can't survive if they put organs from two different people into you at once—your body would never be able to deal with it. I wrote a thank-you note right away that I gave to the nurse. But they don't like you to know who it is; sometimes people feel that their child has been reborn in you and they want to make close contact. That could lead to problems. I still think of it as a different person inside me—yes I do, still. It's not all of me, and it's not all this person either. Actually, I might like some contact with the donor family. . . . You know, I never liked cheese and stuff like that, and some people think I'm joking, but all of a sudden I couldn't stop eating Kraft slices—that was after the first kidney. This time around, the first thing I did was to eat chocolate. I have a craving for chocolate and now I eat some every day. It's driving me crazy because I'm not a chocolate fanatic. So maybe this person who gave me the liver was a chocoholic?! It's funny like that, and some of the doctors say it's the drugs that do things to you. I'm certainly moody these days. You do change whether you like it or not. I can't say that I'm the same person I was, but in a way I think that I'm a better person.

You know, sometimes I feel as if I'm pregnant, as if I'm giving birth to somebody. I don't know what it is really, but there's another life inside of me, and I'm actually storing this life, and it makes me feel fantastic. It's weird, I constantly think of that other person, the donor . . . but I know a lot of people who receive organs don't think about the donors at all.

Awhile ago I saw a TV program about Russia and it seemed as though they were actually killing children in orphanages to take out their eyes and other organs. This disturbed me no end. I hope to God it's not really like that. My parents and my uncles all thought I shouldn't have a transplant, they said you can't be sure that the patient is really dead. Brain-dead is not death, they said. But I know that's not right. I had a friend a few years back who had a bad fall off a bicycle and her husband donated her organs. Once you're brain-dead that's it.

Lock: What do you think happens when people die?

White: I hope I go to heaven! I don't believe in resurrection but I do believe in a heaven and hell and an in-between, you know? I think there's a person up there who knows that I'm carrying a part of her around with me. I always think there's somebody watching me . . . but you know, I don't really believe in religion. . . . I really don't. In a way I wish I could have a pig's liver or kidney—it would be much simpler then.

Despite the power of medical discourse working against animation of organs by patients and the flat rejection of the possibility of any transformation in subjectivity on the part of virtually all doctors, it is clear from numerous interviews carried out independently by Leslie Sharp (1995) and me that a large number of patients in Canada and the United States believe themselves to be "reborn" after a transplant. These patients frequently form affiliations with other transplant recipients, but this newfound group identity is often accompanied by a more substantial transformation; many recipients undergo a profound change in subjectivity and report that they experience embodiment in a radically different way after a transplant.

THE GLOBALIZATION OF SUBJECTIVITY

Not all technologically advanced countries have responded in the same way to transplant technology. By far the majority of organs for transplant are procured from brain-dead bodies. In Japan, a vigorous national debate has taken place for over 30 years in which opponents to the recognition of brain death as the end of human life have effectively blocked almost all organ transplants. Only in 1997 was it finally agreed that the bodies of brain-dead patients could be commodified for use in transplants. Since that time, however, there have been only 17 organ procurements from brain-dead bodies. This situation has meant that for those relatively few patients in Japan who receive transplants, "living related organ donations" is the norm, that is, organs are usually procured from living relatives. One exception was Naka Yoshitomo, 63, a retired school principal who was the recipient of a kidney taken from a 70-year-old American brain-dead donor. The transplant took place between 60 and 70 hours after the kidney was first procured in the United States, having traveled halfway across the world and then languished in a cooler while medical professionals disputed whether it should be used. Exactly one year later, in 1996, when I interviewed him, Naka was experiencing a mild rejection of the kidney, but since that time he has done exceptionally well.

"I've become ten years younger since I had the transplant," he says, "I was on dialysis for 13 years, every Tuesday, Thursday, and Saturday afternoons and evenings."

"How did you feel about having a kidney from such an old donor?"

"My wife was opposed, partly because of the cost. But my son agreed as soon as he understood that I was keen." (Note that Mr. Naka thinks first about the reactions of his family and not about his own feelings.) He goes on,

> I felt really lucky to go right to the top of the list of waiting people just because I happened to be the best match. I didn't want to lose this chance— this seemed really to be a "gift of love and health" (*ui to kenko no okuri-mono*), finally, after all the waiting.

In the event, once the operation was completed, it took only 5 days before the kidney started to function well. In the United States, this kidney would have been thrown out as defective because of its age and the protracted time outside a human body.

One morning shortly after the operation, Naka was completely taken aback when he noticed in the street below the sounds of one of the oppressively noisy military-like vehicles used by the extreme right wing in Japan to stir up nationalistic sentiment. As it crawled back and forth outside the hospital, he gradually became aware of the message being screamed into the loudspeakers: "Bad doctors have taken part in a cover-up. Importation of defective kidneys." On and on they droned, strident and abusive. Lying in his hospital bed, shocked, Naka was plagued by serious doubts and began to believe that in his haste to get a transplant, he had done something wrong. He had been told that the chances of success for the transplant were about 80 percent, but he started to wonder whom he should believe. Time has proved the judgment of the doctors correct—but they do indeed take risks in transplanting aged kidneys into desperate Japanese patients.

Naka and others of his compatriots who have received transplants, as well as transplant surgeons, have been labeled unpatriotic by a few of their bellicose countrymen who have strong nationalistic sentiments. Both the recognition of brain death as the end of human life and the carrying out of organ transplants making use of brain-dead donors, whether the donor is Japanese or foreign, have caused hostile reactions from the extreme Right as "unnatural" acts in which Japan should not participate. After his first shock, Naka had no trouble ignoring the hostility targeted at him and his surgeon. He reported to me that now he lives daily with thoughts about his donor:

> *Naka*: Hopefully I will understand how he felt one day. We must change our ideas in Japan [and be more generous about donation], and that is why I wrote a book about my experience.
>
> *Lock:* Did you write a letter to the donor's family?
>
> *Naka*: Oh yes! I was happy to send that letter. I sent a copy of my book to UNOS (the United States United Network of Organ Sharing) as well. Now I'm working hard on cultural exchange between my hometown and our sister town in America. I go to America all the time arranging visits and events. I can't think of a better way to thank that family for what they did for me.

Naka firmly believes that as a result of the transplant he is able to transcend the boundaries of his former self and has become a citizen of a global community that fosters international cooperation of all kinds.

Although Naka is highly cognizant of the generosity of his donor, this is by no means always the case. Because of the anonymity that has been imposed on donors, many of whom receive nothing more than a brief note of thanks from an organ procurement agency, their altruism has gone virtually unmarked by many recipients and even by some transplant teams. On the contrary, a sense of entitlement to "spare parts" is evident among a good number of people involved with the transplant enterprise.

Ethnographic research has contributed to a growing understanding that public recognition of the indispensable part played by donors in the transplant enterprise is crucial. With increasing frequency, donors' families and organ recipients are brought together, usually at public gatherings at which donors as a group are memorialized. These encounters are not designed for the purpose of bringing donor families together with the recipients of the organs of their relatives but rather to create a community in which both donors and recipients participate. As a result of such gatherings, family members who have already facilitated a donation may well be motivated to encourage other people in their circle of acquaintances to comply with organ donation should a relative of theirs become brain-dead.

To date, because donor families have been pushed into obscurity by a system that requires anonymity, there has been little incentive for them to encourage other people to do what they did; on the contrary, some families retain doubts that can linger for years as to whether they did the right thing in agreeing to donation. Only when donor families are permitted to encounter firsthand the transformation that transplants can have on the lives of so many people will such doubts be dissipated, although even then they may continue for some people. Similarly, the misplaced idea that a few people appear to have, that organ donation is simply a matter of signing an organ donation card, will be dispelled as donor families are increasingly brought into the public domain. Signing a card is, indeed, rather easy, but it is, in the end, the families of donors who, in a state of intense shock, must agree to suppress their own overwhelming feelings of loss and disbelief and permit the procurement of organs.

REFERENCE

Sharp, Leslie. 1995. "Organ Transplantation as a Transformative Experience: Anthropological Insights into the Restructuring of the Self." *Medical Anthropology Quarterly* 9:357–89.

MERATUS EMBRYOLOGY

ANNA LOWENHAUPT TSING

Despite the problematic heritage of ethnographic writing, ethnographies—
with their focus on the specificity of social arrangements and points of view—
are a possible site for drawing attention to both local creativity and regional-
to-global interconnections. In part, this is because ethnographies are messy;
like novels, they tend to include an overabundance of detail, much of it extraneous to the main argument. Rather than condemn this messiness as boring . . . or outdated . . . , I prefer to build from it as a source of analytic
heterogeneity and promise. In this messiness, for example, there is room for
elements that simultaneously draw readers into projects of cultural comparison, regional cultural history, and local/global positionings.

To reemphasize the possibilities for combining such projects, I turn to [a]
Meratus story. . . . This piece is not particularly fragmented or graffiti-like;
rather it creates a coherent set of imagery. Its careful aestheticism makes it
possible to call it a "poem." It is a poem about fetal development and birth; it
tells of individual autonomy amid power and inequality. It is a piece I love
because of its elegance. It would be easy to use the poem to tell of local wisdom.
Unlike stories of police or rock stars, it is challenging to place this poem within
national and global cultural politics. By taking up this challenge, I can illustrate
multiple layers of ethnographic possibility.

Here is the immediate social context in which I heard the poem. A man who
lived in the mountains came to Uma Adang's house for advice. He was in love
with a woman and had decided to divorce his wife and marry the other
woman. Community elders disapproved of this decision and were pressuring
the man to change his mind. Uma Adang counseled him to listen carefully to
the elders and to use mature good sense. At the same time, he should also
follow his heart. She spoke of the individual, unpredictable nature of love
relationships (*juduh*): "There are juduh that last as long as a touch; there are
juduh that last through fondling breasts; there are juduh that last through
becoming lovers; there are juduh that last two or three months. Even if you
already have children, if your juduh is over, then nothing can keep you to-

gether." The statement sounds surprisingly disruptive when one considers the context of the national language, where the term *jodoh* refers to an appropriate match, marriage partner, or mate. It holds some sense of a preordained, and thus durable, couple. In contrast, Uma Adang was making room for more tentative, shifting arrangements.

As one aspect of her advice, Uma Adang began telling the story of the fetus's development in the womb. Fetal development is a familiar Meratus framework for understanding personal agency: Fetal autonomy signals individual autonomy in a world of social demands. Meratus say that the fetus "meditates" (*batapa*) for nine months and nine days in the womb to find its *rajaki*, its future livelihood, luck, health, and wisdom. Because each fetus finds its own rajaki, each individual at some level goes his or her own way in life. Uma Adang's telling also emphasizes the relationship of the fetus with God (*Tuhan*). Indeed, there was considerable talk of God that night. Perhaps it was important that the man had an ambiguous ethnic identity: He was born Banjar but brought up by a Meratus family and currently living in a Meratus community. Uma Adang did not talk to him of her theories of Meratus religion or of adat. Instead, she stressed the regionally transethnic discourse of a mystical God. This was a discourse in which, as Uma Adang and her close friends stressed, God is an aspect of the self "closer than our fingernails."

That night, Uma Adang wove an exceptionally beautiful and haunting version of the story of the fetus. I was so moved that I asked her to repeat the story the next day so that I could record it. She did, in a somewhat abbreviated, but equally poetic fashion. The version I have translated below is the taped version, with a few minor additions from my notes on the earlier telling. The written format is my attempt to give the piece a slower, more formal, oral rhythm.

Beginning with origins.
Water is placed by the father,
    it is entered inside the mother,
    it is enclosed in the mother,
    the water of nine months and nine days.
The water becomes blood,
    the blood becomes eyes,
    the eyes becomes fibered extensions,
    and are filled out into a head.
The fibers are enlivened
    becoming arms
    or becoming legs.

They harden to become bones
    and are blanketed by ligaments,
    wrapped in muscular flesh.
The flesh fills out
    and is covered by skin.
Thus one month, or two months.
At three months we enter into questions.
"What is your desire to eat or drink?"
Everything will be given, every food or drink.
    The mother eats sour fruits and all kinds of sour things
    [for those are the desire of the child].
Our every intention and desire is satisfied.
Thus six months.
We reach our strength and begin to move.
Our food requests follow each other,
    our many desires,
    and all are eaten, whatever is requested.
The vitality of the child in its mother's belly accompanies the
mother's breath.
    For in the belly one cannot breathe, but only go along
    with the mother's breath.
And then it is completed.
All the foods are sufficient in the womb, and at last the child
is ready to lie birthed by the mother,
completing nine months and nine days.
Then it is God the Powerful that asks the human being in its
mother's belly,
    "What is your strength?
    "Where is your power?"
And we answer,
    "I am yet water and blood."
This we answer three times.
And then God asks,
    "What are your desires in life?
    "What will be your strength?
    "How will you form your desires, you
    there in the womb of your mother?
    "How is it that you choose to die?
    "Will you die falling, or seized by a crocodile,

or will you die an ordinary death of illness
and pain?
"Will you be killed by another, what is your
choice?"
These are God's questions that create our vitality in the womb.
And we say,
"Akrana, akrana, I would have my breath."
And God says we must decide what it is we want, coming out
into this world. Only then can we be born, we who were held
by our mother for nine months and nine days. But we first
must acknowledge these questions. And to this humble
subject in the mother's womb, God asks again,
"What will be your death?"
"What are your desires in this world, so broad
and so wide?"
And perhaps we say,
"I will die like this."
Perhaps we say,
"I will live ninety-nine years."
And we ask for our rajaki in this world, as great as it may be.
We ask for peace and perfection in the world. And the whole
world in a tiny drop of water is realized, and we acknowledge
God and we acknowledge the questions.
"'Truly,'" we say, "I ask for the vital essence of God."
And only then are we given vitality.
We are given our rajaki, our life's wisdom, and we hold this
balled up in the palms of our hands.
And only then do we emerge from the womb,
birthed by the mother.
At the moment of birth, we at last receive our breath.
In surprise we cry out,
"A! Owa!"
and we open our fists and our rajaki is scattered. It falls to the
ground with the blood of birth and is scattered across the
earth. And the rest of our lives we must follow that rajaki; we
must work to follow it wherever it takes us. For it may take us
to many villages, and we must always follow to gather it up.
When it is gone, our years are over.

The fetus about to be born chooses its own future life. It chooses its lovers, however few or many; it chooses its travels; it chooses its time and manner of death. Surprised by our passage into the world and the power of our own breath, each of us, as a newborn, lets go of all that wisdom and spends the rest of life working to regain it.

Here, Uma Adang draws from familiar local and regional ideas about fetal development. Her story aims to create a local political effect. Most specifically, it gave a friend permission to follow his own path in choosing a wife. It also aims farther as a statement about personal autonomy; it is an interpretation, a positioning within regional political culture.

Uma Adang's wider ambition was evident in her willingness to record the story for me; she urged me to take her story to the United States. In accepting her challenge and incorporating the poem in my ethnography, I take on the responsibility of explaining her themes. How might these enter an English-oriented conversation?

As a U.S. American, it is difficult for me to consider Uma Adang's story without thinking about United States issues concerning women's reproductive rights, for fetal narratives have been a central element in recent U.S. debates. Since the 1970s, feminists in the United States have worked to make more reproductive options, including abortion, available to women. In the 1980s, an equally passionate movement seeking to criminalize abortion gained increasing institutional support. The antiabortion movement depends heavily on creating verbal and visual imagery with which to show the fetus as separate from the mother. Appreciation of the fetus as a separate entity, a child who happens to be in a womb, is used to activate assumptions that the fetus, like a child, is innocent, vulnerable, and in need of the state's protection. Thus, antiabortionists pass out plastic fetus dolls and show and reshow photographs of sleeping fetuses floating in watery space. One antiabortion film, *The Silent Scream*, has been described as particularly successful because of the way in which it uses ultrasound and fiber optics to create the sense of an independent (yet vulnerable) and responsive (yet threatened) child in the womb.

The success of antiabortion fetal imagery can be measured in part by the inability of feminist critics to respond. Even critics dare not contest assumptions of the transcendent innocence of the fetus-as-child; further assumptions about the necessity for maternal sacrifice and paternal state protection follow easily. Pro-choice activists tend either to ignore fetal imagery or, if they respond to it, to find themselves in the weak position of characterizing abortion as an unfortunate cruelty to innocents in a flawed world. The entrenched interpretation of fetal imagery overwhelms and stifles feminist creativity.

How different a fetus Uma Adang's poem conjures! Her fetus is wise, not innocent; it demands respect, not sacrifice or protection. No dependent being living at the command of paternal law or dying from unregulated, selfish female neglect, this is a fetus who determines its own birth and death. The gaze is shifted from that of a potential parent contemplating an unself-conscious but valued object to that of a once-wise fetus now grown, remembering his or her own origins. Uma Adang's narrative of fetal development stretches the imagination of North Americans, whether feminist or antifeminist. If we could visualize a *wise* fetus, we might not assume that the "rights" of the fetus demand the sacrifice of the mother.

PART IV

Everyday Life, or Exploring the

Body's Times and Spaces

Everyday Life, or Exploring the
Body's Times and Spaces

INTRODUCTION

What could be more mundane, more insignificant, than embodiment? What could be more bodily than everyday life? The daily rhythms of the body—eating, sleeping, excreting, breathing, walking, and so on—seldom demand analysis. Until recently, we have felt no need to assemble a history of the daily lives of our own bodies. So much is taken for granted, and our sights—both as living people and as analysts—are usually set on loftier goals. Whether it is the challenges presented by the day's to-do list or getting closer to understanding the meaning of life that occupies our attention, the concrete embodied practices that enable these important efforts continue to elude our grasp as analysts. One can know the nature of one's own everyday life, though to take it too seriously would be a little odd; but how should we approach the most mundane realities of other individuals, or of whole groups?

Only a few groups of scholars have offered answers to this question. Henri Lefebvre's famous work *The Critique of Everyday Life* (1947) made the point well that everyday life is paradoxically inaccessible to analysis even as it offers the essential medium in which any study of production and consumption—the central concerns of political economy and Marxist cultural studies—must locate itself. Mid-twentieth-century phenomenologists like Maurice Merleau-Ponty (see part II) and Paul Valéry made valiant efforts to turn philosophical attention away from the ideal realms of cognition, while still seeking to characterize some essence of experience beyond the mundane and observable. For these analysts, attention to the ordinary life of the body was a mode of access to deeper and more universal truths; yet their salutary and frequent return to evidence from everyday life is an important legacy for the anthropology of the body. In a more historical mode, the work of the Annales School historians such as Fernand Braudel, Marc Bloch, and Emmanuel Le Roy Ladurie has meticulously documented the material details of collective life in France over periods of many centuries. They have demonstrated fascinating forms of stability and continuity as well as a new way of tracking unexpected changes in material life. Many of these strains of scholarship have converged in the work of our most recent theorists of everyday life, Pierre Bourdieu and Michel de Certeau.

Such recent turns toward the anthropology and history of the body can be seen as attempts to reconsider everyday life and subject it to a gaze that rescues it from our naturalizing routines. Insofar as common sense is embodied and appears to provide an unquestionable foundation for all the arrangements of a particular social world, it is worth reexamining this foundation. As was shown in part III, a denaturalizing reading of the given can call many of the historically contingent and politically problematic formations of bourgeois modernity into question. Euro-American assumptions about everything from breathing to death, when subjected to a relativizing analysis, are difficult to decouple from the racial, class-inflected, and gendered structures that continue to inform collective life.

But a rereading of the everyday can also bring some of the magic back into the mundane. As Michael Taussig argues below, we can simultaneously demystify and re-enchant the given world. The five readings included in this part, while all adopting a critical stance on some previous approaches to embodied life, nevertheless succeed in charging the times and spaces in which we live with unsuspected significance. De Certeau, for example, well known for his vast collaborative study of the practices of everyday life, here considers the phenomenology of walking in the city, contrasting street-level experience with the kind of disembodied knowledge enabled by a gaze from above. In dense, lapidary prose in which every sentence offers fuel for much reflection, he critiques our desire for a kind of knowledge that relies on a totalizing view from above. Contrasting the view from the World Trade Center (New York's tallest buildings before September 11, 2001) with one's experience of the city at street level, he declares the "panorama city"—the one that can be seen from the top of a skyscraper—to be "a 'theoretical' (that is, visual) simulacrum, in short a picture, whose condition of possibility is an oblivion and a misunderstanding of practices." The conceptual picture of the city is of a different logical order than the city that is lived, imagined, practiced, and even visualized while negotiating its streets and encountering its life-size residents. Turning to the practice of walking in the city, he recasts consciousness away from the visible, the legible, and the total: here, the body is "clasped by the streets that turn and return it according to an anonymous law." Walkers "make use of spaces that cannot be seen; their knowledge of them is as blind as that of lovers in each other's arms." The analogies de Certeau employs here seek to remind us that there are worlds of experience that escape the conceptual and elude legibility of the kind we deploy with images and texts. Moreover, he insists that the spatial practices of everyday life, like walking in the city, demand an approach that can place diversity and concreteness at the center of the analysis, avoiding any search for essences or universals. Practices that use and alter social and physical space are important "if one admits that spatial practices in fact secretly structure the determining conditions of social life." The task involves following out these "multiform, resistant, tricky and stubborn procedures that elude discipline without being outside the field in which it is exercised." Here de Certeau reminds us of his important theoretical efforts to go beyond the dyad of power and resistance, finding everyday agency in activities that work on structures of power without opposing them. The goal in this essay, as in much of the anthropology of everyday life, is a "theory of lived space." One could argue that this is no different from an anthropology of the body, especially if abstractions like de Certeau's are matched with empirical history and ethnography.

The cultural historians Peter Stallybrass and Allon White have enriched the anthropology of the body with a wide-ranging "history of bourgeois disgust." In *The Politics and Poetics of Transgression*, they consider all manner of social phenomena—fairs, pigs, authors, coffee, famous psychopaths, domestic servants, sewers, rats, and much more—as contributions to a four-hundred-year evolution of British bourgeois common sense. In the extract provided here, they demonstrate how nineteenth-century urban reform efforts connect to a transformation of the senses of touch and smell, even as bourgeois vision seeks ever more high and remote outposts from which to exercise a condescending and fascinated gaze. Thus they offer a historical perspective on the distinction between the high, abstract position of the (relatively elite) spectator and the practical position of the "low" and the everyday that interests de Certeau.

Stallybrass and White borrow from anthropology (from Mary Douglas, in particular) the idea that what is socially peripheral (the low, the practical, the dirty) is often symbolically central to the elite rulers of proper discourse. In their reading of the records of nineteenth-century urban reform, they find that "the reformers were central in the construction of the urban geography of the bourgeois Imaginary. As the bourgeoisie produced new forms of regulation and prohibition governing their own bodies, they wrote ever more loquaciously of the body of the Other—of the city's 'scum.'" And this organization of experience into high and low, elite self and declassé other, is not just a matter of symbolism or discursive tropes. It is materialized in the emerging form of the city itself as well as in the most intimate and immediate experience of citizens. Depictions of the poor even by sympathetic chroniclers like Charles Dickens and Friedrich Engels involved a representation of filth that was unstable, "sliding between social, moral and psychic domains. At one level, the mapping of the city in terms of dirt and cleanliness tended to repeat the discourse of colonial anthropology." Thus the vast reorganization of cities which saw the building of sewer systems and tidy suburbs forever altered both the cultural constitution of inner cities and the sense of space that was possible for city-dwellers.

Echoing the early Marx (see part II), Stallybrass and White see nineteenth-century disciplines of policing and sanitation as depending upon a transformation of the senses. In their turn to themes of filth and disgust in Freud's cases, they indicate that any such transformation also involves culturally inflected forms of desire: "It is surely no coincidence, though," they point out, "that the zeal for reform was so often accompanied by a prolonged, fascinated gaze from the bourgeoisie." Reading Freud's analyses as an instance of that fascinated bourgeois gaze, they identify a psychic topography specific to the nineteenth century and to Europe. Their work thus amounts to a social and historical (rather than psychoanalytic) deconstruction of "the symptomatic language of the bourgeois body." In this task, they insist, "it is necessary to reconstruct the mediating topography of the city which always-already inscribes relations of class, gender, and race." This accomplishment is more than a critical denaturalization; it can also serve as a joyful rediscovery of the rats, pigs, sewers, and serving girls who inhabit our most unthinking experiences.

In a quite different way (and in a mode critical of studies of "discourse" such as that of Stallybrass and White), Michael Taussig explores the sensuousness of city life, the kinds of experience and embodiment that stem from the chronic state of distraction in which we live. Working through Walter Benjamin's critical observations on modernity, Taussig joins de Certeau in rejecting abstract contemplative vision in favor of "tactility." Like de Certeau, he seeks a deeper recognition of the "plurality in everydayness." He goes on to argue that "this would be an obvious point, the founding orientation of a sociology of experience, were it not for the peculiar and unexamined ways by which 'the everyday' seems, in the diffuseness of its ineffability, to erase difference." This is no dream of an undifferentiated human sameness that rolls along beneath all our debates and contradictions. Rather Taussig seeks to tease out "the trace of a diffuse commonality [that] promises the possibility of other sorts of nonexploitative solidarities which, in order to exist at all, will have to at some point be based on a common sense of the everyday and, what is more, the ability to sense other everydaynesses." This is a fascinating return to a politics of "common sense" that goes beyond critique to re-enchant the everyday. Thus, "a new magic, albeit secular, finds its everyday home in a certain tactility growing out of distracted vision." And because a tactile knowledge "lies as much in the objects and spaces of observation as in the body and mind of the observer," this is truly a form of experience that goes beyond the body proper to propose a disseminated carnal existence.

Taussig refuses, sometimes a bit hastily, dualistic approaches to what he calls a sociology of experience. He sees the old antinomies of emotion and thought, body and mind, as producing only a "tightening of paradox; an intellectual containment of the body's understanding." Instead he aims at "a more accurate, a more mindful, understanding of the play of mind on body in the everyday." Taussig's (and Benjamin's) notions of tactility and distraction, enriched by a materialist concern with the concrete ingredients of daily life, draw our attention to forms of experience for which we thus far have few respectable analytic concepts.

Judith Farquhar, in her consideration of flavors and experience in modern China, draws on esoteric (but, in China, readily accessible) knowledge of herbal medicines to approach an aspect of lived experience. This extract centers on a story about a "medicinal meals" restaurant serving lunchtime tonics—both nutritional and therapeutic—to men from a rural public security bureau. But it also explores possible relations between knowledge and daily experiences of eating. There is extensive classical and systematic lore about herbal medicines in China, though it is mostly specialists who bear it in mind. The language of flavors shared by medicine and cuisines, however, is flexible and can provide idioms for both experience and action, even for nonspecialists. Farquhar suggests that the pleasure and bitterness of oral experience in China are taken to have powers over the body that go little acknowledged in biomedically inflected experience. Embodied theories of oral consumption like these are integral to an everyday life that "mediates and muddles" the dualistic categories of body and mind, emotion and cognition, experience and action, that have informed much Euro-American social science. Yet it could be argued that anyone anywhere can grasp a form of experience in which enjoyment of flavor can have immediate and possibly lasting results throughout the body. Thus Farquhar's article broaches that common terrain in which bodies themselves might be capable of imagining real difference. She suggests that "we should be able to carnally imagine other lifeworlds, or sensory realms, through an ethnographic description that attends to the concrete and the everyday."

Literary studies have also addressed the materialities of everyday life, as exemplified by Nancy Miller's article "Rereading as a Woman: The Body in Practice." Beginning with an image of a literary critic's wife doing the ironing while he reads to her, Miller adopts an approach to reading (and writing) that both invokes and problematizes the bodies that inhabit their texts. She reflects on how literature authored by men and addressed to a male point of view (even at a time when many readers were women) might be read with some suspicion by women, who can "at least . . . imagine the lady's place, imagine while reading the place of a woman's body." In the case of some classic literature renowned for its wit—the example read at greatest length here is Laclos's *Dangerous Liaisons*—women's bodies are multiply enlisted in a male production of language. Miller analyzes a classic scene in *Dangerous Liaisons* in which Valmont pens a letter to the virtuous Mme de Tourvel while he is in bed with the courtesan Emilie. He then invites his friend Mme de Merteuil to appreciate his written wit before she posts the letter on to its female addressee, who is the only one not privy to the joke. Like the male literary critic who laughs at the physical humiliation of a proud woman in Rabelais, the reader-positioned-as-male is expected to enjoy Valmont's double entendre and, one presumes, his sexual prowess. Yet women readers, imagining at least in some cases the lady's place, might laugh a little less heartily at these appropriations of their sex in the production of a thoroughly male-centered narrative.

Miller is careful not to essentialize women as readers or as critics. Her analysis invokes women and men who are "circumscribed by and in the *practice* of reading and writing" rather than pre-given by nature. By treating reading as an everyday embodied activity, she is able to think in league with historically constituted women holding very material texts in their hands. She even suggests, through this kind of analysis, that they—we—might resist or escape some of the class and gender circumscriptions imagined by texts and their original authors. Reading as an ordinary practice is thus opened up to social variation, to the politics of gender, and to the anthropology of embodiment.

All these authors can be thought of as materialists, concerning themselves with concrete practices and the substantive claims of bodies, buildings, foods, and other visible, tangible entities. They tend to treat texts as palpably present in material worlds—words on pages, the shape of a view from above and from street level, the reflection of a neon sign in a puddle—and thus more of an intervention in history than a mere expression of preexisting meanings. This affectionate involvement in the sensuousness of daily life goes beyond the more traditional Marxist materialisms, which tend to look for determinants in the reduced domains of the economic and the productive. A materialist anthropology of this kind has the potential to move beyond the classic dualisms of male/female, capitalist/precapitalist, public/private, modernity/tradition, but not in the direction of a new idealist transcendence. As the articles collected in this part and elsewhere in the volume attest, a new engagement with the concrete life of the body in its necessary dailiness promises theoretical innovation as well as a re-enchanted empirical field for a more sensuous scholarship.

ADDITIONAL READINGS

Garber, Marjorie. 2000. *Bisexuality and the Eroticism of Everyday Life.* New York: Routledge.

Highmore, Ben, ed. 2002. *The Everyday Life Reader.* New York: Routledge.

Willis, Susan. 1991. *A Primer for Daily Life.* New York: Routledge.

WALKING IN THE CITY

MICHEL DE CERTEAU

Seeing Manhattan from the 110th floor of the World Trade Center. Beneath the haze stirred up by the winds, the urban island, a sea in the middle of the sea, lifts up the skyscrapers over Wall Street, sinks down at Greenwich, then rises again to the crests of Midtown, quietly passes over Central Park and finally undulates off into the distance beyond Harlem. A wave of verticals. Its agitation is momentarily arrested by vision. The gigantic mass is immobilized before the eyes. It is transformed into a texturology in which extremes coincide—extremes of ambition and degradation, brutal oppositions of races and styles, contrasts between yesterday's buildings, already transformed into trash cans, and today's urban irruptions that block out its space. Unlike Rome, New York has never learned the art of growing old by playing on all its pasts. Its present invents itself, from hour to hour, in the act of throwing away its previous accomplishments and challenging the future. A city composed of paroxysmal places in monumental reliefs. The spectator can read in it a universe that is constantly exploding. In it are inscribed the architectural figures of the *coincidatio oppositorum* formerly drawn in miniatures and mystical textures. On this stage of concrete, steel and glass, cut out between two oceans (the Atlantic and the American) by a frigid body of water, the tallest letters in the world compose a gigantic rhetoric of excess in both expenditure and production.

VOYEURS OR WALKERS

To what erotics of knowledge does the ecstasy of reading such a cosmos belong? Having taken a voluptuous pleasure in it, I wonder what is the source of this pleasure of "seeing the whole," of looking down on, totalizing the most immoderate of human texts.

To be lifted to the summit of the World Trade Center is to be lifted out of the city's grasp. One's body is no longer clasped by the streets that turn and return it according to an anonymous law; nor is it possessed, whether as player or played, by the rumble of so many differences and by the nervousness of New

York traffic. When one goes up there, he leaves behind the mass that carries off and mixes up in itself any identity of authors or spectators. An Icarus flying above these waters, he can ignore the devices of Daedalus in mobile and endless labyrinths far below. His elevation transfigures him into a voyeur. It puts him at a distance. It transforms the bewitching world by which one was "possessed" into a text that lies before one's eyes. It allows one to read it, to be a solar Eye, looking down like a god. The exaltation of a scopic and gnostic drive: the fiction of knowledge is related to this lust to be a viewpoint and nothing more.

Must one finally fall back into the dark space where crowds move back and forth, crowds that, though visible from on high, are themselves unable to see down below? An Icarian fall. On the 110th floor, a poster, sphinx-like, addresses an enigmatic message to the pedestrian who is for an instant transformed into a visionary: *It's hard to be down when you're up.*

The desire to see the city preceded the means of satisfying it. Medieval or Renaissance painters represented the city as seen in a perspective that no eye had yet enjoyed. This fiction already made the medieval spectator into a celestial eye. It created gods. Have things changed since technical procedures have organized an "all-seeing power"? The totalizing eye imagined by the painters of earlier times lives on in our achievements. The same scopic drive haunts users of architectural productions by materializing today the utopia that yesterday was only painted. The 1,370-foot-high tower that serves as a prow for Manhattan continues to construct the fiction that creates readers, makes the complexity of the city readable, and immobilizes its opaque mobility in a transparent text.

Is the immense texturology spread out before one's eyes anything more than a representation, an optical artifact? It is the analogue of the facsimile produced, through a projection that is a way of keeping aloof, by the space planner urbanist, city planner or cartographer. The panorama-city is a "theoretical" (that is, visual) simulacrum, in short a picture, whose condition of possibility is an oblivion and a misunderstanding of practices. The voyeur-god created by this fiction, who, like Schreber's God, knows only cadavers, must disentangle himself from the murky intertwining daily behaviors and make himself alien to them.

The ordinary practitioners of the city live "down below," below the thresholds at which visibility begins. They walk—an elementary form of this experience of the city; they are walkers, *Wandersmänner*, whose bodies follow the thicks and thins of an urban "text" they write without being able to read it. These practitioners make use of spaces that cannot be seen; their knowledge of them is as blind as that of lovers in each other's arms. The paths that correspond in this intertwining, unrecognized poem in which each body is an element signed by many others, elude legibility. It is as though the practices organizing a bustling city were characterized by their blindness. The networks of these moving, intersecting writings compose a manifold story that has neither author nor spectator, shaped out of fragments of trajectories and alterations of spaces: in relation to representations, it remains daily and indefinitely other.

Escaping the imaginary totalizations produced by the eye, the everyday has a certain strangeness that does not surface, or whose surface is only its upper limit, outlining itself against the visible. Within this ensemble, I shall try to locate the practices that are foreign to the "geometrical" or "geographical" space of visual, panoptic, or theoretical constructions. These practices of space refer to a specific form of *operations* ("ways of operating"), to "another spatiality" (an "anthropological," poetic and mythic experience of space), and to an *opaque and blind* mobility characteristic of the bustling city. A *migrational*, or metaphorical, city thus slips into the clear text of the planned and readable city.

FROM THE CONCEPT OF THE CITY TO URBAN PRACTICES

The World Trade Center is only the most monumental figure of Western urban development. The atopia–utopia of optical knowledge has long had the ambition of surmounting and articulating the contradictions arising from urban agglomeration. It is a question of managing a growth of human agglomeration or accumulation. "The city is a huge monastery," said Erasmus. Perspective vision and prospective vision constitute the twofold projection of an opaque past and an uncertain future onto a surface that can be dealt with. They inaugurate (in the sixteenth century?) the transformation of the urban *fact* into the *concept* of a city. Long before the concept itself gives rise to a particular figure of history, it assumes that this fact can be dealt with as a unity determined by an urbanistic *ratio*. Linking the city to the concept never makes them identical, but it plays on their progressive symbiosis: to plan a city is both to *think the very plurality* of the real and to make that way of thinking the plural *effective*; it is to know how to articulate it and be able to do it. . . .

THE RETURN OF PRACTICES

The Concept-city is decaying. Does that mean that the illness afflicting both the rationality that founded it and its professionals afflicts the urban populations as well? Perhaps cities are deteriorating along with the procedures that organized them. But we must be careful here. The ministers of knowledge have always assumed that the whole universe was threatened by the very changes that affected their ideologies and their positions. They transmute the misfortune of their theories into theories of misfortune. When they transform their bewilderment into "catastrophes," when they seek to enclose the people in the "panic" of their discourses, are they once more necessarily right?

Rather than remaining within the field of a discourse that upholds its privilege by inverting its content (speaking of catastrophe and no longer of progress), one can try another path: one can analyze the microbe-like, singular and plural practices which an urbanistic system was supposed to administer or suppress, but which have outlived its decay; one can follow the swarming activity of these procedures that, far from being regulated or eliminated by panoptic administration, have reinforced themselves in a proliferating illegitimacy, developed and insinuated themselves into the networks of surveillance, and combined in accord with unreadable but stable tactics to the point of constituting everyday regulations and surreptitious creativities that are merely concealed by the frantic mechanisms and discourses of the observational organization.

This pathway could be inscribed as a consequence, but also as the reciprocal, of Foucault's analysis of the structures of power. He moved it in the direction of mechanisms and technical procedures, "minor instrumentalities" capable, merely by their organization of "details," of transforming a human multiplicity into a "disciplinary" society and of managing, differentiating, classifying, and hierarchizing all deviances concerning apprenticeship, health, justice, the army, or work. "These often minuscule ruses of discipline," these "minor but flawless" mechanisms, draw their efficacy from a relationship between procedures and the space that they redistribute in order to make an "operator" out of it. But what *spatial practices* correspond, in the area where discipline is manipulated, to these apparatuses that produce a disciplinary space? In the present conjuncture, which is marked by a contradiction between the collective mode of administration and an individual mode of reappropriation, this question is no less important, if one admits that spatial practices in fact secretly structure the determining conditions of social life. I would like to follow out a few of these multiform, resistant, tricky and stubborn procedures that elude discipline without being outside the field in which it is exercised, and which should lead us to a theory of everyday practices, of lived space, of the disquieting familiarity of the city.

THE CHORUS OF IDLE FOOTSTEPS

The goddess can be recognized by her step. (Virgil, *Aeneid*, I, 405)

Their story begins on ground level, with footsteps. They are myriad, but do not compose a series. They cannot be counted because each unit has a qualitative character: a style of tactile apprehension and kinesthetic appropriation. Their swarming mass is an innumerable collection of singularities. Their intertwined paths give their shape to spaces. They weave places together. In that respect, pedestrian movements form one of these "real systems whose existence in fact makes up the city." They are not localized; it is rather they that spatialize. They are no more inserted within a container than those Chinese characters speakers sketch out on their hands with their fingertips.

It is true that the operations of walking on can be traced on city maps in such a way as to transcribe their paths (here well-trodden, there very faint) and their trajectories (going this way and not that). But these thick or thin curves only refer, like words, to the absence of what has passed by. Surveys of routes miss what was: the act itself of passing by. The operation of walking, wandering, or "window shopping," that is, the activity of passers-by, is transformed into points that draw a totalizing and reversible line on the map. They allow us to grasp only a relic set in the nowhen of a surface of projection. Itself visible, it has the effect of making invisible the operation that made it possible. These fixations constitute procedures for forgetting. The trace left behind is substituted for the practice. It exhibits the (voracious) property that the geographical system has of being able to transform action into legibility, but in doing so it causes a way of being in the world to be forgotten.

PEDESTRIAN SPEECH ACTS

A comparison with the speech act will allow us to go further and not limit ourselves to the critique of graphic representations alone, looking from the shores of legibility toward an inaccessible beyond. The act of walking is to the urban system what the speech act is to language or to the statements uttered. At the most elementary level, it has a triple "enunciative" function: it is a process of *appropriation* of the topographical system on the part of the pedestrian (just as the speaker appropriates and takes on the language); it is a spatial acting-out of the place (just as the speech act is an acoustic acting-out of language); and it implies *relations* among differentiated positions, that is, among pragmatic "contracts" in the form of movements (just as verbal enunciation is an "allocution," "posits another opposite" the speaker and puts contracts between inter-locutors into action). It thus seems possible to give a preliminary definition of walking as a space of enunciation.

We could moreover extend this problematic to the relations between the act of writing and the written text, and even transpose it to the relationships between the "hand" (the touch and the tale of the paintbrush [*le et la geste du pinceau*]) and the finished painting (forms, colors, etc.). At first isolated in the area of verbal communication, the speech act turns out to find only one of its applications there, and its linguistic modality is merely the first determination of a much more general distinction between the *forms used* in a system and the *ways of using* this system (i.e., *rules*), that is, between two "different worlds," since "the same things" are considered from two opposite formal viewpoints.

Considered from this angle, the pedestrian speech act has three characteristics which distinguish it at the outset from the spatial system: the present, the discrete, the "phatic."

First, if it is true that a spatial order organizes an ensemble of possibilities (e.g., by a place in which one can move) and interdictions (e.g., by a wall that prevents one from going further), then the walker actualizes some of these possibilities. In that way, he makes them exist as well as emerge. But he also moves them about and he invents others, since the crossing, drifting away, or improvisation of walking privilege, transform or abandon spatial elements. Thus Charlie Chaplin multiplies the possibilities of his cane: he does other things with the same thing and he goes beyond the limits that the determinants of the object set on its utilization. In the same way, the walker transforms each spatial signifier into something else. And if on the one hand he actualizes only a few of the possibilities fixed by the constructed order (he goes only here and not there), on the other he increases the number of possibilities (for example, by creating shortcuts and detours) and prohibitions (for example, he forbids himself to take paths generally considered accessible or even obligatory). He thus makes a selection. "The user of a city picks out certain fragments of the statement in order to actualize them in secret."

He thus creates a discreteness, whether by making choices among the signifiers of the spatial "language" or by displacing them through the use he makes of them. He condemns certain places to inertia or disappearance and composes with others spatial "turns of phrase" that are "rare," "accidental" or illegitimate. But that already leads into a rhetoric of walking.

In the framework of enunciation, the walker constitutes, in relation to his position, both a near and a far, a *here* and a *there*. To the fact that the adverbs *here* and *there* are the indicators of the locutionary seat in verbal communication —a coincidence that reinforces the parallelism between linguistic and pedestrian enunciation—we must add that this location (*here–there*) (necessarily implied by walking and indicative of a present appropriation of space by an "I") also has the function of introducing an other in relation to this "I" and of thus establishing a conjunctive and disjunctive articulation of places. I would stress particularly the "phatic" aspect, by which I mean the function, isolated by Malinowski and Jakobson, of terms that initiate, maintain, or interrupt contact, such as "hello," "well, well," etc. Walking, which alternately follows a path and has followers, creates a mobile organicity in the environment, a sequence of phatic *topoi*. And if it is true that the phatic function, which is an effort to ensure communication, is already characteristic of the language of talking birds, just as it constitutes the "first verbal function acquired by children," it is not surprising that it also gambols, goes on all fours, dances, and walks about, with a light or heavy step, like a series of "hellos" in an echoing labyrinth, anterior or parallel to informative speech.

The modalities of pedestrian enunciation which a plane representation on a map brings out could be analyzed. They include the kinds of relationship this enunciation entertains with particular paths (or "statements") by according them a truth value ("alethic" modalities of the necessary, the impossible, the possible, or the contingent), an epistemological value ("epistemic" modalities of the certain, the excluded, the plausible, or the questionable) or finally an ethical or legal value ("deontic" modalities of the obligatory, the forbidden, the permitted, or the optional). Walking affirms, suspects, tries out, transgresses, respects, etc., the trajectories it "speaks." All the modalities sing a part in this chorus, changing from step to step, stepping in through proportions, sequences, and intensities which vary according to the time, the path taken and the walker. These enunciatory operations are of an unlimited diversity. They therefore cannot be reduced to their graphic trail. . . .

CREDIBLE THINGS AND MEMORABLE THINGS: HABITABILITY

By a paradox that is only apparent, the discourse that makes people believe is the one that takes away what it urges them to believe in, or never delivers what it promises. Far from expressing a void or describing a tack, it creates such. It makes room for a void. In that way, it opens up clearings; it "allows" a certain play within a system of defined places. It "authorizes" the production of an area of free play (*Spielraum*) on a checkerboard that analyzes and classifies identities. It makes places habitable. On these grounds, I call such discourse a "local authority." It is a crack in the system that saturates places with signification and indeed so reduces them to this signification that it is "impossible to breathe in them." It is a symptomatic tendency of functionalist totalitarianism (including its programming of games and celebrations) that it seeks precisely to eliminate these local authorities, because they compromise the univocity of the system. Totalitarianism attacks what it quite correctly calls *superstitions*: supererogatory semantic overlays that insert themselves "over and above" and "in excess," and annex to a past or poetic realm a part of the land the promoters of technical rationalities and financial profitabilities had reserved for themselves.

Ultimately, since proper names are already "local authorities" or "superstitions," they are replaced by numbers: on the telephone, one no longer dials *Opera*, but 073. The same is true of the stories and legends that haunt urban space like superfluous or additional inhabitants. They are the object of a witch-hunt, by the very logic of the techno-structure. But their extermination (like the extermination of trees, forests, and hidden places in which such legends live) makes the city a "suspended symbolic order." The habitable city is thereby annulled. Thus, as a woman from Rouen put it, no, here "there isn't any place special, except for my own home, that's all. . . . There isn't anything." Nothing "special": nothing that is marked, opened up by a memory or a story, signed by something or someone else. Only the cave of the home remains believable, still open for a certain time to legends, still full of shadows. Except for that, according to another city-dweller, there are only "places in which one can no longer believe in anything."

It is through the opportunity they offer to store up rich silences and wordless stories, or rather through their capacity to create cellars and garrets everywhere, that local legends (*legenda*: what is *to be read*, but also what *can be read*) permit exits, ways of going out and coming back in, and thus habitable spaces. Certainly walking about and traveling substitute for exits, for going away and coming back, which were formerly made available by a body of legends that places nowadays lack. Physical moving about has the itinerant function of yesterday's or today's "superstitions." Travel (like walking) is a substitute for the legends that used to open up space to something different. What does travel ultimately produce if it is not, by a sort of reversal, "an exploration of the deserted places of my memory," the return to nearby exoticism by way of a detour through distant places, and the "discovery" of relics and legends: "fleeting visions of the French countryside," "fragments of music and poetry," in short, something like an "uprooting in one's origins" (Heidegger)? What this walking exile produces is precisely the body of legends that is currently lacking in one's own vicinity; it is a fiction which moreover has the double characteristic, like dreams or pedestrian rhetoric, of being the effect of displacements and condensations. As a corollary, one can measure the importance of these signifying practices (to tell oneself legends) as practices that invent spaces.

From this point of view, their contents remain revelatory, and still more so is the principle that organizes them. Stories about places are makeshift things. They are composed with the world's debris. Even if the literary form and the actantial schema of "superstitions" correspond to stable models whose structures and combinations have often been analyzed over the past thirty years, the materials (all the rhetorical details of their "manifestation") are furnished by the leftovers from nominations, taxonomies, heroic or comic predicates, etc., that is, by fragments of scattered semantic places. These heterogeneous and even contrary elements fill the homogeneous form of the story. Things *extra* and *other* (details and excesses coming from elsewhere) insert themselves into the accepted framework, the imposed order. One thus has the very relationship between spatial practices and the constructed order. The surface of this order is everywhere punched and torn open by ellipses, drifts, and leaks of meaning: it is a sieve-order.

The verbal relics of which the story is composed, being tied to lost stories and opaque acts, are juxtaposed in a collage where their relations are not thought, and for this reason they form a symbolic whole. They are articulated by lacunae. Within the structured space of the text, they thus produce antitexts, effects of dissimulation and escape, possibilities of moving into other landscapes, like cellars and bushes: "*ô massifs, ô pluriels.*" Because of the process of dissemination that they open up, stories differ from *rumors* in that the latter are always injunctions, initiators and results of a leveling of space, creators of common movements that reinforce an order by adding an activity of making people believe things to that of making people do things. Stories diversify, rumors totalize. If there is still a certain oscillation between them, it seems that today there is rather a stratification: stories are becoming private and sink into the secluded places in neighborhoods, families, or individuals, while the rumors propagated by the media cover everything and, gathered under the figure of the City, the masterword of an anonymous law, the substitute for all proper names, they wipe out or combat any superstitions guilty of still resisting the figure.

The dispersion of stories points to the dispersion of the memorable as well. And in fact memory is a sort of anti-museum: it is not localizable. Fragments of it come out in legends. Objects and words also have hollow places in which a past sleeps, as in the everyday acts of walking, eating, going to bed, in which ancient revolutions slumber. A memory is only a Prince Charming who stays just long enough to awaken the Sleeping Beauties of our wordless stories. "*Here*, there used to be a bakery." "*That's* where old lady Dupuis used to live." It is striking here that the places people live in are like the presences of diverse absences. What can be seen designates what is no longer there: "you *see*, here there used to be. . . ." but it can no longer be seen. Demonstratives indicate the invisible identities of the visible: it is the very definition of a place, in fact, that it is composed by these series of displacements and effects among the fragmented strata that form it and that it plays on these moving layers.

"Memories tie us to that place. . . . It's personal, not interesting to anyone else, but after all that's what gives a neighborhood its character." There is no place that is not haunted by many different spirits hidden there in silence, spirits one can "invoke" or not. Haunted places are the only ones people can live in—and this inverts the schema of the *Panopticon*. But like the gothic sculptures of kings and queens that once adorned Notre-Dame and have been buried for two centuries in the basement of a building in the rue de la Chaussée-d'Antin, these "spirits," themselves broken into pieces in like manner, do not *speak* any more than they *see*. This is a sort of knowledge that remains silent. Only hints of what is known but unrevealed are passed on "just between you and me."

Places are fragmentary and inward-turning histories, pasts that others are not allowed to read, accumulated times that can be unfolded but like stories held in reserve, remaining in an enigmatic state, symbolizations encysted in the pain or pleasure of the body. "I feel good here": the well-being underexpressed in the language it appears in like a fleeting glimmer is a spatial practice.

TACTILITY AND DISTRACTION

MICHAEL TAUSSIG

Now, says Hegel, all discourse that remains discourse ends in *boring* man.
—Alexander Kojeve, *Introduction to the Reading of Hegel*

Quite apart from its open invitation to entertain a delicious anarchy, exposing principles no less than dogma to the white heat of daily practicality and contradiction, there is surely plurality in everydayness. My everyday has a certain routine, doubtless, but it is also touched by a deal of unexpectedness, which is what many of us like to think of as essential to life, to a metaphysics of life, itself. And by no means can my everyday be held to be the same as vast numbers of other people's in this city of New York, those who were born here, those who have recently arrived from other everydays far away, those who have money, those who don't. This would be an obvious point, the founding orientation of a sociology of experience, were it not for the peculiar and unexamined ways by which "the everyday" seems, in the diffuseness of its ineffability, to erase difference in much the same way as do modern European-derived notions of the public and the masses.

This apparent erasure suggests the trace of a diffuse commonality in the commonweal so otherwise deeply divided, a commonality that is no doubt used to manipulate consensus but also promises the possibility of other sorts of nonexploitative solidarities which, in order to exist at all, will have to at some point be based on a common sense of the everyday and, what is more, the ability to sense other everydaynesses.

But what sort of sense is constitutive of this everydayness? Surely this sense includes much that is not sense so much as sensuousness, an embodied and somewhat automatic "knowledge" that functions like peripheral vision, not studied contemplation, a knowledge that is imageric and sensate rather than ideational; as such it not only challenges practically all critical practice, across the board, of academic disciplines but is a knowledge that lies as much in the objects and spaces of observation as in the body and mind of the observer. What's more, this sense has an activist, constructivist bent; not so much contemplative as it is caught in *media res* working on, making anew, amalgamating, acting and reacting. We are thus mindful of Nietzsche's notion of the senses as bound to their object as much as their organs of reception, a fluid bond to be sure in which, as he says, "seeing becomes seeing *something*." For many of us, I submit, this puts the study of ideology, discourse, and popular culture in a somewhat new light. Indeed, the notion of "studying," innocent in its unwinking ocularity, may itself be in for some rough handling too.

I was reminded of this when as part of my everyday I bumped into Jim in the hallway of PS 3 (New York City Public School Number Three) where he and I were dropping off our children. In the melee of streaming kids and parents, he was carrying a bunch of small plastic tubes and a metal box, which he told me was a pump, and he was going to spend the morning making a water fountain for the class of which his daughter, age eight, was part. She, however, was more interested in the opportunity for the kids to make moulds of their cupped hands and then convert the moulds into clam shells for the fountain. I should add that Jim and his wife are sculptors, and their home is also their workplace, so Petra, their daughter, probably has an unusually developed everyday sense of sculpting.

It turned out that a few days back Jim had accompanied the class to the city's aquarium in Brooklyn which, among other remarks, triggered the absolutely everyday but continuously fresh insight, on my part as much as his, that here we are, so enmeshed in the everydayness of the city that we rarely bother to see its sights, such as the aquarium. "I've lived here all of seventeen years," he told me, "and never once been there or caught the train out that way." And he marveled at the things he'd seen at the station before the stop for the aquarium—it was a station that had played a prominent part in a Woody Allen film. He was especially struck by the strange script used for public signs. And we went on to complete the thought that when we were living in other places, far away, we would come to the city with a program of things to see and do, but now, living every day in the shadow and blur of all those particular things, we never saw them any more, imagining, fondly, perhaps, that they were in some curious way part of us, as we were part of them. But now Jim and Petra were back from the visit to the aquarium. He was going to make a fountain, and she was going to make moulds of hands that would become clam shells.

"The revealing presentations of the big city," wrote Walter Benjamin in his uncompleted *Passagenwerk*, "are the work of those who have traversed the city absently, as it were, lost in thought or worry." And in his infamously popular and difficult essay, "The Work of Art in the Age of Mechanical Reproduction," written in the mid-1930s, he drew a sharp distinction between contemplation and distraction. He wants to argue that contemplation—which is what academicism is all about—is the studied, eyefull, aloneness with and absorption into the "aura" of the always aloof, always distant, object. The ideal-type for this could well be the worshipper alone with God, but it was the art-work (whether cult object or bourgeois "masterpiece") before the invention of the camera and the movies that Benjamin had in mind. On the other hand, "distraction" here refers to a very different apperceptive mode, the type of flitting and barely conscious peripheral vision perception unleashed with great vigor by modern life at the crossroads of the city, the capitalist market, and modern technology. The ideal-type here would not be God but movies and advertising, and its field of expertise is the modern everyday.

For here not only the shock-rhythm of modernity so literally expressed in the motion of the business cycle, the stock exchange, city traffic, the assembly line and Chaplin's walk, but also a new magic, albeit secular, finds its everyday home in a certain tactility growing out of distracted vision. Benjamin took as a cue here Dadaism and architecture, for Dadaism not only stressed the uselessness of its work for contemplation, but that its work "became an instrument of ballistics. It hit the spectator like a bullet, it happened to him, thus acquiring a tactile quality." He went on to say that Dadaism thus promoted a demand for film, "the distracting element of which," and I quote here for emphasis, "is also primarily tactile, being based on changes of place and focus which periodically assault the spectator." As for architecture, it is especially instructive because it has served as the prototype over millennia not for perception by the contemplative individual, but instead by the distracted collectivity. To the question "How in our everyday lives do we know or perceive a building?" Benjamin answers through usage, meaning, to some crucial extent, through touch, or better still, we might want to say, by proprioception, and this to the degree that this tactility, constituting habit, exerts a decisive impact on optical reception.

Benjamin set no small store by such habitual, or everyday, knowledge. The tasks facing the perceptual apparatus at turning points in history, cannot, he asserted, be solved by optical, contemplative, means, but only gradually, by habit, under the guidance of tactile appropriation. It was this everyday tactility of knowing which fascinated him and which I take to be one of his singular contributions to social philosophy, on a par with Freud's concept of the unconscious.

For what came to constitute perception with the invention of the nineteenth-century technology of optical reproduction of reality was not what the unaided eye took for the real. No. What was revealed was the *optical unconscious*—a term that Benjamin willingly allied with the psychoanalytic unconscious but which, in his rather unsettling way, he so effortlessly confounded subject with object such that the unconscious at stake here would seem to reside more in the object than in the perceiver. Benjamin had in mind both camera still shots and the movies, and it was the ability to enlarge, to frame, to pick out detail and form unknown to the naked eye, as much as the capacity for montage and shocklike abutment of dissimilars, that constituted this optical unconscious which, thanks to the camera, was brought to light for the first time in history. And here again the connection with tactility is paramount, the optical dissolving, as it were, into touch and a certain thickness and density, as where he writes that photography reveals "the physiognomic aspects of visual worlds which dwell in the smallest things, meaningful yet covert enough to find a hiding place in waking dreams, but which, enlarged and capable of formulation, make the difference between technology and magic visible as a thoroughly historical variable." Hence this tactile optics, this physiognomic aspect of visual worlds, was critically important because it was otherwise inconspicuous, dwelling neither in consciousness nor in sleep, but in waking dreams. It was a crucial part of a more exact relation to the objective world, and thus it could not but problematize consciousness of that world, while at the same time intermingling fantasy and hope, as in dream, with waking life. In rewiring seeing as tactility, and hence as habitual knowledge, a sort of technological or secular magic was brought into being and sustained. It displaced the earlier magic of the aura of religious and cult works in a pretechnological age and did so by a process that is well worth our attention, a process of demystification *and* reenchantment, precisely, as I understand it, Benjamin's own self-constituting and contradictorily montaged belief in radical, secular politics *and* messianism, as well as his own mimetic form of revolutionary poetics.

For if Adorno reminds us that in Benjamin's writings "thought presses close to its object, as if through touching, smelling, tasting, it wanted to transform itself," we have also to remember that mimesis was a crucial feature for Benjamin and Adorno, and it meant both copying *and* sensuous materiality—what Frazer in his famous chapter on magic in *The Golden Bough*, coming out of a quite different and far less rigorous philosophic tradition, encompassed as imitative or homeopathic magic, on the one side, and contagious magic, on the other. Imitative magic involves ritual work on the copy (the wax figurine, the drawing or the photograph), while in contagious magic the ritualist requires material substance (such as hair, nail parings, etc.) from the person to be affected. In the multitude of cases that Frazer presented in the 160-odd pages he dedicated to the "principles of magic," these principles of copy and substance are often found to be harnessed together, as with the Malay charm made out of body exuviae of the victim sculpted into his likeness with wax and then slowly scorched for seven nights while intoning, "It is not wax that I am scorching, it is the liver, heart, and spleen of So-and-so that I scorch," and this type of representation hitching likeness to substance is borne out by ethnographic research throughout the twentieth century.

This reminder from the practice of that art form known as "magic" (second only to advertising in terms of its stupendous ability to blend aesthetics with practicality), that mimesis implies *both* copy and substantial connection, *both* visual replication and material transfer, not only neatly parallels Benjamin's insight that visual perception as enhanced by new optical copying technology has a decisively material, tactile, quality, but underscores his specific question as to what happens to the apparent withering of the mimetic faculty with the growing up of the Western child and the world historical cultural revolution we can allude to as Enlightenment, it being his clear thesis that children, anywhere, any time, and people in ancient times and so-called primitive societies are endowed by their circumstance with considerable miming prowess. Part of his answer to the question as to what happens to the withering-away of the mimetic faculty is that it is precisely the function of the new technology of copying reality, meaning above all the camera, to reinstall that mimetic prowess in modernity.

Hence a powerful film criticism which, to quote Paul Virilio quoting the New York video artist Nam June Paik, "Cinema isn't I see, it's I fly," or Dziga Vertov's camera in perpetual movement, "I fall and I fly at one with the bodies falling or rising through the air," registering not merely our sensuous blending with filmic imagery, the eye acting as a conduit for our very bodies being absorbed by the filmic image, but the resurfacing of a vision-mode at home in the pre-Oedipal economy of the crawling infant, the eye grasping, as Gertrude Koch once put it, at what the hand cannot reach.

And how much more might this be the case with advertising, quintessence of America's everyday? In "This Space For Rent," a fragment amid a series of fragments entitled "One Way Street," written between 1925 and 1928, Benjamin anticipated the themes of his essay on mechanical reproduction, written a decade later, claiming it was a waste of time to lament the loss of distance necessary for criticism. For now the most real, the mercantile gaze into the heart of things, is the advertisement, and this "abolishes the space where contemplation moved and all but hits us between the eyes with timings as a car, growing to gigantic proportions, careens at us out of a dim screen." To this tactility of a hit between the eyes is added what he described as "the insistent, jerky nearness" with which commodities were thus hurtled, the overall effect dispatching "matter-of-factness" by the new, magical world of the optical unconscious, as huge leathered cowboys, horses, cigarettes, toothpaste, and perfect women straddle walls of buildings, subway cars, bus stops, and our living rooms via TV, so that sentimentality, as Benjamin put it, "is restored and liberated in American style, just as people whom nothing moves or touches any longer are taught to cry again by films." It is money that moves us to these things whose power lies in the fact that they operate upon us viscerally. Their warmth stirs sentient springs. "What in the end makes advertisements so superior to criticism?" asks Benjamin. "Not what the moving red neon sign says—but the fiery pool reflecting it in the asphalt."

This puts the matter of factness of the everyday on a new analytic footing, one that has for too long been obscured in the embrace of a massive tradition of cultural and sociological analysis searching in vain for grants that would give it distance and perspective. Not what the neon says, but the fiery pool reflecting it in the asphalt; not language, but image; and not just the image but its tactility and the new magic thereof with the transformation of roadway parking-lot bitumen into legendary lakes of fire-ringed prophecy so that once again we cry and, presumably, we buy, just as our ability to calculate value is honed to the razor's edge. It is not a question, therefore, of whether or not we can follow de Certeau and combat strategies with everyday tactics that fill with personal matter the empty signifiers of postmodernity, because the everyday is a question not of universal semiotics but of capitalist mimetics. Nor, as I understand it, is this the Foucauldian problem of being programmed into subjecthood by discursive regimes, for it is the sentient reflection in the fiery pool, its tactility, not what the neon sign says, that matters, all of which puts reading, close or otherwise, literal or metaphoric, in another light of dubious luminosity.

This is not to indulge in the tired game of emotion versus thought, body versus mind, recycled by current academic fashion into concern with "the body" as key to wisdom. For where can such a program end but in the tightening of paradox; an intellectual containment of the body's understanding? What we aim at is a more accurate, a more mindful, understanding of the play of mind on body in the everyday and, as regards academic practice nowhere are the notions of tactility and distraction more obviously important than in the need to critique what I take to be a dominant critical practice which could be called the "allegorizing" mode of reading ideology into events and artifacts, cockfights and carnivals, advertisements and film, private and public spaces, in which the surface phenomenon, as in allegory, stands as a cipher for uncovering horizon after horizon of otherwise obscure systems of meanings. This is not merely to argue that such a mode of analysis is simpleminded in its search for "codes" and manipulative because it superimposes meaning on "the natives' point of view." Rather, as I now understand this practice of reading, its very understanding of "meaning" is uncongenial; its weakness lies in its assuming a contemplative individual when it should, instead, assume a distracted collective reading with a tactile eye. This I take to be Benjamin's contribution, profound and simple, novel yet familiar, to the analysis of the everyday, and unlike the readings we have come to know of everyday life, his has the strange and interesting property of being cut, so to speak, from the same cloth as that which it raises to self-awareness. For his writing, which is to say the very medium of his analysis, is constituted by a certain tactility, by what we could call the objectness of the object, such that (to quote from the first paragraph of his essay on the mimetic faculty) "His gift of seeing resemblances is nothing other than a rudiment of the powerful compulsion in former times to become and behave like something else." This I take to be not only the verbal form of the "optical unconscious," but a form which, in an age wherein analysis does little more than reconstitute the obvious, is capable of surprising us with the flash of a profane illumination.

And so my attention wanders away from the Museum of Natural History on Central Park, upon which so much allegorical "reading," as with other museums, has been recently expended, back to the children and Jim at the aquarium. It is of course fortuitous, overly fortuitous you will say, for my moral concerning tactility and distraction that Jim is a sculptor, but there is the fact of the matter. And I cannot but feel that in being stimulated by the "meaning" of the aquarium to reproduce with the art of mechanical reproduction its watery wonderland by means of pumps and plastic tubes, Jim's tactile eye and ocular grasp have been conditioned by the distractedness of the collective of which he was part, namely the children. Their young eyes have blended a strangely dreamy quality to the tactility afforded the adult eye by the revolution in modern means of copying reality, such that while Jim prefers a fountain, Petra suggests moulds of kids' hands that will be its clam shells.

THE CITY: THE SEWER, THE GAZE, AND

THE CONTAMINATING TOUCH

PETER STALLYBRASS AND ALLON WHITE

In the nineteenth century [a] fear of differences that "have no law, no meaning, and no end" was articulated above all through the "body" of the city: through the separations and inter-penetrations of the suburb and the slum, of grand buildings and the sewer, of the respectable classes and the lumpenproleteriat (what Marx called "the whole indefinite, disintegrated mass thrown hither and thither"). In this chapter, we will trace the transcodings of psychic desire, concepts of the body, and the structuring of the social formation across the city's topography as this was inscribed in the parliamentary report, the texts of social reform, the hysterical symptom of the psychoanalyst's patient, as well as in the poet's journal and the novel.

It was in the reforming text as much as in the novel that the nineteenth-century city was produced as the locus of fear, disgust and fascination. [Sir Edwin] Chadwick's *Report . . . on an Inquiry into the Sanitary Conditions of the Labouring Population of Great Britain* (1842), for instance, was an instant best-seller, and more than 10,000 copies were distributed free. In Chadwick, in [Henry] Mayhew, in countless Victorian reformers, the slum, the laboring poor, the prostitute, the sewer, were re-created for the bourgeois study and drawing-room as much as for the urban council chamber. Indeed, the re-formers were central in the construction of the urban geography of the bour-geois Imaginary. As the bourgeoisie produced new forms of regulation and prohibition governing their own bodies, they wrote ever more loquaciously of the body of the Other—of the city's "scum."

The body of the Other produced contradictory responses. Certainly, it was to be *surveyed*, as [Sir Robert] Southwell and [William] Wordsworth surveyed the fair, from "some high window" or superior position. Chadwick insisted that satisfactory regulation depended upon breaking down those architectural barriers which kept the immoral "secluded from superior inspection and from common observation." At the same time, new forms of propriety must pene-trate and subjugate the recalcitrant body: hence, the insistence upon "regularity of diet," "clean or respectable clothes," even *drill-masters* to restrain "bodily irritability, and thence uncontrollable mental irritability." Chadwick argued that calisthenics, "which to the common eye are expensive and misbefitting luxuries, are in the experience of Sanitary Science, 'formatives,' necessary to impart mobility to all parts of the frame, to get rid of clumsiness and to augment health and productive force:—the objects of an economical administration." But even as the bourgeoisie speculated upon new regimes for the body, they obsessively returned to the "unutterable horrors" of the city, where there were no "architectural barriers or protections of decency and propriety"; to the imaginary place whose empirical existence a Scottish police superintendent asserted was a place where there lived "a thousand children who have no names whatever, or only nicknames, like dogs."

SLUM AND SEWER

The relation of social division and exclusion to the production of desire emerges with great clarity in the nineteenth-century city. New boundaries between high and low, between aristocrat and rag-picker, were there simultaneously established and transgressed. On the one hand, the slum was separated from the suburb: "the undrained clay beneath the slums oozed with cesspits and sweated with fever; the gravelly heights of the suburbs were dotted with springs and bloomed with health." On the other hand, the streets were a "minglemangle," "a hodge-podge," where the costermonger, the businessman, the prostitute, the clerk, the nanny and the crossing-sweeper jostled for place.

Henry Mayhew's *London Labour and the London Poor* (1861) is traversed and fractured by contradictory formulations of these relations of high to low. His work begins with a chapter "Of wandering tribes in general" in which he separates out two distinct "races": "the wanderers and the civilized tribes." Mayhew's definition of the nomadic is a demonized version of what Bakhtin later defined as the grotesque. The nomad, Mayhew writes, is distinguished from the civilized

> by his repugnance to regular and continuous labour—by his want of providence in laying up store for the future . . .—by his passion for stupefying herbs and roots, and, when possible, for intoxicating fermented liquors . . .—by his love of libidinous dances . . .—by the looseness of his notion as to property—by the absence of chastity among his women, and his disregard of female honour—and lastly, by his vague sense of religion. (Mayhew 1861–62: I, 2)

Mayhew constructs the nomad in terms of his desires ("passion," "love," "pleasure") and in terms of his rejections or ignorance ("repugnance," "want," "looseness," "absence," "disregard"). In the slum, the bourgeois spectator surveyed and classified *his own antithesis*.

The nomads are improvident: "like all who make a living as it were by a game of chance, plodding, carefulness, and saving habits cannot be reckoned among their virtues." Their habits are "not domestic": for those who inhabit the markets, streets, beer shops, dancing-rooms and theaters, "home has few attractions." They are indifferent to marriage: "[only] one-tenth—at the outside one-tenth—of the couples living together and carrying on the costermongering trade are married. Of the rights of 'legitimate' and 'illegitimate' children the costermongers understand nothing, and account it a mere waste of time and money to go through the ceremony of wedlock." They are opposed to constituted authority and above all to the police: "The hatred of a costermonger to a 'peeler' is intense, and with their opinion of the police, all the more ignorant unite that of the governing power"; "in their continual warfare with the force, they resemble many savage nations, from the cunning and treachery they use." They are ignorant of religion: "not 3 in 100 costermongers had ever been in the interior of a church, or any place of worship, or knew what was meant by Christianity," whilst a 9-year-old mud-lark "did not know what religion was. God was God, he said. He had heard he was good, but he didn't know what good he was to him." Above all, the "nomads" confront the bourgeois observer as a spectacle of filth.

As the nomads transgress all settled boundaries of "home," they simultaneously map out the area which lies beyond cleanliness. However much Mayhew attempts to separate "moral wickedness" from "physical filthiness" the very categories of his work (excluding as they do the railway man, the factory worker and the domestic servant) foreground the connections between topography, physical appearance and morality. The emergent proletariat is displaced by the lumpenproleteriat whom Marx describes in *The Eighteenth Brumaire*:

> Alongside decayed *roués* with dubious means of subsistence and of dubious origins, alongside ruined and adventurous off-shoots of the bourgeoisie, were vagabonds, discharged jailbirds, escaped galley slaves, swindlers, mountebanks, *lazzaroni*, pickpockets, tricksters, gamblers, *maquereaus*, brothel-keepers, porters, *literati*, organ-grinders, ragpickers, knife-grinders, tinkers, beggars,—in short, the whole indefinite, disintegrated mass thrown hither and thither, which the French call *la bohème*. (Marx 1852 [1951], I: 267)

Like Mayhew, Marx here concentrates on those who are marginal to the forces of production. And, paradoxically, it is this very group which stimulates his linguistic productivity. Marx ransacks French, Latin and Italian in his attempt to grasp this "indefinite, disintegrated mass." Marginal in terms of production, the lumpenproletariat are yet central to the "Imaginary," the object of disgust and fascination.

Mayhew's *London Labour*, then, covers not *all* forms of labor but those forms which, lying on the margins of the nameable ("dubious," "indefinite," "disintegrated"), characteristically embody the carnivalized picturesque. Mayhew fixates upon bone-grubbers, rag-gatherers, "pure"-finders (collectors of dog shit), sewer-hunters, mud-larks, dustmen, scavengers, crossing-sweepers, rat-killers, prostitutes, thieves, swindlers, beggars and cheats. And his attempt at social analysis is inseparable from his scopophilia.

The emphasis upon dirt was also central to the discourse which traced the concealed links between slum and suburb, sewage and "civilization." In *The Condition of the Working Class in England*, for instance, Engels analyzed the planning of a city whereby the "dirty" was made invisible to the bourgeoisie. Manchester, he argued, was divided into three circles: an inner ring of commerce, where warehouses and offices were already partially concealed behind the expensive shops of the main thoroughfares; a second ring of working-class housing; and an outer ring of suburbs inhabited by the bourgeoisie. The inner and outer rings were joined by main thoroughfares along which a "good" class of shops developed so as to service the bourgeoisie on their way to and from work. In the process, the working-class housing "disappeared" behind a respectable front.

Engels represented the second ring, where the poor lived, in grim detail. In a Salford cow-shed, he found a man "too old for regular work" who "earned a living by removing manure and garbage with his handcart. Pools of filth lay close to his shed." Off Long Millgate, Engels visits a court where "the privy is so dirty that the inhabitants can only enter or leave the court if they are prepared to wade through puddles of stale urine and excrement." From Ducie Bridge, he observes the River Irke, "a narrow, coal-black, stinking river full of filth and rubbish" and of "the liquid and solid discharges" from factories as well as of "the contents of the adjacent sewers and privies." (One might compare Dickens's vision of London in *Little Dorrit*: "Through the heart of the town a deadly sewer ebbed and flowed, in the place of a fine river," while in "fifty thousand lairs . . . the inhabitants gasped for air" (Dickens 1857 [1979]:29).)

But although Engels and Dickens attempt to analyze the city by tracing the sewer back to the suburb, the representation of filth which traverses their work is unstable, sliding between social, moral and psychic domains. At one level, the mapping of the city in terms of dirt and cleanliness tended to repeat the discourse of colonial anthropology. In 1881, Captain Bourke, a cavalry officer in the U.S. Army, observed the "characteristic dances" of the Zuni Indians in New Mexico. He watched the "filthy brutes" drinking urine; he heard one dancer relate how they normally "made it a point of honor to eat the excrement of men and dogs"; finally, when the dance, which had taken place indoors, was over, he ran from the room, which had become "foul and filthy" from the presence of a hundred Indians." Ten years later, Bourke published *Scatologic Rites of All Nations* (1891) in which he obsessively dwelled upon the filth of "savages," contrasting them with Christians and Hebrews, who were "now absolutely free from any suggestion of this filth taint" (quoted in Greenblatt 1982:2). In Bourke's work the division between cleanliness and filth, purity and impurity, is that between Christian and pagan, the civilized and the savage. But the nineteenth-century sanitary reformers mapped out the same division across the city's topography, separating the suburb from the slum, the respectable from the "nomad" along the same lines. Chadwick, the leading exponent of "the sanitary idea" in Britain, asked, "How much of rebellion, of moral depravity and of crime has its root in physical disorder and depravity?" and he argued that "[the] fever nests and seats of physical depravity are also the seats of moral depravity, disorder, and crime with which the police have most to do." Chadwick connects slums to sewage, sewage to disease, and disease to moral degradation: "adverse circumstances" lead to a population which is "short-lived, improvident, reckless, and intemperate, and with an habitual avidity for sensual gratifications." Like most of the sanitary reformers, Chadwick traces the metonymic associations between filth and disease: and the metonymic associations (between the poor and animals, between the slum-dweller and sewage) are read at first as the signs of an imposed social condition for which the State is responsible. But the metonymic associations (which trace the *social* articulation of "depravity") are constantly elided with and displaced by a metaphoric language in which filth stands in for the slum-dweller: the poor *are* pigs.

In Mayhew, we can observe the same sliding between the metonymic and the metaphoric. At the beginning of Mayhew's work, the street-people are remarked upon for their "greater development of the animal than of the intellectual or moral nature of man . . . for their high cheeks and protruding jaws," and this vision of innate animality permeates even a sympathetic account of an old woman who had been a "pure-finder." She lies like "a bundle of rags and filth stretched on some dirty straw" in a place "redolent of filth and pregnant with pestilential diseases, and whither all the outcasts of the metropolitan population seem to be drawn." That last phrase is troubling, implying that the "filthy" are *drawn* to the filth (as pigs were said to be drawn to the mire). To the extent that the poor are constituted in terms of bestiality, the bourgeois subject is positioned as the neutral observer of self-willed degradation.

Transgressing the boundaries through which the bourgeois reformers separated dirt from cleanliness, the poor were interpreted as also transgressing the boundaries of the "civilized" body and the boundaries which separated the human from the animal. Even Engels, despite his desire to demonstrate the culpability of the bourgeoisie for the slums, retains an essentialist category of the sub-human "nomad": the Irish. Engels, indeed, works within a colonial discourse which had been formed in the late sixteenth century and early seventeenth century when the Irish had been constructed as a race living "beyond the pale": they were said to be "more hurtfull and wilde" than "wilde Beastes"; they were accused of "uncleannesse in Apparrell, Diet and Lodging"; they were said to live in a "foul dunghill" in their "swinesteads," "snatching food like beasts out of ditches." Engels particularly develops the association between the Irish and "swinesteads":

> the Irishman allows the pig to share his own living quarters. This new, abnormal method of rearing livestock in the large towns is entirely of the Irish origin. . . . The Irishman lives and sleeps with the pig, the children play with the pig, ride on its back, and roll about in the filth with it.

Engels was, in fact, quite wrong: there was nothing specifically Irish about keeping pigs in town, which was a common English practice. But by condensing the "abnormal" practices of the slum into the figure of the savage Irishman, Engels attempted to purify the English proletariat. He insisted upon the contingent, metonymic relation between the English poor and filth, while simultaneously establishing a fixed and "natural" metaphoric relation between the Irish and animality.

Once the metaphoric relations were established, they could be reversed. If the Irish were like animals, animals were like the Irish. One of the sewer workers who talked to Mayhew described the sewers (which Irish laborers had helped to build) as full of rats "fighting and squeaking . . . like a parcel of drunken Irishmen." More generally, the links which associated the poor with animals and disease could be traced backwards; disease itself could be read as a member of the dangerous classes. In 1864, William Farr of the Registrar-General's Office wrote that "zymotic poisons, as dangerous as mad dogs, are still allowed to be kept in close rooms, in cesspools and in sewers, from which they prowl, in the light of day, and in the darkness of night, with impunity, to destroy mankind."

However "close" and confined the slums were, they were not confined enough. As the orifices of the poor opened to contaminate the purity of bourgeois space (at the turn of the century 44 percent of poor Glasgow children were defined as "mouth breathers"), so in the bourgeois imagination the slums opened (particularly at night) to let forth the thief, the murderer, the prostitute and the germs—the "mad dogs" which could "destroy mankind." The discursive elision of disease and crime suggested an elision of the means with which to cope with them: like crime, disease could be policed. In 1843, Farr argued that "the Legislature" should enact "the removal of known sources of disease, and, if necessary, trench upon the liberty of the subject and the privilege of property, upon the same principle that it arrests and removes murderers." The notorious Contagious Diseases Acts of 1864, 1866, and 1869 allowed special policemen to arrest women, subject them to internal examination and incarcerate them in lock-hospitals if they were suffering from gonorrhoea or syphilis. Justifying police regulations, W. R. Greg claimed that "the same rule of natural law which justifies the officer in shooting a plague-stricken sufferer who breaks through a *cordon sanitaire* justified him in arresting and confining the syphilitic prostitute who, if not arrested, would spread infection all around her."

In 1901, Charles Masterman published *The Heart of the Empire* in which he described the battle between "the forces of progress" and "social diseases." In his introductory essay on the "Realities at home," he described how town authorities were "pushing their activities into the dark places of the earth; slum areas are broken up, sanitary regulations enforced, the policemen and the inspector at every corner." As the Empire shed light upon the "darkness" of Africa, so the sanitary regime would shed light upon the city's "dark places." The connection between sanitation, light and policing can be seen in a Hudson's soap advertisement of the 1890s. In the picture, a policeman stands in the night holding up his lantern to illuminate a poster of Hudson's "extract of soap." At the top of the poster "PUBLIC HEALTH" is written, and underneath: "Dirt Harbours Germs of Disease." But the "source of Danger" will be removed by using "Hudson's" (in huge letters, occupying the center of the poster). The bottom half of the poster is given over to the miraculous powers of Hudson's and concludes: "Home, Sweet Home! The Sweetest, Healthiest Homes are those where HUDSON'S EXTRACT OF SOAP is in Daily use." The policeman and soap are analogous: they penetrate the dark, public realm with its disease and danger so as to secure the domestic realm ("Sweet Home") from contamination. The police and soap, then, were the antithesis of the crime and disease which supposedly lurked in the slums, prowling out at night to the suburbs; they were the agents of discipline, surveillance, purity.

The discipline of policing and sanitation depended in turn upon a transformation of the senses. As Foucault has argued, nineteenth-century policing found its privileged form in Bentham's Panopticon, which ensured the "permanent visibility" of the inmate. The gaze is structured in the Panopticon so that "in the peripheric ring, one is totally seen without ever seeing; in the central tower, one sees everything without ever being seen" (Foucault 1979:202). Throughout the nineteenth century, the "invisibility" of the poor was a source of fear. In Britain, the Select Committee of 1838 noted that there were whole areas of London through which "no great thoroughfare passed" and, as a consequence, "a dense population of the lowest classes of persons" were "entirely secluded from the observation and influence of better educated neighbours" (quoted in Stedman Jones 1971:166). The "laboring" and "dangerous" classes would be transformed, it was implied, once they became visible. On the one hand, there would be surveillance by *policing*; on the other, the inculcation of *politeness* through the benign gaze of the bourgeoisie.

But the bourgeoisie's organization of the gaze was always problematic. If the dominant discourses about the slum were structured by the language of reform, they could not but dwell upon the seductions for which they were the supposed cure. It was, perhaps, as a remedy for the ambivalence of the gaze that there was an increased regulation of *touch*. For even if the bourgeoisie could establish the purity of their own gaze, the stare of the urban poor themselves was rarely felt as one of deference and respect. On the contrary, it was more frequently seen as an aggressive and humiliating act of physical contact.

Thus even as the separation of the suburb from the slum established a certain class difference, the development of the city simultaneously threatened the clarity of that segregation. The tram, the railway station, the ice rink, above all the streets themselves, were shockingly promiscuous. And the fear of that promiscuity was encoded above all in terms of the fear of being touched. "Contagion" and "contamination" became the tropes through which city life was apprehended. It was impossible for the bourgeoisie to free themselves from the taint of "the Great Unwashed" (an English expression which emerged in the 1830s). Even money bore their stain. One government official paid Freud in paper florins which had been ironed out at home. It was a matter of conscience with him, he explained, not to hand anyone dirty paper florins: "for they harboured all sorts of dangerous bacteria and might do some harm to the recipient." The capitalist commodity itself permitted, and even encouraged, alarming conjunctions of the élite and the vulgar. In late nineteenth-century Holland, the bourgeois Versehoors would clandestinely burn the packages which *Jürgens Solo Margerine* came in, so that the neighbors would not discover them in the garbage can.

If the vulgar commodity could contaminate the home, the sorties of the home into the street were even more dangerous. In her book on good manners (*Goede Manieren*), Mrs. Van Zutphen van Dedem devoted a whole chapter to the "act of avoiding and excluding." She listed places to be avoided, which included slums, local trains and streetcars, third-class pubs, cheap seats at movie theaters, and crowds or celebrations in the streets. But since the promiscuity of public space was unavoidable, one must make all "the greater effort not to *touch* any 'undesirable.' " The "more refined person" was to avoid even

> the slightest contact, so far as possible, with the bodies and garments of other people, in the knowledge that, even greater than the hygienic danger of contamination, there is always the danger of contact with the spiritually inferior and the repugnant who at any moment can appear in our immediate vicinity, especially in the densely populated centres of the cities, like germs in an unhealthy body.

The "healthy" body is *refined*, uncontaminated by the "germs" of "the spiritually inferior," yet it is constantly assailed by them.

The gaze/the touch: desire/contamination. These contradictory concepts underlie the symbolic significance of the *balcony* in nineteenth-century literature and painting. From the balcony, one could gaze, but not be touched. The gentleman farmer who presided over a harvest feast would commonly arrange the table so that he could sit at its head *inside the house*, distributing hospitality through an open window or door. Similarly, the bourgeoisie on their balconies could both participate in the banquet of the streets and yet remain separated.

The flâneur, on the contrary, appalled by the "horror of one's home," sought out the urban carnival. "Even when he flees from town," wrote Baudelaire, "he is still in search of the mob" (Baudelaire 1983:71). Yet when he mingles with the crowd, he does not feel one of them. Indeed, Baudelaire sneered at George Sand's "love for the working classes," and argued that "it is indeed a proof of the degradation of the men of this century that several have been capable of falling in love with this latrine" (Baudelaire 1983:67). Preferring "harlots to Society women," Baudelaire nevertheless talked of "contaminated" women and wrote obsessively of "Hygiene Projects," "Hygiene Morality," "Hygiene. Conduct. Morality" (Baudelaire 1983:70, 96–102). He tried to abstain from "all stimulants" by obeying "the strictest principles of sobriety," yet his sobriety at home was the spur to ever greater excesses in the city. The silenced desires of the bourgeois citizen ("A summary of wisdom: Toilet/Prayer/Work" [Baudelaire 1983:99]) found their loquacious expression through the topography of Paris.

Within this social and psychic economy, a key figure was the prostitute. "There is, indeed, no exalted pleasure which cannot be related to prostitution," wrote Baudelaire (1983:21). It was above all around the figure of the prostitute that the gaze and touch, the desires and contaminations, of the bourgeois male were articulated. From the perspective of the righteous patriarch, every young man was "meeting with, and being accosted by, women of the street at every step":

> His path is beset on the right hand and on the left, so that he is . . . exposed to temptation from boyhood to mature age, his life is one continued struggle against it. (Quoted in Walkowitz 1980:34)

And the "contamination" of the prostitute seeped into the respectable home. In 1857, a writer in the *Lancet* complained that

> The typical Pater-familias, living in a grand house near the park, sees his son allured into debauchery, dares not walk with his daughters through the streets after nightfall, and is disturbed in his night-slumbers by the drunken screams and foul oaths of prostitutes reeling home with daylight. If he look from his window he sees the pavement—his pavement—occupied by the flaunting daughters of sin, whose loud, ribald talk forces him to keep his casement closed. (Quoted in Trudgill 1973:694)

In the 1850s, the fears of the "respectable" increasingly concentrated upon "the great social evil," prostitution. But through the discourse on prostitution they encoded their own fascinated preoccupation with the carnival of the night, a landscape of darkness, drunkenness, noise and obscenity.

This is not, of course, to deny the *regulatory* aspect of the construction of prostitution as "the great social evil." Following the Contagious Diseases Acts of the 1860s, there was undoubtedly an increased categorization and surveillance of the "unrespectable" poor. A new disciplinary regime could be inscribed upon the bodies of prostitutes once they had been classified and confined. In 1873, Inspector Sloggett recorded that the women in the Royal Albert Hospital in Southampton were "clad alike in blue serge dresses, their hair neatly draped and wearing muslin caps" and seemed "rather like a number of respectable young women in domestic service than registered prostitutes." And William Acton claimed that the women in Aldershot Lock Hospital were "most respectful; there was no noise, no bad language, no sullenness, no levity" (quoted in Walkowitz 1980:223). (Undoubtedly, these were idealized accounts of "reformation." In Portsmouth, women in the lock-hospital rioted and smashed windows, and one commentator complained that they were given to "insane frenzy," "singing, dancing, swearing, or destroying the blankets and rugs given them to sleep in" [quoted in Walkowitz 1980:216, 224]).

The "prostitute," though, was just the privileged category in a metonymic chain of contagion which led back to the culture of the working classes. The social-purity crusade of Ellice Hopkins in Plymouth aimed not only at inculcating new standards in young men but also at establishing legal and institutional programs to combat working-class "immorality." During the attempt to suppress Plymouth's Fancy Fair in 1886, Sergeant-Major Young claimed that numbers of young girls "were being ruined in that place every week, and afterwards bringing contamination into the homes of the well-to-do as nurse girls, and servants" (Walkowitz 1980:242, 244). When a girl disappeared at the fair, a migratory fiddler, Henry Greenslade, was prosecuted, but the trial was almost entirely devoted to "the immoral influence of the fair," which, it was claimed, had destroyed the girl's character. Purity crusaders set up the Girls' Evening Home Movement (which emphasized reading, music, and cooking lessons) in opposition to the "aimless street saunters" of working girls who could too easily stray "to such places as the fancy fair" (Walkowitz 1980:244). Similarly, Trinity Fair in Southampton was limited to one day while the Above-Bar Fair was abolished because of its "moral delinquencies" and "customary origins" (Walkowitz 1980:245). Like the prostitute, the fair was conceptualized as the breeding ground of physical and spiritual germs which, through the mediation of servants, would bring "contamination into the homes of the well-to-do."

It is surely no coincidence, though, that the zeal for reform was so often accompanied by a prolonged, fascinated gaze from the bourgeoisie. In the 1830s, for instance, the plans to construct "great thoroughfares," by means of which the "civilized" would, by their mere visibility, improve the "normally degraded," coincided with a flood of books which titillated the middle-class reader with tales of "a hardened, semi-criminal race of outlaws, safe from public interference within ancient citadels of crime and vice" (Stedman Jones 1971:180). Pierce Egan's *Life in London* (1812), for instance, was avidly consumed both as a book and as a play in the 1820s and 1830s. In the book, Tom and Jerry find "life" (i.e., drinking, dancing, swearing) in the East End of London, where "lascars, blacks, jack-tars, coal heavers, dustmen, women of colour, old and young, and a sprinkling of the remnants of once fine girls, etc., were all *jigging* together." It was common during this period for young bloods, sometimes protected by detectives, to visit Ratcliffe Highway ("a Babel of Blasphemy") to gaze at the sailors and prostitutes. Similarly in the 1880s, a time of crisis for the poor and of renewed moral panic among the bourgeoisie, there was "an epidemic of slumming" (Stedman Jones 1971:285). And again, there was a flood of writing *about* the slums which could be consumed within the safe confines of the home. Writing, then, made the grotesque *visible* while keeping it at an *untouchable* distance. The city however still continued to invade the privatized body and household of the bourgeoisie as *smell*. It was, primarily, the sense of smell which engaged social reformers, since smell, while, like touch, encoding revulsion, had a pervasive and invisible presence difficult to regulate.

Chadwick, the great sanitary reformer of the early nineteenth century, worked in Benthamite circles in the 1820s and from 1830–32 worked closely with Bentham himself. In 1846, Chadwick wrote: "all smell is, if it be intense, immediate disease, and eventually we may say that, by depressing the system and making it susceptible to the action of other causes, all smell is disease" (quoted in Schoenwald 1973:681). Smell was organized above all around *disgust*. George Buchanan, a Medical Officer of Health, attributed to "the influence of stink" not only loss of appetite, nausea, but also "a general sense of depression or malaise"; another Medical Officer, John Liddle, found the smell of the poor's linen, even when just washed, "very offensive." The Great Stink of 1858 only focused more intensively the bourgeoisie's obsessive concern with "the unmistakable and most disgusting odour of living miasm" (Wohl 1983:81).

At one level smell was re-formed as an agent of class differentiation. Disgust was inseparable from refinement: while it designated the "depraved" domain of the poor, it simultaneously established the purified domain of the bourgeoisie. The process is similar to that which we have already observed in Mayhew. Depicting the "nomad," Mayhew was able to construct by back-formation the "civilized": "regular and continuous labour," "providence," "property," "chastity," "religion" (Mayhew 1861–62: I, 2). Yet the imagined pleasures of the nomadic (including the smells) remained to undermine the "civilized." Mayhew's text, like the sanitary reports, testifies to one of the ways by which the nomad and the slum made their way into the bourgeois study and drawing room, to be *read* as objects of horror, contempt, pity, and fascination. Texts which were structured by antithetical thinking became gaps in the domestic scene through which contaminating desires leaked.

Like Chadwick, Mayhew was aware of the practical problems of sanitation. The subject of "London Sewerage and Scavengery" was, he wrote, "vast," concerning "the cleansing of the capital city, with its thousands of miles of streets and roads *on* the surface: and its thousands of miles of sewers and drains *under* the surface" (Mayhew 1861–62: II, 179). But in describing the functional process of cleaning, Mayhew articulates the sewers as a symbolic system. Indeed, he repeats one of the dominant tropes of western metaphysics: truth lies hidden behind a veil. But "truth" is now conceived materially, as excrement. In *Les Misérables*, in what might be called, without irony, one of the most brilliant explorations of the semantics of the sewer, Victor Hugo wrote that there could be "no false appearance" in the "vast confusion" of the "ditch of truth": "[the] last veil is stripped away. . . . This sincerity of filth pleases us and soothes the spirit" (Hugo 1980:2, 369). What Dickens called "the attraction of repulsion" is developed and analyzed in Hugo's text. Here the attraction is constructed as simply the revelation of the bodily functions, hidden by "the last veil."

Curiously, Freud's "Wolf Man" conceptualized reality through the same image of the veil:

> The world, he said, was hidden from him by a veil. . . . The veil was torn, strange to say, in one situation only; and that was at the moment when, as a result of an enema, he passed a motion through his anus. He then felt well again, and for a very short time he saw the world clearly. (Freud 1918:340)

The basic constitutive elements of the symbolic system of both "Wolf Man" and Hugo are the same: the veil, excrement, the "truth," and pleasure ("he felt well again"). But whereas Freud articulated the *psychic* formation of that system, Hugo represented its *social* formation. The sewer, Hugo wrote, was

> the conscience of the town where all things converge and clash. There is darkness here, but no secrets. . . . Every foulness of civilization, fallen into disuse, sinks into the ditch of truth wherein ends the huge social downslide, to be swallowed, but to spread. No false appearances, no whitewashing, is possible; filth strips off its shirt in utter starkness, all illusions and mirages scattered, nothing left except what is, showing the ugly face of what ends. (Hugo 1980: II, 369)

The sewer here represents a *non plus ultra* of naturalist reason, truth itself which, unimaginable "*on* the surface," can only subsist "*under* the surface":

> the spittle of Caiaphas encounters the vomit of Falstaff, the gold piece from the gaming house rattles against the nail from which the suicide hung, a livid foetus is wrapped in the spangles, which last Shrove Tuesday danced at the Opera, a wig which passed judgment on men wallows near the decay

which was the skirt of Margoton. It is more than fraternity, it is close intimacy. (Hugo 1980:II, 369)

The melodramatic coercion of extreme opposites into close intimacy here becomes the ultimate truth of the social. For indeed the signs of the sewer could not be confined "under the surface." The sewer—the city's "conscience" —insisted, as Freud said of the hysterical symptom, in "joining in the conversation." Hugo imagines a social "return of the repressed" in terms of the city's topography:

the cloaca at times flowed back into the town, giving Paris a taste of bile. . . . The town was angered by the audacity of its filth, and could not accept that its ordure should return. (Hugo 1980: II, 371)

Hugo, though, was writing about a past when "the sewerage was opposed to any discipline" and in its "confusion of cellars" mirrored the "confusion of tongues": the sewer had been "the labyrinth below Babel." But at the moment when Hugo wrote, the sewers had been cleaned up:

Today the sewer is clean, cold, straight, and correct, almost achieving that ideal which the English convey by the word "respectable." (Hugo 1980: II, 375)

Before the cholera outbreaks of the 1830s, the "excremental crypt" had asserted itself by flooding, and through it had entered "crime, intelligence, social protest, liberty of conscience, thought and theft, everything that human laws pursue" (Hugo 1980: II, 368). But the sewer had been transformed:

The sewer today has a certain official aspect. . . . Words referring to it in administrative language are lofty and dignified. What was once called a sluice is now a gallery, and a hole has become a clearing. (Hugo 1980: II, 376)

As the sewer was more rigorously segregated from the city above, it was linguistically reformed, absorbed into the discourse of respectability. "A good sewer," Ruskin declared, was a "far nobler and a far holier thing . . . than the most admired Madonna ever printed" (quoted in Wohl 1983:101). The nobility of the Victorian sewer was nowhere more dramatically confirmed than in the opening ceremonies of Bazalgette's intercepting sewers, south and north of the Thames, which were attended by the Prince of Wales, Prince Edeward of Saxe-Weimar, the Lord Mayor, the Archbishop of Canterbury and the Archbishop of York (Wohl 1983:107). Yet paradoxically the sewer's improved status depended upon its invisibility. In 1865, the *Illustrated Times* depicted "the Prince of Wales starting up the main-drainage works at Crossness": the pumping station is portrayed as a striking architectural monument and in the foreground of the picture a police sergeant holds a large flag. The sewer was becoming acceptable because it was locked and patrolled to prevent contamination, "the keyhole and the bolt securely in place," as Hugo wrote, with added protection from "one of those prison locks" (Hugo 1980: II, 394). The only remaining trace of the sewer was "a vaguely suspect odour, like Tartuffe after confession" (Hugo 1980: II, 376).

The sewer, however, like all the low and grotesque systems we have here examined, could not entirely be closed off from above. Passing between the sewer ("the conscience of the town" which was now blocked off) and the city (with its "noble buildings" [Hugo 1980: II, 367]) were the rats: "and here, in the foetid darkness, the rat is to be found, apparently the sole product of Paris's labour" (Hugo 1980: II, 368). Indeed, Hugo claimed that despite the rebuilding of the sewers the "immemorial rodent population" was "more numerous than ever" (Hugo 1980: II, 376). Rats had, of course, been the objects of hatred before the nineteenth century (Zinsser 1985). But just as the meaning of the grotesque body was transformed by its diacritical relation to the emergent notion of the bourgeois body, so the symbolic meaning of the rat was re-fashioned in relation to the sanitary and medical developments of the nineteenth century. As the connections between physical and moral hygiene were developed and redeployed, there was a new attention to the purveyors of physical and moral "dirt." The rat was no longer primarily an economic liability (as the spoiler of grain, for instance): it was the object of fear and loathing, a threat to civilized life. Hence, the stories which Mayhew recorded of sewer rats attacking men "with such fury that the people have escaped from them with difficulty" (1861–62: II, 151).

The rat, then, furtively emerged from the city's underground conscience as the demonized Other. But as it transgressed the boundaries that separated the city from the sewer, above from below, it was a source of fascination as well as horror. In one of Freud's case studies, Frau Emmy von N. spoke of "a case of white rats" while "[she] clenched and unclenched her hand several times." "Keep still!—Don't say anything!—Don't touch me!—Suppose a creature like that was in the bed!" (She shuddered.) "Only think when it's unpacked! There's a dead rat in among them—one that's been gn-aw-aw-ed at!" (Freud 1893:107). It is true that the rat was only one of various animals about which Frau Emmy hallucinated, but her particular fascination with rats is suggested by her "extreme horror" at the story of Bishop Hatto, who was supposedly eaten by them (Freud 1893:131). (The story implies a dramatic contrast of high and low: the bishop, who preaches of a transcendent heaven, is destroyed by rats, which live in a physical "hell").

Elsewhere Freud named one of his patients after the rat, and his case permits us to analyze "the attraction of repulsion" in greater detail. Freud called his patient the "Rat Man" because of his "great obsessive fear" which was triggered by a story told him by an army officer about the punishment of a criminal in the East:

> "a pot was turned upside down on his buttocks . . . some *rats* were put into it . . . and they . . ." [the patient] had again got up, and was showing every sign of horror and resistance—". . . . bored their way in. . . ."—Into his anus, I helped him out. (Freud 1909:47)

But even as "Rat Man" recalled this story, Freud observed "a very strange, composite expression" on his face, "one of *horror at pleasure of his own of which he himself was unaware*" (Freud 1909:48, Freud's italics). The pleasure was derived, Freud argued, from the "anal erotism" (Freud 1909:93) which his patient had repressed: the pleasure reappeared in the form of a negation and with the eroticism represented by the rat which bored into his anus. Thus, a bourgeois "of irreproachable conduct" (Freud 1909:40) found his way back down the axis of his body to the censored realm of excremental ambivalence. Freud, to be sure, analyzes the "rat" as a sliding signifier within the domain of the psyche, but he nevertheless treats the concept of "rat" as unproblematically *given*, the "natural" symbol of his patient's repression. But the process of symbolization is in need of social as well as psychic explication. We would argue that, although symbolic systems are never entirely reducible to each other, one cannot analyze the psychic domain without examining the processes of transcoding between the body, topography and the social formation. . . .

The vertical axis of the bourgeois body is primarily emphasized in the education of the child: as s/he grows up/is cleaned up, the lower bodily stratum is regulated or denied, as far as possible, by the correct posture ("stand up straight," "don't squat," "don't kneel on all fours"–the postures of servants and savages), and by the censoring of lower "bodily" references along with bodily wastes. But while the "low of the bourgeois body becomes unmentionable, we hear an ever increasing garrulity about the *city's* low"—the slum, the ragpicker, the prostitute, the sewer—the "dirt" which is "down there." In other words, the axis of the body is transcoded through the axis of the city, and while the bodily low is "forgotten," the city's low becomes a site of obsessive preoccupation, a preoccupation which is itself intimately conceptualized in terms of discourses of the body. But this means that the obsessional neurosis or hysterical symptom can never be immediately traced back through the psychic domain. To deconstruct the symptomatic language of the bourgeois body it is necessary to reconstruct the mediating topography of the city which always-already inscribes relations of class, gender, and race.

We would argue, then, that "Rat Man" "speaks" his body through the topography of the city, a topography which is in turn shaped and controlled by the divisions of social formation. Body and social formation are inseparable. "Rats," "sewage," "filth" are not transparent signifiers which lead directly back to some primal moment. If they speak the unconscious, it is only through the mediation of the slum. The vertical axis of the body's top and bottom is transcoded through the vertical axis of the city and the sewer and through the horizontal axis of the suburb and the slum or of East End and West End. Furthermore the topography of the city . . . is represented within the bourgeois household itself through the relation of the family to its servants, through the relation of "upstairs" to "downstairs."

Indeed, an analysis of Rat Man's sociolect requires an examination of the relation between the topography of the city and that of the household. As a child, "Rat Man" crept up his governess's skirt or stared in fascination at Fraülein Lina's abscesses (Freud 1909:41–42). But to the analysand, such delights can only be thought of as the obsessions of a rat. Ergo, children themselves must be rats (Freud 1909:97). This enables the patient with one part of his psyche to adopt the position of his father: he must be punished:

> The notion of a rat is inseparably bound up with the fact that it has sharp teeth with which it gnaws and bites. But rats cannot be sharp-toothed, greedy and dirty with impunity: they are cruelly persecuted and mercilessly put to death by man, as the patient had often observed with horror. (Freud 1909:96)

As an adult "of irreproachable conduct," he must shun not only rats but also those elements with which he associated them. Hence, his disgust at prostitutes (Freud 1909:39). Knowing that rats were carriers of dangerous infectious diseases, "he could . . . employ them as symbols of his dread . . . of *syphilitic infection*" (Freud 1909:94, Freud's italics). And since the penis was itself a carrier of syphilis, "he could consider the rat as a male organ of sex" (Freud 1909:94).

But if the symbolization of rats positioned "Rat Man" with his father as the censor of his own childhood pleasure, it also determined his phantasies of rebelling. When visiting his father's grave, he saw what he took to be a rat (Freud believed that it was really a weasel) and he imagined that it "had actually come out of his father's grave, and had just been having a meal off his corpse" (Freud 1909:96). He associated the punishment in which a rat bored its way up the criminal's anus both with his father and with the woman whom he was thinking of marrying (Freud 1909:48). The German word for "to marry" ("heiraten") was associated both with "*Ratten*" (rats) and with "*Raten*" (installments): "so many florins, so many rats," "Rat Man" had told Freud. So in his fantasy, the middle-class fiancée was elided with the rat and the prostitute, the sewer and the slum. The rat then was a phobic mediator between high and low, a kind of debased coinage in the symbolic exchange underpinning the economy of the body. The symbolic figure of the rat overran not only the boundaries between city and sewer: it gnawed away at the distinctions which separated patriarch from child, bourgeois beloved from prostitute, mother from abscessed maid, the pure from the contaminated.

Just as the rat was one of the dominant signs through which the bourgeoisie imagined the passage between "the noble buildings" and "the foetid darkness," so too the pig was to be transvalued. On the one hand, it was conspicuously present in the cities. In the middle of the wealthy suburb of North Kensington lay the Potteries, a seven-acre slum with open sewers, stinking ditches, and a stagnant, poisonous lake. The 1851 census revealed three pigs for every human there; the pigs provided bacon for the surrounding suburb while the inhabitants provided the servants, prostitutes, chimney-sweeps, and night-soil men to "service" the bourgeois households. While the pig moved up the social scale to the middle-class table, the swill of the suburbs passed down into the slums.

A man "who had moved in good society" told Mayhew that

[when] a man's lost caste in society, he may as well go the whole hog, bristles and all, and a low lodging house is the entire pig. (Mayhew 1861–62: I, 255)

Indeed, the pig could appear in more troubling shapes than "the entire pig" of the lodging house. Mayhew heard a strange tale from the sewer-hunters

of a race of wild hogs inhabiting the sewers in the neighbourhood of Hampstead. The story runs, that a sow in young by some accident got down the sewer through an opening, and, wandering away from the spot, littered and reared her offspring in the drain, feeding on the offal and garbage washed into it continually. Here, it is alleged, the breed multiplied exceedingly, and have become almost as ferocious as they are numerous. (Mayhew 1861–62: II, 754)

This surreal narrative perfectly embodies the phobic inversion of the carnivalesque icon. It participates in the formation of a "carnival of the night" which was to trouble the dreams of the bourgeoisie. The pig, reared in the slums, is displaced by an imaginary race of sewer pigs, living in darkness, multiplying like rats, eating garbage, threatening the high with the ferocity of the low.

In the symbolic formation of the city, the pig too, like the rat, could figure as recalcitrant Other to trouble the fantasy of an independent, separate, "proper" identity. It would surely be mistaken to see the pig and the rat here as merely the residual signifiers of a pre-capitalist formation. On the contrary, the reformation of the senses produced, as a necessary corollary, new thresholds of shame, embarrassment and disgust. And in the nineteenth century, those thresholds were articulated above all through specific *contents*—the slum, the sewer, the nomad, the savage, the rat—which, in turn, remapped the body. It is important to emphasize that this "manifest content" was no incidental and contingent metaphor in the structuring of the bourgeois Imaginary. It was not a secondary over-coding of some anterior and subjective psychic content. Indeed it participated in the *constitution* of the subject, precisely to the degree that identity is discursively produced from the moment of entry into language by such oppositions and differences as we have explored here.

REFERENCES

Baudelaire, Charles. 1983. *Intimate Journals*. Tr. Christopher Isherwood. San Francisco: City Lights.

Chadwick, Edwin. 1965 (1842). *Report on the Sanitary Condition of the Labouring Population of Great Britain*. Edinburgh: University Press.

Dickens, Charles. 1979 (1857). *Little Dorrit*. Oxford: Clarendon Press.

Engels, Friedrich. 1971 (1844). *The Condition of the Working Class in England*. Oxford: Blackwell.

Foucault, Michel. 1979. *Discipline and Punish*. New York: Random House.

Freud, Sigmund. 1974 (1893). "Studies on Hysteria." In A. Richards, ed., *The Pelican Freud Library 3*. Harmondsworth: Penguin.

——. 1979 (1909). "Notes Upon a Case of Obsessional Neurosis (The 'Rat Man')." In A. Richards, ed., *The Pelican Freud Library 9*. Harmondsworth: Penguin.

——. 1979 (1918). "From the History of an Infantile Neurosis (The 'Wolf Man')." In A. Richards, ed., *The Pelican Freud Library 9*. Harmondsworth: Penguin.

Greenblatt, Stephen. 1982. "Filthy Rites." *Daedalus* 3, 3: 1–16.

Hugo, Victor. 1980 (1862). *Les Misérables*. Tr. N. Denny. 2 vols. Harmondsworth: Penguin.

Marx, Karl. 1951 (1852). "The Eighteenth Brumaire of Louis Bonaparte." In *Marx-Engels Selected Works*. London: Lawrence and Wishart.

Mayhew, Henry. 1967 (1861–62). *London Labour and the London Poor*. London: Frank Cass.

Schoenwald, Richard L. 1973. "Training Urban Man: A Hypothesis About the Sanitary

Movement." In H. J. Dyos and Michael Wolff, eds., *The Victorian City: Images and Realities*. London: Routledge and Kegan Paul. Vol. 2, 669–92.

Stedman Jones, Gareth. 1971. *Outcast London: A Study in the Relationship Between Classes in Victorian Society*. Oxford: Clarendon Press.

Trudgill, Eric. 1973. "Prostitution and Paterfamilias." In H. J. Dyos and Michael Wolff, eds., *The Victorian City: Images and Realities*. London: Routledge and Kegan Paul. Vol. 2, 693–705.

Walkowitz, Judith. 1980. *Prostitution and Victorian England: Women, Class, and the State*. Cambridge: Cambridge University Press.

Wohl, Anthony S. 1983. *Endangered Lives: Public Health in Victorian Britain*. London: Methuen.

MEDICINAL MEALS

JUDITH FARQUHAR

A LOGIC OF THE CONCRETE

What is the nature of the systematic knowledge used in medicinal meals? This is an interesting question because in everyday life knowledge is not held apart from other kinds of experience. Far from being a merely conceptual thing, knowledge has its own efficacies. One hardly needs to go further to understand this than to recall the clinical commonplace that diagnostic procedures often have therapeutic efficacy. An elderly but still fairly healthy relative of mine, for example, periodically develops chest and arm pains and begins to worry a great deal about heart disease. Thus far, his symptoms have been markedly alleviated by the administration of a full cardiac workup; the knowledge he thereby gains that his cardiovascular system is still in good working order holds his symptoms at bay for long periods of time. This is not just an example of psychosomatic medicine; it is also testament to the powers that knowledge wields in experience.

In the 1960s Lévi-Strauss, continuing his investigations into the logic of the concrete, inaugurated his *Mythologiques* series with *The Raw and the Cooked*. Arguing that "there is a kind of logic in tangible qualities" (1969:1), he justified his decision to "operate at the sign level" as follows:

> Even when very restricted in number, [signs] lend themselves to rigorously organized combinations which can translate even the finer shades of the whole range of sense experience. We can thus hope to reach a plane where logical properties, as attributes of things, will be manifested as directly as flavors or perfumes; perfumes are unmistakably identifiable, yet we know that they result from combinations of elements which, if subjected to a different selection and organization, would have created awareness of a different perfume. Our task, then, is to use the concept of the sign in such a way as to introduce these secondary qualities—on the plane of the intelligible and not only the tangible—into the operations of truth. (1969:14)

The way in which he then proceeded to introduce the "secondary qualities" of sense experience into the operations of truth performed the particular reduction for which structural analysis is well known: he treated the matter of the myths he analyzed (the cooked and rotten meat, the jaguars and the maize trees) as "an instrument, not an object, of signification" (341; 346–47). In order for tangible qualities to have a logic, the analyst needed to show how myth narratives arranged concrete things in significant relationships; objects and qualities became operators reflecting more abstract contrastive pairs. Once the logic of this signifying arrangement was appreciated—promoted, as it were, to the cognitive level—the matter through which it had been made could be forgotten.

But the tangible qualities of food, like those of perfume, are not so readily forgotten. Nor are they easily disciplined within the confines of a "logic." The analytic power and tidiness of structuralist analysis in the Lévi-Straussian manner gratifies me as an anthropologist even as it annoys the eater in me; explanatory power about signification seems to be gained at the expense of the poetry—the flavors and pleasures—inherent in everyday reality. The structuralist analyst works through the concrete to reach the logical, leaving the charms of mundane experience far behind. The contradiction is not only Levi-Strauss's; ethnographic description, with its local cultural commitments, often seems to work at cross-purposes with anthropological analysis, which compares and generalizes.

My point here, that there is a big difference between the mundane pleasures evoked by Lévi-Strauss and the cerebral gratifications of his structuralist reasoning, is neither original nor new: almost from its inception structural analysis stirred a "humanist" response in a debate that posed poetry against science, experience against cognition, politics against "mentalities," even "body" against "spirit." These dualisms may be impossible to overcome, but perhaps they can be made less foundational, less taken-for-granted, by noting the many ways in which everyday life mediates and muddles them. In its detailed attention to the qualities of things eaten and sensed, Lévi-Strauss's project—and a few that have followed it in this respect—still inspires my own reading of medicine and embodiment in contemporary China. If bodies are capable of imagining, we should be able to carnally imagine other lifeworlds, or sensory realms, through an ethnographic description that attends to the concrete and the everyday.

This description and this reading are no simple transfer of the most "real" parts of one lifeworld into the foreign context of another, however. (If you have never eaten lichees, you could hardly guess from a description what they taste like.) Rather they resemble a translation process that must take the specificity of terms and entities in both "source" and "target" languages very seriously. Direct sensory experience, the material attributes of concrete things and mundane activities, can be invoked, and thereby imagined, but only by way of language and images, and only in the context of times, places, and habitus that impose constraints on what can be experienced or imagined. If ethnography searches for a "logic of the concrete" or a "science of the tangible," then, it must always work on both sides of these somewhat oxymoronic terms. That is, both logical-scientific forms of knowledge and concrete characteristics of social life must be kept in play.

For this contradictory task, Chinese medicine offers much, assuming as it does a world, a consciousness, and a physiology that unite knowledge and experience, power and knowledge, in a single system of simultaneous understanding and action. Unlike those scientific empiricisms that seek to draw aside the curtain of (mere) appearances to reveal the (truer) underlying processes knowable only to the analyst, Chinese medicine relies heavily on many sorts of experience (*jingyan*). This experience is not a direct, naive immediacy, rather it is a heavily emphasized and ideologically fraught category that links a great many clinical and scholarly practices together. Nevertheless, even the act of positing experience as a central resource and authority for the field makes a strong contrast with experimental science. The perceived qualities of symptoms and of drugs with which Chinese pharmacy often concerns itself would be extraneous factors—needing to be controlled before a clean study could be designed—in most laboratories.

Chinese pharmacy meets anthropology's science of the concrete with what appears at first to be an elaborate science of tangible qualities. Drug attributes such as flavor, warmth, directionality, and speed both classify (abstractly) medicinal substances and name (concretely) their sensory, material characteristics. Moreover, as I will show below, the qualities of herbal drugs and of bodily states are not mere passive reflections of hidden causes at work; these known and experienced attributes themselves have power. In this they resemble the odors and colors that figure in the myths analyzed by Lévi-Strauss; at the level of the narrative or the therapeutic process, the "hotness" or "sweetness" of a drug makes a difference. But unlike structuralist analysis of myths, Chinese medical theory does not abandon these manifest qualities in favor of an articulation of the abstract minimal contrasts through which they might signify. As will be seen below, Chinese medicine does not even employ its "native" dualism of yin and yang to this end. Rather, doctors dwell on the concrete attributes of thousands of herbal, mineral, and animal drugs, weaving together known qualities into potent cocktails of flavor, heat, and positional tendency. At the same time, the flavors of sweet, bitter, pungent, salty, and sour (and the attributes of warming and cooling, and the positional tendencies that affect the five great visceral systems) offer classificatory rubrics from which principles of combination can be derived. Recall Lévi-Strauss's perfume example: a distinctive aroma produced from diverse elements by methodical combination is quite like a Chinese herbal prescription. Except that the latter has the power to act directly on whole bodies. And this is an important difference.

MEDICINAL MEALS IN A COUNTY TOWN

Early in 1993, the brother of a doctor I knew opened a small "herbal medicine meals" (*yao shan*) restaurant in a county town in Shandong Province. At that time there were very few indoor restaurants in town, though street-corner noodle stands had begun to appear to meet the needs of short-term visitors to town—truck drivers, villagers selling or contracting agricultural products, purchasing agents for smaller communities or rural enterprises, customers for the town's two periodic markets. A few formal restaurants had opened, mainly in hotels and guest houses; most of the trade for these establishments consisted of banquets for visitors hosted by county government officials or the town's few business leaders. . . . Mr. Wu's Yao Shan Restaurant was somewhere between these two levels of public eating. He and his wife, assisted by his brother who had a small private Chinese medical practice, rented a shop-house and converted the front into space for two tables and four banquet rooms, each of which seated about eight. The living space in the back became the kitchen. Though the appearance of the place was not charming (it was quite low, dark, and cramped), the food was good and so was business. Mr. Wu attributed their success, especially during the lunch hour, to their proximity to the county's main Public Security Agency complex. Police and cadres often ate and entertained there.

In fact, Mr. Wu and his family were quite surprised—almost alarmed—when I walked in one evening with my friend Sarah, another foreign researcher. Their surprise was not simply attributable to the rarity of foreigners in town; it had more to do with our gender and our anomalous status as women unattached to any known household. The family was very hospitable and happy to serve us an excellent meal; but they insisted we eat in a private room, and they were glad we had come for dinner rather than lunch. They later suggested that any return visits be in the evening. They apparently did not like the idea of us mixing with their lunchtime clientele, middle-aged men who sometimes drank a great deal of white liquor and became rather rowdy at mid-day. (I didn't relish the idea myself.)

More puzzling was the fact that our waitress, Mr. Wu's daughter, did not see the point of serving us any medicinal foods, despite the prominence of the restaurant's specialty. (It was their large "herbal medicine meals" sign that had brought me to them, after all, not their relationship—which I learned later—with a doctor I had already interviewed.) She said that Sarah and I were not especially good candidates for this restaurant's herbal meals; the dishes they prepared daily for customers were meant to improve (*bu*, or bolster and tonify) the health of middle-aged men.

Dr. Wu, the proprietor's brother, later gave me a list of the herbal substances used most often in the restaurant: ginseng, astragalus, schisandra, chrysan-themum, fennel, hawthorn, and Asian cornelian cherry. Boiled up in soups (*tang*, the same word used for Chinese medical herbal decoctions), these drugs show a definite tendency toward supplementing qi-energy, preventing the de-pletion of jing-essence (sometimes translated as seminal essence), and improv-ing the functions of the visceral system of which the kidneys are the central locus. (The Kidney Visceral System not only governs aspects of the fluid econ-omy of the body, it is central to sexual and reproductive functions.) In other words—and somewhat over-simplifying—these were tonics for men; not quite aphrodisiacs, but designed to improve the physiological basis for a fully func-tional masculinity. No wonder the Wu family's pokey little restaurant was such a hit with the Public Security Bureau.

I imagine everyone concerned understood the gendered character of these herbal medicine meals better than I did at the time. Though there must have been some conversation and explanation—when the restaurant was opening, for example—about the specific virtues of the soups, much certainly went without saying. For one thing, everyone knows without dwelling on it what a tonic for middle-aged men is; apart from the brothers Wu, I may have been the only person who cared about the names and efficacies of the specific drugs in use. Still, even while I remained ignorant of these details it was not hard to perceive the strongly masculine orientation of the service in this establish-ment. This was a quality that was conveyed to Sarah and me from the moment we stepped across the threshold, especially in the family's surprise and in their kind but rather abashed service. (Being foreigners, of course, we were ac-customed to being attributed with—and forgiven for—many small transgres-sions; it was only later that I began to think about the reasons why this place was not for us.) No one really explained it, but where spaces and practices are strongly gender marked, there is perhaps no need to explain. Several ill-assorted forms of habitus confronted each other there, and we got the message.

Not all "medicinal meals" restaurants are so strongly gendered, and this one was perhaps a little unusual in its insistence on specializing. The particular soups served by Mr. Wu in his restaurant were not unlike the commercially packaged tonics and high-priced herbal drugs given by people of both sexes to their male seniors. Despite their association with the physiological substrate of sexual powers, these objects carry no improper innuendo; they are no more embarrassing or unspeakable than beauty products for women would be. Moreover, despite Miss Wu's protests, I can find no suggestion in the materia medica literature that the soups she served would not also benefit my aging female physiology. . . . Women too must harbor seminal essence (jing), and nourishing qi and improving Kidney System function should be important for all aging bodies. Why were the medicinal specialties at the Yao Shan Restaurant more appropriate for the bureaucrats from the Public Security Bureau than for me?

Technically speaking, perhaps, they were not, but that is how they were billed. Men who ate regularly there must have enjoyed the experience at several levels. One level is certainly that of a change in subjective bodily state. At least some of the substances used in the soups reliably deliver noticeable effects, the stimulant activity of ginseng being only the most obvious. Especially if customers consumed these things expecting a little energy boost, smoother digestion, and an improvement in mood, they would have walked away feeling stronger and more in command of at least a few things. At another level, it is worth bearing in mind that the early 1990s was not the best of times for many large government agencies. Many aspects of government in this Shandong county town had been removed in the 1980s from party control and assigned to a newly organized civil government; in parallel with this change, it had begun to be clear that technical expertise was more valued in local government than the "redness" on which an older generation of administrators had built their careers. Only a year or so later a national campaign for early retirement began to banish men 55 and older from the offices (and chauffeured vehicles and mid-day banquets) where they had long served as loyal party members. At least some of Mr. Wu's patrons, I suggest, must have felt that their jobs were shaky, their loyalties outmoded, and their skills devalued. The physical boost of a medicinal meal—reaching both their mood and their masculinity—would have been a welcome addition to their daily lives.

Most of the drugs used in Chinese medicine are not like ginseng, chrysanthe-mum, and hawthorn—their physiological effect is not immediately noticeable. Their taste, however, is hard to miss. The languages of Chinese food and traditional herbal therapy share specialized terms that classify flavors and summarize elaborate technologies of cooking. Though considerations of ap-pearance and texture, so important in the world of food, count for little in the preparation of medicine, both kinds of work are family affairs that can domi-nate daily life. Further, both cooking and herbal medicine draw on a wide variety of substances, which come to these domestic spaces bringing connota-tions of place, seasonality, and textual elaboration.

Some of these connotations are carried by flavor words. Take *ku*, bitter, for example. The most common term in modern Chinese for suffering is *chiku*, eating bitterness. As Gang Yue argues in his discussion of Wang Ruowang's *Hunger Trilogy*, this ordinary word is strongly invested with a historical sense—it is easy to solicit rueful comments from people about how much bitterness they or their family members have swallowed in the past. And whatever per-sonal bitterness they may be thinking of, to refer to this suffering as *chiku* is to link one's own difficulties with those of the nation. *Ku* is widely used in other compounds as well, with a denotative scope ranging from the relatively literal referents of bitterness and pain to more figurative notions of earnest serious-ness and painstaking effort. Hard work, for example, is often described as *xinku*, a compound translatable literally as pungent-bitter.

The technical uses of the flavor terms in Chinese medicine appear at first to be on the literal end of this range. Not only are these five commonly known flavor terms—sour (*suan*), bitter (*ku*), sweet (*gan*), pungent (*xin*), and salty (*xian*)—used to classify herbal medicines, these words often (but not always) refer to the actual tastes of particular herbal drugs. It is often said among Chinese medical people that Shen Nong, the sage-king and mythical founder of herbal medicine, "tasted 100 herbs," and on the basis of this experience produced the first materia medica texts. When I began to study Chinese medi-cine I assumed that the five flavors served solely as classificatory rubrics, so this use of the verb "to taste" puzzled me; why wasn't Shen Nong said to have used, or tested, or classified 100 herbs? Once I turned to the concrete qualities of medicines that interest doctors and patients—encouraged more by the exten-sive descriptive information in pharmacopeia reference works than by the flavor terms per se—the word looked much more essential. Thus, though patients often complain that herbal decoctions are "too bitter" (*ku*), they also admit that they can detect the characteristically sweet taste of licorice root (*gancao*) in most prescriptions. A refined palate can no doubt also distinguish amidst the bitterness some tastes that are more sour, salty or pungent. Considering that individual drugs of diverse flavors are usually boiled together, it must be difficult to sort out all the tastes of a complex prescription. But there's no doubt, I think, that for a medicine to do anything very complicated it must assault the sufferer with a strong and complex flavor.

I am no Shen Nong, and my palate is not educated for discriminating Chinese medicine's "five flavors." Every herbal decoction I have ever swallowed has tasted simply horrid (i.e., *ku*) to me, and many Chinese patients I have checked with agree with me on this point. Still, the fact that drugs in the classic decoction form have flavor, i.e., both an experiential quality and a classificatory function in a system of pharmaceutical effects, raises the question, what is the efficacy of a "flavor"? Isn't it rather odd, at least for those of us steeped in the subject/object divide of Euro-American common sense, to think of a personal experience such as a flavor acting directly on a biological condition? Lévi-Strauss, when he invoked the notion of a perfume, did not accord much power to the sensory qualities of aroma beyond, perhaps, the ability to invoke memories or set in train other more or less mental processes. When we think of the experiences that accompany therapies (the pain with physical therapy, the nausea with chemotherapy), we are more likely to think of them as side-effects of a primary action which is—almost by definition—sub-experiential. Muscles mend and bones knit, tissues or microorganisms die or proliferate, outside of our direct awareness.

In biomedical discourses, the causes of illness or healing certainly don't *need* to be perceived by the sufferer. This is why we consult a doctor, to learn the actual causes of our discomforts, which could never be figured out "for sure" from our experience. (It is not too many hours spent squinting at small print in archives that's causing our headaches, but a pinched nerve; not a lack of vigorous exercise at the root of this back pain, but a collapsing disc. The experiential reasoning with which we connect our ailments to the rest of our lives is more often than not dismissed or bypassed in biomedical diagnosis.) In addition, the metaphorical billiard balls of the standard materialist model of causation still bounce around in biomedical explanations of etiology and efficacy, and action at a distance—such as the effects achieved by acupuncture—remains difficult to explain. Muscles, tissues, and microorganisms are entities familiar to an anatomically based medicine of structures, and can be invoked in etiological narratives that accord with a materialist logic of causes. But Chinese medicine cares little about anatomy. Rather it has usually been characterized as a functional medicine that reads the manifestations of physiological and pathological changes without resorting to models of fixed structural relations. Moreover, Chinese medicine is said to be "holistic"—it links manifestations of illness to causal narratives not with reference to an underlying anatomical field but in relation to temporal emergence. The appearance of symptoms in sequence or at the same time suggests that they are related phenomena, and the physician's job is to identify the process that could produce all these symptoms in precisely this temporal relation. Proximity in anatomical space counts for little in this holism. If these descriptions of Chinese medicine are correct at all, then some other form of causality—and efficacy, even power—must also be entailed.

The following excerpt from a 1978 pharmacy textbook indicates both how complex and how direct the relationship between drug flavors and drug powers is conceived to be:

> The five flavors are the five types of pungent, sweet, sour, bitter, and salty. Some drugs have a clear or an astringent flavor, so in reality the types are not confined to five; but they are customarily still called the five flavors. The five flavors are also an expression of the roles of drugs, different flavors having different functions. . . . Generalizing from the historical experience of using drug flavors, their functions are as follows:
>
> *Pungent* has the function of spreading and disseminating, moving qi, moving Blood, or nourishing with moisture. . . . *Sweet* has the function of replenishing and supplementing, regulating the activity in the Middle *Jiao*,[1] and moderating acuteness. . . . *Sour* has the functions of contracting and constricting. . . . *Astringent* has functions similar to those of sour drugs. . . . *Bitter* has the function of draining and drying. . . . *Salty* has the function of softening hardness, dispersing lumps, and draining downward. . . . *Clear* has the function of condensing Dampness and causing urine to flow.

The functions listed here (disseminating, constricting, etc.) are both technically inflected theoretical terms for normal physiological activities of the human body and words that could, with a little reflection, be readily applied to personal embodied experience. In other words, the logic that connects the flavors to the powers of medicines has room for the sensed responses of the lived body. What seems to require explaining, however, is the *connection* between the experience of a flavor and the experience of a bodily change. How can "sour" bring about a general contraction and constriction? Why does "bitter" cause fluids to drain? Though this passage introduces a textbook in which the classificatory functions of "pungent, sweet, sour, bitter, and salty" are emphasized, it also makes it clear that these are not entirely arbitrary rubrics. Shen Nong *tasted* 100 herbs when he founded Chinese pharmacy. Pharmaceutical classification is presented as reflecting the actual tastes of substances in the materia medica corpus. That these tastes are then correlated with particular efficacies is a fact that, in Chinese discourses, requires no explanation beyond the usual reference to accumulated historical experience.

English does not offer a language for whole-body responses to tastes, or a theory of flavor causation, of this kind. Perhaps the closest we come is to the notion of "heavy" or "light" meals affecting our alertness, or learning that certain foods "disagree" with our stomachs. The idea that "flavors" could have powerful physiological efficacies is odd enough to have been politely ignored by most of the English-language literature on Chinese herbal medicine. In North American nutritional lore, we tend to relegate tastes to that domain in which the (relatively isolated) human subject receives sensory input, registering pleasure or revulsion in response to food. We think of those forces and entities that actually alter our bodies as properties of the food that are quantifiable (e.g. fat, vitamin, or protein content) and which inhere in the food whether we eat it or not. The body in question is a part of nature, while the cultural enjoyments of which it is capable don't really belong to a biological discourse. The rationally known efficacies of things cancel the relatively ephemeral experience of ingesting them, and our carnal tastes, when they are invoked, drift upward toward the cultural domain where subjective experience is stored. Apparently pleasure, when spoken of in English, has weak causal force.

In modern China, people who suffer from disease and seek relief in the use of herbal medicine tend to link the hungry or over-worked, exciting or relaxing past experiences of their bodies to the present state of their pleasures and pains. This is a well-known feature of patients' subjective accounts of illness the world over. But in the idioms made available by Chinese language and Chinese herbs, we can see bitter experiences rendered treatable with sweet drugs, pungent substances used to set the stagnant fixities of old pathology into more wholesome motion. This experiential side to Chinese medicine encourages a personal micro-politics, as patients seek to govern themselves and their immediate environment using techniques that fuse thinking and feeling, forming habits that make sense to their own senses.

NOTE

1. The Upper, Lower, and Middle *Jiao*, sometimes translated as the Three Burners, are classified together as a visceral system in Chinese medicine. The standard modern accounts present them as three spatial regions of the functional body, especially responsible for managing the movement of fluids in the body.

REFERENCES

Lévi-Strauss, Claude. 1969. *The Raw and the Cooked*. New York: Harper and Row.
Yue, Gang. 1999. *The Mouth that Begs: Hunger, Cannibalism, and the Politics of Eating in Modern China*. Durham: Duke University Press.

NANCY K. MILLER

"Would it not stand to reason that men and women read differently, that there must be a fundamental disparity between what they bring to, do to, demand from and write about texts? And if we do read differently, is it not necessary to figure out how and where we do so? To what extent is the 'body of scholarship' on any given author really *two* bodies?" These questions, formulated by Susan Winnett as the rationale for a special session at MLA [the Modern Language Association] (1983) called "Reading and Sexual Difference," are the pretext and context of what follows. I have chosen to address the question of sexual difference literally (to the letter, as it will emerge), taking its terms, for the purposes of argument, as a question of identity tied to a material body circumscribed by and in the *practice* of reading and writing.

Feminists have no sense of humor, as Wayne Booth has shown.

In a celebrated moment at a symposium organized by *Critical Inquiry* in 1981 on the "Politics of Interpretation," Wayne Booth, then President of MLA, came out as a male feminist; or at least, as an admitted "academic liberal," embraced the "feminist challenge" to *reread*, recognizing what "many feminist critics have been saying all along: our various canons have been established by men, reading books written mostly for men, with *women as eavesdroppers*" (1982:74; emphasis added). Taking Rabelais as the authorial example to illustrate his point, Booth addressed the question of ideology and aesthetic response; of sexual difference and the pleasure of the text:

> When I read, as a young man, the account of how Panurge got his revenge on the Lady of Paris [as you recall, he punishes the lady for turning him down by sprinkling her gown with the pulverized genitals of a bitch in heat; he then withdraws to watch gleefully the spectacle of the assembled male dogs of Paris pissing on her from head to toe] I was transported with delighted laughter; and when I later read Rabelais aloud to my young wife, as she did the ironing(!), she could easily tell that I expected her to be as fully transported as I was. Of course she did find a lot of it funny; a great

deal of it *is* very funny. But now, reading passages like that, when everything I know about the work as a whole suggests that my earlier response was closer to the spirit of the work itself, I draw back and start thinking rather than laughing, taking a different kind of pleasure with a *somewhat* diminished text (68).

Booth's account of his conversion experience is of course important and gratifying for a feminist critic; but what I want to take off from here is not so much his self-conscious rereading *as a man*, as his staging, within the scene of his own discourse of a woman's point of view, his placing of a "voice," however muted, that he finds absent from Rabelais's, and Bakhtin's, dialogics (61, 65). Mr. Booth does not tell us how he knew what parts his wife did not find funny, or why. But let us guess that listening to this story from a book "written mostly for men," Mrs. Booth felt a bit like an eavesdropper, like a reluctant voyeur called upon to witness a scene of male bonding (a man and his best friends) which excluded her. Or perhaps, as a woman (reader), she put herself, for a moment, in the Lady's place.

To reread as a woman is at least to imagine the lady's place; to imagine while reading the place of a woman's body; to read reminded that her identity is also remembered in stories of the body.

Like the young Mrs. Booth, the ironing notwithstanding, I too can easily tell when I am supposed to find something funny. I can tell, read to or reading, when I am expected to be transported, as "fully transported" as those we might think of as the *dominant responders* (or "dr's")—one's husband, one's teachers, the critics; though clearly I am not nearly so good a sport. So let us turn now to a less "carnivalesque" but equally witty example of the canon; to an epistolary novel of the Enlightenment in which every reader is perforce an eavesdropper, or rather a voyeur reduced to the scopic delights that may be derived from reading someone else's mail.

In letter 47 of *Les Liaisons Dangereuses* the Vicomte de Valmont explains to his correspondent, Mme de Merteuil, that while making love to Emilie, a courtesan, he writes a passionate love letter to Mme de Tourvel—the woman in the novel he is trying to seduce—literally, or so he would have the Marquise believe, *on* Emilie.[1] Writing vividly "to the moment," Valmont codes his "undressed" physical exertions in the rhetoric of "dressed" amorous discourse:[2] "I have scarcely enough command of myself to put my ideas in order. I foresee already that I shall not be able to finish this letter without breaking off," etc. Double meanings follow upon each other like a rigorously expanding metaphor; and the foreseen interruption of the (in)scription occupies (neatly) the blank space between paragraphs. The performance over, Valmont forwards the letter—48 in the volume—to Merteuil, leaving it open for her to read, seal, and mail (he wants, he says, to have the letter postmarked from Paris). Emilie, the woman as desk, and the first reader of the love letter, Valmont recounts, "a ri comme une folle," she "split her sides laughing." "I hope you will laugh too," Valmont adds, by way of preparing Merteuil's response to his text and its context: his one-man show.

Readers, on the whole, have responded to the optative and been amused. In an essay entitled, "The Witticisms of M. de Valmont," a critic observes:

> Doubtless many readers will find this instance of Valmont's sense of humor particularly intolerable, and will rebel at this display of poor taste. To them this long series of double meanings will not endear the Vicomte, nor make him in any way more charming. But *willy-nilly*, as they read this letter, *they will become a party in his conspiracy*, for the simple reason that they will understand the hidden though transparent meaning of the letter, to which the addressee, Mme de Tourvel, cannot but remain blind. Moreover, these morally inclined readers who, to be sure, will condemn Valmont on elementary grounds of decency, *will also have to appreciate*, on a *strictly intellectual or even perhaps esthetic level*, Valmont's remarkable though devilish intelligence and his enviable mastery of stylistic devices (May 1963:182–83; emphasis mine).

This model of double reading, which opposes the "morally inclined" to the "strictly intellectual or even perhaps esthetic," and valorizes the victory of the latter position—since it implies a narcissistic alliance between two hegemonic modes ("enviable mastery")—describes very nicely the plight of the feminist critic of whom it is regularly said, as Booth imagines it will be said of him: "I've lost my sense of humor or I don't know how to read 'aesthetically' " (68).

In a more recent work on Laclos, the double reading becomes triple and the aesthetic grounds, hermeneutic. The critic, glossing letter 48 and in particular the coding of scriptus interruptus—"I must leave you for a moment to calm an excitement which mounts with every moment, and which is fast becoming more than I can control"—concludes: "This is a blatant example *of double entendre*, amusing and well done; it demands a rereading, and maybe even a third reading, at first to laugh, and then to wonder" (Rosbottom, 1979:108).

The third reading then wonders about the transmission of information in the novel, the ways in which the novel insists upon "the process of signifying," and takes as its subject, "how things mean" (109). But it might also wonder whether it is really the perception of the double register in itself—"the very table on which I write, never before put to such a use, has become in my eyes an altar consecrated to love"—that compels complicity. It seems, rather, that the appeal of the letter derives from a masculine identificatory admiration for Valmont's ability to do (these particular) two things at once. This is of course not to deny that to conjoin in representation activities not normally conjoined is textually arresting—like that of Aesop's master, who, to save time, pissed as he walked, as Montaigne relates in "Of Experience," or, *a contrario*, Gerry Ford, of whom it was said, in the cleaned-up version, that he could not walk down the street and chew gum at the same time. But I do want to insist (at the same time) on the specific, almost textbook phallogocentric reference (referents) of Valmont's conjunction.[3]

Let us move now from the ways in which sexual difference can be said to structure the scene of production, the actual production of reading material, to the scene of reception, the reading of the letter, and the glossing of its text.

We should recall that the dominant trope of the act of novel-reading in the eighteenth century is the figure, or allegory, perhaps even the fact, of the *lectrice*, the woman reader reading. *Les Liaisons* provides us both with the standard model, Mme de Tourvel—the beleaguered heroine in a story she does not understand, reading, as fortification against the plotting hero, volume two of *Christian Thoughts* and volume one of *Clarissa*; and the model ironized, Mme de Merteuil. The super woman reader who would be (male) author (early in the novel she proposes to write Valmont's memoirs in his place), the marquise instead rereads from Crébillon, Rousseau and La Fontaine to prepare for her part in the fiction she embodies but cannot represent in letters; or rather, which she at the end re-presents as truth at the cost of disfiguration.

The tropology of the woman reader reading is also evoked in the novel's frame as the Editor rehearses his "anxiety of authorship" with a particular emphasis on the reception of the work by a female readership.[4] Thus in his prefatory moves, the Editor invokes the sanction of a "real" woman reader to authorize his publication: a good mother, who, having read the correspondence in manuscript, declares her intention (in direct quotes) to give her daughter the book on her wedding day. This singular "already read" fantasized wishfully as a collective imprimatur—"if mothers in every family thought like this I should never cease to congratulate myself for having published it"— consoles the Editor for the more limited readership he "realistically" imagines for this collection. Like Emilie, the bad daughter *(la fille)*, the first woman reader, the good mother, has a sense of humor, even wit: *de l'esprit*. But are they reading as women? Put another way, what would it mean to place oneself, to find one's place as a woman reader within a phallic economy that doubly derives female identity through the interdigitated productions of the penis and the pen. Thus, on the one hand, the body-letter receives the sperm-words, she is the master's piece, the masterpiece;[5] and on the other, she figures as metaphorical "woman," disembodied because interchangeable: I hope you will laugh too.

If we follow out the privileged hermeneutic chain, Emilie, Merteuil, me, you (all of us narratees doubled into whores by the act of reading itself), it seems fair to claim that the female reader of this novel is expected to identify with the site/sight, topographical and visual, of her figured complicity—like that of the classical *lectrice*, Mme de Tourvel—within the dominant order. Tourvel is the "immasculated" reader, who is "taught to . . . identify with a male point of view. . . . Intellectually male, sexually female, one is in effect no one, no where, immasculated" (Fetterley, 1978:xx, xxii). What is it possible for a woman to read in these conditions of effacement and estrangement; in a universe, to return to our previous discussion, where the rules of aesthetic reception and indeed of the hermeneutic act itself are mapped onto a phallomorphic regime of production?

The third reading I evoked earlier might not so much ponder the (im)possibility of fixing meaning, or admire the text's "modernity," as interrogate the particular kind of phallic modeling that attaches itself to the very process of encoding and decoding, or vice versa. A third reading might wish instead to reevaluate the reciprocities of sexual and scriptional practices, and rethink a metaphorics of writing and reading that figures, and at the same time paradoxically *grounds*, "woman" as material support for a masculine self-celebration; a metaphorics that specularizes *double entendre*—which we might freely understand as the trope of interpretation itself—as the couple man/woman (man over woman) and fetishizes the superscription of the masculine. (This might also be thought of as a working definition of the canon.)

But even a third reading produces for me not only a *"somewhat* diminished text," and a "different kind of pleasure," to return to Wayne Booth's formulation, but finally an acute desire to read something else altogether and to read for something else.

In closing, therefore, I want to evoke another (almost) eighteenth-century author, and another—for me more congenial—staging of the scene of reading, writing, and sexual difference. I am referring to the famous moment in Jane Austen's last novel, *Persuasion*, when Captain Wentworth drops his pen. "While supposed to be writing only to Captain Benwick," (239) and while eavesdropping on a conversation between his friend Captain Harville and the heroine Anne Elliot about men's and women's inconstancy and constancy, Captain Wentworth, in a room full of people, writes a love letter.

The pen falls as Anne sets forth what we have come to call the sex/gender arrangements of our culture: the division of labor that grants men, among other things, the professions, and women, the private world of feelings. Here a woman figuratively picks up the pen, as Austen's heroine decorously but specifically protests against that troping of the spheres, against the penmanship of the hegemonic culture: "men have had every advantage of us in telling their own story. The pen has been in their hands. I will not allow books to prove anything." The "histories" from which one quotes to prove one's point were all, as Captain Harville is quick to grant, written by men. In *Persuasion*, which recycles the epistolary and retrieves its earlier (more Rousseauian) desire for the proper destination, Wentworth delivers his letter by hand its "direction hardly legible, to 'Miss A.E.-.'" And we read *with* her: "While supposed to be writing only to Captain Benwick, he had been also addressing her!" (239).

To miss this letter is to run the risk of believing that the pen is nothing but a metaphorical penis, even though we know, as we stand here ironing, that that has never been entirely true. For if Valmont writes to (and on) women in order, ultimately to be read by men—other libertines, literary critics etc. (women are, we know, always and merely the cover and site of exchange for this founding homosocial contract)—then Wentworth, "at work in the common sitting room," as Gilbert and Gubar have persuasively imagined him, "alert for inauspicious interruptions, using his other letter as a kind of blotter to camouflage his designs," much like "Austen herself" (1979, 179)—, writes, I would argue, *as a woman* to a woman in order to be read by women. (Paradoxically, this shift in *address* is made possible by the historical development of the female readership in the eighteenth century, whose authorization, we saw, the Editor of *Les Liaisons* anxiously and ironically hypothesized in his preface.) By having Captain Wentworth both drop his pen because of the sound of a woman's voice, and write a love letter in dialogue with the claims of her discourse, Austen, I think, operates a powerful revision of the standard account: she puts the pen in the hand of a man who, unlike Valmont (or Lovelace), wishes to have his feelings for the other "penetrated" (240). This wish for transparence, however nostalgic, may be said to deconstruct the familiar logic of the body as parts, and by that move writes the possibility of a body beyond parts; a body, therefore, through which the stories of sexual difference would have to be figured otherwise.

1. There is, despite the insistence on the physical aspect of the action—producing "a letter written in bed, in the arms, almost, of a trollop"—a peculiar effacement of (the) woman's body in and as representation. It is curious to note that critics do not seem to agree, or rather do not seem to be able to *say* exactly what goes on here. Peter Brooks, in his presentation at MLA, imag(in)ed the letter written on Emilie's "bare backside." Jacques Bourgeacq argues (against "the literary tradition that limits Emilie's role in the composition of the letter to the use of her back") that Emilie actively participates in the action; that the "action external to the letter" as he euphemizes it, "organizes the text itself" (1980:185–86; my translation). Thomas Fries, reading deconstructively the modeling of sexual difference in the novel, seems to assimilate the ambiguity implicit in the positioning of the woman's body to the general problematics of the novel, to the "Figurality of language and more precisely in the figure Latin rhetoric calls *dissimulatio*" (1976:1308). For him this ambiguous manipulation of knowledge is the source of the pleasure of the text; he cites "one of the supposedly most scandalous passages of this book, the famous letter *written from and on some lower part of Emilie's back*," as a typical example (1309, emphasis mine). The blockage of mimesis is, I think, tied to the narrative framing. By the use of a figure which is either periphrasis, or catachresis, or both—"in the arms, almost"—interpretation is both forced and inhibited. Paradoxically, this "undecidability" does not interfere with the reading effect which derives from the reading *process*, from the semiosis: one knows that a woman's body is being enlisted in a male production of language.

2. I allude to and play with the distinctions in letter writing styles established in Altman (1978).

3. A conjunction, and therefore sexualization, that, as I will argue, dangerously privileges the letter writer over the letter reader; and conflates, predictably, virility and authority: "the juxtaposition of the lover's desire for sexual gratification and the writer's penetration into his subject is too strong to be overlooked" (Fries, 1976:1309). And Michel Butor, enlisted by Fries, supplies a depressing and acute parenthesis on the subject of the lover as writer (1964:50): "what is most surprising in Valmont's famous letter to the Presidente written by using Emilie as a desk, is not that he interrupted it to 'commit a downright infidelity,' it's the transformation of the woman into a desk, into a means of writing to another [woman]; it's that in her bed and almost in her arms, he began with the letter, and the infidelity committed, he had nothing more pressing to do than to start writing again" (my translation). Male desire, then, originates not in the "arms" of a woman present, but in the desired absence of the woman to be written to. Would it not be possible to argue that the effacement of the woman's body in representation is here due precisely to its transformation into instrumentality? Emilie must remain invisible since her function is merely and classically to facilitate the exchange of women and/or signs.

4. I am playing here with the phrase Sandra Gilbert and Susan Gubar have constructed—against Harold Bloom's model of a male "anxiety of influence"—to account for the psychological posture of the female artist: her "need for sisterly precursors and successors, her urgent sense of her need for a female audience" (1979:50).

5. In " 'The Blank Page' and the Issues of Female Creativity," Susan Gubar summa-rizes the history of this figuration: "This model of the pen-penis writing on the virgin-page participates in a long tradition identifying the author as a male who is primary and the female as his passive creation—a secondary object lacking autonomy, endowed with often contradictory meaning but denied intentionality" (1981:247). It is interesting to note that for "virgin" we can here substitute "whore" and leave the meaning of the argument intact. (Though one might also want to read a desire for the virgin in Valmont's hieratic acrobatics: "the very table on which I write, *never before put to such use*, has become in my eyes an altar consecrated to love.")

REFERENCES

Altman, Janet. 1978. "Addressed and Undressed Language in *Les Liaisons Dangereuses*." In *Laclos: Critical Approaches to Les Liaisons Dangereuses.* Madrid: Studia Human-itatis.

Austin, Jane. 1975 (1818). *Persuasion.* Harmondsworth, Middlesex: Penguin.

Booth, Wayne. 1982. "Freedom of Interpretation: Bakhtin and the Challenge of Feminist Criticism." *Critical Inquiry* 9:1 (September): 45–76.

Fetterley, Judith. 1978. *The Resisting Reader: A Feminist Approach to American Fiction.* Bloomington: Indiana University Press.

Gilbert, Sandra, and Susan Gubar. 1979. *The Madwoman in the Attic: The Woman Writer and the 19th Century Literary Imagination.* New Haven: Yale University Press.

Gubar, Susan. 1981. " 'The Blank Page' and the Issues of Female Creativity." *Critical Inquiry* 8:2 (Winter), 243–64.

Laclos, Choderlos de. 1951 (1782). *Les Liaisons Dangereuses.* Paris: Gallimard.

May, Georges. 1963. "The Witticisms of Monsieur de Valmont." *L'Esprit Créateur* 3:4, 181–87.

Rosbottom, Ronald. 1979. *Choderlos de Laclos.* Boston: Twayne.

PART V

Colonized Bodies, or Analyzing the

Materiality of Domination

Colonized Bodies, or Analyzing the

Materiality of Domination

INTRODUCTION

The authors of the essays that follow are not content with decontextualized arguments about the repressive effects of colonial and postcolonial regimes. Through fine-grained historiography and ethnography they draw attention to body representations and practices as key sites for intense examination of the embodied effects generated by many forms of colonization. Here the fraught relationships among colonial and postcolonial regimes, their regional representatives and administrators, and local communities and individuals are carefully described. Hybrid bodies proliferate under colonial regimes in ways that surprise and frustrate the imperial impulse. Gone are older portrayals of unmediated domination by those in power over the health, illness, comportment, and reproduction of peoples cast as though they have no culture and little or no agency.

During the eighteenth and nineteenth centuries the extension of colonial control to remote territories was billed at home as a humanistic endeavor—a means of saving primitive peoples from their ignorance, dirt, and lack of godliness. Colonized bodies were understood as "symbolic inversions" of European bodies, as "diseased, lazy, and grotesque." But it was also believed that "strict regimes of personal and domestic hygiene" would naturally bring about improvements.[1] It was clear from the outset, however, that concern about disease in the colonies was far from benevolent. Fears of contagion led to efforts to partition off "native" from expatriate populations, and colonial governments reorganized local populations so new forms of labor could be extracted.[2] In hindsight, it is evident that colonized populations responded to the impositions of colonizers in a variety of ways. Mestizo elites emerged in new social strata whose access to cosmopolitan resources was real but still limited by "race." Class, cultural and gender hierarchies already present in these regions were sometimes exacerbated, with effects that continue to undermine social stability long after formal decolonization.

In his book *Colonizing the Body*, for example, David Arnold takes up the relationship among the British colonizers, the Indian State, indigenous medical practitioners in nineteenth-century India, and the management of the infectious diseases of smallpox, cholera, and the plague. Arnold uses historical documents to show how the Indian body became a "site of colonizing power and of contestation between the colonized and the colonizers" (1993:8). Arnold shows that colonization is very much a corporeal matter—the body is made into a site where those with power act out their authority, legitimacy, and control—but, he argues, equally important was the knowledge that could be garnered from these newly encountered exotic bodies in health and disease. Colonized bodies were in effect transformed into scientific gauges by means of which the environment, with its pox, miasmas, and toxins, could now be calibrated. In this project tropical medicine was born, and knowledge obtained from this new specialty gradually modified medical wisdom in the capitals of Europe.

Arnold shows how a vast array of ideological discourses, administrative mechanisms, and health-related practices associated with the beginnings of modern medical knowledge became insinuated into Indian society. But he goes further and demonstrates the wide range of Indian responses to the situation, ranging from resistance to accommodation, participation, and appropriation of medical discourses and powers. As he observes, "Medicine was too powerful, too authoritative, a species of discourse and praxis to be left to the colonizers alone" (1993:10).

In an equally subtle history, Jean Comaroff, writing about nineteenth-century representations of southern Africa, argues that Europeans imagined Africa as a "hothouse of fever and affliction," filled with suffering and degenerate peoples to which they brought a mission of "humane imperialism."[3] In Africa, in particular, colonization went hand in hand with the introduction of European medical practices carried out by Christian missionaries: activities that mutually reinforced and legitimized each other. Relegated to the bottom of the Enlightenment "chain of being," Africans were depicted as being sorely in need of both physical and spiritual help. In her book *Body of Power, Spirit of Resistance*, Comaroff discusses at length the effects that colonization had on everyday life and the mixed responses of resistance and assimilation on the part of local peoples. In the millennial Zionist or Spirit churches that sprang up everywhere in opposition to both mission Protestantism and secular medical management, reconstrual of the lived body was a crucial activity involving rites of healing, dietary taboos, and dress codes. Comaroff documents the transformation brought about in local consciousness through participation in such churches. Significantly, these were sites where liberal use was still made of earlier, indigenous symbols and ritual. These cults of affliction dealt with more than physical ills; they united members in a reconstructed social and moral order designed to bring about individual cleansing and increased practical control over their everyday lives, perceived to be under siege by the colonizers.[4]

Comaroff and Arnold contribute prominently through these works to a colonial historiography that has recently exploded with richly empirical studies. The essay by Susan Pedersen reproduced here, set in British-controlled Kenya of the early twentieth century, shortly before independence was declared, is an example of this kind of history. Hers is a cautionary tale about the importance of attending to actual practice, rather than settling simply for an analysis of official discourse to be found in archived documents. In researching what became known as the "female circumcision controversy," Pedersen is in part concerned, like Comaroff, with colonial missionary activity. She effectively shows the enormous gaps that existed between the law passed in Great Britain making it illegal to carry out female circumcision in Africa and several disputatious positions taken by people working in the Colonial Office in London. These views in turn diverged from those of local colonial authorities in Kenya, and what actually devolved in practice was most influenced by a patchwork of mediating social groups who sometimes applied the law and sometimes altered or deviated from it entirely. Missionaries were among those who attempted to enforce the law against circumcision, but in doing so they brought on the largest outbreak of popular protest among the Kikuyu that the British government had experienced until that time.

Pedersen's essay exposes the stubborn practical dilemmas that all contenders in this dispute had to face. She recounts how divisive political arguments in London were evidence of a concern for the careful management of an emerging nationalism associated with the impending transition to self-government. And she documents how ingrained ideas about the sexuality of African women were deployed—often tacitly—to shore up this debate. Unlike most of the politicians, leading feminists of the day, together with some noted humanitarians, were concerned above all about the actual welfare of African women; but because they could discuss circumcision only in euphemistic terms, their objectives more often than not appeared confused. Like the missionaries but for different reasons, these activists wanted circumcision to be stopped. White settlers, on the other hand, wanted no interference with the "savage black race" in order that they, the settlers, could assume political leadership in Kenya at independence. They contended that it was dangerous to interfere with local custom, and that white rule was justified in part *because* of barbaric practices such as circumcision. Pedersen notes that the absent voices in this dispute were, of course, those of African women.

Bitter arguments about female genital surgery intimately associated with nationalism, religious fundamentalism, feminism, and postcolonial discourse continue to this day, but now African women, although by no means united among themselves, make a vociferous contribution to the debate.

The next three essays are graphic examples of the impact on everyday life of the colonization of bodies and consciousness. Stuart Cosgrove's essay on zoot suits, set in America of the 1940s, offers a remarkable example of the use of style in clothing and body adornment to make powerfully subversive statements. Zoot suits spoke of subcultural identities and forms of minority belonging, while also displaying strong feelings of ambivalence and dissent toward the dominant culture. Cosgrove argues that this form of display is best understood in historical context, in light of the legacy of colonialism. The African American and Mexican-American youth (*pachucos*), who paraded in zoot suits while taking pleasure in the effects their eccentric outfits had on others, at the same time exhibited a form of resistance to the subservient positions in which they inevitably found themselves. In flaunting their own style in this way, these young men knew they would be associated by the dominant culture with delinquency and draw the attention of unfriendly authorities. In this way they challenged power in a domain—dress—that could not be regulated by law.

To what extent zoot suiters were consciously aware of the political ramifications of their wardrobe choices is a moot point, but the riots that took place between white youth and black and Latino zoot suiters in Los Angeles in 1943 must have left little doubt in anyone's mind about the political potency of dress style. Many similar examples of the political uses of style in dress and bodily adornment have been documented (see, for example, Dick Hebdige's now-classic book on the subject [1979]). One striking example is the patriotic dress adopted by many Americans after the terrorist attacks of September 11, 2001, in which national symbols were highly visible. Terence Turner's essay in part I of this book is, as we have seen, also concerned with style—in the Amazon of the 1960s. As in contemporary America, in the Kayapo case it was affirmations of belonging and not those of dissent that were on display.

Following Michel Foucault, David Arnold argues that there is a "sense in which all modern medicine is engaged in a colonizing process."[5] In the first of these essays, Janice Boddy, juxtaposing accounts of birth in the 1980s in Sudan and Toronto, makes this claim vividly real. While doing fieldwork in a village on the east bank of the Nile over the course of more than a decade, Boddy was sufficiently integrated into village activities that she was invited to participate in birthing practices. She opens her essay with a moving account of her return to the village in 1994 and her visit with the family of Amal, a friend who had died in childbirth during Boddy's absence. Reflecting on Amal's death, Boddy realizes that she is unable to account for the problems that arose simply in terms of a lack of biomedical resources or poor local conditions. She thus weaves a complex account in which her participation in the stressful, surgically induced delivery of a friend in Toronto is contrasted with deliveries she witnessed in the Sudan. She notes that contemporary Sudanese practices are indebted to the work of a pair of formidable British sisters who, in the early twentieth century, took it upon themselves to retrain Sudanese midwives, partially transforming local birthing practices. Their approach was one in which birth was made into pathology, in keeping with modern obstetrical attitudes, rather than being understood as a natural event requiring patience on the part of all participants. As an instance of the resultant hybrid public health of childbirth, we meet Sheffa, a Sudanese midwife, whose childbirth practices are informed by local custom and a few simple modern technologies. The means by which Sheffa manages relatively safe births by women who have been infibulated—a form of genital surgery—is particularly interesting. Boddy's description invites the reader to compare surgical childbirth in a Toronto hospital with a no more violent or traumatic birth in a Sudanese village house. Boddy turns to Foucault to remind readers that, under the modern state, power confers benefits even as it disciplines; she ultimately argues that the majority of women in countries where medical technology is readily available have actively participated in the general move throughout the twentieth century toward "androcentric" birthing practices. This widely desired biomedicalization of birth represents, she suggests, an "impressive colonization of consciousness—of selfhood." Boddy laments that medical management of birth is now more commonly understood in wealthy countries as palliative, rather than therapeutic; the difference between suffering in the course of a normal reproductive life and seeing one's body pathologized in the clinic should not be forgotten. But she notes the terrible irony that, in the Sudan, where poverty, malnutrition, economic dependency of women on men, and infectious and parasitic diseases are widespread, and where female circumcision is practiced, an androcentric and colonizing medical intervention is probably the only way to avoid deaths like that of her friend Amal.

Patricia Kaufert and John O'Neil also write about birth. They document the historical and political context in which medical records were first created and then made use of in the Canadian north during the 1960s and 1970s, specifically in order to justify medical control over Inuit birth. This control was exerted in a particularly coercive manner: a policy was enacted that required all but a very few pregnant Inuit women living in the Keewatin region of the Northwest Territories to be evacuated by plane to the cities in the south to give birth. Kaufert and O'Neil examine how biased epidemiological data and sparse obstetrical records were drawn on to justify this policy, with the objective of lowering an assumed but unsubstantiated high infant mortality rate.

These authors make it clear how standard risk assessments made in connection with pregnancy were and are inappropriately applied to Inuit women; they critique the dual assumptions present in modern obstetrics: that giving birth is everywhere the same basic process and that the risks associated with birth are thus universal. A brief discussion of traditional birth practices described by Inuit elders and midwives reveals the moral, symbolic, and political importance assigned to a private and silent labor among Inuit women, demonstrating prized values of independence and self-sufficiency. As Kaufert has elsewhere argued, Inuit villagers often think of risk quite differently from officials, whose perceptions of risk are inseparable from statistical models. Thus, this article shows that during the colonial period, few if any government officials were concerned with birthing or public health among the Inuit. It has been only over the past thirty years, when the Inuit gradually gave up a nomadic lifestyle and started to settle in communities, that they became visible to those in authority as anything other than an exotic Other of the far north. As northern populations were more closely integrated into the Canadian polity, their health status became an issue for statisticians. Anxiety about a relatively poor national infant mortality rate in Canada as compared to most parts of Europe ensured that a "beneficence of power" would be the response to "improving" Inuit health and well-being.

The policy of evacuating pregnant women who were ready to give birth was easily justified in the minds of bureaucrats in terms of what was obviously safer, cleaner, and the right thing to do. During the 1990s, however, in the face of heightened concern in Canadian politics about past injustices against indigenous peoples during and after colonization, coupled with activism on the part of Inuit themselves, policies were revised. Formal training of Inuit midwives who could remain in their rural communities was instituted. In part as a result of official reports made by Kaufert and O'Neil, the evacuation policy was radically modified, thereby decolonizing the bodies of Inuit women seeking more control over local childbirth.

The short essay that concludes this section turns attention to India. This excerpt is taken from Jean Langford's book *Fluent Bodies: Ayurvedic Remedies for Postcolonial Imbalance*. Like Boddy, Langford makes clever use of juxtaposition, in this case of the "dosic" bodies of Ayurvedic medicine with the "docile" modern bodies classically discussed by Foucault. The dosic body that appears in Ayurvedic texts is above all fluid and polyvalent, it "bears the imprint of a social matrix. . . . In this body the heart can be both a center of circulation and a center of consciousness, while warmth can be simultaneously temperature, temperament, and a somato-environmental agent that eludes modern categories." Citing Frances Barker, Langford argues with respect to dosic bodies that "it would be better to speak of a certain 'bodiliness' than of 'the body.'" However, in modern Ayurvedic anatomy departments the bodies represented are almost without exception depictions taken from biomedical texts. Such depictions are designed to create a naturalized "homogenous" space in which body parts can easily be located. Depictions of the Ayurvedic body, in contrast, leave much to the imagination, so that the specificities associated with each patient can be appreciated. While it is possible to make comparisons between the grotesque body of Rabelais as described by Bakhtin and dosic bodies, Langford is quick to point out that the dosic body is not transgressive; nor is it a body reacting to bourgeois boundaries. The dosic body continues to undergo postcolonial transformations up to the present time, as Ayurvedic practitioners seek to equal and even surpass European medical practices; like Sudanese and Inuit midwives, many practitioners exhibit hybridity in their knowledge and practice and draw on ideas about both docile and dosic bodies. Several of the essays included below make it vividly clear that the legacy of colonialism persists today. In postcolonial situations, with the rise of ardent nationalisms and of open conflict over ethnic and cultural differences (many of which stem directly from colonial restructuring of local social practice), tensions created as a result of body politics are, if anything, exacerbated, even as the technologies associated with modernity and Westernization are selectively embraced. Above all, it is clear that pluralism and hybridity abound; stark oppositions between tradition and modernity, indigenous and biomedical bodies, colonizer and colonized, colonial and postcolonial regimes are no longer valid, if indeed they ever were.

NOTES

1. Anderson 2003:235.
2. Bush 2000.
3. Comaroff 1993.

4. Comaroff 1985.

5. Arnold 1993:9.

ADDITIONAL READINGS

Adams, Vincanne. 2001. "Establishing Proof: Translating 'Science' and the State in Tibetan Medicine." In *New Horizons in Medical Anthropology: Essays in Honour of Charles Leslie*, edited by M. Nichter and M. Lock. New York: Routledge.

Briggs, Charles, and Clara Mantini Briggs. 2003. *Stories in the Time of Cholera: Racial Profiling During a Medical Nightmare*. Berkeley: University of California Press.

Burke, Timothy. 1996. *Lifebuoy Men, Lux Women: Commodification, Consumption, and Cleanliness in Modern Zimbabwe*. Durham: Duke University Press.

Greenblatt, Stephen. 1991. *Marvelous Possessions: The Wonder of the New World*. Chicago: University of Chicago Press.

Stoler, Ann Laura. 1995. *Race and the Education of Desire: Foucault's History of Sexuality and the Colonial Order of Things*. Durham: Duke University Press.

Vaughan, Megan. 1991. *Curing Their Ills: Colonial Power and African Illness*. Stanford: Stanford University Press.

REMEMBERING AMAL: ON BIRTH AND
THE BRITISH IN NORTHERN SUDAN

JANICE BODDY

NOVEMBER 1994

The ochre light of dusk, birds clamoring in the ne'em tree, a muezzin's call to prayer. I open the familiar blue door and enter the courtyard without knocking. A sense of home engulfs me. Everything seems as it was: to my left, the breezeway sheltering water jars, its earth floor damp and cool; to my right, the gallery off the storeroom where we sat to pound spices, wash clothes, brew coffee, and talk. Ahead is the saysaban Salima planted all those years ago, alive still, miraculously grown, marking time. But the far verandah has been walled for rooms. There is a new garden with morning glories, sunflowers, and henna, gifts of piped water one hour a day.

Salima emerges from the goat pen at the back. She sets her milk bowl on the kitchen sill, embraces me without a word. At once she starts to tremble, weeping. She wails a rhythmic mourning phrase whose drawn out words I cannot catch. Now I am crying too, holding her shoulders as she holds mine.

Moments later we sit and dry our eyes in silence. Then, abruptly, she smiles. "You've been a long time away," she says. "Much has happened. Awad has two more daughters, both very young. Would you like to meet his new wife? Would you like to see Amal's little girl? She's ten years old now, and very smart, first in her class at school."

Awad and his wife appear, each with a freshly bathed toddler in arms. Amal's daughter, tall for her age, hangs back, hiding behind her father's legs. He places a hand on her shoulder and gently draws her forth. "Her name is Nura," he says. "We named her for our grandmother." The girl is a miniature of her mother, over half as old as she was at her death.

For several months in the mid-1980s, I lived with Salima, Awad, their parents, and Amal, Awad's bride, on the east bank of the Nile in an Arab Sudanese village I call Hofriyat. I had spent over a year there in the mid-1970s doing ethnographic research and had returned to continue the work. Amal was pregnant and understandably apprehensive. She had been ill with malaria, anemia, and a bladder infection. The clinician had detected proteins in her urine; she seemed always to be in pain. Whole days she lay groaning on an *angareeb* (rope bed) in the gallery. Other days she insisted on doing her share of the work, hauling water home from the well, scrubbing Awad's clothes.

From the moment she knew she was expecting, Amal wore her wedding gold to thwart capricious spirits that might seize her womb and loosen its captive seed. Amal's body had become a protected domestic space, a figurative house wherein mingle the male and female contributions that shape and sustain human life (Boddy 1989). Her blood nurtured a kindred being who would remain an intimate part of herself even after birth.

Amal ate well when she could, "clean," expensive foods—eggs, grapefruit, guavas, tomatoes, meat fava beans, lentils—that Awad supplied to help build her blood. She went to the village clinic for weekly prenatal checks, but the government midwife was never there—out on call Amal was told. Only the male nurse's aide in charge of the tiny, woefully understocked dispensary had examined Amal—or so she said.

Later she revealed that he had not. She did not trust the man; he is not related to the people of Hofriyat. Worse, he's rumored to be possessed by *zar* spirits that are known to play havoc with women's fertility. At his office he is regularly found listening to recordings of *zar* songs to appease his spirits. We suspect he is possessed by the European Doctor *zar*, Hakim-bi-Dor (Doctor by Turns), because both man and spirit are least discontent when patients are lined up waiting to see them. The clinician often roams the village in character: stethoscope slung round his neck, blood pressure cuff strapped to his arm. Neither implement is properly functional, as his watch lacks a second-hand. The spirit-possessed clinician is a parody of Western medicine, a stinging, if suitable, comment on its efficacy in impoverished postcolonial Sudan. Amal would not let him palpate her belly; she was too shy, she said, too ashamed. Besides, he was always trying to sell her more vitamins to "fix her body" and had once advised a lengthy course of tetracycline which Awad vetoed, aware of the harm it could cause the unborn child.

Awad and I begged her to see a proper doctor in Shendi, the closest town, thirty kilometers away. We offered countless times to accompany her there on the bus. I would approach the doctor, I said, and I would stay with her; she need not feel ashamed, pregnancy was natural, she was a married woman, it was important to take care of herself and the baby. Awad even ordered her to go, but she refused, and left to stay with her mother a few doors away.

Amal's stubbornness was not without reason; she confided that she could not bear the thought of being examined by strangers, all of them men. Hofriyati women conceal their pregnancies in public, something that modesty wraps and a preference for corpulent figures make it possible to do. Even close female kin refer to the condition euphemistically, for pregnancy is unmistakable evidence of sexual activity, and it is shameful for women to evince their sexuality. It is doubly improper to do so in front of men.

So we went for Miriam, the busy district midwife, who questioned Amal and examined her eyes and mouth but was not alarmed. When the pains returned I suggested we call for Sheffa, the midwife from a neighboring district, who had impressed me with her cool professionalism and kindly, nononsense approach. Sheffa was a qualified nurse's aide—well beyond Miriam's achievements. She had a confidence Miriam lacked, and a bedside manner that would, I was sure, convince Amal to seek the medical help she clearly required. But the family objected that Miriam would take offense, placing their good relations with her, hence the welfare of all its women, in jeopardy. Midwives are powerful people; their prompt response can mean life rather than death for mothers and infants alike.

I left Hofriyat shortly before Amal's baby was due. Hugging her good-bye on boarding the rickety bus for Khartoum, I feared I would not see her again. I think she knew that I would not.

I have since had the luxury to ponder Amal's response and my insistence, my undeterred faith in biomedical technology, a faith that increasingly resides with Hofriyati too. I am less certain now that Amal's death in childbirth was readily preventable, and less convinced that she died for the sake of shame, or if she did, that it was a meaningless or, for her, avoidable act. I have no answers, only half-figured questions and a weight of accumulating ironies for consolation. This essay juggles and tries to make sense of these by exploring some politics of female reproductive experience.

A few months before leaving for Hofriyat in 1983, I played labor coach to my friend Melanie, then in the last trimester of pregnancy with a husband stationed abroad. Together we attended prenatal classes, practiced breathing, saw films of normal and cesarean births that made our blood run cold. Melanie had ceased being herself when she learned that she was expecting: she stopped smoking, drank no alcohol, ate a diet rich in protein, calcium, iron, folic acid, all the things a fetus needs. Pregnancy consumed her. She consulted books to learn what her body was doing at every stage. She stoically eschewed medication: nary an aspirin passed her lips the whole nine months. She came to regard herself as an incubator, nature's perfect instrument for growing human life, and was confirmed in her assessment when we at last saw her baby on the monitor of an ultrasound machine. The sonogram ratified the internal evidence of her body, offering objective corroboration that motherhood was imminent and real. Melanie's fetus now became a person, someone distinct from herself. And in her culture, persons are identified by the things that surround them, by what they possess.

So Melanie devoted her prenatal work-leave to fixing up a nursery. She papered and painted in primary colors to stimulate the baby's mind; hung cheerful, clinky mobiles; put an array of stuffed animals on a bookshelf placed within sight of the crib. She bought tiny outfits, baby blankets, sheets, sleepers, vests, bibs, box upon box of disposable diapers.

Her due date arrived but the baby did not. The next check-up came, and the one after that. She was agitated driving back from her latest appointment, so we decided to stop at the mall; maybe window shopping would persuade the baby to be born.

When we got home there was a message on the answering machine. Melanie must return to hospital right away; she would deliver by cesarean section that afternoon.

....

In Canada maternal mortality rates are among the lowest in the world: four deaths per one hundred thousand live births in 1985. In Ontario the rate that year was seven. But only about two-thirds of that province's births were normal vaginal deliveries and few of these took place at home. Fully one in five was a cesarean section and a further one in ten was operative, performed with forceps or vacuum extraction.

In 1983 Ontario's perinatal (still birth plus early neonatal) mortality rate was 6.7 per thousand births. The rate for early neonatal deaths alone was under five.

For Sudan such finely calibrated statistics are impossible to get. Infant mortality (live births surviving less than a year) is estimated to be somewhere between 110 and 120 per thousand based on imperfect census reports from

1973. The time of Amal's pregnancy (1983–84) was one of hardship; the north was on the brink of famine and even bread was scarce. Since nutrition plays a major role in determining pregnancy outcome, rates were likely higher then and continued so in the difficult years that ensued. Figures from the 1920s, though of questionable reliability, hint at deteriorating conditions since the heyday of colonial rule: in the northern provinces in 1926, infant mortality was recorded at seventy-one per thousand. By 1929 that rate had dropped to sixty.

To put this in perspective, the infant mortality rate in England and Wales between 1896 and 1900, when the British were engaged in conquering Sudan, was 156 per thousand. By the early 1930s, with improved hygiene and nutrition especially for the poor, it had fallen to sixty-two, on par with figures for northern Sudan.

Rates of maternal mortality in Sudan are comparatively high, with figures ranging from 655 to 2,270 per 100,000 live births. Because this figure is based on live births, it does not, of course, include women who died during pregnancy, as a complication of pregnancy following still birth, miscarriage, or induced abortion, or as a long-term consequence of childbearing. Since deaths are seldom reported at all, it is subject to gross distortion. Even so, it means that in Sudan, where the number of live births a woman can expect to experience averages seven, a conservative estimate of the lifetime risk of dying as a result of pregnancy is one in twenty-one. A more realistic figure could well be one in nine. By contrast, the lifetime risk in developed regions is as low as one in ten thousand. I would stress that these are actuarial figures and do not convey risks to individual women. Several studies have shown that in the developing world maternal mortality rates for girls who become pregnant before age fifteen are five to seven times higher than for those aged twenty or more; for those in their late teens, the rates are still two to three times as high.

By 1983 a handful of Hofriyati had delivered in hospital, at least two by cesarean section. And in 1994 there was widespread talk in the village about being able soon to "reserve" a hospital bed for childbirth, as is increasingly done in the city.

....

Life for women in Hofriyat is undeniably precarious, and it is understandable that many seek to minimize childbearing risks by all available means. Most marry young and experience early and multiple pregnancies. All in prepubescence undergo infibulation, a form of "female circumcision" in which the midwife excises the girl's clitoris and inner labia, then pares her outer labia and stitches them together so as to "cover" or "veil" the vaginal meatus, leaving a pinhole opening for the elimination of urine and menstrual blood. A girl's virginity thus becomes a revisable social assertion rather than one whose expression is natural and discrete. At marriage many infibulated women must be surgically opened in order to have sexual intercourse; in labor all require the presence of a midwife to perform one or more episiotomies before the baby can be born. After delivery a woman is resewn, and thereby regains a measure of virginal status.

While infibulation is designed to safeguard reproductive capacity, sadly, it often precipitates its distress. This frustrates the cultural imperative that married women demonstrate fecundity, and may lead to devastating loss of status and economic support through divorce or co-wifery. Infibulation is clearly hazardous to health. Most of its complications are long term, and derive from improper drainage of urine and menstrual blood. Childbirth is problematic, as the inelastic scar may prolong the second stage of labor, the contraction phase when the baby is leaving the womb, and, even when a midwife is present to cut through the fibrous tissue, the outcome can be death or brain damage to the child. For the mother, lengthy delivery can cause fistulae (passages) to develop between the vagina and bladder or rectum, resulting in incontinence for which she may face ostracism and divorce. Dermoid cysts have been reported to develop in the line of the scar, and keloids sometimes form that complicate an anterior episiotomy. Maternal death can result from blood loss and exhaustion or a revisitation of unsanitary conditions attending the original operation. In order to avoid traumatic delivery some women in their third trimester cut back on their food, resulting in low birth weights, a problematic start for the child, and slow recovery for the mother. Still, whatever its risks to individuals, infibulation seems not to have had a negative effect on overall fertility levels in Sudan, where the annual rate of increase hovers around 3 percent. Infibulation transforms a woman's natural fertility into a moral asset that defines her sense of self; despite growing awareness of its harmfulness, no Hofriyati woman I spoke to in the mid-1970s or early 1980s was willing to give up the practice.

Soon after arriving in Hofriyat in 1976, I was taken to witness a birth. The laboring woman lay flat on her back on an *angareeb* that had been spread with a mattress topped by an auspiciously "red" fibrous mat linked to demonstrations of female fertility. The lower end of the mattress was slightly elevated and covered with a waterproof sheet. This supported the woman's hips and heels. Her knees were raised in a local equivalent of the stirruped birthing posture customary in Western societies. To deliver, she would have to push the baby uphill.

The prone or "lithotomy" birthing posture was introduced to Sudan by British sisters, Mabel and Gertrude Wolff, who, in the 1920s and 1930s, established the Midwives Training School in Omdurman adjacent to Khartoum. British motives for initiating midwifery reform were philanthropic, but also political and pragmatic. Administrators hoped thereby to remedy Sudan's underpopulation, seen as a legacy of famine and displacement caused by the preceding century of unrest, and aggravated by ignorance. They were eager to secure a native workforce for prospective cotton plantations like the Gezira Scheme, created to feed the Lancashire mills.

More subtle, biopolitical transformations were left to British health and educational staff. As two decades of Pax Britannica had effected little growth in the northern population, officials became convinced that infibulation held the key. They blamed the practice for sustained low birth weights and high infant mortality, though, ironically, their own figures for the latter rivaled Britain's at the time. When the Director of the Sudan Medical Service disagreed, suggesting that malaria and venereal diseases were largely to blame, his opinion fell on disbelieving ears. Efficiently reproductive women were crucial to the colonial venture in Sudan, but their bodily reform required diplomacy, tact, and a flexible approach. This the formidable Wolff sisters—whom officials dubbed "the Wolves"—would arguably provide.

JANUARY 1984

An angry wind and pale, cold sun. Salima and I shiver as we walk the elevated railway track laid down by the British at the turn of the century and still the only excavated road in the north. We are on our way to visit Sheffa, the senior midwife who lives in a village several miles away.

We find her at work in an outlying hamlet. Sheffa seems delighted to see us and suggests I accompany her on her rounds. An eighteen-year-old is now in labor with her first pregnancy. The birth is likely to be prolonged and Sheffa offers to arrange transport if we get late. But first we must check on the laboring woman, then return to the clinic for Sheffa's regular prenatal appointments.

Sheffa astonishes me; she is unlike any rural woman I have met. Her clothes —a jeans skirt and tight fitting t-shirt—are urban and, apart from the white *towb* worn over top, more Western than Sudanese. Over her shoulders she had slung a sheepskin coat in case the day stays cold; instead of a midwife's traditional red and gold box, she used a battered leather briefcase to carry her equipment. Sartorial iconoclasm enhances her commanding presence: she is tall, handsome, and radiates health, moving with unbridled energy, not the torpid gait considered feminine in these parts. And though only in her mid-thirties, what Sheffa says clearly goes; few would dare to oppose her.

We make our way to her patient's natal home where, as is the custom, she has returned to give birth. There we are served coffee to warm us, though the chill has left the morning and a strengthening sun has begun to shimmer the air. In the dark, still-cool reception room, Sheffa and I are directed to metal chairs set before the laboring girl, Nemad, who reclines on an *angareeb*. We are met by a junior midwife, Foziya, who reports that Nemad's contractions are still well spaced. As we sip they discuss the examination they will do in a few moments' time. Now and then Nemad moans and as she does she grabs my hand. Sheffa conforts her, "It's all right, the pains are good. Soon it will be finished."

Gently, Foziya manipulates Nemad's belly to check the baby's location. The first position, she reports, head pointing down, face to the back, buttocks on mother's right side. But the body is high—common for a first delivery. Foziya inspects Nemad's mouth and gums for signs of illness, her eyes for anemia, her hair and scalp for lice, which if present must be dealt with promptly lest they infect the baby after birth. All is well.

Sheffa scrubs her hands. Nemad is told to hoist herself aboard some rolled up mattresses on top of another *angareeb*. These have been spread with a vinyl cloth—blue to counter the evil eye, and laced with images of Pepsi bottle caps. Sheffa cautiously measures Nemad's dilation. "I circumcised her," she says. "Not too tight." Nemad's eyes widen and she cries out in pain." "I'd like to use surgical gloves for this, but we can't always get them in Sudan." One and a half fingers' width. Still the latent phase of the first stage of labor; there are several more hours to go. If the baby's head remains high at five o'clock, she must go to hospital in Shendi. Nemad is encouraged to walk; we leave her to it and move on to the clinic.

There, a fourteen-year-old girl is waiting to be seen. "I delivered her," says Sheffa. "She was my first local baby." The girl, she explains, was married at eleven, before she'd begun to menstruate; she's now five months pregnant. Natural delivery will be impossible, as her pelvis is not yet fully formed and the baby quite large. This, Sheffa continues, is the tragedy of a recent trend: increasingly, parents in straitened circumstances are marrying their prepubescent daughters to suitors twice their age or more, who have lucrative jobs in Arabia. The men obtain compliant, partly educated, but indisputably virgin wives unlikely to question their domestic role. Since consummation takes place soon after menarche, the girls may conceive before their bones have matured.

If so, their bodies channel energy from their own growth toward that of their babies, leaving stunted pelves prone to "disproportion"—too small for the baby's head to pass. Sheffa predicts a sharp rise in the number of cesareans over the next few years.

Years later I read in a World Health Organization piece that precocious childbearing and grand multiparity are "cultural practices" more pernicious even than female circumcision. The latter condemns women to lives of suffering but is seldom lethal; early and frequent births regularly kill.

SEPTEMBER, 1983

A bright late summer evening. Little traffic clogs the lake shore as Melanie and I drive to the hospital in downtown Toronto. We are excited but apprehensive —I, over whether I can fulfill my promise to bear witness. As an undergraduate I'd volunteered to be a "droplette" at a blood donor clinic but had fainted at the sight of my first customer.

At the hospital Melanie is quickly braceleted and whisked upstairs to the maternity floor. Moments later I find her lying gowned and strapped to a fetal monitor attended by a nurse in surgical uniform. This is not a good sign. Though Melanie is hardly romantic about experiencing vaginal birth, things are not turning out as we imagined. The room is a soulless box filled with gadgets and dials that have apparently detected fetal stress. Melanie is nervous, on the verge of tears. Another nurse enters the room and prepares to shave her. The anesthetist arrives to start an epidural block; the doctor and his team are scrubbing. I leave to phone Melanie's husband overseas.

JANUARY, 1984

As Sheffa changes back into work clothes a woman bursts into the courtyard. Nemad's labor is hot, she says. We must come quickly. Now!

We find Nemad lying on her red fertility mat. Again Sheffa washes and examines her. Two and a half fingers. Not quite there. Nemad's mother helps her walk back and forth across the room to bring the baby's head into position. Within half an hour she had dilated to three fingers, the head is down, and her pains are truly "hot." This will take time, cautions Sheffa, it will be several hours yet.

We drink syrupy tea and converse. I am asked how women give birth in Canada, and tell them about being Melanie's coach. "But normally," I say, "the woman's husband attends and helps." My companions are aghast, but think it rather sweet. In Sudan a woman's husband cannot set foot in the birth chamber the entire forty days of her confinement. He may not even hold his child until it is five months old. But this does not apply to all men, only the baby's father. Nemad's brothers are here today, impatient to welcome the baby.

Ninety minutes later Sheffa measures Nemad's pelvis relative to the width of the baby's head. She will be ready to deliver in an hour or two. By 5:15 Nemad is close to giving birth. Sheffa readies the "delivery room." Working at a table near the south-facing windows—for now our only source of light—she places scissors, cotton gauze, plastic tubing and a length of string in a large enamel bowl and pours boiling water over all. She fills a clean syringe with anesthetic, then directs that an *angareeb* be placed in the center of the room. A lawn chair is put at the end of the bed; this is where Sheffa will sit. Immediately behind it to the left is the *angareeb* where I am to observe and hold aloft the kerosene lamp when required. Beside me is another chair, with arms, that will serve as the baby's crib. Mattresses are piled up at one end of the birth bed and behind them in its center is placed a rope-strung stool. Over the mattresses the Pepsi mackintosh is laid as before.

Nemad is led into the room accompanied by four middle-aged women. The door is shut. Her mother's sister lifts herself onto the stool atop the *angareeb*, then extends her legs to either side of the mattress roll. The other aunts help Nemad onto the sheet before her. Nemad will give birth in her kinswoman's arms, her back leaning against her aunt's chest, her torso and hips angled upright. Sheffa, unlike Miriam, has modified the supine posture to better resemble rope delivery with its gravitational advantage.

The room is hot; its mud-brick walls, having absorbed the heat of the day, now emit it with a vengeance. A hint of breeze wafts through the north windows; we are lucky the wind has dropped and there is no blowing sand. The room is shadowy, muted in the fading light and alive with the humming of flies.

Nemad's legs are spread and held back, an aunt on either side of the *angareeb* to perform this task. The one nearest me is wearing Bruce Lee "Kung Fu" plastic sandals. Nemad's skirt is stretched between her elevated knees to provide modesty and prevent her from seeing what Sheffa might do. On instruction Nemad lifts her arms and locks them backward round her aunt's neck. Her mother stands by her side, fanning her face.

....

When I return to the monitored cubicle, Melanie has disappeared. A nurse escorts me to an antechamber where I don booties, hospital gown, and cap, and place a surgical mask over my nose and mouth. "They're about to begin," I'm told. "Go through the blue doors to your left."

The operating theater is large, brightly lit, and chill. Figures in aphid green are clustered round an object ahead; they turn toward me and by their surprised expressions I know I've come through the wrong door. Now I see Melanie's body, stretched out and divided in two by a massive screen; people, lights and equipment are focused on her lower half, nearest me. The surgeon has already made a bright red incision through her belly and is about to clamp it back. I take a deep breath and walk to the chair by Melanie's head.

....

Sheffa tells Nemad to breathe; she wants her to cry out with the pain, rather than observe the custom of silence lest she hold her breath and deprive the baby of oxygen. A lamp is lit and passed to me through the window. I hold it high over Sheffa's shoulder. She inserts the syringe into the vaginal opening and makes several small injections around the area to be cut, much like a dentist freezing one's gums. Nemad is bearing down; Sheffa feels inside her and punctures the amniotic sac; fluid gushes onto the vinyl sheet and into a basin below. I fan away the flies. Soon the top of the baby's head shows through the opening. Sheffa pours water over the area to clean away some blood. She washes Nemad with carbolic soap then inserts two fingers between the head and the perineum. She waits until the head is crowning well, then quickly cuts through the muscle to the left and down. There is a spurt of blood. Kung Fu auntie swoons and leaves the room; Nemad's mother takes her place. The flies are growing thick; hand rhythmically brushes them away. After several more contractions the baby still cannot pass; Sheffa cuts a further inch or so. Next push, she eases back the muscle, gently grasps the baby's head and glides him into the world.

....

Melanie is flat on her back with her right arm strapped to a blood pressure cuff. Her free hand, attached to an intravenous drip, grabs my arm as I sit down. "I want to know when they're going to cut," she says. "Relax," I tell her, "you're already wide open." Beyond the screen the doctor's eyes crinkle in a grin. Melanie sinks back and closes her eyes; tension ebbs from her hand.

....

Sheffa sucks mucous from the baby's mouth with a small plastic pipe. He gives a hearty cry. Nemad's kinswomen are elated and congratulate her on the birth of a son. Sheffa ties the cord close to the baby and cuts it, then wraps him in a clean cloth and places him in the chair next to me. He whimpers and tests his legs, then puts his hand to his mouth and sucks. He is small but extremely alert. Sheffa closes the window and returns her attention to Nemad. The afterbirth is delivered without mishap; now she threads a needle with suture and prepares to sew up the wound. I bring the lamp closer and wave away more flies.

....

The doctor tells Melanie she'll feel a moment of pressure as the baby is massaged from her womb. The anesthetist checks her instrument panel and adjusts a dial or two. She has oxygen ready should Melanie faint. Melanie moans with the oddity of the sensation, and a few seconds later she has a baby girl. A team of nurses shuttle her to an incubation table. They weigh her, put drops in her eyes, wrap her in a flannel blanket, and give her to me to hold. I lift her up for Melanie to see. "Hello, Julia." I hear myself say. "Wait till you see your room."

I telephone Melanie's husband as she and Julia are wheeled upstairs to a private room. When I rejoin them Melanie is in pain and about to receive medicine for its relief. The baby is taken to the nursery for the night, to be bathed and thoroughly examined. Melanie is attached to an intravenous drip. I leave as a large bouquet of roses arrives.

....

Nemad is helped down and supported by her aunts as Sheffa and her mother rearrange the room. The *angareeb* is moved back against the wall and Nemad soon put to bed. Sheffa uses oil to clean the baby of vernix and blood, then swaddles him in clean soft cotton and gives him to his mother to feed. I am taken by the simplicity of it all, and hope that when Amal's time comes it will be like this.

Nemad's mother stays with her as the rest of the delivery team go out onto the verandah for a breath of air. Incense is lit and placed beneath an *angareeb*. Nemad's brothers arrive, overjoyed that the child is a boy; her husband's family are heard cheering a few doors away. As we sip the celebratory coffee, women ask how birth in my country differs from what I've just seen. I am tired and emotional, and want to avoid a lengthy discussion. I reply that it's similar, but that we do not circumcise. "Baraka," someone says, "blessing," and the others murmur assent.

....

How different is giving birth in rural Sudan from doing so in urban Toronto? On one level the answer is obvious; on another it is not. In both places the biomedicine shapes the process, yet in one more thoroughly than in the other. In Sudan, where persons are unthinkable save as members of a social whole, and spirit possession is a conventional event, the body and its parts may be less reified, less readily viewed as objects in themselves. Perhaps. The extent to which the natural process is respected may be greater there as well. But if Sudan is less interventionist, this seems more for want of means than by design. Moreover, it is hardly anti-interventionist: infibulation, after all, is underwritten by the view that the female body (like the male) is naturally flawed and must be perfected by human hand. In a deeper vein, we have much in common, it seems.

It is by now a truism that human bodies are everywhere subjectively informed by cultural meanings, including those labeled "scientific" and deemed unencumbered by ideology. In so far as such meanings are backed by the power of the state and alternatives suppressed, bodies are "colonized" and "normalized" as well. Moreover, as Foucault has pointed out, power in the modern state is masked and diffuse: it invests bodies, confers benefits as it disciplines, makes us subjects in the dual sense of that word. The gradual replacement of gynocentric birthing practices by those informed by assumptions of a tacitly androcentric biomedicine has taken place with the participation of women themselves. It has been an impressive colonization of consciousness— of selfhood—that of Western subjects most of all.

Aggressive biomedicine in the West unarguably saves the lives of mothers and babies who would otherwise die; but the cost to the majority can be dear. In the novel *Surfacing*, Margaret Atwood's protagonist rages famously at the indignities of modern birth:

> They shave the hair off you and tie your hands down and they don't let you see, they don't want you to understand, they want you to believe it's their power, not yours. They stick needles into you so you won't hear anything, you might as well be a dead pig, your legs are up in a metal frame, they bend over you, technicians, mechanics, butchers, students clumsy or sniggering practicing on your body, they take the baby out with a fork like a pickle out of a pickle jar. After that they fill your veins up with red plastic, I saw it running down through the tube. I won't let them do that to me ever again.

Yet these words, we may presume, are those of a healthy, well-fed Canadian woman, who is unlikely to have been nutritionally compromised as a child, and whose privileged material position enables her to eschew technological intervention that she legitimately sees as intrusive. For her, home birth attended by an empathetic midwife is a reasonable alternative. Where it is the norm the options are different, and few.

In Sudan, biomedicine's advance has been uneven and slow. Yet its promise is alluring, and few who live and give birth under the country's parlous conditions want to resist what offers so much hope. But Sheffa's emphatic decision to deliver in Shendi hospital must be set against Amal's modest refusal to seek out the expertise available there. And both Sheffa and Miriam have forged different syntheses of local and biomedical techniques in their respective midwifery practices. Like the women they attend, Sudanese midwives are not all alike. Moreover, to view them as remnants of a golden age of women's power, in which women controlled the process of birth and delivered safely in their own or their mothers' homes, is to ignore both history and social context, and the inordinate risks and responsibilities of midwifery work.

I remain convinced that Amal's chances of survival would have been better —not certain—had Sheffa rather then Miriam attended her from the start. But could she have been saved had she been whisked off to Canada for the moment of birth? I cannot say. While the duality of body and mind at the heart of biomedical science has enabled countless advances in knowledge, it has also encouraged an implicit confidence in the omnipotence of technology. We seem convinced that the more elaborate the procedure, the surer the positive outcome. Yet the assumption that institutional birth results in lower rates of maternal death is not wholly borne out even here.

It is true that throughout the twentieth century maternal mortality has fallen in the West as rates of hospital delivery and technical intervention have increased. Counterintuitively, however, the drop in maternal mortality owes more to the rising standards of nutrition, overall health, and sanitation, than to ever-expanding institutional interventions. The effect is cumulative over the generations: a woman's physical fitness to reproduce depends on her mother's nutritional health while pregnant, as well as on her own from the time of her birth.

In Sudan, standards of living are low, and female circumcision plainly adds to reproductive distress. But in a country where material want is great, where women remain economically dependent on men and are expected to produce large families—in part so as to diversify their means of survival—and where male offspring are preferred, malnutrition alone likely rivals, perhaps exceeds, circumcision in this role. It is estimated that a third of Sudanese children between the ages of two and five suffer from stunted growth. In one study in Khartoum, 44 percent of children under five were infested with parasites, attributable to poor living conditions. From my experience in the rural north, children consume few vegetables or fruits and very little meat, even in plentiful years; as people abandon traditional breads made from protein-rich sorghum or whole wheat flours, and adopt more "modern" foods like refined white baguettes, this situation is likely to worsen. Moreover, malnutrition compromises immunity, and is therefore linked to the prevalence of illnesses like dysentery, malaria, and hepatitis which add to childbearing risks and for which infibulation may be performed as a childhood cure. It is also responsible for rickets, which causes bone malformation leading to pelvic disproportion during pregnancy and, failing cesarean delivery, death. As Amal's country sinks deeper into poverty with the rest of the developing world, the nutritional status of its womenfolk is increasingly threatened. The economically motivated early marriage of daughters is becoming more common as well, adding precocious pregnancy to the litany of hazards women face. Yet there as elsewhere, technological fixes are offered for problems that are political and economic at heart. The costly medicalization of birth continues to be seen as remedial, not palliative, touted as the appropriate response to high rates of maternal and infant mortality. Paradoxically, given the present inequitable distribution of power and global wealth, that view may be correct. And this, I think, is the true irony and tragedy of Amal's untimely death.

NATIONAL BODIES, UNSPEAKABLE ACTS:
THE SEXUAL POLITICS OF COLONIAL
POLICY MAKING

SUSAN PEDERSEN

Government may claim and sometimes exercise its monopoly over the legitimate use of violence, but much of the time political rule can seem an elaborate charade, in which leaders only ratify policies decided elsewhere and fictions of compliance mask more complex networks of domination, inaction, or protest. In no area was authority so erratic as in British-controlled Kenya, where a "dignified" rhetoric of "native paramountcy" masked an "efficient" interest in labor control and economic growth on the part of settlers and (usually) the government. Indirect rule could thus devolve in practice into either arbitrary rule or no rule at all, as a patchwork of mediating social groups enforced, altered, or ignored the laws proceeding from an impervious Government House.

The fictions of rule were exposed in 1926, when missionary pressure and some level of official concern about the prevalent practice of clitoridectomy led the governors of the East African dependencies to meet together to devise a strategy to combat the practice. Noting that one Kikuyu "native council" had passed a law requiring the licensing of all operators and restricting the extent of cutting to "simple clitoridectomy" (as opposed to the usual practice of removing the entire external genitalia), the governors endorsed this shift to the "less brutal" form. The new guidelines were adopted without protest by a range of councils and with a gratifying circulation of regulations and papers. Imagine the surprise of the "native commissioners," then, when the decision by some missions in 1929 to give teeth to the rulings by refusing communion to all Christians unwilling to forswear the practice led to the largest outbreak of popular protest among the Kikuyu that the British government had yet faced. The 1926 law, it seems, had been a dead letter all along. Only the missions, controlling as they did access to worship and education, could provide real sanctions against clitoridectomy; only their intervention exposed the elaborate dance of theoretical government concern and equally theoretical compliance.

When Sir Edward Grigg, governor of Kenya during the disturbances of 1929–30, explained to Sidney Webb at the Colonial Office precisely why it was difficult for the colonial administration to dissuade young girls from undergoing the operation of clitoridectomy, he tended to blame an omnipotent "tradition" which made such customs virtually unquestionable. "The chief opponents to any reform are the victims themselves," he wrote, "and their attitude is due to the fact that they are bound by custom." Yet the main white agitators against clitoridectomy, the Church of Scotland Mission, offered a very different explanation for their failure. "Except for the championship of the rite as a national symbol by the Kikuyu Central Association," they wrote, "it probably would by now have been widely abandoned. This faction, in spite of its professed aims as to progress, education, and enlightenment, has, through its ill-advised attitude, done more than anything else to delay the emancipation of Kikuyu womanhood."

Embedded though they are in the teleology of Western enlightenment, both of these statements contained partial truths. Clitoridectomy was, as Grigg noted, certainly a custom of central significance to Kikuyu life. Like the male circumcision undergone by Kikuyu boys, the public ritual of clitoridectomy marked for women not only the passage from childhood to marriageable adulthood but also the moment of entry into the full life of the community. Young women initiated together became members of the same age group, which was the basic unit of Kikuyu social organization. Age groups acted as associations for mutual aid; they also served to enforce the distinctions of status and behavior prescribed for their members at different stages of life. Women thus understandably often identified clitoridectomy as the most important—if also the most painful—experience of their lives. Wanjiku, a Kikuyu woman born around 1910 and initiated before the controversy discussed here, recalled in an interview with Jean Davison: "Anybody who has not felt the pain of *Irua* [the 'circumcision' ceremony] cannot abuse me. . . . From *Irua* I learned what it meant to be grown-up, with more brains. . . . Also from *Irua*, I learned what it means to be pure Mûgîkûyû—to have earned the stage of maturity when, being a circumcised person, one no longer moves about with those not yet circumcised." "*Irua* was like being given a degree for going from childhood to adulthood," one Kikuyu midwife recalled. Whereas uncircumcised girls were not considered marriageable, "the minute you got circumcised, no one would stand in your way."

If clitoridectomy remained central to Kikuyu identity and social structure, however, it was not because the Kikuyu were, as Grigg and his colonial civil servants liked to put it, "just emerging from a state of barbarism." Rather, clitoridectomy persisted in spite of (or perhaps partly because of) the stress and dislocation borne by the Kikuyu during the rapid pace of colonization in Kenya. Historians have delineated in detail how the alienation and settlement of "White Highlands," the establishment of adjoining "native reserves," the construction of a system of "tribal" authority often at odds with earlier political structures, and the deliberate use of taxation and legislation to restrict Africans' freedom to grow cash crops and to coerce them into wage labor left many Kikuyus with little of their "traditional" life, except perhaps the still-powerful rituals of clitoridectomy and male circumcision. Tabitha Kanogo, in her excellent book on Kikuyu "squatter" farmers in the White Highlands, shows how community norms were maintained through the transportation of the rites central to life on the reserves.

But if the missionaries were quite wrong in their contention that prior to 1929, clitoridectomy was an atavistic custom in terminal decline, several careful historical studies have substantiated their claim that the protonationalists of the Kikuyu Central Association (KCA) found the missionaries' censure of the practice a useful catalyst for organization and resistance. The intervention of the KCA catapulted the controversy into the category of a full-scale political revolt and endowed the practice itself with new meaning. As a defense of clitoridectomy became entangled with long-standing Kikuyu grievances about mission influence and access to land, clitoridectomy, always the sign of the "true Kikuyu," also came to be seen as a mark of loyalty to the incipient, as yet imaginary, nation.

A brief account of the course and content of the revolt will substantiate these points. Roman Catholic, Presbyterian, Anglican, and American missions were active in the Kikuyu areas from the turn of the century, and a group of African converts committed to education, "progress," or simply personal advancement soon grew up around the mission schools. Although the extent of mission interference with "traditional" customs varied, all of the Protestant missions looked with disfavor on clitoridectomy—quite as much for the "pagan ritual" that accompanied it as for its "brutality"—and attempted to combat it through teaching and the establishment of dormitories for young girls. Probably the most vehement opponent was Dr. John Arthur, an outspoken leader in the Church of Scotland Mission and the nominated representative for African interests on the government's Executive Council. From the mid-twenties, both the African Inland Mission and the Church of Scotland Mission campaigned actively against the practice; Arthur even provided legal support for girls who contended that they had been circumcised against their will.

African responses to the missions' teachings on clitoridectomy were not monolithic but varied with age, locale, and economic position. David Sandgren, studying the American-run African Inland Mission, found that the older converts who had found new lifestyles, opportunities, and status through the missions could support the campaign against clitoridectomy, while younger converts in more remote areas resented not only mission attacks on accepted ritual practices but also the older converts' dominance of church decision making. Nevertheless, the missions' insistence that one could not both be a good Christian and respect traditional customs struck many African Christians as hypocritical and baffling, especially when the vernacular translations of the Bible prepared by the missions themselves often used the terminology of initiation rites to describe Christian rituals. Quiet disregard rather than overt rejection of such teachings could prevail until March of 1929, however, when a conference representing the churches of all Protestant missions active among the Kikuyu endorsed the suspension of church members unwilling to abandon the practice of clitoridectomy. The Church of Scotland Mission could now claim that the decision to excommunicate those who condoned the practice was supported by missionaries and African members alike.

Such claims notwithstanding, the decision to make clitoridectomy a punishable offense was unpopular among most African converts from the beginning, especially since Arthur had stated his goal to be the abolition of *Irua*—the entire practice of initiation—and hence also the age-grade system on which Kikuyu social life was based. African support dwindled further in the face of draconian mission enforcement and KCA countermobilization. The Kikuyu Central Association, formally organized in the Fort Hall area in 1924, had concentrated prior to 1929 on attempts to secure the return of lands appropriated for European settlement and the release from detention of the earlier political leader, Harry Thuku. The outbreak of controversy over female circumcision furnished them with their first chance for real popularity, and in August the KCA Council—except for the young Johnstone (later, Jomo) Kenyatta, who had been sent to London to represent the KCA's views—signed a "Lament for the Abolition of Female Circumcision," which protested against the "law" introduced by Arthur and urged all Kikuyu chiefs to mobilize in defense of tribal customs. On September 11, Grigg informed Sidney Webb in a worried private letter that the state of Kikuyu province was causing him great anxiety, with the KCA "engaged in a widespread campaign for the revival of the most brutal form of female circumcision, which they declare to be necessary to true Kikuyu nationality." . . .

THE QUARREL IN BRITAIN

. . . Behind the Labour government stood an articulate collection of activists for "native rights." Loosely organized around the party's Advisory Committee on Imperial Questions, this group included Leonard Woolf and the noted humanitarians and M.P.s Josiah Wedgwood and Charles Roden Buxton, as well as Dr. Norman Leys and William McGregor Ross, both of whom had lived in Kenya before the war and had become well-known critics of British solicitude toward the settler community. Although Webb himself paid little attention to these anticolonialist radicals, they kept in contact with his Parliamentary undersecretary, Thomas Drummond Shiels, an energetic Scottish doctor who had participated in their meetings before Labour took office and who from December of 1929 oversaw much of Kenya policy.

On December 11, 1929, these critics of empire initiated a debate on "colonial policy in relation to coloured races" in the House, in which Labour's commitment to "native paramountcy" was questioned and reaffirmed. The resolution was moved that:

> The Native population of our dependencies should not be exploited as a source of low-grade labour; no governmental pressure should be used to provide wage-labour for employers, due care should be taken of Native social well-being; the Native demand for land should be adequately and satisfactorily met and their rights therein properly safeguarded; where the Native population is not yet fitted for self-Government direct imperial control of Native policy should be fully maintained; Native self-governing should be fostered; and franchise and legal rights should be based upon the principle of equality for all without regard to race or colour.

This statement could have been written as a direct critique of past colonial policy in Kenya, and C. R. Buxton, in seconding the amendment, made no bones about the fact that he had Kenya particularly in mind. "Matters have come to a head in East Africa," he said, "and something has to be done there." Under no conditions, however, should powers be turned over to the settlers clamoring for greater constitutional rights; rather, any movement toward self-government must be on nonracial lines. In his reply for the government, Drummond Shiels accepted the resolution and stressed Labour's commitment to work along similar lines.

If the debate over colonial policy had stopped there, all would have been in agreement. But Buxton's speech was followed by two other speeches relevant to Kenya, both of which raised new problems for anticolonialist critics and shocked and outraged some members of the House. While broadly agreeing with the doctrine of "native paramountcy," these speeches proposed that the British government be held responsible not merely for equal rights between races but also for guaranteeing equal rights between Black women and Black men and for "protecting" women from "barbaric" practices. The speakers were Katherine, the Duchess of Atholl, and Eleanor Rathbone,[1] and with their intervention clitoridectomy became, astonishingly, a subject for debate in the House of Commons. . . .

THE FIRST PROBLEM: NATIONALISM CUTS BOTH WAYS

The circumcision controversy left the Labour government on the horns of a dilemma. On the one hand, they wished to recognize the legitimacy of African political aspirations; on the other, they hoped to conduct foreign policy along humanitarian and moral lines. But what were they to do if the two aims came into conflict, if representative African organizations won popularity by championing rituals that "humanitarians" found indefensible, and if the British enforced "humanitarian" reform through draconian political repression? Labour politicians tried to solve this problem by persuading the Kikuyu Central Association to give up its support of clitoridectomy; in other words, they attempted to dissociate popular nationalism from the defense of prevailing cultural or sexual practices. Drummond Shiels made this strategy explicit when he proposed that he should perhaps meet Kenyatta himself, "since his organisation is evidently making this a question of Kikuyu patriotism." Kenyatta's defense of clitoridectomy could, he felt, "be sublimated into a willingness for abolition in the interests of the race, if it were properly put."

Shiels did in fact see Kenyatta at the House of Commons in late January of 1930 and attempted to win him over. As a medical doctor, he emphasized that the growth of scar tissue following clitoridectomy would make childbirth difficult and dangerous. He also explicitly stated that adherence to clitoridectomy was unlikely to further the KCA's nationalist ambitions. "You must take care," he told Kenyatta, "that the expressions of your attachment to your country do not bring discredit upon it and you." Kenyatta—who had the previous day warned against "any attempt to coerce my people by 'force majeure' "— responded mildly that, while people had resented the moralism of the missionaries' campaign, they would probably be willing to hear about the medical hazards of the practice from government doctors. Shiels and Webb both considered the meeting a considerable success and wrote to Grigg of their hopes for both a gradual abolition of clitoridectomy and the taming of the KCA. Shiels insisted that with "careful handling" Kenyatta and the KCA could become responsible participants in the constitutional process, and he urged Grigg to pursue such a strategy.

Grigg, however, had no intention of nursing the ambitions of the KCA. Just about the time Shiels saw Kenyatta, Grigg telegraphed Webb, warning him that British sympathy with the KCA was being exaggerated in Kenya for seditious purposes and requesting unequivocal support for his own government and the constituted Native Councils. Grigg's request was echoed more bluntly by the community of Kenyan settlers, who in letters and visits to the Colonial Office made known their opinion that the root of the problem was, quite simply, the interference of a lot of "ignorant" and "mischievous" missionaries and politicians, who had no notion of the danger and futility of interfering with the deeply held customs of a "savage black race." Put simply, many committed to white hegemony in Kenya had little desire either to combat clitoridectomy or to purge nationalism of its sexual content. Rather, they read both the practice and the KCA's support as further evidence of Kikuyu "barbarism" and hence as another justification of white rule. They also worried that, should the government attempt to interfere with Kikuyu women, Kikuyu men would vent their rage on white women. Any attempt to put down the practice, they contended, would quite possibly lead to a "serious native rising," with violence directed against white women. An elderly woman missionary known for her opposition to clitoridectomy had been found murdered on January 3, 1930, with vaginal wounds suggesting an attempt at "circumcision," and Grigg warned Webb that further violence could only be prevented if there were a "clear demonstration that in no quarter of Parliament is there any sympathy with [the] general challenge of Kikuyu Central Association to [the] authority of the chiefs and Government."

If Grigg and the settlers found the prospect of transforming the KCA into a legitimate "civilized" national body less than appealing, the KCA's leadership in Kenya did not seem particularly interested in the prospect either. The circumcision issue was "useful" to the KCA precisely because it could be read as a political conflict uniting all Kikuyu against the combined forces of British imperialism. The KCA thus ironically shared Grigg's propensity to minimize the religious and social aspects of the revolt and played an important role in constructing the "dominant" political interpretation of the conflict. Their description of the mission rules as "laws" reflected their refusal to accept

British claims that the missions were distinct from the government; similarly, they dismissed any African sympathy for the missions' campaign as pro-British lackeyism. Their strategy was aimed less at altering Grigg's policy than at convincing the London government to break with the political status quo in Kenya entirely. Thus they interpreted Shiels's interview with Kenyatta not as the admonition and "guidance" Labour had intended, but as a sign that the British government recognized the Kikuyu Central Association as the legitimate representative of the Kikuyu people and intended to repudiate the institutions of "indirect rule" managed by Grigg. Furthermore, although the KCA's acting president defended initiation as a "deeply ingrained" custom of the Kikuyu, with "a definite place in their social, economic and moral life," the championing of the custom by those identified with "modernity" rather than "tradition" endowed this ostensibly timeless and immutable practice with new, and highly political, meanings. "Tradition" was redefined; and, in a final ironic twist, some of the non-Christian Kikuyu leaders recognized by the British as the defenders of Kikuyu customs supported the missions' attack on "pagan rituals" in the hopes of defeating the KCA's challenge to their authority.

Perhaps a second reason for the difficulty of disentangling nationalism from sexuality, and hence for the failure of Labour's strategy, was that the KCA's defenders in Britain, quite as much as their opponents in Kenya, found it far easier to support their politics by appealing to sexual customs or preying on sexual fears than to dissociate the politics of nation from questions of women's rights. The usual left-leaning defenders of the political rights of African peoples—Leonard Woolf, William and Isabel McGregor Ross, and Norman Leys—mobilized early in 1930 to defend the Kikuyu from interference in their customs; in doing so, however, their scrupulous rejection of claims to cultural superiority soon shaded into a trivialization of the practice itself. Thus Woolf described clitoridectomy only as an "eminently religious custom," while Norman Leys insisted that the medical case against it had been "ridiculously exaggerated." They also questioned the good faith of feminists and humanitarians campaigning against the practice, charging them with a prurient desire for "atrocities to wallow in."

The particular purchase threats of sexual violence had even for anti-imperialist men was vividly displayed in the letter written by William McGregor Ross to Shiels in early February, especially in comparison with his wife's letter of that same month. While Isabel Ross questioned the depth of knowledge of British women campaigners and pointed out that African women suffered "in the long run much more" from poor education and the forced labor system, she made no attempt to minimize the seriousness of the practice of clitoridectomy. In McGregor Ross's letter, by contrast, clitoridectomy was described as a "time-honoured practice," while his defense of "native rights" quickly became a defense of African men's rights to control Black women. A real government attack on the practice would entail "the wholesale examination of the sexual organs of Kikuyu young women," and he warned that "the Kikuyu are not such worms as to tolerate wholesale interference in their women by Government." Ross did not seem aware that this sentence identified "the Kikuyu" exclusively with Kikuyu *men;* he also seemed to feel that men's defense of "their women" against government interference was a sign of their manliness and, hence, of their worthiness as citizens of the incipient Kikuyu nation.

Yet Ross's definition of national self-determination as constituted through men's control over women was not applied only to Kenya. Just as he defended African men's control of "their women," so too his critique of British policy expressed itself in fears that government indiscretion would undermine white men's ability to protect and control white women. In terms indistinguishable from those used by the Kenyan settlers, Ross warned that, should the government proceed with its campaign, "there is at least a possibility that the local Government might be faced with a rising on a savage scale, with possibilities of rape, murder or circumcision of white women on scattered farms. No white woman or girl would feel safe." The safety of white women, he implied, was contingent on recognition of Black men's rights over Black women. Since clitoridectomy was widely supported by African women as well as men, Ross's characterization was more than a little inaccurate. It was enough, however, to frighten Sir Cosmo Parkinson, then the head of the East African Department in the Colonial Office, who noted on Ross's letter that the government was not contemplating anything more than some propaganda by medical officers.

The controversy over clitoridectomy divided the Kenyan settlers and politicians from the Labour government, but it also split Labour loyalists among themselves. Although the political aims of groups hostile to the government's proposed campaign could not have been further apart, we have seen that they shared an understanding of the issues at stake in the attempt to dissociate nationalist ambitions from issues of sexual order. Clitoridectomy, and the British campaign against it, were contentious issues partially because some men divided by political allegiance and even race nevertheless accepted that the right to speak for women of one's own race and culture was a central and ineradicable component of male political rights. And even those opposed to the political goal of self-rule were willing to concede this sexual prerogative—provided, of course, that recognition of rights over women of one's own race was understood to preclude any "poaching" among women in the brother's camp.

Attention to the similarities in the rhetoric employed by settlers and anti-colonialist radicals can show how men's anxieties about sexual control were exploited to sustain both imperialist and anti-imperialist politics. In leaving aside the opinions and activities of both Kikuyu and British women, however, such a focus could imply—wrongly—that the campaign against clitoridectomy failed because of masculine opposition alone. For nationalism is not only a concern of men. As Cora Ann Presley has shown, many Kikuyu women in the interwar period identified their interests firmly with those of Harry Thuku and other protonationalist leaders; the identification of clitoridectomy with anti-colonial resistance could only have made the practice more meaningful for women opposed to the government's labor and pass laws. Similarly, and even more problematically for our story, the call of national interest and imperial stability fragmented the loyalties of British women activists. Rathbone and Atholl had little sympathy for the KCA's political claims and none at all for its defense of Kikuyu "custom." At times, then, it was not altogether clear whether the demise of clitoridectomy or the defeat of the KCA was perceived as the greater good. . . .

THE SECOND PROBLEM: PROTECTING THE SUPERFLUOUS

If the first problem faced by British campaigners was, in a sense, the difficulty of resisting a seductive identification with the "national body," the second was the entirely different dilemma of imagining a way to defend a "thing"—the clitoris—that was, in the august corridors of the Colonial Office, "unspeakable." Yet this problem took some time to become apparent and to immobilize their campaign, for Atholl, Rathbone, and their colleagues did have an available rhetoric of maternalism and "racial hygiene" in which to couch their disapproval of clitoridectomy. In time, however, and when they confronted an alleged distinction between the "brutal" radical operation and "simple clitoridectomy," the limits of maternalist rhetoric became apparent. How could one oppose "simple clitoridectomy" without discussing sex? The failure to resolve this dilemma reveals both the difficulties of feminist activism in the interwar period and a virtually unfathomable degree of ignorance and silence about women's sexual response.

When Atholl and Rathbone first raised the issue of sexual mutilation in the Commons, they did not attack clitoridectomy by name but rather focused on a "pre-marriage initiation rite," which amounted to "actual mutilation" only in its "brutal form." In doing so, they implicitly endorsed a reproductive framework for the clitoridectomy issue that dated back at least four years. In its ineffectual circular against the practice in 1926, the Kenyan government distinguished clearly between a "brutal form" of clitoridectomy, which also involved the removal of the labia minora and half of the labia majora, and clitoridectomy itself. While the former was claimed to cause such severe scarring as to interfere with childbirth, sexual intercourse, and even menstruation, the latter was dismissed as "a simple operation unlikely to be followed by any serious effects." Prohibiting both practices would lead to "sullen and resentful opposition," the circular predicted, but the population could perhaps be induced, "in the interests of humanity, native eugenics, and increase of population, to revert to the milder form of the operation, which is indeed more in keeping with ancient tribal usage." No evidence was given of the existence of any "milder form"—much less of its "traditional" character—but this distinction between a "brutal form" harmful to reproduction and "the race" and a "simple" form damaging to neither, patterned all subsequent discussion of the government's role.

This maternalist rhetoric was effective, at least in a limited sense. It was the graphic descriptions of the effects of the "brutal form" on childbirth—of children released only by cutting through the scar tissue around the vagina— that convinced extraparliamentary feminist organizations to raise the matter publicly and that galvanized the British government into contemplating state intervention in "native customs." The government of Kenya was about to replace the old Indian Penal Code—surely a propitious moment to move against the practice of "female circumcision." In a detailed memorandum to Webb, Grigg reported that the Supreme Court had interpreted the code in such a way as to make it impossible to convict any operator of grievous hurt even in a radical operation, provided the girl consented. Yet should a person be permitted to consent to his or her own maiming? Grigg thought not, and recommended that the new code be amended so that the "brutal operation" would fall within the definition of a "maim." Consent, then, would no longer be a defense against prosecution for such maiming. In April of 1930, Passfield agreed to this course of action.

Yet while a concern for reproduction and for public health convinced Grigg of the need to limit the extent of cutting, it did nothing to combat clitoridectomy itself. After all, the simple removal of the clitoris did not affect a woman's capacity to give birth, and the medical establishment in Britain seemed unable to imagine any other arguments against the practice. Even Dr. Gilks, the director of medical services in Kenya, told Atholl's committee that although he himself would not countenance the performance of clitoridectomies by his staff, the operation itself was "relatively innocuous." In a statement prepared for the government he wrote that simple clitoridectomy "can hardly be followed by undesirable results other than might occasionally result from sepsis." . . .

The clitoris has no functions save that of sexual pleasure; as Gayatri Spivak writes, it "escapes reproductive framing."[2] Its excision is not hazardous for childbirth; indeed, if the subject of sexual pleasure is inadmissible, clitoridectomy seems no different or worse than scarification. Bluntly, there is no way to oppose clitoridectomy without discussing sexual response. Yet this was precisely what Atholl was unable to do, except in the most euphemistic terms. Even the women doctors she interviewed, both at this point and later, were willing to state only that the clitoris was "an extremely sensitive part" and "very richly supplied with sensory nerves," so that "its excision must be accompanied by intense pain" and would result in "a severe mental shock and sexual trauma."

Nothing so debilitated the campaign against clitoridectomy as this silence about women's sexual response. Neither the women M.P.s, nor the Labour intellectuals, nor the British medical establishment thought it possible or relevant to mention the effects of the operation on women's sexual pleasure. Indeed, in the entire debate over clitoridectomy, I have found only one such statement, and even in this case this transgressive opinion was surrounded by a formidable protective casing denying much of its power. It was Dr. Arthur and his fellow male mission doctors, in their blunt "Brief Statement in Nontechnical language regarding the Medical aspects of Female and Male Circumcision and Clitoridectomy," who wrote: "Although this organ is not, as is erroneously supposed, the main seat of sexual gratification, it certainly contributes to this, which makes its unnecessary excision all the more indefensible." Weak as this statement was, it was the only defense of women's sexual pleasure made during the entire controversy.

Hesitation and euphemism on the part of the committee left the Colonial Office free to counterattack with ridicule and "expertise." When Atholl requested that the opinions of the Women's Medical Federation (WMF) be heard on the issue, Dr. Stanton, the medical advisor to the Colonial Office, retorted that "the technical aspect of the matter is I think quite clear and there is no need to seek advice from the BMA [British Medical Association] or the WMF who know nothing whatever about it." And in discrediting the seriousness of the practice, colonial administrators exploited a rhetorical invention even older than the distinction between the "minor" and "brutal" forms. This was the term "female circumcision" itself.

Few analogies are more spurious than that of clitoridectomy to male circumcision, yet civil servants and even doctors out to justify inaction deliberately invoked this false analogy. Norman Leys, concerned to defend the validity of Kikuyu customs, thus expressed doubts about whether the practice was any more likely to produce "bad result[s]" than male circumcision had been in the days before antiseptic. Similarly, when the Duchess of Atholl sent on to the Colonial Office her reservations about the government's tolerance of "simple clitoridectomy," Dr. Stanton claimed that "it is as though she were advised that the rite of male circumcision among Jews in this country should be made a penal offense." The very use of the term "female circumcision" condemned its opponents—then as now—to the absurd task of refuting a presumed medical and physiological commensurability with male circumcision.

The divided and (on some questions) inarticulate committee members were no match for self-confident experts wielding a useful lie. As a result the "official" campaign continued to tie a genuine revulsion against the "brutality" of the "major operation" to a blithe dismissal of clitoral excision as an innocuous practice. The central "silence" holding this curious blend of opinions in harmony was the exclusion of all knowledge or consideration of women's sexual response. Only with the erasure of the very idea of women's sexual pleasure could clitoridectomy be described as analogous to the circumcision—rather than the amputation—of the penis. Since a positive statement of female sexual pleasures and rights was beyond the powers, and possibly the knowledge, of most of those campaigning against clitoridectomy, at no point did this linkage fracture and the sexual function of the clitoris become "speakable." . . .

In the end, however, the inability of women to find a name for their knowledge proved irrelevant. In September of 1930, the Colonial Office received a telegram announcing unexpected problems with the government's plans to legislate against even the "major" operation. Within six months, this strategy had been abandoned and the proposed campaign against the practice entirely contained. The final stage of the controversy exposed the essential absurdity of the British strategy, but at the same time it ended the government's interest in doing anything at all.

RESOLUTION: THE CAMPAIGN CONTAINED

Not that Grigg had been dragging his feet: he had already begun to prepare the legislation against the "brutal" operation. In the early months of 1930, he met separately with leading missionaries, his own senior commissioners, and fellow governors of the East African dependencies and carefully explained the government's position. He told the missionaries: "It was the duty of Government to put down brutality and cruelty, and in so far as the operation known as female circumcision was brutal it should be made an offence against the law and put down accordingly. It was also the duty of the Government to protect unwilling victims against an operation of any kind. Any operation of this nature, therefore, would be illegal, even of the milder form, if it were proved that it had been carried out without consent. Beyond that, however, Government could not go." The missionaries agreed that they would consider it "entirely satisfactory" if the simple removal of the clitoris were allowed but all additional cutting made a penal offence. Grigg's commissioners and fellow Governors concurred, feeling that such a compromise would ensure that "the ordinary customs of the people will not be interfered with," while at the same time putting down "brutality."

Unfortunately, this felicitous resolution—and indeed the entire campaign—hinged not only on the problematic claim that "simple clitoridectomy" was not brutal but also, and more fundamentally, on the presumption that more extensive cutting was relatively rare. In the fall of 1930 Grigg discovered to his surprise what the missionaries had known all along: that "simple" clitoridectomy was virtually unknown among the Kikuyu, and that what had been dubbed the "brutal form" was in fact "the ordinary custom of the people." It was this discovery that sparked his warning telegram, which was quickly followed by an official dispatch with the facts. Of 374 Kenyan women examined by the Medical Department, fully 370 were "circumcised"; a mere three of these were "simple clitoridectomies."

The proposed strategy of criminalizing the "brutal form" would thus involve the government not in putting down an unusual and excessive variant of an accepted "native custom" but, rather, in stamping out the custom itself. "Such a situation was never contemplated," Grigg's deputy H. M. Moore wrote to Webb, "and I feel sure Your Lordship will agree with me that in the circumstances it would be unwise further to proceed with the proposal at present." The revelation of the seriousness of genital mutilation in Kenya thus ironically lessened government interest in doing anything about it. The Kikuyu Reserve was finally quiet, but authorities feared legislative action would set off further unrest. In an argument reviving the linkage between peaceful coexistence and a recognition of male rights, Moore argued that nothing at all should be done to feed suspicions among "the tribe" that the government intended to interfere with "their women." . . .

The story of the clitoridectomy controversy, as I have told it here, is a diffi-cult but illuminating one. Whatever their political and religious differences, the most knowledgeable and sophisticated participants in the controversy—especially Kenyatta, Leakey, and Isabel Ross—all agreed that British interven-tion was riddled with problems. Most participants expressed some concern about the reproductive—if not the sexual—consequences of clitoridectomy, and even Norman Leys agreed that the British should ensure that girls were not operated on against their will. Yet the proposal to introduce a general legisla-tive prohibition aroused apprehension: even some missionaries recognized the danger of using "the lever of political repression to achieve a moral end." It would, Kenyatta warned, "have the very opposite of the desired effect," as people would attach "accentuated importance to the maintenance of this custom."

He was precisely right. Issues of sexual practice and sexual order proved useful vehicles for both nationalist and imperialist agendas; in the end, the linkage proved indisseverable. Kikuyu patriots resistant to British rule found the defense of "traditional" rituals to be a basis for popular mobilization; white male radicals in Britain eager to support their political claims had little diffi-culty excusing clitoridectomy as a "time honoured practice." Since the Colo-nial Office's concern for African women's health was strictly secondary to their interest in the stability of imperial rule, the British government quickly backed down. This process of contestation, far from eradicating the practice, gave it new meaning, as initiation came to symbolize an attachment to an evolv-ing "nation."[3] And this definition outlasted the controversy itself: Wanjiku, a Kikuyu woman interviewed by Jean Davison, recalled that while mission-educated young women refused to undergo initiation, the practice was revived during Mau Mau as a way of demonstrating loyalty to Kikuyu traditions. Only direct repudiation of the practice by the post-independence leaders has gone some distance to breaking the link between sexual mutilation and nationalism.

British men involved in the clitoridectomy controversy also assumed that political order could be maintained only if the sexual order remained stable, although their vision of "stability" was quite different. Whether defending the rights of African men to control "their women" (as in the case of McGregor Ross) or subjecting men to church discipline for failing to protect their daugh-ters from "pagan ritual" (as in the case of John Arthur), white men construed women's bodies as largely under the control of men. In doing so, they betrayed their own assumptions about the familial and sexual prerogatives of men but showed themselves incapable of imagining that African young women could resist or support clitoridectomy through their own volition. Their rhetoric shaped the course of government action, but it obscured the central role of the women whose bodies were, after all, the bone of contention.

For clitoridectomy had always been largely a woman's issue. The identification of clitoridectomy as a loyalty test by both the missions and the KCA subjected young women to new pressures, but it did not make them into mere dupes. When Jocelyn Murray conducted her research in the early seventies, she found that in one area the story of the young daughter of a church elder who defied her father and ran away from the mission dormitories to be initiated in 1932 had become a local legend—but so too had the story of her brother, who in 1940 became the first local man to marry an uninitiated woman.[4] Thus while most young women willingly held fast to rituals they found central to their identity as women, as "true Kikuyu," and as patriots, other girls defied the taunts of their schoolmates and the pressures of their relatives to resist practices they found incompatible with their faith. These few were sorely tested by a political movement that identified them as traitors, by a judicial system reluctant lo impose significant penalties for abduction and forcible circumcision, and by a British officialdom that justified its own inaction with the racist and cavalier statement that "the average native girl has not a will of her own—she is a puppet of tradition." Nor could they find real help from British women who, distant and isolated, found their own attempt to "protect" African women limited by the ridicule or apathy of civil servants and fellow politicians, by a culturally patterned reticence in speaking about sex, and by their own reluctance to jeopardize the national interest. . . .

We know too little about how such cross-cultural campaigns reshaped Western feminism and affected British women's attitudes toward their own culture. The clitoridectomy campaign exposed the degree to which fear and denial of female sexual response suffused British as well as Kenyan society, but British women remained blind to such parallels. They were unable to see government treatment of their campaign as a comment on their own tenuous place in political life.

NOTES

1. By this time, Katherine Atholl was already well known in the House of Commons as its most prominent woman conservative. By no means a self-confessed feminist, Atholl had opposed women's suffrage before the war, but also had served on various local government committees. Eleanor Rathbone was equally prominent, and, as the

ten-year president of the National Union of Societies for Equal Citizenship, was perhaps the nation's most strenuous defender of women's rights. For further biographical information, see S. J. Hetherington, *Katherine Atholl, 1874–1960: Against the Tide* (Aberdeen, 1989); Mary D. Stocks, *Eleanor Rathbone: A Biography* (London, 1949).

2. Gayatri Chakravorty Spivak, "French Feminism in an International Frame," in *In Other Worlds: Essays in Cultural Politics* (New York: Methuen, 1987), 51. The difficult and sensitive question of initiated women's sexual response is best approached through the writings of African women health professionals; two excellent studies are those by Olayinko Koso-Thomas, *The Circumcision of Women: A Strategy for Eradication* (London: Zed Books, 1987); and Raqiya Haji Dualeh Abdalla, *Sisters in Affliction: Circumcision and Infibulation of Women in Africa* (London: Zed Press, 1982). Koso-Thomas questioned 140 circumcised women in Sierra Leone and found that although some experienced sexual desire and pleasure, none experienced orgasm. She also found a high incidence of medical difficulties traceable to clitoridectomy. Abdalla's conclusions were similar; on an encouraging note, however, she found that while "everybody is reluctant to break through the established institution and tradition connected with this custom and no one dares to be the first to abolish it," a majority nevertheless thought it should be abolished (94). By contrast, Hanny Lightfoot-Klein's recent study of the Sudan, Kenya, and Egypt underscores the extreme medical and sexual problems endured by initiated women but contends (on the basis of the author's extensive interviews) that "circumcised" women could and did experience orgasm (*Prisoners of Ritual: An Odyssey into Female Genital Circumcision in Africa* [New York: Haworth Press, 1989]).

3. Harriet Lyons, in defending anthropological studies of genital mutilations from feminists' criticisms, also notes that the racism embedded in attacks on Kikuyu circumcision "serve[d] only to instil the custom with validating meaning within a nationalist context" ("Anthropologists, Moralities, and Relativities: The Problem of Genital Mutilations," *Canadian Review of Sociology and Anthropology* 18, no. 4 [1981]: 512).

4. Jocelyn Murray, "The Kikuyu Female Circumcision Controversy, with Special Reference to the Church Missionary Society's 'Sphere of Influence' " (Ph.D. dissertation, University of California, Los Angeles, 1974). Murray is unique among the historians of the controversy in Kenya in integrating women fully into her account; the question of African women's interests and action during the controversy is best approached through her work.

THE ZOOT SUIT AND STYLE WARFARE

STUART COSGROVE

INTRODUCTION: THE SILENT NOISE OF SINISTER CLOWNS

What about those fellows waiting still and silent there on the platform, so still and silent they clash with the crowd in their very immobility, standing noisy in their very silence; harsh as a cry of terror in their quietness? What about these three boys, coming now along the platform, tall and slender, walking with swinging shoulders in their well-pressed, too-hot-for-summer suits, their collars high and tight about their necks, their identical hats of black cheap felt set upon the crowns of their heads with a severe formality above their conked hair? It was as though I'd never seen their like before: walking slowly, their shoulders swaying, their legs swinging from their hips in trousers that ballooned upward from cuffs fitting snug about their ankles; their coats long and hip-tight with shoulders far too broad to be those of natural western men. These fellows whose bodies seemed—what had one of my teachers said of me?—"You're like one of those African sculptures, distorted in the interest of design." Well, what design and whose?[1]

The zoot suit is more than an exaggerated costume, more than a sartorial statement: it is the bearer of a complex and contradictory history. When the nameless narrator of Ellison's *Invisible Man* confronted the subversive sight of three young and extravagantly dressed blacks, his reaction was one of fascination, not of fear. These youths were not simply grotesque dandies parading the city's secret underworld, they were "the stewards of something uncomfortable,"[2] a spectacular reminder that the social order had failed to contain their energy and difference. The zoot suit was more than the drape-shape of 1940s fashion, more than a colorful stage-prop hanging from the shoulders of Cab Calloway; it was, in the most direct and obvious ways, an emblem of ethnicity and a way of negotiating an identity. The zoot suit was a refusal: a subcultural gesture that refused to concede to the manners of subservience. By the late 1930s, the term "zoot" was in common circulation within urban jazz culture. "Zoot" meant something worn or performed in an extravagant style, and since many young blacks wore suits with outrageously padded shoulders and trousers that were fiercely tapered at the ankles, the term "zoot suit" passed into everyday usage. In the subcultural world of Harlem's nightlife, the language of rhyming slang succinctly described the zoot suit's unmistakable style: "a killer-diller coat with a drape-shape, reat-pleats and shoulders padded like a lunatic's cell." The study of the relationships between fashion and social action is notoriously underdeveloped, but there is every indication that the zoot-suit riots that erupted in the United States in the summer of 1943 had a profound effect on a whole generation of socially disadvantaged youths. It was during his period as a young zoot-suiter that the Chicano union activist César Chávez first came into contact with community politics, and it was through the experiences of participating in zoot-suit riots in Harlem that the young pimp "Detroit Red" began a political education that transformed him into the Black radical leader Malcolm X. Although the zoot suit occupies an almost mythical place within the history of jazz music, its social and political importance has been virtually ignored. There can be no certainty about when, where or why the zoot suit came into existence, but what is certain is that during the summer months of 1943 "the killer-diller coat" was the uniform of young rioters and the symbol of a moral panic about juvenile delinquency that was to intensify in the postwar period.

At the height of the Los Angeles riots of June 1943, the *New York Times* carried a front-page article which claimed without reservation that the first zoot suit had been purchased by a black bus worker, Clyde Duncan, from a tailor's shop in Gainesville, Georgia.[3] Allegedly, Duncan had been inspired by the film *Gone With the Wind*, and had set out to look like Rhett Butler. This explanation clearly found favor throughout the United States. The national press forwarded countless others. Some reports claimed that the zoot suit was an invention of Harlem nightlife; others suggested it grew out of jazz culture and the exhibitionist stage-costumes of the band leaders; and some argued that the zoot suit was derived from military uniforms and imported from Britain. The alternative and independent press, particularly *Crisis* and *Negro Quarterly*, more convincingly argued that the zoot suit was the product of a particular social context.[4] They emphasized the importance of Mexican-American youths, or *pachucos*, in the emergence of zoot-suit style and, in tentative ways, tried to relate their appearance on the streets to the concept of *pachuquismo*.

In his pioneering book *The Labyrinth of Solitude*, the Mexican poet and social commentator Octavio Paz throws imaginative light on *pachuco* style and indirectly establishes a framework within which the zoot suit can be understood. Paz's study of the Mexican national consciousness examines the changes brought about by the movement of labor, particularly the generations of Mexicans who migrated northward to the USA. This movement, and the new economic and social patterns it implies, has, according to Paz, forced young Mexican-Americans into an ambivalent experience between two cultures:

> What distinguishes them, I think, is their furtive, restless air: they act like persons who are wearing disguises, who are afraid of a stranger's look because it could strip them and leave them stark naked. . . . This spiritual condition, or lack of a spirit, has given birth to a type known as the pachuco. The pachucos are youths, for the most part of Mexican origin, who form gangs in southern cities; they can be identified by their language and behavior as well as by the clothing they affect. They are instinctive rebels, and North American racism has vented its wrath on them more than once. But the pachucos do not attempt to vindicate their race or the nationality of their forebears. Their attitude reveals an obstinate, almost fanatical will-to-be, but this will affirms nothing specific except their determination . . . not to be like those around them.[5]

Pachuco youth embodied all the characteristics of second-generation working-class immigrants. In the most obvious ways they had been stripped of their customs, beliefs and language. The *pachucos* were a disinherited generation within a disadvantaged sector of North American society; and predictably their experiences in education, welfare and employment alienated them from the aspirations of their parents and the dominant assumptions of the society in which they lived. The *pachuco* subculture was defined not only by ostentatious fashion, but by petty crime, delinquency and drug-taking. Rather than disguise their alienation or efface their hostility to the dominant society, the *pachucos* adopted an arrogant posture. They flaunted their difference, and the zoot suit became the means by which that difference was announced. Those "impassive and sinister clowns" whose purpose was "to cause terror instead of laughter,"[6] invited the kind of attention that led to both prestige and persecution. For Octavio Paz the *pachuco*'s appropriation of the zoot suit was an admission of the ambivalent place he occupied. "It is the only way he can establish a more vital relationship with the society he is antagonizing. As a victim he can occupy a place in the world that previously ignored him: as a delinquent, he can become one of its wicked heroes."[7] The zoot-suit riots of 1943 encapsulated this paradox. They emerged out of the dialectics of delinquency and persecution, during a period in which American society was undergoing profound structural change.

The major social change brought about by the United States' involvement in the war was the recruitment to the armed forces of over four million civilians and the entrance of over five million women into the wartime labor force. The rapid increase in military recruitment and the radical shift in the composition of the labor force led in turn to changes in family life, particularly the erosion of parental control and authority. The large-scale and prolonged separation of millions of families precipitated an unprecedented increase in the rate of juvenile crime and delinquency. By the summer of 1943 it was commonplace for teenagers to be left to their own initiatives while their parents were either on active military service or involved in war work. The increase in night work compounded the problem. With their parents or guardians working unsocial hours, it became possible for many more young people to gather late into the night at major urban centers or simply on the street corners.

The rate of social mobility intensified during the period of the zoot-suit riots. With over fifteen million civilians and twelve million military personnel on the move throughout the country, there was a corresponding increase in vagrancy. Petty crimes became more difficult to detect and control; itinerants became increasingly common, and social transience put unforeseen pressure on housing and welfare. The new patterns of social mobility also led to congestion in military and industrial areas. Significantly, it was the overcrowded military towns along the Pacific coast and the industrial conurbations of Detroit, Pittsburgh and Los Angeles that witnessed the most violent outbreaks of zoot-suit rioting.[8]

"Delinquency" emerged from the dictionary of new sociology to become an everyday term, as wartime statistics revealed these new patterns of adolescent behavior. The *pachucos* of the Los Angeles area were particularly vulnerable to the effects of war. Being neither Mexican nor American, the *pachucos*, like the black youths with whom they shared the zoot-suit style, simply did not fit. In their own terms they were "24-hour orphans," having rejected the ideologies of their migrant parents. As the war furthered the dislocation of family relationships, the *pachucos* gravitated away from the home to the only place where their status was visible: the streets and bars of the towns and cities. But if the *pachucos* laid themselves open to a life of delinquency and detention, they also asserted their distinct identity, with their own style of dress, their own way of life and a shared set of experiences.

The zoot-suit riots sharply revealed a polarization between two youth groups within wartime society: the gangs of predominantly black and Mexican youths who were at the forefront of the zoot-suit subculture, and the predominantly white American servicemen stationed along the Pacific coast. The riots invariably had racial and social resonances, but the primary issue seems to have been patriotism and attitudes to the war. With the entry of the United States into the war in December 1941, the nation had to come to terms with the restrictions of rationing and the prospects of conscription. In March 1942, the War Production Board's first rationing act had a direct effect on the manufacture of suits and all clothing containing wool. In an attempt to institute a 26 percent cutback in the use of fabrics, the War Production Board drew up regulations for the wartime manufacture of what *Esquire* magazine called "streamlined suits by Uncle Sam."[9] The regulations effectively forbade the manufacture of zoot suits, and most legitimate tailoring companies ceased to manufacture or advertise any suits that fell outside the War Production Board's guidelines. However, the demand for zoot suits did not decline, and a network of bootleg tailors based in Los Angeles and New York continued to manufacture the garments. Thus the polarization between servicemen and *pachucos* was immediately visible: the chino shirt and battledress were evidently uniforms of patriotism, whereas wearing a zoot suit was a deliberate and public way of flouting the regulations of rationing. The zoot suit was a moral and social scandal in the eyes of the authorities, not simply because it was associated with petty crime and violence, but because it openly snubbed the laws of rationing. In the fragile harmony of wartime society, the zoot-suiters were, according to Octavio Paz, "a symbol of love and joy or of horror and loathing, an embodiment of liberty, of disorder, of the forbidden."[10]

The zoot-suit riots, which were initially confined to Los Angeles, began in the first few days of June 1943. During the first weekend of the month, over sixty zoot-suiters were arrested and charged at Los Angeles County jail, after violent and well-publicized fights between servicemen on shore leave and gangs of Mexican-American youths. In order to prevent further outbreaks of fighting, the police patrolled the eastern sections of the city, as rumors spread from the military bases that servicemen were intending to form vigilante groups. The *Washington Post*'s report of the incidents, on the morning of Wednesday 9 June 1943, clearly saw the events from the point of view of the servicemen:

Disgusted with being robbed and beaten with tire irons, weighted ropes, belts and fists employed by overwhelming numbers of the youthful hood-lums, the uniformed men passed the word quietly among themselves and opened their campaign in force on Friday night. At central jail, where spectators jammed the sidewalks and police made no efforts to halt auto loads of servicemen openly cruising in search of zoot-suiters, the youths streamed gladly into the sanctity of the cells after being snatched from bar rooms, pool halls and theaters and stripped of their attire.[11]

During the ensuing weeks of rioting, the ritualistic stripping of zoot-suiters became the major means by which the servicemen reestablished their status over the *pachucos*. It became commonplace for gangs of marines to ambush zoot-suiters, strip them down to their underwear and leave them helpless in the streets. In one particularly vicious incident, a gang of drunken sailors rampaged through a cinema after discovering two zoot-suiters. They dragged the *pachucos* on to the stage as the film was being screened, stripped them in front of the audience and, as a final insult, urinated on the suits.

The press coverage of these incidents ranged from the careful and caution-ary liberalism of the *Los Angeles Times* to the more hysterical hate-mongering of William Randolph Hearst's west coast papers. Although the practice of stripping and publicly humiliating the zoot-suiters was not prompted by the press, several reports did little to discourage the attacks:

> Zoot suits smoldered in the ashes of street bonfires where they had been tossed by grimly methodical tank forces of service men. . . . The zooters, who earlier in the day had spread boasts that they were organized to "kill every cop" they could find, showed no inclination to try to make good their boasts. . . . Searching parties of soldiers, sailors and Marines hunted them out and drove them out into the open like bird dogs flushing quail. Pro-cedure was standard: grab a zooter. Take off his pants and frock coat and tear them up or burn them. Trim the "Argentine Ducktail" haircut that goes with the screwy costume.[12]

The second week of June witnessed the worst incidents of rioting and public disorder. A sailor was slashed and disfigured by a *pachuco* gang; a policeman was run down when he tried to question a car-load of zoot-suiters; a young Mexican was stabbed at a party by drunken Marines; a trainload of sailors were stoned by *pachucos* as their train approached Long Beach; streetfights broke out daily in San Bernardino; over four hundred vigilantes toured the streets of San Diego looking for zoot-suiters, and many individuals from both factions were arrested.[13] On 9 June, the *Los Angeles Times* published the first in a series of editorials designed to reduce the level of violence, but which also tried to allay the growing concern about the racial character of the riots:

> To preserve the peace and good name of the Los Angeles area, the strongest measures must be taken jointly by the police, the Sheriff's office and Army and Navy authorities, to prevent any further outbreaks of 'zoot suit' rioting. While members of the armed forces received considerable provocation at the hands of the unidentified miscreants, such a situation cannot be cured by indiscriminate assault on every youth wearing a particular type of costume.

> It would not do, for a large number of reasons, to let the impression circulate in South America that persons of Spanish-American ancestry were being singled out for mistreatment in Southern California. And the incidents here were capable of being exaggerated to give that impression.[14]

THE CHIEF, THE BLACK WIDOWS, AND THE TOMAHAWK KID

The pleas for tolerance from civic authorities and representatives of the church and state had no immediate effect, and the riots became more frequent and more violent. A zoot-suited youth was shot by a special police officer in Azusa; a gang of *pachucos* were arrested for rioting and carrying weapons in the Lincoln Heights area; twenty-five black zoot-suiters were arrested for wrecking an electric railway train in Watts; and one thousand additional police were drafted into East Los Angeles. The press coverage increasingly focused on the most "spectacular" incidents and began to identify leaders of zoot-suit style. On the morning of Thursday 10 June 1943, most newspapers carried photographs and reports on three "notorious" zoot-suit gang leaders. Of the thousands of *pachucos* that allegedly belonged to the hundreds of zoot-gangs in Los Angeles, the press singled out the arrests of Lewis D. English, a twenty-three-year-old black, charged with felony and carrying a "16-inch razor sharp butcher knife"; Frank H. Tellez, a twenty-two-year-old Mexican held on vagrancy charges; and another Mexican, Luis "The Chief" Verdusco (twenty-seven years of age), allegedly the leader of the Los Angeles *pachucos*.[15]

The arrests of English, Tellez and Verdusco seemed to confirm popular perceptions of the zoot-suiters widely expressed for weeks prior to the riots. First, that the zoot-suit gangs were predominantly, but not exclusively, comprised of black and Mexican youths. Second, that many of the zoot-suiters were old enough to be in the armed forces but were either avoiding conscription or had been exempted on medical grounds. Finally, in the case of Frank

Tellez, who was photographed wearing a pancake hat with a rear feather, that zoot-suit style was an expensive fashion often funded by theft and petty extortion. Tellez allegedly wore a colorful, long drape coat that was "part of a $75 suit" and a pair of pegged trousers "very full at the knees and narrow at the cuffs" which were allegedly part of another suit. The caption of the Associated Press photograph indignantly added that "Tellez holds a medical discharge from the Army."[16] What newspaper reports tended to suppress was information on the Marines who were arrested for inciting riots, the existence of gangs of white American zoot-suiters, and the opinions of Mexican-American servicemen stationed in California, who were part of the war effort but who refused to take part in vigilante raids on *pachuco* hangouts.

As the zoot-suit riots spread throughout California, to cities in Texas and Arizona, a new dimension began to influence press coverage of the riots in Los Angeles. On a day when 125 zoot-suited youths clashed with Marines in Watts and armed police had to quell riots in Boyle Heights, the Los Angeles press concentrated on a razor attack on a local mother, Betty Morgan. What distinguished this incident from hundreds of comparable attacks was that the assailants were girls. The press related the incident to the arrest of Amelia Venegas, a woman zoot-suiter who was charged with carrying, and threatening to use, a brass knuckleduster. The revelation that girls were active within *pachuco* subculture led to consistent press coverage of the activities of two female gangs: the Slick Chicks and the Black Widows.[17] The latter gang took its name from the members' distinctive dress: black zoot-suit jackets, short black skirts and black fish-net stockings. In retrospect the Black Widows, and their active part in the subcultural violence of the zoot-suit riots, disturb conventional understandings of the concept of *pachuquismo*.

As Joan W. Moore implies in *Homeboys*, her definitive study of Los Angeles youth gangs, the concept of *pachuquismo* is too readily and unproblematically equated with the better known concept of *machismo*.[18] Undoubtedly, they share certain ideological traits, not least a swaggering and at times aggressive sense of power and bravado; but the two concepts derive from different sets of social definitions. Whereas *machismo* can be defined in terms of male power and sexuality, *pachuquismo* predominantly derives from ethnic, generational and class-based aspirations, and is less evidently a question of gender. What the zoot-suit riots brought to the surface was the complexity of *pachuco* style. The Black Widows and their aggressive image confounded the *pachuco* stereotype of the lazy male delinquent who avoided conscription for a life of dandyism and petty crime, and reinforced radical readings of *pachuco* subculture. The

Black Widows were a reminder that ethnic and generational alienation was a pressing social problem and an indication of the tensions that existed in minority, low-income communities.

Although detailed information on the role of girls within zoot-suit subculture is limited to very brief press reports, the appearance of female *pachucos* coincided with a dramatic rise in the delinquency rates among girls aged between twelve and twenty. The disintegration of traditional family relationships and the entry of young women into the labor force undoubtedly had an effect on the social roles and responsibilities of female adolescents, but it is difficult to be precise about the relationships between changed patterns of social experience and the rise in delinquency. However, wartime society brought about an increase in unprepared and irregular sexual intercourse, which in turn led to significant increases in the rates of abortion, illegitimate births and venereal diseases. Although statistics are difficult to trace, there are many indications that the war years saw a remarkable increase in the numbers of young women who were taken into social care or referred to penal institutions, as a result of the specific social problems they had to encounter.

Later studies provide evidence that young women and girls were also heavily involved in the traffic and transaction of soft drugs. The *pachuco* subculture within the Los Angeles metropolitan area was directly associated with a widespread growth in the use of marijuana. It has been suggested that female zoot-suiters concealed quantities of drugs on their bodies, since they were less likely to be closely searched by male members of the law enforcement agencies. Unfortunately, the absence of consistent or reliable information on the female gangs makes it particularly difficult to be certain about their status within the riots, or their place within traditions of feminine resistance. The Black Widows and Slick Chicks were spectacular in a subcultural sense, but their black drape jackets, tight skirts, fish-net stockings and heavily emphasized make-up, were ridiculed in the press. The Black Widow clearly existed outside the orthodoxies of wartime society: playing no part in the industrial war effort, and openly challenging conventional notions of feminine beauty and sexuality.

Toward the end of the second week of June, the riots in Los Angeles were dying out. Sporadic incidents broke out in other cities, particularly Detroit, New York and Philadelphia, where two members of Gene Krupa's dance band were beaten up in a station for wearing the band's zoot-suit costumes; but these, like the residual events in Los Angeles, were not taken seriously. The authorities failed to read the inarticulate warning signs proffered in two separate incidents in California: in one a zoot-suiter was arrested for throwing gasoline flares at a theater; and in the second another was arrested for carrying a silver tomahawk. The zoot-suit riots had become a public and spectacular enactment of social disaffection. The authorities in Detroit chose to dismiss a zoot-suit riot at the city's Cooley High School as an adolescent imitation of the Los Angeles disturbances.[19] Within three weeks Detroit was in the midst of the worst race riot in its history.[20] The United States was still involved in the war abroad when violent events on the home front signaled the beginnings of a new era in racial politics. . . .

THE MYSTERY OF THE SIGNIFYING MONKEY

> The pachuco is the prey of society, but instead of hiding he adorns himself to attract the hunter's attention. Persecution redeems him and breaks his solitude: his salvation depends on him becoming part of the very society he appears to deny.[21]

The zoot suit was associated with a multiplicity of different traits and conditions. It was simultaneously the garb of the victim and the attacker, the persecutor and the persecuted, the "sinister clown" and the grotesque dandy. But the central opposition was between the style of the delinquent and that of the disinherited. To wear a zoot suit was to risk the repressive intolerance of wartime society and to invite the attention of the police, the parent generation and the uniformed members of the armed forces. For many *pachucos* the zoot-suit riots were simply high times in Los Angeles when momentarily they had control of the streets; for others it was a realization that they were outcasts in a society that was not of their making. For the black radical writer Chester Himes, the riots in his neighborhood were unambiguous: "Zoot Riots are Race Riots."[22] For other contemporary commentators the wearing of the zoot suit could be anything from unconscious dandyism to a conscious "political" engagement. The zoot-suit riots were *not* "political" riots in the strictest sense, but for many participants they were an entry into the language of politics, an inarticulate rejection of the "straight world" and its organization.

It is remarkable how many postwar activists were inspired by the zoot-suit disturbances. Luis Valdez of the radical theater company El Teatro Campesino allegedly learned the "chicano" from his cousin the zoot-suiter Billy Miranda.[23] The novelists Ralph Ellison and Richard Wright both conveyed a literary and political fascination with the power and potential of the zoot suit. One of Ellison's editorials for the journal *Negro Quarterly* expressed his own sense of frustration at the enigmatic attraction of zoot-suit style:

A third major problem, and one that is indispensable to the centralization and direction of power, is that of learning the meaning of myths and symbols which abound among the Negro masses. For without this knowledge, leadership, no matter how correct its program, will fail. Much in Negro life remains a mystery; perhaps the zoot-suit conceals profound political meaning; perhaps the symmetrical frenzy of the Lindy-hop conceals clues to great potential powers, if only leaders could solve this riddle.[24]

Although Ellison's remarks are undoubtedly compromised by their own mysterious idealism, he touches on the zoot suit's major source of interest. It is in everyday rituals that resistance can find natural and unconscious expression. In retrospect, the zoot suit's history can be seen as a point of intersection between the related potential of ethnicity and politics on the one hand, and the pleasures of identity and difference on the other. It is the zoot suit's political and ethnic associations that have made it such a rich reference point for subsequent generations. From the music of Thelonious Monk and Kid Creole to the jazz-poetry of Larry Neal, the zoot suit has inherited new meanings and new mysteries. In his book *Hoodoo Hollerin' Bebop Ghosts*, Neal uses the image of the zoot suit as the symbol of Black America's cultural resistance. For Neal, the zoot suit ceased to be a costume and became a tapestry of meaning, where music, politics and social action merged. The zoot suit became a symbol for the enigmas of Black culture and the mystery of the signifying monkey:

But there is rhythm here
Its own special substance:
I hear Billie sing, no Good Man, and dig Prez, wearing the Zoot
suit of life,
the Porkpie hat tilted at the correct angle; through the Harlem
smoke of beer
and whisky, I understand the mystery of the Signifying Monkey.[25]

NOTES

1. Ralph Ellison, *Invisible Man*, New York, 1947, 380.
2. Ellison, *Invisible Man*, 381.
3. "Zoot Suit Originated in Georgia," *New York Times*, 11 June 1943, 21.
4. For the most extensive sociological study of the zoot-suit riots of 1943, see Ralph H. Turner and Samuel J. Surace, "Zoot Suiters and Mexicans: Symbols in Crowd Behaviour," *American Journal of Sociology*, 62, 1956, 14–20.
5. Octavio Paz, *The Labyrinth of Solitude*, Harmondsworth, Penguin, 1967, 5–6.
6. Paz, *Labyrinth of Solitude*, 8.

7. Ibid.

8. See K. L. Nelson, ed., *The Impact of War on American Life*, New York, 1971.

9. O. E. Schoeffler and W. Gale, *Esquire's Encyclopedia of Twentieth-Century Men's Fashion*, New York, 1973, 24.

10. Paz, *Labyrinth of Solitude*, 8.

11. "Zoot-Suiters Again on the Prowl as Navy Holds Back Sailors," *Washington Post*, 9 June 1943, 4.

12. Quoted in S. Menefee, *Assignment USA*, New York, 1943, 24.

13. Details of the riots are taken from newspaper reports and press releases for the weeks in question, particularly from the *Los Angeles Times*, *New York Times*, *Washington Post*, *Washington Star*, and *Time Magazine*.

14. "Strong Measures Must be Taken Against Rioting," *Los Angeles Times*, 9 June 1943, 4.

15. "Zoot-Suit Fighting Spreads On the Coast," *New York Times*, 10 June 1943, 23.

16. Ibid.

17. "Zoot-Girls Use Knife in Attack," *Los Angeles Times*, 11 June 1943, 1.

18. Joan W. Moore, *Homeboys: Gangs, Drugs and Prison in the Barrios of Los Angeles*, Philadelphia, 1978.

19. "Zoot Suit Warfare Spreads to Pupils of Detriot Area," *Washington Star*, 11 June 1943, 1.

20. Although the Detroit Race Riots of 1943 were not zoot-suit riots, nor evidently about "youth" or "delinquency," the social context in which they took place was obviously comparable. For a lengthy study of the Detroit riots, see R. Shogun and T. Craig, *The Detroit Race Riot: A Study in Violence*, Philadelphia and New York, 1964.

21. Paz, *Labyrinth of Solitude*, 9.

22. Chester Himes, "Zoot Riots are Race Riots," *The Crisis*, July 1943; reprinted in Himes, *Black on Black: Baby Sister and Selected Writings*, London, 1975.

23. El Teatro Campesino presented the first Chicano play to achieve full commercial Broadway production. The play, written by Luis Valdez and entitled *Zoot Suit*, was a drama documentary on the Sleepy Lagoon murder and the events leading to the Los Angeles riots. (The Sleepy Lagoon murder of August 1942 resulted in twenty-four *pachucos* being indicted for conspiracy to murder.)

24. Quoted in Larry Neal, "Ellison's Zoot Suit," in J. Hersey, ed., *Ralph Ellison: A Collection of Critical Essays*, New Jersey, 1974, 67.

25. From Larry Neal's poem "Malcolm X: An Autobiography," in L. Neal, *Hoodoo Hollerin' Bebop Ghosts*, Washington, D.C., 1974, 9.

COOPTATION AND CONTROL:

THE RECONSTRUCTION OF INUIT BIRTH

PATRICIA LEYLAND KAUFERT AND JOHN D. O'NEIL

At one level in this article we deal with the impact that record keeping, evaluation procedures, and statistical techniques have had on policies relating to childbirth in the Canadian Arctic. At another level, we take the forms and rituals of epidemiological knowledge, not as givens but as a legitimate field for anthropological inquiry.

We took our material from a research project examining the delivery of obstetric care to Inuit women from the Keewatin Region of the Northwest Territories. The epidemiological component of this study included an audit of the obstetrical records for all women from the area who gave birth between 1979 and 1985. As part of the same project, we reviewed government publications, policy documents, and academic reports dealing with childbirth in the Canadian Arctic. These activities brought us into close contact not only with the raw data and standard procedures of epidemiological research but also with the ways in which the products of such research are used by policy makers. In an earlier paper, we discussed some of the methodological problems involved in the use of data taken from obstetrical records (Kaufert et al. 1990). Our concern here is less with the ways in which the use of these data violate standard procedures of epidemiological research and more with the historical and political context in which the records were created and used. We are particularly concerned with the use (and sometimes misuse) of rates, numbers, and statistics in the extension of medical control over Inuit childbirth.

The Keewatin Region of the Northwest Territories has an area of approximately 225,000 square miles and lies along the western coast of Hudson Bay. Five thousand five hundred Inuit live in eight communities in the region. Ranging in size from two hundred to eleven hundred people, these communities are between two hundred and eight hundred miles from Churchill, Manitoba, which is located some six hundred miles north of Winnipeg, the provincial capital. Flying is the main mode of travel between the eight communities and between the Keewatin Region and Churchill or Winnipeg.

Until 1988, when responsibility for health-care services was transferred to the Government of the Northwest Territories, provision of medical care in the area was the direct responsibility of the Canadian Federal Government as exercised through the Medical Services Branch of Health and Welfare, Canada. Medical Services in turn had contracted with the University of Manitoba for the provision of physician and obstetrician services. From the late 1960s through the early 1980s, obstetric policy allowed a woman whose pregnancy was defined as low risk to give birth (attended by a nurse-midwife) at a health center (known as a nursing station) within her own community. This option has now been withdrawn, and all but a few women from the Keewatin are evacuated for childbirth to hospitals in either Churchill or Winnipeg. Protests against the new policy from the communities resulted in the formulation of this research project to investigate the impact of evacuation on women and their families.

METHODS

Unusual for research in the anthropology of childbirth, this article is not based on observation of women in labor or on stories told by women about their birthing experiences. Rather than trying to capture the actuality of the individual birth, our interest is in how Inuit childbirth has been represented through the medium of obstetrical records, official statistics, and epidemiological reports. For this reason, the main source of our data is obstetrical records themselves, plus an assortment of government reports, administrative archives, and research papers (both published and unpublished).

The review of obstetrical records includes every birth to women who had a child between 1979 and 1985. We have only the written record for each of these births. The birth is past and cannot be observed. We have not gone to the woman herself and asked about the birth; neither have we talked with the physician or nurse who cared for her. Though such interviews might be logistically difficult, they would not be impossible. The barrier to doing them lies with hospitals, which own and control access to the records. To protect confidentiality, hospitals permit no information to be abstracted which might identify the individual patient. Given this constraint, contacting a woman in order to inquire into the relationship between her experience and the data recorded in her chart was impossible.

However, material for this article is not limited to hospital records. As part of the project, a series of in-depth interviews were completed with traditional Inuit midwives and elders, women in their childbearing years, health professionals, and administrators. We held meetings in each of the communities to discuss the problems of evacuating women for childbirth and organized a workshop which brought together community representatives, health professionals, and researchers to discuss obstetric policy in the North (O'Neil and Gilbert 1990). (All these interviews, meetings, and the workshop were tape recorded and transcribed.) The problems created by the policy of evacuating women for childbirth is of concern to many different groups in the North, but particularly to Inuit women, the nurses, physicians, and administrators who work in the Keewatin, Inuit politicians, Ottawa bureaucrats, and other researchers. In addition to our formal interviews, we have participated in numerous informal discussions and from them have gained insight into many of the issues involved in the delivery of obstetric care in the Keewatin.

BACKGROUND

THE HISTORY OF MEDICAL CARE IN THE KEEWATIN

Present childbirth policy in the Keewatin cannot be understood without some knowledge of the political and historical background of the area, including the contact and relationship between Inuit and the central government. Early in this century, the Inuit sustained sporadic contacts with whalers and explorers, who were followed by the Hudson's Bay Company (the first trading post in the Keewatin was set up in 1911), and then by missionaries, who established the first mission in 1912. Small numbers of Canadian and U.S. military personnel came to the area during World War II and again later to set up sites for the Distant Early Warning radar line. According to Williamson (1974), the administration of the Keewatin by the Canadian Government was relatively minimal until 1953, when the Federal Department of Northern Affairs was established. Before 1953, the main government presence was a scattering of the Royal Canadian Mounted Police (RCMP) and the staff of Department of Transport weather stations. The primary reason for a sustained government presence was the construction of a nickel mine at Rankin Inlet in 1953.

Williamson describes how Inuit life changed as hunting for subsistence was modified in response to the fur-trapping economy, new but fluctuating wage employment opportunities, and the impact of Anglican and Catholic missionaries. It was famine and disease, however, which nearly brought destruction to some inland groups and severely disrupted the whole Keewatin. Epidemic disease, particularly tuberculosis and poliomyelitis, spread gradually throughout the area, but medical care was scant. Except for a mission hospital at

Chesterfield Inlet, formal Western medicine was only introduced in the Canadian North in the 1950s, when physicians began to accompany ships supplying the missions and trading posts scattered across the Arctic coast. In the middle and late 1950s, a shift in the migration patterns of the caribou brought starvation to many inland groups and decimated population in the south Keewatin Plain and northern Central Keewatin.

Although the Inuit had once been discouraged officially from congregating around trading posts and radar stations, once the Department of Northern Affairs was established in 1953, the government became actively involved in resettling the remaining inland peoples to coastal communities, which thus became an amalgam of groups, many of which had previously had little contact with each other, had practiced different subsistence patterns, and had spoken different dialects. Though well intentioned, the resettlement policy was often ill-conceived and poorly executed. According to an environmental health officer writing in the mid-1970s, communities were

> fostered or have been allowed to grow where drainage and waste disposal is impossible. Migration has been allowed to occur into communities already overcrowded and without hope of catching up with housing demand . . .— Water is in chronically short supply in many communities. . . . To these basic problems have been added those caused by individual indifference to garbage and hazard strewn surroundings, by the demands of financial economy in food marketing.

Housing conditions were poor. Levels of infectious disease (particularly tuberculosis and diphtheria) were high. Families were broken apart by death or separated by the government policy of sending the sick south for medical care (Williamson 1974). People "disappeared" into hospitals and sanatoria, unable to contact their families, who knew little of where their sick kin were or how they fared (Hodgson 1982). Disrupted by resettlement, famines, and epidemics, communities in the Keewatin were thus disorganized politically and socially. By the 1960s, people were dependent on government services and vulnerable to government control, including control over childbirth.

TRADITIONAL BIRTH

The elders (including a few surviving traditional midwives), from whom we collected accounts of childbirth before the building of the nursing stations, spoke about the births of their own children, births at which they had assisted, and births about which they had been told by their mothers and grandmothers. (The midwives often said that they had been trained by their mothers who had also been midwives.) Their stories always included details about where and when the birth occurred (in a summer camp, traveling in a storm), who was or was not present (birthing alone, with the midwife, with their husband only, with other women), what happened at the birth (whether the labor was long or short, easy or hard). Birth was depicted as a normal rather than a pathological process, although the midwives recognized that things could go devastatingly wrong. Talking about past births, they told of their fears of postpartum hemorrhage or a breech delivery. They described how they turned a malpositioned infant or dealt with a retained placenta or hastened a prolonged labor. Each birth story was presented as if it were a unique happening. The elders were careful never to talk as if a single Inuit model of childbirth existed, and they always emphasized the specificity of their own ways or experiences.

Indeed, many spoke of important regional variations and emphasized the importance of these differences in contemporary communities where women have migrated from different regions. Although the particularity of each birth was emphasized, an analysis of the stories uncovers a common emphasis on obedience, virtue, courage, and stoicism. The old women recalled how they were told as young girls that a quick birth and a small but lively baby could be ensured by working hard, being constantly active, not overeating, being respectful to elders, and being not just obedient but promptly obedient. If a short labor in childbirth was the reward for virtue, the threat of a long and painful labor was used as a form of sanction. While stoicism through a long labor was sometimes described with admiration, the preference was for a quick labor. Crying out in pain was discouraged, because it showed a lack of both courage and concentration. A woman was expected to focus on pushing hard to ensure a speedier, and thereby safer, birth.

Whenever the elders spoke about traditional ways of childbirth (whether with us or at public gatherings), they talked about the importance of passing on their knowledge to younger women. Yet there was also a political and moral purpose to their stories: particular pride was attached to births which women had deliberately managed alone, without help from anyone. Whether or not this circumstance was common, the woman laboring silently and in isolation constituted a powerful contemporary symbol of women's earlier independence and self-sufficiency. Praise of the past was a criticism of a present in which women had lost control over the place and conditions of childbirth and had become dependent on Southern medical care.

Some of the traditional births described by elders may have occurred in the 1930s and 1940s, but many took place in the 1950s and 1960s, the years marked by major upheaval and disruption. Outside influences on traditional ways of childbirth may date back even earlier than these stories, for there are reports of missionaries, Hudson's Bay employees, and RCMP officers having served occasionally as midwives. A few women also gave birth in the mission hospital in Chesterfield Inlet, and later, the military hospital in Churchill provided some obstetric care to Inuit women. The impact of these earlier influences is unknown, but it was probably spasmodic and unevenly disseminated throughout the area.

The full-scale medicalization of childbirth dates to the building of nursing stations by the federal government during the 1960s. By 1970, each of the eight communities had its own nursing station, which was staffed by one to four nurses, including at least one nurse midwife. Schaefer, a physician who worked for the federal government in the Canadian Arctic, has described the process of bringing childbirth into the nursing stations.

> There was a push from above for more hospital deliveries and deliveries in nursing stations instead of in tents and igloos. I do not say that I was one hundred percent convinced that there was a great need for it, but eventually it was inevitable.

Criteria for determining which women should stay in the communities to have their babies in the nursing station and which should be evacuated to hospitals were established by the Medical Services Branch of the Department of Health and Welfare in Ottawa (the "above" in the quotation from Schaefer). Until 1977, the annual reports by government on health conditions in the Northwest Territories always included some version of the following statement:

> We have continued the policy that sees all primagravida and grand multiparae (fifth or subsequent infants) evacuated to a hospital for delivery as are all complicated pregnancies or anticipated complications. Provided no complications ensued at the birth of the first infant or if all else is well, second, third or fourth babies are delivered in nursing stations.

According to these annual reports, the criteria for evacuation remained unchanged. but our data show a gradual but steady decline in the number of nursing station births throughout the 1970s. Physicians and nurses who worked in the Keewatin during those years have told us that the pressure to evacuate increased steadily until the early 1980s, when official policy became one of evacuating all women to hospitals.

Many of the women we have asked about the experience of being evacuated for childbirth have complained about being separated from their community and family, about the loneliness of childbirth in a strange, medical setting, about loss of control over the ways in which they gave birth, about being alienated through language and culture from those providing care. Their complaints are familiar, for the same themes appear and reappear in anthropological research and feminist critiques of the medicalization of childbirth (Annandale 1988; Davis-Floyd 1988; Lazarus 1988; Michaelson 1988). The transition for Inuit women from birth within their own homes under the care of traditional midwives to hospital birth under physician control can be seen as a repetition of a process which happened earlier for women living in Southern Canada. One of the differences for Inuit women, however, was that these changes were compressed into less than twenty years. In our review of obstetrical records, for example, we have data on women who had their first birth in a hunting camp, attended by a traditional midwife, and their last birth in a tertiary care hospital a thousand miles south of their home community.

Inuit women who have delivered in hospitals have been subject to the same type of interventions in the birth process as other North American women over the past twenty years. Data from our review of obstetrical records from 1979–85 indicate an increasing use of ultrasound and fetal monitors and rising rates of induction and cesarean section.

When we talked with health professionals in general about evacuation, the majority were hesitant about supporting a return to childbirth in the nursing stations attended by midwives (O'Neil and Kaufert 1990). Those who had worked in the North acknowledged the problems that evacuation creates, but their belief that birth should only be in a hospital with specialist care, services, and machinery close at hand was much stronger. Their attitudes in this regard were reminiscent of the American obstetricians interviewed by Davis-Floyd (1987), probably reflecting similarities in education and professional socialization across North America.

THE OBSTETRICAL RECORD AND THE BIRTH EXPERIENCE

In the literature on the medicalization of childbirth, the impact that such innovations as the creation of records, the codification of data, and the processing of statistics have had on women's childbirth experiences has been largely ignored, partly because these changes are less obvious than the new machinery of obstetrics. Fetal monitors, for example, are visually more dramatic than a perinatal risk scoring form, but it is this form, completed as part of an Inuit woman's prenatal care, that determines to which hospital she is sent for childbirth. (High risk women are evacuated to Winnipeg; low risk women remain in Churchill.) Similarly, the immediate decision-making power of the attending physician is highly visible to a woman in labor (or an observing anthropologist taking notes). It is much less obvious that the actions of this physician are constrained by decisions relating to obstetric policy and practice that are outside his control and may be based on statistics and epidemiological data. The earlier quotation from Schaefer, for example, shows that he had no choice but to comply with an administrative decision to move childbirth into nursing stations or hospitals. This decision had been based on (questionable) statistics indicating a very high mortality rate.

The impact of record keeping and the compilation and analysis of data on the medicalization of childbirth has been ignored in the literature partly because researchers who work with obstetrical records tend to be epidemiologists uninterested in such questions. Anthropologists and others interested in the medicalization of childbirth prefer to work in the ethnographic tradition by observing the women in labor or talking with them about their experiences of hospital birth (e.g., Lazarus 1988), or, like Annandale (1988) or Davis-Floyd (1988), by observing and talking with those who provide care to women in childbirth (midwives, nurses, physicians). The result is a rich ethnographic account of hospital birth, but it is one which largely ignores the role of information gathering and epidemiological analyses on the childbirth experience.

We are very aware, however, of the limitations of obstetrical records when viewed from an anthropological perspective. We can reconstruct what happened at nearly every birth over a period of six years but only within the limitations of the data entered into the record. We can tell, for example, whether fetal monitors were used and show that usage increased, but not what the experience of being monitored meant to Inuit women. We can see that some babies were born on airplanes while women were being evacuated, but we cannot talk to the women involved about the experience of being in labor so publicly and so dangerously. We know the annual rates of cesarean section but not the psychological impact of undergoing this operation instead of being able to give birth naturally like a traditional Inuit woman.

Rather than dismissing the review of obstetric records as inadequate substitutes for the usual forms of anthropological research, we have treated them as a field of inquiry. Records are a source of information in the same sense as any set of historical documents. They can reveal, for example, the creeping process of medicalization as measured by increased rates of induction and cesarean section. They can also be examined to explore the same sort of questions historians ask about the documents they use, such as why they were created, how they were used, and what were their political and other implications.

OBSTETRICAL RECORDS AS A MEANS OF COMMUNICATION

Working occasionally with records for women whose first birth occurred in the early 1970s, we know that record keeping in the Keewatin became more and more detailed, more and more formal, and more and more standardized over time. Over the period of approximately fifteen years, record keepers became more meticulous in completing records. They also had to deal with an increasing array of requests for information. In corroboration of this greater workload, we have found letters in the administrative files for the 1970s from the nursing stations to the nursing supervisor complaining about an increasing volume of paper work. We also read letters sent from administrative officers for Medical Services in Ottawa to local administrators in the Keewatin demanding more thorough attention to record keeping or discussing the introduction of new forms.

A bureaucratic concern in Ottawa with forms and statistics may have intensified the concern with record keeping in the Keewatin, but the increasing importance of records was a general phenomenon in obstetric care in the 1970s and 1980s, partly reflecting a preoccupation with malpractice suits, for which documentation is seen as a critical element in defensive medicine. As the use of tests increased with the increasing complexity of prenatal and obstetric care in this period, records also expanded to accommodate the results of procedures and the reports of an expanding team of specialists. And as the amount of information increased, its reduction into a compressed and standardized format became more important. As a result, standard forms developed. Finally, criticism of the poor research reputation of obstetrics was a factor in expanding and standardizing obstetrical record keeping. In the Keewatin, reviews for research purposes were a relatively frequent phenomenon and may well have influenced the growing standardization of information.

Copies of a woman's prenatal record are sent with her when she goes south for childbirth, and a copy of the forms relating to labor and delivery return with her as the official story of what happened at the birth. In a sense, her obstetrical record is the medical equivalent of the stories of childbirth we have been told by the traditional midwives. Like birth stories, records have an underlying narrative form; their purpose is to allow others to reconstruct what happened, who did what, why, and when. Like all medical records, an obstetrical record is "the intervening document between one practitioner and another" (Pettinari 1988:18). It is a means of communication intended for use solely by the initiated, the health professional trained in its language.

Although we cannot talk to those who completed records, we can see how these documents have passed from one professional to another, accruing pieces of paper reporting test results, collecting comments and diagnoses from each person who examined the patient. Most records show that a woman saw the nurse at the nursing station (possibly more than one nurse), the physician(s) visiting the community, and a varying assortment of physicians, nurses, and obstetricians when evacuated to hospital. Occasionally we can recognize a signature, but most entries into obstetrical records have been made by an unknown hand. Some are members in a constantly changing group of physicians and nurses working in the Keewatin; others never went north but worked in Winnipeg seeing Inuit patients as they would any other woman having a baby on the labor and delivery ward.

Obstetrical records are not simply a neutral medium for storing information but are constructed in ways which predetermine what information is collected and in what form. Record keeping has its own vocabulary and standard sets of questions. Forms vary in the amount of detail, but spaces are always allotted for recording maternal age and parity, date of the last menstrual period, expected dates of confinement, blood pressures and weight gains, survival of previous infants, exact timing of each stage of labor, Apgar scores, and notations on the use of analgesics, episiotomy, induction, fetal monitors, and forceps. These are the standard elements seen as important for clinicians to know about a pregnancy and birth. The focus is clearly on the mechanical elements of birth, not the emotional and experiential; we find occasional references in nursing notes to a woman being lonely, worried, or afraid, but no formal place exists for such comments within the record of labor and delivery. The rich and complex experience of birth is broken into parts and packed into categories.

The forms that comprise obstetrical records in the Keewatin usually derive from Manitoba, the province to which most women are sent for obstetric care. The advantage of using common forms lies in communication among practitioners. The same information is in the same place on the same form, regardless of whether the patient is from Winnipeg or a thousand miles away in Repulse Bay. The records coming from the Keewatin can be read and understood by people working in Winnipeg hospitals; the records sent from hospitals to the nursing stations are in the same common language of Western obstetrics familiar to both nurses and physicians from their training. No one using the forms need have any knowledge of the North, of Inuit culture, or of the work of nursing stations.

Allowing no space for information particular to the North has clear disadvantages in making policy decisions. For example, the current pregnancy risk status assessment form used with Northern women is based on criteria developed elsewhere. A consultant working for Medical Services in 1978 concluded that

> at least three of the currently used systems for scoring risk factors in pregnancy are not appropriate for use in the eastern Arctic; that regional and ethnic differences can significantly affect the importance of some variables: that scoring systems, if they are to be used, should be individualized to the population being cared for.

Among other criticisms, [this consultant] noted that tuberculosis therapy was a significant factor in predicting risk but that parity was not. Yet the risk status assessment used in the Keewatin is the one used in Manitoba. Parity is weighed highly in the scoring system for risk, and gestational diabetes (rare among Inuit women) is included but not tuberculosis therapy. Obstetrical forms, like the one for risk assessment, represent a perspective on birth which is based on the experience of those who construct them. Tuberculosis is not a risk factor in the South and was excluded; the effects of diabetes are quite frequently seen in the South, and it was included. The underlying assumption is that birth is a universal process and so also are its risks. The assumption is in fact false, and what is treated as universal is particular to Southern women and Southern clinical experience.

RECORDS AS A MEANS OF SURVEILLANCE

The nurse-midwives working in the nursing stations were at the periphery of an administrative hierarchy with its headquarters in Ottawa. Travel conditions have improved greatly over the past twenty years, but geography and climate continue to make flying difficult and expensive. Since direct supervision of nurses is therefore impossible, the written record is an important source for monitoring performance. Anyone adding to a record, particularly the nurses, knows that they are open to repeated scrutiny by peers and superiors. In prenatal care, administrative regulations require that physicians on their monthly visits to the communities should review the records of all women who are pregnant, and the record of prenatal care from both physicians and nurses is sent out with each woman who is evacuated. If anything goes wrong in a pregnancy (particularly if a woman goes into labor while still in the community), the record is reviewed for problems missed earlier.

The cases that go wrong, rather than births successfully accomplished, preoccupy nurses, obstetricians and other physicians, and administrators. Conversations with health professionals about problems of obstetric care in the North often switched to the medical horror story of the birth which ended in death and disaster or the one in which death was narrowly averted. The more formal expression of this preoccupation is found in the detailed investigation and reporting of all deaths under one year, established in 1971 with the following purpose: "In this way we may be able to ascertain what other preventive measures are necessary to reduce infant mortality" (National Health and Welfare Canada [Medical Services] 1971:65). Since no woman died in childbirth in the Keewatin between 1970 and 1985, maternal mortality rates were an issue only as an event which was feared, not as a reality. The concern was rather with stillbirths and deaths within the first month of life. A perinatal mortality committee was established in 1971 and marked the beginning of a discourse which continually examined, evaluated, and commented not so much on birth, but on death in birth.

The function of the committee is to determine what happened around each death, why, and what to avoid in the future. The meetings of the committee are confidential but presumably follow the usual practice of reviewing the obstetrical record, the official history of what has happened. The activities of the perinatal mortality committee may be seen simply as an exercise in critical self-evaluation, such as one finds in any other setting following the death of a newborn infant. But the setting of the Keewatin has added a particular concern with the location of birth and death. Published reviews of perinatal mortality rates in the Keewatin during the 1970s evaluate each death in a nursing station according to whether it might have been avoided had the woman been in hospital. Most of the deaths were due to premature labor or other causes impervious to changes in evacuation policy (such as severe congenital anomaly).

Our data suggest that practice shifted toward hospitalization and away from births in nursing stations long before an official policy was declared in favor of all women being evacuated. Though evacuation did not become standard policy until after 1980, the proportion of Inuit births occurring in Manitoba hospitals rose from 65 percent in 1971, to 77 percent in 1975, to 88 percent in 1980. Moreover, one element in this change appears to have been the practice of constantly reviewing the contribution of birth in a nursing station to the death rates of newborns. The existence of a climate of fear associated with surveillance is suggested by Mason in writing about the period immediately before the decision to require evacuation: "As the policy of total evacuation became more pronounced, the nurses became more nervous about doing low risk deliveries, even when their results were excellent. . . . There came to be an atmosphere of crisis around birth" (Mason 1987:226). This "crisis" atmosphere existed despite the fact that formal published reviews of perinatal mortality acknowledged that shifting births from nursing stations to hospitals would have minimal impact on rates. Rather than data, it seems to have been an ideological commitment to the hospital as the safest place for birth which influenced not only physicians and nurses in making decisions about the care of individual women but also administrators in Ottawa in developing obstetric policies for the North.

RECORDS, STATISTICS, AND GOVERNMENT POLICY

Yet records are not simply a method of communication or surveillance; they are a source of information. The reliance placed on them as the location of "facts" seems always to run contrary to actual clinical experience. At one level, health professionals know that records are inevitably the product of a series of qualitative judgments and decisions transposed into quantitative terms and that their appearance of precision and objectivity is misleading. At another level, written data appear to be reassuring to both the clinical and the epidemiological mind. In medical care evaluation, records are generally considered more reliable than information retrieved directly from a patient.

From an epidemiological perspective, the value of record keeping lies in the reduction of each birth to a series of common units that can be summed, compared, or otherwise related one to another. By translating each category into a numeric code, a researcher can look at the relationship between gestational age and birthweight, the age of the mother and the survival of the infant, or Apgar scores and birth by cesarean section. Perinatal mortality rates are the customary measure of epidemiological discourse. By turning raw numbers into rates, comparisons can be made from one period to another, from one place of birth to another, from one area to another. Taking the formula for calculating mortality rates and plugging in the numbers for the Keewatin has an appealing simplicity, as well as an appearance of appropriate scientific method.

Reviews of obstetrical records have been a constant feature of medical care in the Keewatin, although most have been completed by health professionals with some epidemiological training rather than by professional epidemiologists. One of the earliest reviews was a project known as the "Perinatal and Infant Mortality Study," which included a review of the obstetrical records of every child born in the Keewatin in 1973. Funded by the federal government as a response to high rates of both perinatal and infant mortality, the study was carried out by university researchers and was followed by a series of less extensive reviews over the next seventeen years by an array of obstetricians, general practitioners, medical students, government bureaucrats, and nurse administrators. . . .

Reading through both the annual government reports and the published reviews leaves one with an impression that births and deaths are items in a system of Northern bookkeeping, in which the balance of profit and loss is based on comparisons with rates for previous years, rates for other groups living in the North, and rates for Canada as a nation. A report for 1966, for example, includes a graph showing an infant mortality rate for the Inuit of approximately 240 per 1,000 in 1956 (relative to a Canadian rate of approximately 30 per 1,000), which fell to 90 per 1,000 in 1963 (relative to a Canadian rate of 28 per 1,000) (National Health and Welfare Canada [Medical Services] 1967). The unstated implication is that prior to the introduction of medical services, traditional Inuit ways of childbirth and child-rearing were dangerous and often deadly. The steeply declining line presents a striking image of the beneficial impact of government intervention.

The presumed basis for such a graph is statistics and mathematical formulae. The graph has an air of scientific legitimacy, but in actuality this one is based on guesswork rather than data. The starting point for the graph is located in the 1950s, when the collection of vital statistics was still the responsibility of a handful of RCMP (Williamson 1974). Many of the inland Inuit would still have been living in scattered groups; patterns of settled life were not yet fully established. The likelihood of government having an exact tally of the children who died in the first year is extremely low. The slope of the graph is a fiction, though the trend it depicts may not be untrue. The death rate among young infants was probably much higher in the Keewatin during the 1950s than elsewhere in Canada, and as socioeconomic conditions in the settlements improved, mortality may even have fallen as steeply as the graph suggests.

The tendency to claim more than is justified by the available statistics is common. In an earlier review of the use of perinatal mortality rates in the debate over obstetric policy in the Keewatin, we showed similarly that the numbers of births and deaths were simply too low to have any epidemiological meaning. Between 1970 and 1985, the number of births fluctuated annually from 120 to 168; the number of deaths within the first month of life ranged from 0 to 7, but in most years only 2 or 3 deaths occurred. When numbers are this low, it is statistically impossible to tell whether a difference in the rate from one year to another is due to chance or reflects some real change. Furthermore, the assumption that the figures prove the dangers of allowing births in nursing stations is based on misinterpretation of a statistical association. A properly epidemiological comparison of hospital and nursing station would require random assignment of women to one place of birth or another and a minimum of 7,700 births. These numbers are impossible to achieve in the Keewatin. Thus, the debate over evacuation is statistically meaningless, since without such numbers neither side can prove its point.

Why, then, have perinatal mortality rates assumed such importance in the debate over evacuation and the discussion of obstetric policy for the Keewatin? References to a high perinatal mortality rate are made almost ritualistically by health professionals, as if they were the ultimate scientific defense against a political and ideological clamor for change. Moreover, preoccupation with rates and measures of mortality dates back to the earliest introduction of medical care in the area.

> In matters of health, vital statistics are the measure of our success or failure and for this reason this year's report will follow the same format so that we may compare our figures with those of the past and rejoice in our achievements and scrutinize our failures so that we may discover the reasons for the same.

There is a biblical ring to the phrasing of this report and a strong hint of a sense of civilizing mission. We suggest that in the relationship of the Canadian government with the Inuit, declining mortality rates became a metaphor for the success and moral virtue of Canadian colonial penetration into its own northern fringe. Preoccupation with perinatal mortality rates transformed the death of a baby from a problem only for the individual woman, her family, and community, into a concern for the Canadian Government. The impact of poor housing, infectious disease, tuberculosis, the demoralization of famine, and relocation was transferred to the body of the Inuit woman and read and reacted to through the ability of government to ensure that she carried an infant to term and that it survived the passage of birth.

CONCLUSION

Obstetric policies and practices in the Keewatin are not uniquely vulnerable to information systems and those who collect, analyze, and manipulate the data they contain. The linkages between policies and information systems are simply more visible in this setting, partly because the government is directly involved in the provision of health care and the bureaucracy needs reports, numbers, and statistical evidence. In addition, the university-based physician in the Keewatin brings to clinical practice a concern for research, the collection and analysis of data, and the publication of results. Yet one of the more intriguing aspects of the link between policy and research in the Keewatin is that the data so often point in a direction different from the policies enacted.

Inuit childbirth could remain outside the control of medicine only as long as Inuit remained a scattered and mobile society in which most births were unseen by the official eye. With settlement, each individual was a unit to be counted. Each pregnancy became open to scrutiny and documentation. Discussing the production of an infant mortality rate in Britain in 1877, Armstrong writes, "The creation of a specific mortality rate for infants at that time suggests both the emergence of a social awareness of these young deaths and, more importantly, the social recognition of the infant as a discrete entity" (1986:212). In the Keewatin, the preoccupation with perinatal mortality rates became an expression of a new "awareness" of the Inuit by the Canadian Federal Government. Control over childbirth became essential to ensure constant proof of the beneficence of power. When talking with Inuit women, a recurrent theme is the loss of control by the individual woman over the place of birth, its timing, and process. This loss has become a metaphor for the loss of political control by Inuit over their lives and their communities. Hence, an ongoing campaign to recapture choice in childbirth has become as much an item on the Inuit political agenda as it is an issue for Inuit women. The two causes are not separate.

REFERENCES

Annandale, Ellen C. 1988. "How Midwives Accomplish Natural Birth: Managing Risk and Balancing Expectations." *Social Problems* 35, 2: 95–110.

Armstrong, David. 1983. *Political Anatomy of the Body: Medical Knowledge in the Twentieth Century*. Cambridge: Cambridge University Press.

Davis-Floyd, Robbie. 1987. "Obstetric Training as a Rite of Passage." *Medical Anthropology Quarterly* (n.s.) 1, 3: 288–318.

——. 1988. "Birth as an American Rite of Passage." In Karen Michaelson, ed., *Childbirth in America*, 153–177. South Hadley, Mass.: Bergin and Garvey.

Hodgson, Corinne. 1982. "The Social and Political Implications of Tuberculosis Among Native Canadians." *Canadian Review of Sociology and Anthropology* 19, 4: 502–512.

Lazarus, Ellen. 1988. "Theoretical Considerations for the Study of the Doctor-Patient Relationship." *Medical Anthropology Quarterly* (n.s.) 2,1: 34–59.

Mason, Jutta. 1987. "A History of Midwifery in Canada." Appendix I in *Report of the Task*

Force on the Implementation of Midwifery in Ontario. Toronto: Ontario Ministry of Health.

Michaelson, Karen. 1988. "Childbirth in America." In Karen Michaelson, ed., *Childbirth in America: Anthropological Perspectives*, 1–33. South Hadley, Mass.: Bergin and Garvey.

Pettinari, Catherine Johnson, ed. 1988. *Task, Talk, and Text in the Operating Room: A Study in Medical Discourse.* Advances in Discourse Processes, Vol. 33. Roy O. Freedle, ed., 1–169. Norwood N.J.: Ablex.

Williamson, Robert G., ed. 1974. "Eskimo Underground: Socio-Cultural Change in the Canadian Central Arctic." *Occasional Papers II.* Uppsala: Almqvist and Wiksell.

DOSIC BODIES/DOCILE BODIES

JEAN LANGFORD

Has the introduction of modern hospital and educational disciplines transformed dosic bodies into docile bodies? Can the anatomical body with its well-
defined boundaries, its division into discrete regions, its militaristic defense
against viral invasion be mapped onto the body of dosa [illness forces], dhātu
[tissues], agni [fire], and mal [waste products]?[1] . . . In this chapter I consider
the disciplining of bodies and illness in contemporary Ayurvedic hospitals. I
am especially concerned with those moments of slippage when the dosic body
overflows the anatomical borderlines and slides out from under modern somatic disciplines. For even at the bodily level, modern Ayurvedic practitioners
negotiate a narrative of disease that resists assimilation by a biomedical univer-
salism. Even in the daily activities of diagnosis and treatment, many physicians
employ the biomedical techniques that make Ayurveda parallel to biomedicine
while circumventing the biomedical epistemologies that would make Ayurveda subordinate to biomedicine. . . .

The dosic body is what I have called . . . a fluent body, coursing with
climates and appetites, messages and passions, winds and tempers. The dosic
body spans the divide between text and world. It is inscribed with signs that are
more productively understood as versatile signifiers than visualized as definite
objects. To say that it is a fluent body is to say that it is overflowing not only
with dosic currents, but also with polyvalent syntax. The dosic body bears the
imprint of a social matrix, the somato-psychic consequences of living with or
against dharma. In this body the heart can be both a center of circulation and a
center of consciousness, while warmth can be simultaneously temperature,
temperament, and a somato-environmental agent that eludes modern categories. One M.D. student who was completing a dissertation in English on
katiúúl, or lower back pain, told me that one of the words he had left in the
original Sanskrit was *usna*. People call it heat, he said, but it is not exactly heat
in the sense of temperature. For example, usna is one of six *guna* that are
variously enhanced by six flavors (*rasa*). Usna is enhanced by both pungent
(*katu*) and sour (*amla*) flavors. Since sourness increases usna, one professor

explained, sour medicines and foods are good for poor digestion. Since sourness also enhances the quality of oiliness (*snigdh*) and, to a lesser extent, lightness (*laghu*), sour medicines and foods are also good for angina: the oily quality decreases the dryness (*ruskśa*) that causes the arteries to constrict, and the light quality helps to open up the arterial space. It is for this reason, he said, that pomegranates are good for the heart. The body implicit in such a narrative of heat and sourness, dessication and oiliness, bears more resemblance to a weather pattern than to a biological entity. Dosic embodiment makes more obvious what might be said of any kind of embodiment: "The body in question is not a hypostatized object . . . but a relation in a system of liaisons. . . . It would be better to speak of a certain 'bodiliness' than of 'the body' " (Barker 1995:10).

How are we to understand the discourses of dosic bodiliness in relation to the discourses of docile bodies? In *Śivanātha Sāgar* and other vernacular texts of its era, the bodies depicted are not yet the passive bodies of modern anatomy. They are certainly interiorized, with organs and bones exposed. They are also, however, living bodies, not frozen in anatomical poses, but gesturing and gazing off the page. In modern Ayurvedic departments, on the other hand, almost the only representations of the body are borrowed from biomedicine—plastic figurines in rigid postures, with inner organs painted in different colors. In one anatomy department, however, I noticed among these items a life-sized model of the ideally proportioned human male, poised in midaction, one arm bent and raised in a fist and one leg stepping forward, eyes open and focused, bones and organs hidden. . . . The dimensions of each part of his body, I soon learned, are in a particular prescribed proportion to the central section of the middle finger. A perfectly proportioned body is an indicator of a long life. . . . Contemporary students still learn the *swanguli pramān* (self-finger measurement), although neither they nor their professors use it clinically. Swanguli pramān is one of the few topics in the Ayurvedic anatomy class that does not correlate to biomedical anatomy. The body of swanguli pramān is not measured out in units of neutral space like an anatomical body, but in units of a self-referential space that mathematically multiplies the length of the body's own middle finger. In contrast to the body parts all around it, this plastic statue represents a body that is ideal rather than normative, dynamic rather than docile, as much engaged in seeing as in being seen and in acting as being acted upon. In fact, despite being an abstract body, a sign of longevity, this body is reminiscent of what Romanyshyn has called the "pantomimic body" in that it gestures to a world and embodies an emotional situation, if only the situation of glorying in its own health.

As Romanyshyn has noted, the modern anatomical body is sectioned as if by a grid imposed on a neutral space—as if seen, in other words, through a *camera obscura*. The perfectly proportioned Ayurvedic body is sectioned also, according to a pattern of *marmas* or vulnerable areas, which are also measured according to swanguli pramān. Susruta, recognized as the ancient authority on Ayurvedic surgery, identified one hundred seven marmas, which are classified according to both the area of the body in which they appear and the consequences of their injury. The injury of marmas in one category, for example, leads to instantaneous death, while the injury of those in other categories leads to death after seven or fourteen days, to deformity or disability, or simply to pain. Unlike the anatomical body, then, the marmic body is spatialized not so much by a cameralike vision as by possible crises. Romanyshyn speaks of how the anatomical body is actually "lived" only in moments disruptive of the body's ordinary engagement with the world, when, for example, torn muscles, raw bone, and severed veins become visible or painful. Such moments of disruption are, of course, precisely the concern of the physicians and their very motivation for positing marmic or anatomical bodies. However much the anatomical body is projected as a neutral space, independent of anyone's intentions, it is, like the marmic body, an instrumental space in which a surgeon's or anatomist's hands can maneuver. In biomedicine, however, this instrumental space is imagined not in relationship to an injury or treatment, but as "real," independent of any situation of injury or treatment. In this way, biomedicine creates the naturalized *effect* of an empty homogeneous space in which we locate national borders and territories. In marmic embodiment, on the other hand, space is imagined, as in pantomimic embodiment, in relation to possible situations, especially situations of bodily harm.

It is tempting to assimilate dosic bodiliness to the nondisciplinary bodies of premodern Europe, especially the grotesque Rabelaisian body, spilling out of its boundaries, described by Mikhail Bakhtin. Indeed, Bakhtin tells us that the grotesque body was inspired partly by accounts of fabulous oriental bodies, with disarranged and half-animal anatomies, that were pictured in popular medieval compilations of stories about India under the name of "Indian Wonders." Is it possible that in such texts Europeans begin to look to other peoples for the counterpoint, the point of release for their own disciplines, in this case a budding anatomical discipline? For the "Indian Wonders" include not only magic herbs, fountains of youth, and devils spitting fire, not only unicorns, phoenixes, and griffins, but also satyrs, centaurs, Cyclopes, headless "leumans" with faces on their chests, and so on. "This," as Bakhtin notes, "is an entire gallery of images with bodies of mixed parts . . . All this constitutes a wild anatomical fantasy"(1968:345). In the "Indian Wonders," this vision of anatomy run wild is projected away from the bourgeois tidiness of European medicine onto an Indian landscape.

Bakhtin suggests that the grotesque body encompasses every nonmodern embodiment. He assimilates the grotesque Rabelaisian body to the humoral body of Hippocratic medicine, which is in some respects similar to the dosic body, insofar as "the confines of the body and the world are effaced." Yet precisely here and in other passages, as when Bakhtin writes that the "Indian Wonders" primed the medieval imagination for the "*transgression* of the limits dividing the body from the world," it is possible to glimpse important differences between grotesque and dosic embodiment. For the grotesque body, at least of Rabelais's time, seems to take its wondrous force from its transgressiveness. It seems to owe its significatory impact to an already intimated separation between body and world and an already initiated parceling of the body into discrete parts. Is this grotesque body not related to the body of the public dissections that marked the transition from spectacular punishments to anatomical discipline? After all, one of these public dissections was performed by Rabelais himself. I suggest, therefore, that the grotesque body plays out the suppressed horrors implied by that public dismemberment that is both entertainment and scientific undertaking. The dosic body is not, however, transgressive or transgressed in the same way: there is nothing horrible about its ebb and flow. It is not a body reacting to bourgeois boundaries. If the intercourse between the grotesque body and the world is envisioned anatomically through a hyperbolic displacement and replication of organs and orifices, then the intercourse between the dosic body and the world is imagined dynamically through the continuous flow of qualities such as warmth and coolness; flavors such as sweetness and astringency; or elements such as earth, water, and ether. The grotesque body is fragmented wildly, as befits a body that is almost rebelliously and riotously opposed to a newly dominating anatomical order. The marmic body of swanguli pramān, on the other hand, is fragmented according to an elaborate somatic geometry. This body appears as a harmonious text of interrelated equations.

In Europe it is possible to trace historical development from predisciplinary bodies, which are one with text or world, to docile bodies. In postcolonial India a shift to docile bodies is still being contested. Ayurvedic practitioners, in their need to both parallel and surpass European medicine, necessarily work with bodies that are both dosic and docile, both fluent and fixed, both text- and material-referent, both coextensive with and apart from the world.

NOTE

1. Definitions for these Ayurvedic medicine terms are simplified from the glossary in *Fluent Bodies*. Eds.

REFERENCES

Bakhtin, Mikhail. 1968. "The Grotesque Image of the Body," 303–67. In Bakhtin, *Rabelais and His World*. Trans. Helene Iswolsky. Cambridge, Mass.: MIT Press.

Barker, Francis. 1995. *The Tremulous Private Body: Essays on Subjection*. Ann Arbor: University of Michigan Press.

Romanyshin, Robert D. 1989. *Technology as Symptom and Dream*. New York: Routledge.

PART VI

Desires and Identities, or

Negotiating Sex and Gender

Desires and Identities, or

Negotiating Sex and Gender

INTRODUCTION

Identity has proven a useful, almost indispensable concept in late twentieth-century cultural anthropology. By examining the genesis of a sense of identity (or identities or identification) in practice, it has been possible to replace the over-rigid divide between individual and society inherited from classic social theory with a more dynamic approach to social experience. Few anthropologists would now depict the social construction of personal experience as a simple imposition of collective imperatives over the unconstructed "nature" of a biological individual. Though the language of social inscription on the blank slate of the body is sometimes still used, newer analyses seek to show how individuals manifest many forms of agency and resistance as part of complex ideological contexts. Some of this resistant activity takes the form of seeking recognition as an identifiable member of a social group or community of practice. Especially where resistance does not seem to be directly addressed to overcoming an oppressor, it has proven useful to see identity-work as one important goal of human constructive activity. And the character of experience, constructed in social life and suffused with desires, can be seen as both cause and consequence of the identity positions that result.

The classic feminist distinction between sex and gender—the former referring to biological difference and the latter to its varying social expressions—has been consistent with discourses on identity. Identity construction is very often understood as necessarily involving the adoption of a sexual orientation that is part of one's social persona. As Simone de Beauvoir famously said, "One is not born a woman, but rather becomes one." It is conventionally expected that sexual identity in the domain of gender will "normally" follow the biological "sex" of the individual, so female children will tend to craft gender identities as women who "naturally" desire men. To account for the many empirical variations in the social practice of gender, however, it was useful for mainstream social scientists to hold sex constant—usually seeing it as biological and consequently unproblematic—while focusing on the social field in which variable gender roles—men, women, homosexuals, etc.—were determined.

Historically and philosophically, however, this tidy divide between biology and the social, sex and gender, has not held up. Nor has it served feminists particularly well, in that the more "real" domain of biology still seems to charter a vision of females as the weaker or more lacking sex, the marked category that can only deviate from the male norm. In this respect even much of psychoanalysis has clung to a medical understanding of the invariant biological base. But Michel Foucault's *History of Sexuality*, volume 1, has been very influential in relativizing sex. His argument that sexuality has a history has by now made it difficult to stick to the standard biological sense of the word. The philosopher Judith Butler (see part II) elaborated Foucault's argument with specific reference to the sex–gender distinction in her work of the 1980s; she showed how, once sex itself is seen as varying historically and biology no longer provides a culture-free basis for identity work, it makes no sense to hold the biological and social domains of sex and gender apart.

Among the historical critiques of the sex–gender distinction that emerged in the 1980s was a study of the history of reproductive biology by Thomas Laqueur. In an influential article entitled "Orgasm, Generation, and the Politics of Reproductive Biology," Laqueur argued that premodern anatomical and physiological accounts of human reproduction tended to depict male and female genital organs as more alike than different—essentially identical structures but differently disposed on the insides and outsides of bodies. He also showed how the biology of sex was held by Galenic authorities to be contingent on degrees of heat in the body, such that cooler natures matured differently as sexual beings than warmer ones. This now-forgotten vision of the sexed and gendered body allowed for fine gradations and variations in sexual behavior and had little difficulty in accounting for such phenomena as hermaphrodites.

Though Laqueur's history of reproductive physiology is partial and exploratory, it nevertheless suggests that a strictly binary sex distinction, providing biological females and males to ground entirely discrete female and male identities, is a fairly recent development in biology, one that has recast anatomical and physiological "facts" that once were contestable and flexible. This and other closer looks at the archive make it difficult to see biological sex as an unproblematic foundation on which social gender is built. To demonstrate that sex is no more natural and no less social than gender is a task not only for theorists but also for historians.

The articles collected here all contribute to a similar relativization and denaturalization of sex/gender. The historian John Boswell's study of the efforts of Thomas Aquinas to denounce homosexual acts as unnatural goes to the heart of the issue: the difficulty of Aquinas and his peers was precisely in defining *nature* in a way that was consistent with their preexisting attitudes toward homosexual acts. Popular morality dictated that homosexual behavior be seen in opposition to nature, but Aquinas ran into logical and empirical difficulties when he tried to explain exactly why this was the case. Boswell's close reading of the theologian's careful arguments, showing the lengths to which he went to sustain a circular notion of homosexuality as unnatural, unsettles the ground on which theological and civil dogma about the propriety of various sexual acts has rested for hundreds of years. Aquinas himself admitted that he could craft no philosophical arguments persuasive enough to make homosexual behavior a greater sin than, for example, gluttony. Yet the theological conclusion, fixed in canon law and incorporated into emerging secular law codes that remain influential today, was that homosexuality is prima facie unnatural.

Questions of sexual propriety are prominent in quite a different way in the Japanese discourses examined by Gregory Pflugfelder in his book *Cartographies of Desire: Male-Male Sexuality in Japanese Discourse, 1600–1950.* Tokugawa (or Edo) era writings on *nanshoku* (male love of males) and *shudō* (the way of [loving] youths) flourished between the seventeenth and nineteenth centuries, and they were very concerned with questions of moral and conventional sexual relations. But they did not regard male-male eroticism as unnatural. Rather, premodern Japanese "mappings" of sexual behavior, found in popular writing and visual culture, in legal discourses, and in medical texts, were both formed by social and cultural forces and themselves fashioned an array of desires. Eroticism between males was not denounced as a problem in itself. Rather, excessive involvement in an increasingly commercialized world of same-sex erotic services and violation of codes of interpersonal propriety in the context of sexual relationships were the target of critical and parodic writing. The excerpt from Pflugfelder's book that we have included here follows the structure of an early eighteenth-century satire by Nankai no Sanjin in laying out the social cartography of male-male eroticism in Tokugawa terms. Important distinctions were made among samurai, priests and acolytes, townspeople, and peasants, not only in terms of social status but in terms of sexual characteristics and preferences. Pflugfelder makes it clear that there was no unitary (or even dual) sexual "nature" to which either proper or improper variations in practice could be referred.

A modern tendency to define nature for purposes of grounding moral propriety is thus prefigured in medieval theology and given the lie by comparative history. The odd relationship between nature and morality continues to affect our lives, however, as is evidenced in the research of Emily Martin. This anthropologist's study of the rhetoric of contemporary reproductive biology troubles North American sex/gender regimes, tackling naturalizations of the (culturally specific) sexual in scientific knowledge. She analyzes metaphors that are in play even in the technical representations of reproductive physiology: menstruation is consistently depicted as a monthly failure and ovulation as wasteful, while the extraordinary volume of sperm production by males is unrelentingly seen as some kind of miracle. Apparently there is no talk of waste or failure where semen is concerned. She also shows how the images used to describe the activities of egg and sperm in the process of conception are strongly gendered, and she provides an interesting survey of some recent research that has required a shift in metaphors. Martin reminds us that maps and metaphors of human reproduction, even when they are less conventionally gendered than in the past, are a form of knowledge that is never innocent of social and political agendas. Thus it behooves us to be more conscious of the power of metaphor in naturalizing historical sex/gender regimes and thus in seeming to place them beyond the reach of history, moral debates, and deliberate human action.

The final extract in this section, from Gilles Deleuze's and Felix Guattari's influential book *Anti-Oedipus*, pursues the possibility of experience, sexuality, desire, and pleasure beyond gender identities and sexual metaphors. The work of these two theorists is famously difficult to read, partly because, in an effort to escape many of the assumptions built into ordinary language, they adopted new terms to reference highly original concepts. Examples in the pages that follow are molar/molecular, desiring machine, and body without organs. A careful tracing of these usages through the Deleuze and Guattari corpus rewards the reader with a startling vision of experience and knowledge beyond the dualisms and foundational arguments identified and critiqued elsewhere in this book.

This short extract cannot open the whole world of Deleuze and Guattari to the reader, but it does offer tantalizing glimpses of the dispersed, active, and enjoying body their thinking offers us. Taken as it is from the authors' extended critique of psychoanalysis and returning to Marx in a rereading influenced by a kind of radical phenomenology, this text asks readers to recall earlier medical, psychoanalytic, and speculative treatments of sexuality. One need not know much about Wilhelm Reich's "Orgonomics," for example, to see how the work of this mid-twentieth-century sex researcher can be reread to liberate the dispersed intensities of a "body without organs" (that is, a lived body thought of as potential, "poised," as Brian Massumi says, "for any action in its repertory"). Nor need one be fully conversant with Freudian uses of the metaphor of castration to see its connection to a notion of fantasy that has informed standard psychologies of desire. When Deleuze and Guattari challenge Jean-François Lyotard's suggestion that "only fantasies are truly desired," we can see how a phallocentric psychoanalytic logic may have prevented an understanding of the "microscopic transsexuality" that is possible in the flows and multiplicities of actual experience.

Deleuze and Guattari distinguish between "molar" and "molecular" modes of composition of individuals (not necessarily human individuals). Brian Massumi explains: "Molecular and molar do not correspond to 'small and large,' 'part' and 'whole,' 'organ' and 'organism,' 'individual' and 'society.' There are molarities of every magnitude (the smallest being the nucleus of the atom). The distinction is not one of scale, but of mode of composition: it is qualitative not quantitative."[1] Thus when Deleuze and Guattari say that "desiring machines" (assemblages that are the contingent site of desiring) are either "assigned to the molecular order that is their own, or they are assigned to the molar order where they form the organic or social machines," they are making a distinction between both kinds and levels of activity. The molar and the molecular are contingent forms of integration: a molar body might be contrasted with and accompanied by molecular dispersed flows of desire. "Desiring machines" might conventionally be seen as individual human bodies or particular populations or institutions; but by drawing attention to these sites of desire as "assemblages," this approach allows us to imagine desiring without a fixed basis in pregiven sociological or biological units. Here we might recall Foucault's famous metaphor from his preface to *Herculine Barbin*: in these pleasures, we can see the grin hanging about without the cat.[2]

In the introduction to *Anti-Oedipus* the authors recommend that we read their work as we listen to recorded music, not resisting the distractions and the varying attention that is characteristic of this kind of appreciative activity. We think this is good advice for much reading of theoretical scholarship. It is often productive to focus on those points that encourage a departure from the text, toward empirical and personally significant examples, shedding new light on the powers of abstraction to connect old realities in unexpected combinations. Particularly when our attention turns to sexuality and embodiment, domains in which much has been invested in the fixed identities of male and female, homosexual and heterosexual, innocence and sexuality, activity and passivity, an overcommitment to the standard categories is a limitation of our understanding. In their critique of "the pitiful little familialist secret" that lies at the heart of psychoanalysis and its Oedipal logic, Deleuze and Guattari invite our imaginations to return to the molecular flows and multiplicities of attachment that can open up experience far beyond the body we thought we were stuck with.

NOTES

1. Massumi 1992:54.
2. Foucault 1980.

ADDITIONAL READINGS

Barlow, Tani. 1994. "Theorizing Woman: Funü, guojia, jiating." In *Body, Subject, and Power in China*, edited by T. Barlow and A. Zito. Chicago: University of Chicago Press.

Grosz, Elizabeth. 1994. *Volatile Bodies: Toward a Corporeal Feminism*. Bloomington: Indiana University Press.

Stoler, Ann. 1995. *Race and the Education of Desire: Foucault's* History of Sexuality *and the Colonial Order of Things*. Durham: Duke University Press.

JOHN BOSWELL

By the middle of the thirteenth century, as the church began the synthesis of theology and canon law which was to stand almost unchallenged into the twentieth century, and as most European states were incorporating theological principles into secular law codes, opposition between "nature" and homosexual behavior was a common assumption of Europeans. Few questioned exactly what this "nature" was, and even fewer were able to explain it; but the average thirteenth-century reader was apt to encounter it in so many different guises that it probably came to seem self-evident. In part "nature" was a beneficent and lovable goddess appearing in the most popular literature of the day and generally speaking for the sexual prejudices and desires of the majority; in part it was the source of a law which was assumed to be universal and which appeared to provide the foundation for all civil and much canon law; in part "nature" was a complex philosophical construct inherited from Boethius and the twelfth-century naturalists and now being expanded with knowledge gained from new translations of Plato and Aristotle.

Albertus Magnus was the first of those responsible for the final synthesis to comment extensively on homosexual behavior. Almost inevitably, his writings evince a certain confusion and inconsistency in regard to the "naturalness" of homosexual acts. In his *Summa theologiae* Albertus condemned homosexual acts as the gravest type of sexual sins because they offended "grace, reason, and nature." (Next after these would be those acts that offended "only grace and reason," e.g., adultery.) Albertus offered no explanation of the precise way in which *sodomia*—which he defined as the carnal union of persons of the same gender—violated nature, but he did cite Romans I: 26–27 as an authority for his opinion.

In other writings, however, Albertus described homosexuality as a contagious disease which passed from one person to another and was especially common among the wealthy. In his commentary on Luke he cited a biblical text suggesting that it was innate and observed that those who had it scarcely ever got rid of it, but in his treatise on animals he described a relatively easy cure: the fur from the neck of an Arabian animal he called "alzabo," burned with pitch and ground to a fine powder, would "cure" a "sodomite" to whose anus it was applied. (Note that this suggests that a "sodomite" is a homosexual male who engages in anal intercourse.) . . .

The moral authority of "ideal" nature reached its most influential and in many ways its final development at the hands of Albertus's most famous pupil, Saint Thomas Aquinas (d. 1274), whose *Summa theologiae* became the standard of orthodox opinion on every point of Catholic dogma for nearly a millennium and permanently and irrevocably established the "natural" as the touchstone of Roman Catholic sexual ethics.

Since Aquinas's teachings represent to a large extent the final synthesis of high medieval moral theology, they merit particularly detailed attention. It may be worth repeating here that the aim of such analysis is not to engage in polemics on moral issues but to investigate the extent to which such positions reflect logical or consistent application of traditional Christian principles, and where they do not, to suggest other ways of accounting for their development. It is difficult to see how Aquinas's attitudes toward homosexual behavior could even be made consonant with his general moral principles, much less understood as the outgrowth of them. Despite his absolute conviction in every other context that humans were morally and intellectually superior to animals and therefore not only permitted but obliged to engage in many types of activity unknown or impossible to lower beings, Aquinas resorted again and again to animal behavior as the final arbiter in matters of human sexuality. Even granting the illogic of the premise, such an undertaking was no mean task for a mind of Thomas's acuity. In condemning promiscuity ("fornication"), for instance, he had to come to grips with the fact—well known in spite of the recent ascendance of animals as models of sexual propriety—that promiscuous sexuality was common among familiar animals like dogs and cats; so common, in fact, that even the most devoted adherents of "nature" compared humans given to obsessive or wanton venereal pursuits to animals. If animals could "naturally" pursue lives of such carefree and expansive sexuality, why could not humans "naturally" do likewise?

To answer this question, Aquinas had to stretch considerably the Platonic tradition of selective inference from birds, arguing that there is some inherent distinction among animals on the basis of postnatal requirements for the offspring:

> We see in fact that among all those animals for whom the care of a male and
> a female is required for the upbringing of the offspring, there is no promis-

cuity [*vagus concubitus*] but only one male with one female, or several females: this is the case among all birds. It is different, however, among those animals for whom the female alone is sufficient for the upbringing of the offspring, among whom there is promiscuity, as is evident in the case of dogs and other similar animals.

Although judged by the standards of the time this argument evinces remarkable biological insight, it presents a great many moral and philosophical difficulties. Aside from the fundamental paradox of this whole line of reasoning, that man, the paragon of the great chain of being, should have to follow the example of lower animals in matters of morality; and overlooking the factual error in the premise that all birds are monogamous, one is still struck by the many crucial questions left unanswered in Aquinas's "answer."

It is difficult to believe, moreover, that animal behavior actually suggested this position to Saint Thomas: he can only appeal to birds, a tiny minority, as monogamous—elsewhere he qualifies his example even further as only some birds—and the analogy between the parental duties of birds and those of humans is questionable, to say the least. The invocation of "nature" is significant, however, as an indication of the lengths to which Scholastic apologists for Christian ethics would go to demonstrate that "nature" was at the foundation of Christian society's sexual taboos. Even granting the selective inference from monogamous species of birds, sexual promiscuity ought to have been no more reprehensible in "natural" ethics than gluttony, which also prescinds from the supposedly "natural" tendency of animals to eat only what is necessary for sustenance. Indeed, Aquinas concedes, heterosexual promiscuity would be no more serious than gluttony if it were not for its potentially harmful effects. While one excessive meal has no permanent consequences, a single act of heterosexual fornication may ruin the life of a human being: that of the illegitimate and (Thomas assumes) uncared-for child produced by it (*Summa theologiae* 2a.2ae.154.2 ad 6).

One would surmise from this argument that Aquinas would regard homosexual acts as no more serious than gluttony. He could argue that they did not fulfill any requirement of nature, but hardly that they produced unwanted or neglected children. The only argument which prevented his "natural" ethics from accepting heterosexual promiscuity as mere intemperance could not be applied to homosexual acts. This left gay sexuality in the position it had occupied in the minds of earlier theologians like Burchard, i.e., at the very worst comparable to drunkenness, and considerably less serious than heterosexual fornication.

But Aquinas could not pursue his logic this far out of the mainstream of thirteenth-century popular morality and public intolerance, and he struggled instead to construct a philosophical justification for classifying homosexual acts as not only serious but worse than comparable heterosexual ones; in fact he promoted them to a position of unique enormity unparalleled since the time of Chrysostom.

In an early work (*Summa contra gentiles* 3.122) Saint Thomas had predicated his objection to homosexual activity not on animal sexuality but on an argument which many later theologians were to seize upon in regard both to contraception and "unnatural" sex acts—that semen and its ejaculation were intended by "nature" to produce children, and that any other use of them was "contrary to nature" and hence sinful, since the design of "nature" represented the will of God. Unlike later writers, however, Saint Thomas realized that this argument had fatal flaws. He himself raised the question of other "misuses" of "nature's" design. Is it sinful for a man to walk on his hands, when "nature" has clearly designed the feet for this purpose? Or is it morally wrong to use the feet for something (e.g., pedaling an organ) which the hands ordinarily do? To obviate this difficulty, he shifted ground and tacitly recognized that it was not the *misuse* of the organs involved which comprised the sin but the fact that through the act in question the propagation of the human species was impeded.

This line of reasoning was of course based on an ethical premise—that the physical increase of the human species constitutes a major moral good—which bore no relation to any New Testament or early Christian authority and which had been specifically rejected by Saint Augustine. Moreover, it contradicted Aquinas's own teachings. Nocturnal emissions "impede" the increase of the human race in precisely the same way as homosexuality—i.e., by expending semen to no procreative purpose—and yet Aquinas not only considered them inherently sinless but the result of "natural" causes. And voluntary virginity, which Aquinas and others considered the crowning Christian virtue (*Summa theologiae* 2a.2ae.151,152), so clearly operated to the detriment of the species in this regard that he very specifically argued in its defense that individual humans are *not* obliged to contribute to the increase or preservation of the species through procreation; it is only the race as a whole which is so obligated. Because of this, Aquinas found it necessary to shift ground again in formulating theological opposition to sexual nonconformity in his major and most influential moral treatise, the *Summa theologiae*.

There are three substantive comments on homosexuality in the *Summa*. In the last and best known of these Aquinas discusses under two headings

(1) whether "vices against nature" constitute a species of lust (he concludes they do) and (2) whether they are the most sinful species of lust (they are). "Vices against nature" include masturbation, intercourse with animals, homosexual intercourse, and nonprocreative heterosexual coitus.

Although nature is defined elsewhere in the *Summa* in many different, sometimes conflicting ways, ranging from "the order of creation" to "the principle of intrinsic motion," no definition is provided here for the "nature" these sins are against, and all common conceptions of "nature" are missing from or excluded by the particulars of the discussion. "Animal" sexuality is opposed to the "natural" at one point, and no other sense of "nature" suggested would apply any more to homosexual acts than to procreative extramarital sexuality. Although at one point he does remark that the potentially procreative types of lust discussed earlier under "fornication" and "adultery" do not "violate human nature," this is directly contradicted by his assertion in the treatment of "fornication" that "it is against human nature to engage in promiscuous intercourse." Indeed, as he subsequently admits, not only are all sexual sins "unnatural," but all sins of any sort are "unnatural." The "natural" in this section is in fact simply the "moral"; and it seems circular, to say the least, to argue that homosexual acts are immoral because they are immoral.

In an earlier part of the *Summa*, however, in a discussion of whether there can be "unnatural" pleasures (the answer is yes), Aquinas does offer more explicit ideas about "nature" and "natural" in relation to homosexuality. In fact, he provides some surprising definitions. "It should be observed that a thing is called 'natural' when it is according to 'nature.' . . . 'Nature,' in the case of man, may be taken in two senses. On one hand the 'nature' of man is particularly the intellect and reason, since it is in regard to this that man is distinct as a species." This first definition appears to refer to the "nature" *of* something, in this case man, but its use is paradoxical because what Aquinas here takes to comprise the "nature" of man is exactly what most adherents of "ideal" nature exclude: his reason. It is indeed very difficult to see how homosexuality violates "nature" in the sense of man's reason. It was precisely the reason of man which proponents of gay sexuality had recently used to defend themselves against "ideal" nature, arguing that it is man's "nature" to rise above what is "natural" to animals and to love regardless of the physical compulsions of procreation (e.g., in "Ganymede and Helen," which Aquinas might have known). Aquinas would not have had to alter his commitment to procreation as the function of sexuality in any way to have recognized that "natural" affection, which in animals exists of "necessity" between mates and relatives, is transferred by "human nature" to relations where there is no "necessity" for affection—e.g., voluntary friendships—without moral defect, and that an analogous argument could be used to justify sexual relations among humans where no "necessity" compels. In the immediately preceding section, for instance, he distinguishes between "natural" and "unnatural" desires: the former are those which animals experience as a consequence of necessity. The latter are unique to humans, "who alone can recognize as good and fitting something which is beyond the requirements of nature." These desires "beyond the natural" are characterized by Aquinas not only as "rational, individual, and acquired" but as appertaining to things which are "good and fitting," despite their exceeding the "natural" and being unknown to animals.

As his second definition of "nature" Saint Thomas then offers a meaning which directly contradicts the first: "On the other hand, 'nature' in man may be taken to mean that which is distinct from the rational, i.e., that which is common to men and other beings, particularly that which is not subject to reason." This appears to be the ever popular concept of "animal nature," a meaning not just peripheral to but ostensibly rejected in the treatment of homosexuality above, but here providing the only substantiation for the claim that homosexual acts are "unnatural." Things are "natural" to both men and animals when they pertain to the preservation of the individual or the species: the examples of food, drink, sleep, and sex are cited. They have nothing to do with thought but are the responses *necessary* for the existence of either the individual or the species. Homosexuality and celibacy might be "against nature" in this sense if one took the simplistic view that indulgence in them somehow precluded the reproduction of the human; certainly they do not diminish the existence of the individual. Aquinas does not, however, show that homosexuality would preclude the reproduction of the race; he could only do so if there were a logical compulsion that if any humans were to engage in homosexual acts, all would then be exclusively homosexual. Otherwise the position of homosexuality could be considered the same as "unnatural" desire (or celibacy): unnecessary, but not evil.

This difficulty pales, however, beside the startling revelation following the second definition that homosexuality may in fact be quite "natural" to a given individual, in either sense of the word. "Thus it may happen that something which is against human nature, in regard either to reason or to the preservation of the body, may become natural to a particular man, owing to some defect of nature in him." The "defect" of nature mentioned here should not be taken as implying some contravention of "natural laws." Aquinas compares this sort of "innate" homosexuality to hot water: although water is not "naturally" hot, it may be altogether "natural" for water under certain circumstances to become hot. Although it may not be "natural" for humans in general to be homosexual, it is apparently quite "natural" for particular individuals. This circumstantial etiology of homosexuality cannot be taken as indicating in itself moral inferiority. Aquinas also believed that women were produced by "defective" circumstances (Ia.92.1): if conception took place under completely "natural" circumstances, males would always result ("for the active force of the male seed intends to produce something similar to itself, perfect in its masculinity"), but if some peculiarity intervened—a defect in sperm or seed or the prevalence of a moist south wind at the time of conception—females would be born (here quoting Aristotle). Although Aquinas did believe that females were in this sense "defective" males and although he certainly considered women inferior to men in many practical ways, it cannot be argued that he considered the condition of femaleness to be morally reprehensible, nor would he have argued that behavior which is the result of a female "nature" is morally inferior to behavior dictated by the more "natural" male condition. Neither homosexuality nor femaleness can be shown to be "immoral" simply because it does not represent the primary intent of "nature," and both are in fact "natural" to the individuals in question.

In Aquinas's view, moreover, everything which is any way "natural" has a purpose, and the purpose is good: "Natural inclinations occur in things because of God, who moves all things. . . . Whatever is the end of anything natural cannot be bad in itself, since everything which exists naturally is ordained by divine providence to fill some purpose. Since both homosexuality and femaleness occur "naturally" in some individuals, neither can be said to be inherently bad, and both must have an end. The *Summa* does not speculate on what the "end" of homosexuality might be, but this is hardly surprising in light of the prejudices of the day.

If, then, the "nature" of man in general is to desire some things which are not "naturally" required or enjoyed by animals (Ia.2ae.30) and if the "nature" of some individuals is to desire homosexual intercourse (Ia.2ae.31), in what sense could homosexual acts be "unnatural"? In his third comment on homosexuality, although he refers again to the "animal nature" which he has admitted should not limit human behavior, Aquinas gives a clue to the real origin of his attitude toward the "unnaturalness" of homosexual behavior:

It must be noted that the nature of man may be spoken of either as that which is peculiar to man, and according to this all sins, insofar as they are against reason, are against nature (as is stated by Damascene); or as that which is common to man and other animals, according to which certain particular sins are said to be against nature, as intercourse between males (which is specifically called the vice against nature) is contrary to the union of male and female which is natural to all animals.

In the end Aquinas admits that his categorization of homosexual acts as "unnatural" is a concession to popular sentiment and parlance. Since theologically sins are necessarily "unnatural," it is simply redundant to argue that homosexuality is sinful because it is "unnatural"; homosexual acts would have to be shown to be sinful *apart from* their "unnaturalness" to be immoral from a theological point of view; but Aquinas could bring to bear no argument against homosexual behavior which would make it more serious than overeating and admitted, moreover, that homosexual desire was the result of a "natural" condition, which would logically have made behavior resulting from it not only inculpable but "good."

....

But homosexual acts "are called the unnatural vice," he observes, because they do not occur among animals, and he bows to the speech patterns and zoological notions of his contemporaries. Aquinas was not an innovator; the *Summa*'s position, in this as in many matters, was a response to, not the origin of, popular attitudes. The arguments Aquinas and his contemporaries used to justify categorizing homosexual acts as the gravest of sexual sins (of all sins, according to some) cannot be shown to derive from the previous Western moral tradition, but they had all been brought to bear on the subject of homosexuality in civil legislation and popular diatribe before the *Summa* was written. The popular satirical poem "Quam pravus est mos" predated by more than a century each of Aquinas's major arguments against gay sexuality, derogating it as unknown among animals, a violation of "nature," a departure from reason, and an impediment to the reproduction of the human race. Albertus Magnus, Vincent of Beauvais, and Saint Thomas Aquinas were all writing in societies which had already passed laws against homosexual behavior and in which popular hostility toward gay people was becoming a literary commonplace. The *Summa* was not begun until 1265, after antigay provisions had been incorporated in law codes in Castile, France, and parts of Aquinas's native Italy. And although opposition to homosexual acts based on the "natural" necessity of procreation should have applied to all nonprocreative sexuality, in fact most theologians, like Albertus Magnus, applied their condemnations only to gay people; others, like Saint Thomas, although admitting that all vices were "unnatural," proceeded to use "unnatural" as specifically referring to homosexuality.

The positions of Aquinas and other high medieval theologians regarding homosexuality appear to have been a response more to the pressures of popular antipathy than to the weight of the Christian tradition, but this is not to suggest that the *Summa* itself did not affect subsequent attitudes. It must be recognized that the context of an accusation is often as damning as the charge itself: Aquinas played to his audience not simply by calling on popular concepts of "nature" but also by linking homosexuality to behavior which was certain to evoke reactions of horror and fear. He compared homosexual acts not with other instances of exceeding what is necessary, like overeating or drunkenness, nor with other behavior of which animals are supposed to be incapable, such as telling lies or counterfeiting currency, but with violent or disgusting acts of the most shocking type, like cannibalism, bestiality, or eating dirt. Indeed, by suggesting subliminally to his thirteenth-century readers that homosexual behavior belonged in a class with actions which were either violently antisocial (like cannibalism) or threateningly dangerous (like heresy), Aquinas subtly but definitively transferred it from its former position among sins of excess or wantonness to a new and singular degree of enormity among the types of behavior most feared by the common people and most severely repressed by the church.

Moreover, it was particularly significant for gay people that Thomas's ideas about homosexuality triumphed just at the moment when the church began to enforce orthodoxy more rigorously than ever before and to insist that everyone accept in every detail not just the infallible pronouncements of popes and councils but every statement of orthodox theologians. Although the intent was not to eradicate acceptance of homosexuality in particular, the effect was to eliminate all opinion in the church which did not accord with accepted theology on every matter, and since it was Aquinas's authority which ultimately became the rule, acceptance of homosexuality ceased to be a safe option for Catholics liable to prosecution for heresy.

Because of the extraordinarily conservative nature of Catholic theology and the persistence of the prejudices which animated the hostile theological developments of the thirteenth century, the popular opposition to homosexuality given official expression in the writings of Aquinas and his contemporaries continued to influence religious and moral attitudes well into modern times. It must be remembered, however, that intellectual responses to homosexuality generally reflected rather than caused intolerance. It is instructive to note in this regard that there was, by any objective standard, a much more powerful medieval *moral* tradition against usury than against homosexual behavior. Unlike homosexuality, usury had been condemned almost unanimously by philosophers of the ancient world as uncharitable, demeaning, and contrary to "nature," both because it violated the kindness which humans ought to extend to each other in times of need and because it represented an "unnatural" growth of money (the usurer did nothing to earn the increase which accrued to him, and the money therefore increased "unnaturally"). Because they were thought to exploit the poor, who were most in need of loans and least able to afford interest, usurers were looked upon everywhere with disgust. Cicero mentions them in the same breath with child molesters. Early theologians universally regarded Jesus's command to "lend hoping for nothing again" (Luke 6: 35) as an extension of Levitical prohibitions of usury among Jews to the entire Christian community.

The ethical case against usury was considerably stronger than that against homosexuality. Many more biblical passages could be claimed to relate to it, including, with only a little stretching, Jesus's constant condemnations of the rich. "Natural law" forbade it. The fathers of the church forbade it. The very same theologians influential in condemning homosexuality forbade absolutely and in no uncertain terms lending money at interest: Peter Cantor, Albertus Magnus, and Saint Thomas Aquinas. Many more church councils had condemned it, beginning with Nicea, the most famous of all, and including dozens of others before the steady and severe proscriptions of the First, Third and Fourth Laterans.

By the fourteenth century usury incurred more severe penalties in church law than "sodomy" did and was derogated in exactly the same terms. The most famous of the commentators on canon law, Panormitanus, equated it explicitly with "unnatural" sexuality: "Whenever humans sin against nature, whether in sexual intercourse, worshiping idols, or any other unnatural act, the church may always exercise its jurisdiction. . . . For by such sins God Himself is offended, since He is the author of nature. This is why Jean Lemoine felt . . . that the church could prosecute usurers and not thieves or robbers, because usurers violate nature by making money grow which would not increase naturally."

Because usurers were almost necessarily well-to-do, they were at first even more eagerly prosecuted under civil law than gay people. The same thirteenth-century laws which penalized gay people—the Coutumes of Touraine-Anjou, the Etablissements, etc.—stipulated that the property of anyone who had practiced usury within a year of his death was to be confiscated to the king automatically. Many local statutes empowered nobles to exact the same lucrative penalty. Less judicious proceedings were also employed: the crusade against the Albigensians named usurers as well as heretics as the objects of its enmity. The former were presumably even more tempting to northern nobles short of cash.

But theology, ethics, law, and even crusades were powerless against a practice which increasingly met the needs of the age and which soon ceased to derive support from widespread popular antipathy. As long as most usurers were Jews, prejudice provided a visceral impetus to prosecution for usury, but by the fourteenth century interest banking more and more frequently involved the Christian majority as well, and the emotional basis of opposition to the practice was steadily eroded by its manifest utility and increased familiarity. As a part of the everyday life of the majority culture, its erstwhile objectionableness eventually came to seem so distant that the ethical tradition against it was sidestepped altogether by the ingenious expedient of declaring ancient prohibitions against it to apply only to the demanding of *excessive* interest.

There were few popular reasons for reinterpreting thirteenth-century strictures against gay people, Jews, witches, or other groups who remained objects of suspicion or hatred on the part of the general population. The prejudices which had been largely responsible for ecclesiastical condemnations continued to animate them, and most of them stood unchallenged at least through the Reformation. There was of course great variation in the fortunes of such groups and their individual members in varying locales and times; this story remains to be written. But there was little change, for a very long time, in public and institutional attitudes toward them. . . . Religious sanctions and intellectual support created by later medieval theology crystallized public and official expression of such attitudes in the thirteenth century and prolonged their effects for centuries thereafter; such expressions both inspired and drew life from the vehement antipathy of the masses. Only when and where the latter abated did such groups experience a general amelioration of their fortunes. In the case of gay people, such changes were relatively rare and lie far beyond the scope of this study.

CARTOGRAPHIES OF DESIRE:

MALE-MALE SEXUALITY

IN JAPANESE DISCOURSE

GREGORY M. PFLUGFELDER

The human record suggests that not only sexual understandings but erotic desires themselves are less uniform across time and space than most people ordinarily suppose. The biologistic reasoning that informs much of late twentieth-century thinking on sexuality insists, to the contrary, that sexual desire, and perhaps even sexual "orientation," are inscribed in the very deepest structures of the human body (for example, in our hypothalamuses or in our genes). In this formulation, which derives cultural authority from various contemporary institutions of medicine and science, the transformation of biological givens or "imperatives" into actual practices of sexuality and in turn the representation of these practices through language (which inevitably involves sexual categories) are epiphenomenal to the essentially pre-social and pre-linguistic, hence pre-historical, reality of sexual desire itself. Even those who recognize that designations for various forms of sexuality differ according to place and speaker—just as "Germany" appears in certain atlases as "Deutschland" and in others as "Allemagne," or is sometimes shaded in yellow, sometimes in red—seldom consider that the underlying geography of desire is anything but universal.

A growing body of scholarship on the history of sexuality challenges this biologistic view. One of the premises of the present study, like others that follow an interpretive tradition commonly labeled "constructionist," is that desire, sexual or otherwise, is not a constant or a given, but is shaped in crucial ways by the very manner in which we think and speak about it. As David Halperin has put it, there is "no orgasm without ideology." From such a perspective, mappings of sexuality do not reflect an unchanging reality so much as participate centrally in its construction, helping to engender the very desires and subjectivities that they purport merely to represent. In this sense, maps of the conventional and metaphorical variety are in the end not so very different, for few would deny the crucial role that cartographers and surveyors play in determining the boundaries and thereby maintaining the integrity of such an artificial entity as "Germany," or "Japan," or any other nation-state. Indeed, if political borders were physically inscribed on the earth's surface for all to see and recognize, many wars might never have been fought. . . .

AUTHORIZING PLEASURE: MALE-MALE SEXUALITY IN EDO-PERIOD POPULAR DISCOURSE

FINDING THE WAY

For Japanese of the Edo period, "homosexuality" was an unfamiliar and perhaps unimaginable concept. I do not mean to suggest, of course, that people of this era never engaged in sexual practices with nor experienced erotic desires toward individuals of the same sex. A wealth of contemporary sources testifies to the fact that they did, often with great relish. Nevertheless, if a twentieth-century American text on "homosexuality" were to find its way by some miraculous means to seventeenth-century Japan, a perplexing task would face the translator who wished to transpose it into the vernacular. To begin with, how to render the term "homosexuality"? Although the period's sexual vocabulary offered various expressions that could be used to refer to erotic activities between males or between females, there was no single word that signified both, so that even finding an appropriate name for the work would pose a considerable challenge. Indeed, outside this hypothetical instance, such a term would hardly have been necessary, since male-male and female-female varieties of sexual behavior were unlikely to appear even in the course of the same discussion.

How, then, might our imaginary translator have conveyed the subject of his text to contemporary readers? I use the masculine pronoun here intentionally, for it is reasonable to assume that the translator would have been male. Literary production in most of its forms (with important exceptions, such as *waka* poetry) was gendered chiefly as a masculine pursuit, requiring an education in Chinese characters that far more seventeenth-century men than women possessed. A substantially larger portion of the female population may have been able to read all or parts of the finished text, particularly if it featured phonetic glosses in the native *kana* syllabary. Nevertheless, given the uneven distribution of reading abilities, readership, too, was likely to be predominantly male. The question of literacy was compounded by that of economic power, which put disposable income, and hence access to the products of a burgeoning publishing industry, disproportionately into the hands of men.

Such factors considered, our translator might even have decided to ignore the female-female content of the work altogether. Such an approach would have offered several advantages. First of all, as a male himself, the translator was more likely to have firsthand knowledge of male-male erotic practices than female-female, and therefore stood on more familiar ground in writing about them. The same would be true of most of his audience, and even female readers would not have been entirely ignorant of the former. More importantly, however, writings on male-male sexuality already possessed an indigenous textual tradition, with its own vocabulary and commercial market. During the seventeenth century, this field of cultural production was in fact undergoing a rapid expansion. By assimilating the twentieth-century treatise to such contemporary works, our translator might even have been able to make himself a pretty penny.

Let us return, though, to the original question: what equivalent would our translator have selected for "homosexuality"? One plausible candidate would be *nanshoku*, a term found, for example, in the title of Ihara Saika-ku's best-selling story collection of 1687, *Nanshoku Ōkagami*, which Paul Schalow has translated into English as *The Great Mirror of Male Love*. This choice would have been apt in at least one sense, since both nanshoku and "homosexuality" derive from classical languages—in the case of the former, Chinese, and in the latter, an awkward mixture of Greek and Latin—and hence carried a certain amount of scholarly cachet. Furthermore, just as "homosexuality" is often conceptually paired with "heterosexuality," nanshoku had a companion term in *joshoku* (also pronounced nyoshoku), an expression likewise denoting male-female eroticism. Here, though, the resemblances ended. Nanshoku and joshoku could convey only half of what "homosexuality" and "heterosexuality" convey, both these former terms being predicated upon an implicitly male erotic subject. Thus, while joshoku was written with ideographs meaning "female" and "love," it referred specifically to a male "love of females," and never the love of one female for another. Similarly, although nanshoku contained only one character for "male," it actually signified an erotic interaction between two (or more) males, rather than a "love of males" by women. The nanshoku/joshoku dichotomy, in other words, mapped the universe of sexual possibilities from an exclusively male perspective, and neither female-female eroticism nor female sexual agency vis-à-vis males enjoyed any place within its signifying system.

Nor was the "sexuality" of "homosexuality" quite the same as the "love" of "male love." Such categories as "homosexuality" and "heterosexuality" are deeply embedded in a medico-scientific model of erotic behavior that, in Japan as in the West, gained general currency only in the present century. Nanshoku and joshoku, on the other hand, were closely linked with Buddhist thought, which passed to Japan by way of China and Korea nearly a millennium and a half earlier. The ideograph that Schalow translates as "love" (Chinese: *se*; Japanese: *iro*) literally meant "color," referring in Buddhist philosophy to the world of visually perceptible forms toward which lower beings, including humans, experienced desire, thus hindering their progress along the path of enlightenment. More specifically, it came to denote the realm of erotic pleasure, which, again from the perspective of the masculine subject, could be divided into nanshoku and joshoku hemispheres depending upon the nature of the distracting form. The pleasures of this realm sprang neither from purely physiological processes, as the medico-scientific model of "sexuality" suggests, nor from a lofty spiritual source, as the term "love" often implies; instead, more akin to the Greek "eros," they partook equally of physical and emotional elements, both of which were understood to pose a similar degree of threat to the unenlightened soul.

It should not be imagined, however, that nanshoku always carried such overtly religious overtones. During the Edo period, nanshoku appears in popular writings as one of the two most widely used expressions for male-male eroticism, featuring prominently in the titles of many works that, far from warning of its dangers, extolled its pursuit. The resident of seventeenth-century Edo (today's Tokyo) was less likely to associate it with religious teachings than with a lively and increasingly commercialized culture that articulated itself not only through such prose texts as Saikaku's, but also in poetry, song, dance, drama, woodblock prints, and, of course, the various pleasures of the flesh. If she or he made a connection with Buddhism, it was probably with the figure of the priest, whom many writers depicted as one of that culture's prime connoisseurs. It is this realm of popular discourse, and within it primarily the media of prose and poetry, that serves as the chief source of materials for this chapter's discussion.

But perhaps our translator would have chosen an alternative expression. Another term for male-male eroticism that frequently appears in popular discourse of the era is *shudō*, which entered into use around the beginning of the Edo period. Nanshoku and shudō shared certain lexical characteristics, and were used in practice almost interchangeably, yet differences of nuance existed as well. Although shudō, like nanshoku, could be written in Chinese characters, it would have made little sense to a resident of the continent, since it possessed an entirely indigenous etymology. The fuller form of the word was *wakashudō* or the "way of youths," *wakashu* denoting in vernacular Japanese an adolescent male. More frequently, however, the wakashu element was abbreviated into one of its two constituent characters, producing the compounds *nyakudō* (alternately pronounced *jakudō*), whose first surviving appearance in a written text dates from 1482, and shudō, which became the preferred combination during the seventeenth century.

In shudō, too, a masculine erotic subject lurked somewhere beneath the surface of the ideographs. The "way of youths" was not the possession of youths themselves, as the characters might literally suggest, but existed instead from the perspective of their male admirers, specifically those old enough to perceive a contrast with the former's adolescence. Shudō, in other words, was not so much the "way of youths" as the "way of loving youths," an erotic path that younger males traveled only in their capacity as sexual objects, and females could not tread at all. True, one may find isolated instances in which women are described as pursuing the "way of youths," such as a late eighteenth-century comic verse that pokes fun at a widowed head of household because of her fondness for male prostitutes: "The audacious [literally, obese] widow chiefly travels the path of shudō." It was the very fact, however, that shudō was understood to be a male prerogative that made this verse humorous, while rendering "audacious" (*futoi*)—the word hints, incidentally, that she may also be pregnant—the economically independent female who attempted to usurp the privilege. . . .

On a more quotidian plane, how did Edo-period writers map shudō onto the geography of their own archipelago? . . . Nankai no Sanjin's story ["Nanshoku Dew on a Mountain Path"] is suggestive. The author describes Tagasuke, who had exhausted the pleasures of "Japanese shudō," as being familiar not only with the "stage boys" of the three metropolises, but also with the non-professional youths of "warrior houses, temple precincts, towns and rural districts"—in other words, the inhabitants of every social sphere. For Nankai no Sanjin, the most conspicuous regional distinction divides metropolis from hinterland—that is to say, the "three ports" of Edo, Kyoto, and Osaka on the one hand, and the rest of the country on the other. Because the publication industry was based in these metropolitan centers, the gaze of the shudō text tended to extend outward from the former to the latter. Although urban authors portrayed the "way of youths" as penetrating into the farthest reaches of the country, . . . their works generally implied that its most elaborate and sophisticated forms, personified in Nankai no Sanjin's story by the figure of the "stage boy" or actor-prostitute, were to be found in a metropolitan setting. Fujimoto Kizan writes, for instance, of a young man he met during a 1674 visit to Kumamoto that the provincial lad was pitifully ignorant of the "ways of eros," in contrast to his peers fortunate enough to be born in Kamigata, the highly urbanized region around Kyoto and Osaka, who were without exception versed in the intricacies of both joshoku and nanshoku.

What is especially striking about Fujimoto's observation is the fact that the very area in which he traveled—namely, the southwestern island of Kyushu—was one that in the Meiji period . . . would gain a notoriety not for the poverty of its male-male erotic culture, but rather for an unseemly abundance. Edo-period writers, by contrast, did not conventionally associate male-male sexual practices with any particular region of the archipelago. For them, the most significant boundaries on the erotic landscape were social rather than geo-graphic, as Nankai no Sanjin's litany of "warrior houses, temple precincts, towns, and rural districts" illustrates. Each of these spaces was the preserve of a distinct social class—warrior, priest, townsman, and peasant—and Nankai no Sanjin's enumeration of them as archetypal sites of shudō suggests that he regarded the "way" not only as broadly distributed throughout Japanese so-ciety, but as differing subtly in character among the various groups. As a way of exploring the class dimensions of shudō discourse, let us examine Nankai no Sanjin's four constituencies in order.

The prominent position that Nankai no Sanjin accords "warrior houses" is by no means accidental. Samurai, who made up somewhere between 5 and 10 percent of the population of the archipelago, formed the dominant social elite of the Edo period, and held a tight grip on the reins of government. Their place in Nankai no Sanjin's list mirrors the official ideology of "four estates," a similarly quadripartite division of society, which asserted the precedence of samurai over, in descending order, peasants, artisans, and merchants. Nankai no Sanjin's privileging of warriors was a reflection not only of their elite status, but of a link that popular discourse commonly made between the "way of youths" and the ways of this particular class. Samurai are a conspicuous pres-ence in the pages of the shudō text (Saikaku devotes nearly the entire first half of his *Great Mirror* to them), while conversely, accounts of warrior life make frequent reference to male-male eroticism. One of Nankai no Sanjin's charac-ters, the townsman Takujūrō, goes so far as to describe nanshoku as the "flower of the military estate."

Yet just as Takujūrō's remark falls from the lips of a townsman, the image of warrior culture that shudō texts helped to disseminate was to a large extent a product of the commoner's brush. Some authors, including Hiraga, the poet Basho, and the dramatist Chikamatsu Monzaemon, sprang from samurai roots, yet the majority wrote about a class to which they did not themselves belong, and toward which many in their audience felt a certain amount of awe.

Shudō texts therefore tell us more about the perceptions and expectations of non-elites than about the lifestyles and values of samurai themselves, which varied greatly with individual circumstance and underwent significant changes over the course of the Edo period. For a more nuanced picture of samurai society, and of the place that male-male sexuality held within it, it is necessary to compare popular writings with other types of documentary evidence, including diaries, letters, and other texts produced outside the framework of the publishing industry. . . . Here, the discussion will focus on the world of the warrior as it was constructed in popular texts, keeping in mind that representation did not always correspond to lived reality.

Popular representations of samurai shudō implicated its practice in a class-specific code of masculine honor. The central idea of this code was the notion of *giri*, which Caryl Callahan has defined as the "complex of obligations which the individual samurai was honor-bound to fulfill, even at the cost of his own life." Although townsmen and other social classes had their own versions of giri, the demands of honor were seen to fall most heavily upon male members of the warrior estate. As portrayed in popular texts, the giri ideal informed shudō practices among samurai in several ways. First of all, shudō ties provided a means of transmitting the values and skills essential to the pursuit of giri from the senior to the junior party, thereby serving a pedagogical function. According to Yoshida, it was the *nenja*'s task to instruct the samurai youth in the military arts, which included not only such practical accomplishments as fencing, lancemanship, archery, and horseback riding, but, most importantly, the mental attitude and etiquette befitting of a warrior. At the same time, the shudō bond was supposed to stimulate the cultivation of honor on the part of the nenja by encouraging him to strive in order to prove himself worthy of his partner; in this sense, shudō was believed to have a mutually ennobling effect. The wakashu and nenja were expected, furthermore, to assist each other in fulfilling the various obligations that made up the social fabric of giri, ranging from their primary responsibility as retainers to a lord to such private matters as duels and vendettas, all of which furnished favorite themes of popular fiction. Finally, the shudō tie was itself construed as a form of personal obligation subject to the codes of giri, demanding a loyalty no less exacting than that of the lord-vassal bond itself.

Popular discourse often linked warrior shudō with violence and death. A mid-eighteenth-century verse thus warns: "Nanshoku leads to dangerous obligations." Certainly this is the impression one would get from reading popular accounts of samurai shudō, in which swordplay is frequently a key element of the drama, and death, whether at the hands of another or through ritual disembowelment, an all too familiar denouement. Although the reaction of contemporary audiences is a difficult matter to judge, it is likely that such representations appealed to samurai and townsmen readers in different ways. For townsmen, who were prohibited as a class from wearing the long sword that was the most prized possession of the samurai male, they no doubt underscored the "otherness" of warrior culture, offering a comfortable glimpse into a world where the pursuit of shudō involved greater (or at least different) risks than in their own. But even for the samurai among their readers, the world portrayed in such works stood at a considerable remove from daily experience, not least because the setting for such narratives lay often in the historical past. For all their blood and gore, the days in which their ancestors had routinely braved death for the sake of honor (the historical accuracy of this portrait is, of course, open to question) may have evoked a sense of nostalgia among urbanized warriors living in an era of peace where the sword was found more often in the pawnshop than on the battlefield, giri was a largely formulaic notion, and male-male eroticism was increasingly a commodity for sale.

In this masculine world of honor and violence, women enjoyed only a marginal place. The social obligations that constituted giri were directed primarily toward other samurai males, while the sense of pride that fueled the pursuit of honor, and which Yoshida defined as the very essence of shudō, was understood to be a fundamentally masculine trait. Such legendary warrior women as Tomoe Gozen notwithstanding, females not only were excluded as a rule from direct participation in warfare, but were perceived as a threat to the martial valor of their menfolk. The samurai male who formed ties with women, wrote Yoshida, would be less inclined to risk death on the battlefield. (Never mind that such an argument testified more to the power of men's attraction to women than to the latter's unworthiness, or that by the time that Yoshida wrote these words, Japan's battlefields had been silent for many decades.) The obverse of warrior homosociality and androcentrism was thus a profound misogyny. It is no coincidence that many of the "woman-haters" portrayed in shudō texts are warriors, their abhorrence for the female sex at times assuming hyperbolic proportions. Urushiya, for example, describes one samurai male who would not eat a certain type of rice dumpling because it was associated with the annual Girls' Festival, while in *Fūryū hiyokudori*, a masterless swordsman proclaims that, had he known then what he knows now of women, he would never have suckled at his mother's breast. Among his rationales for avoiding women, the latter also cites the danger of menstrual pollution, a Shinto-related belief that was by no means limited to the warrior class.

The stalwart warrior-misogynist represents only one of several samurai types to appear in the pages of the shudō text. Within the samurai class, there existed a wide range of status groups, the hierarchy among them more minutely graded and their socioeconomic circumstances no less varied than the "four estates" themselves. At the top of the pyramid stood the warrior aristocracy, including the shogun, his higher-ranking vassals, and the several hundred daimyo. Among this elitest of elites, popular discourse held male-male erotic ties not only to be widespread, but to assume a highly conspicuous and even ostentatious form. The object of affection here was usually represented as a page boy, as in the case of the hairy-shinned attendants we saw favored by the "eccentric" daimyo of the 1708 story. The figure of the page, bedecked in finery and accompanying his lord on flower-viewing expeditions and other refined pursuits, holds an attraction even for the "boorish" proponent of the "way of women" in *Denbu monogatari*, who concedes that he, too, would take pleasure in surrounding himself with such elegant minions had he been born of a station that permitted it. Not all samurai could afford to maintain a private retinue of youths, however, nor did all live in such a rarefied atmosphere. Among the characters depicted as clients at houses of male prostitution, for instance, was the stock figure of the rustic samurai, whose shabby dress and crude manners make him an object of derision even among those who are theoretically his social inferiors.

Unlike samurai, the second constituency in Nankai no Sanjin's list, the Buddhist clergy, fell outside the Confucian-based taxonomy of "four estates." The same was true also of various other groups populating the shudō text, including the court nobility, Shinto priests, Confucian scholars, physicians, the blind, and outcasts. What distinguished the Buddhist clergy from these other outsiders, guaranteeing it such a prominent place in Nankai no Sanjin's social topography, was not official ideology but textual tradition, for popular discourse had long associated Buddhist monasteries and temples with male-male eroticism. Some of the earliest male-male erotic texts, we have seen, emerged from this environment, and even with the advent of commercial publishing, clerical settings remained commonplace in shudō literature. Just as nanshoku represents for Nankai no Sanjin's Takujūrō the "flower of the military estate," a character in *Fūryū hiyokudori* describes it as the "delight of the Buddhist clergy."

As in the case of samurai, priestly shudō was popularly associated with an ideological hostility toward women. An episode from *Taihei hyaku monogatari* (Hundred Tales from an Age of Great Peace), a collection of ghost stories dating from 1732, helps to illustrate this point. In the course of a pilgrimage, a priest named Guzen encounters a beautiful young widow—in actuality an evil spirit—who wishes to give him not only lodging, but her own hand in marriage as well. Guzen's response summarizes some of the religious considerations that discouraged him and his colleagues from pursuing erotic relations with women, while drawing them conversely toward youths. The priest reminds his temptress that the Buddha has forbidden those who joined the religious orders from engaging even in the briefest of flirtations, much less entering into marriage; indeed, the very word for monk (*shukke*) literally signifies one who has "left the household." "Priests love beautiful boys," the widow retorts, "so why should women present any more of a problem?" Much like the speaker in an erotic debate, Guzen lists three aspects of shudō that make it a lesser obstacle to Buddhist discipline than the "way of women." First, the affection one feels for a youth gradually fades as he grows older, so that desire for him is only a temporary distraction. Second, an affair with a youth does not result in offspring and the karmic attachments that this entails. Finally, to associate with women is to consort with morally inferior creatures fated to suffer in this life from the "five obstacles" and the "three obediences," the former a Buddhist doctrine holding females to be incapable of attaining enlightenment and the latter a no less misogynistic Confucian notion. Evidently persuaded by his logic, the widow transforms herself obligingly into a comely youth, upon which the priest realizes her true identity and rescues himself from being cannibalized.

The "beautiful boys" of whom the widow so matter-of-factly assumes priests to be fond appear in popular writings in one of two guises. The first is that of the acolyte or "temple page," an adolescent male, technically a layperson, often sent to serve in a temple or monastery in order to receive an education. The figure of the acolyte had played a central role in medieval writings on male-male eroticism, ranging from the Book of Acolytes picture scroll, to such didactic treatises as *Chigo kyōkun*, to the genre of the *chigo monogatari*, in which a priest's infatuation with a youth, in many cases the incarnation of a Buddhist deity, leads him to attain enlightenment. In popular discourse of the Edo period, however, priest-acolyte relations were less likely to occasion spiritual salvation than earthy humor, whether in such forms of comic verse as *senryū* or in the pages of the anecdote book. Acolytes feature prominently in *Seisuishō* (Rousing Laughter; completed in 1623), one of the progenitors of the latter genre, whose humorous stories the author Anrakuan Sakuden had gathered over many decades of monastic life. Anrakuan typically depicts the acolyte as a less than willing sexual partner to the senior inhabitants of the temple: one youth curses the "hateful" Kukai for bringing such a "bothersome" thing as shudō to Japan. The monastic culture of male-male eroticism in which the acolyte found himself the object of such unwanted attentions possessed a set of conventions and specialized vocabulary that were not always familiar to those who lived beyond its walls, so that the youths in several of Anrakuan's stories are hard-pressed to explain to their parents the meaning of such terms as *nyake* (referring to an anus) or *subari* (signifying a particularly constricted orifice).

A second object of the priest's affections was the male prostitute. Just as the lay youth, upon becoming an acolyte, entered—if only temporarily and often unwillingly—into a centuries-old intramural culture of male-male eroticism, so did the clergy venture out into the arena of commercialized shudō that had arisen more recently outside the cloister. In the *Great Mirror*, Saikaku portrays the supply-demand imbalance that resulted when clerics from around the country gathered in Kyoto in 1659 for religious services in commemoration of a famous Zen master. Visiting the entertainment district of Shijogawara, they "fell in love with the handsome youths there, the likes of which they had never seen in the countryside, and began buying them up indiscriminately without a thought for their priestly duties." As a result, reports the market-savvy townsman Saikaku, the fee for actor-prostitutes rose overnight, so that "their extravagance continues to cause untold hardship for the pleasure-seekers of our day." While Saikaku's account can hardly be taken as historical fact, it serves as literary testament to the entry of the Buddhist clergy into the burgeoning shudō market of the seventeenth century. Saikaku's tone of coy resentment should be read less as malicious anticlericalism than as good-natured irony: beneath the ascetic facade of the clergy, he and his fellow authors imply, lie some very human desires.

Nankai no Sanjin's third constituency, that of townsmen, comprised two separate groups in the "four estates" schema—namely, artisan and merchant households. Popular usage, however, made little distinction between the two. Many of the authors of the shudō text, including Saikaku, belonged to this social stratum, along with a substantial portion of their readership. Because a distinctly townsman culture of male-male eroticism had received little articulation prior to the Edo period, the ideals of shudō that popular texts enunciate —honor, pride, sincerity, compassion, and so forth—borrowed heavily from those of other classes. With the exception of compassion, however, these ideals drew less upon the conceptual vocabulary of the Buddhist clergy than upon that of the samurai elite. Saikaku himself suggests something of this indebtedness by devoting the first half of his *Great Mirror* to stories of warrior shudō before proceeding in the second to describe the world of the townsman. It was by measuring itself against the standards of the samurai, as it were, that the townsman culture of shudō maintained a sense of its own distinctiveness.

We see a metaphorical representation of this relationship in the figure of Takujūrō, whom Nankai no Sanjin, it will be remembered, had made to praise the "flower of the warrior estate." Despite his townsman lineage, the wealthy merchant Takujūrō is described as cultivating a "warriorlike spirit." Honor, violence, and misogyny are all part of his adopted persona: he lectures his wakashu on the need to "polish manliness," he practices the martial arts, and bans all females—canine as well as human—from his household. Nevertheless, when a visit from some robbers puts his mettle to the test, Takujūrō proves himself a helpless coward. Challenged by his wakashu to redeem his masculine honor by disemboweling himself, as a samurai would do, Takujūrō can only exclaim: "It's times like this that one is grateful to be a townsman, for life itself is the source [literally, the seed] of all things."

Takujūrō's remark would surely have chagrined Yamamoto Tsunetomo, who held that "The Way of the Samurai is found in death," or at least confirmed his sense of class superiority. Although Nankai no Sanjin's account, too, conveyed the message that townsmen, no matter how wealthy, must not presume to imitate their social betters, it implied that the "way of the townsman" (although Nankai no Sanjin does not use this term) had rewards of its own— an opinion unlikely to issue from the Kyushu warrior. Whereas Yamamoto valorized death, Takujūrō's experience causes him to take stock in the value of life: although death, in Yamamoto's paradigm, might lead to honor, it is life— the "seed of all things"—that makes possible the generation of wealth and opportunities so essential to merchant activity. Popular representations of townsman shudō often echo this affirmation of life and focus on the material, presenting an implicit contrast to the culture of death and preoccupation with honor seen as characteristic of its samurai counterpart.

The role of money was a key factor distinguishing townsman and samurai versions of shudō culture. Warrior ideology held commercial dealings to be base—a prejudice confirmed by the merchant's lowly status in the official class hierarchy. For townsmen, on the other hand, monetary exchange was the essence of their livelihood—the catalyst, as it were, that served to transform "life" into "things." In the commercialized milieu of the townsman, even sexual pleasure and erotic knowledge commanded a price, as the entrepreneurs of the prostitution and publishing trades were well aware. Although samurai numbered among the consumers of both these industries, the commodification of male-male sexuality troubled many warrior ideologues, who viewed erotic ties between males as in principle a matter of loyalty and honor. Even Yamamoto, who professed his admiration for Saikaku's prose, could scoff in the same breath that a samurai youth who pledged himself to more than one male lover in the course of a lifetime was no better than a *yarō* or *kagema*—two types of male prostitute that Saikaku glorifies in the very work that Yamamoto praised. It was one thing, evidently, for a warrior to sample the products of townsman shudō culture, and another to endorse, much less embody in his own person, its mercenary ethic.

For samurai observers like Yamamoto, the commodification of male-male sexuality was most concretely personified in the figure of the prostitute. The yarō, or kabuki actor, and his colleague the kagema, with his more tenuous theatrical credentials, stood at the heart of an urban culture of male prostitution that Yamamoto had seen expand phenomenally in the course of his lifetime, and that provides the thematic focus for the second half of Saikaku's *Great Mirror*. Although townsmen (and townswomen) were not the only clients for its services, the male prostitution industry was a quintessentially townsman phenomenon in that both entrepreneurs and prostitutes were, legally speaking, members of this class, and because the kabuki theater that served to advertise their wares formed one of the central institutions of urban culture. It was in this milieu, and in the popular texts that represented it, that sexuality and cash came to be most blatantly equated. "Both boys," writes Saikaku, for example, of two yarō, are "easily worth 1,000 pieces of gold"; or elsewhere, with respect to their clients, "merchants are well aware that you get what you pay for."

The outlay of cash served to measure not only the worth of the prostitute, but also the quality of the buyer. The well-heeled townsman patron and the elaborate entertainments that he provided for fellow revelers form a favorite subject of popular fiction, whose readers might thereby partake vicariously of pleasures that their economic circumstances did not otherwise afford. The epitome of townsman extravagance is to be found in Nishiki Bunryū's 1703 tale *Karanashi daimon yashiki* (Mansion of the Great Quince Gate), which describes a "rivalry in love" between two fabulously rich Osaka merchants over the favors of a kabuki actor: when one bribes the city's night watchmen to prevent the other from freely passing through its gates, the latter retaliates by hiring up its entire fleet of palanquins so as to immobilize his rival. Such extreme examples of conspicuous consumption, however, did not necessarily provide an ideal for emulation, whether from the standpoint of the shogunate, which had earlier in the same year confiscated the riches of a prosperous merchant house that served as the inspiration for Nishiki's work, or from that of shudō esthetics, which dictated that the display of wealth be accompanied by a commensurate degree of taste. From the perspective of shogunal officials, the ostentation of affluent merchants like those described by Nishiki represented a challenge to the class hierarchy; from the latter, it was a "boorish" breach of etiquette. Yet paradoxically, even taste was something that was difficult to acquire without a certain expenditure of money, whether in order to imbibe the consummate refinement of the pleasure quarters or to peruse the texts that served to codify its canons. The courtier Konoe Nobuhiro had recognized as much in 1619 when he wrote in his didactic work *Inu tsurezure* (Mongrel Essays in Idleness; published in 1653) that poverty, far from being "stylish," as some claimed, was in fact a "hindrance to all ways," and that the indigent were best advised not to pursue the "way of youths."

Not all forms of townsman shudō, of course, involved an exchange of cash. Nankai no Sanjin himself distinguishes between "stage boys," whose services were available for a nominally fixed price, and amateur wakashu of the town, who, apart from an occasional gift perhaps, did not ordinarily require remuneration. According to Yoshida's *Lucid Mirror*, it was uncouth for a youth of the latter variety to speak of such monetary matters as commodity prices, market conditions, or the laying-in of stock, even if his townsman upbringing had made him clever with the abacus. Male-male erotic ties also took place between fellow employees of merchant houses, as in the *senryū* stereotype of the adult store clerk who is loath to scold by day the shop boy whose favors he enjoys at night. Clerks at clothing stores, often predominantly male establishments, seem to have been especially notorious in this respect: it is to a clothing store, suggestively, that the subject of a 1789 verse chooses to pass on a used copy of a guidebook on actor-prostitutes. Senryū and other forms of popular discourse also associated male-male sexual practices with the figure of the delivery boy at the sake shop, whose peripatetic rounds taking orders and collecting empty containers exposed him on a regular basis to the advances of lecherous customers, and sometimes to outright rape. From the perspective of the author of *Gengenkyō*, however, such forcible assaults upon shop boys and other preadolescents were the work not of a true lover of shudō, but of undiscriminating brutes whose need for a "hole" might be just as easily satisfied by a wolf in forelocks.

Townsman ideology displayed far less hostility toward erotic ties between men and women than that of samurai or of the Buddhist clergy. This is not to say that there were no townsman "woman-haters": indeed, we have already encountered literary representations of two such figures, Ejima's Jūgorō and Nankai no Sanjin's Takujūrō. It is significant, however, that in both cases the authors imply that the character's misogyny is more typical of other social classes. Ejima presents Jūgorō as a foil to the hypocritical womanizing of his priestly brother, while Nankai no Sanjin connects Takujūrō's disdain for women with his warrior-like pretensions. In a townsman setting, the "way of youths" and the "way of women" were more apt to be portrayed as equivalent options for male pleasure, as with the townsman revelers who debate, in Saikaku's story, whether "courtesans are better" or "actors are better." The erotic-debate genre itself primarily addressed a townsman and lower-ranked samurai audience, as reflected in the fact that its speakers sometimes bracket the erotic practices of the higher warrior aristocracy and of Buddhist priests as belonging to a world with different standards than their own. The pursuit of one or the other erotic path rested, in this townsman-centered construction, largely on esthetic rather than ideological considerations, and occasionally on little more than a whim: "Waking from a drunken stupor," a late eighteenth-century senryū reads, "I find myself embracing a *kagema* (a male prostitute)."

Let us turn last to the final constituency in Nankai no Sanjin's list, the peasantry, who made up the overwhelming majority of the Japanese population. In the official class hierarchy, peasants ranked second only to the ruling elite, who deemed agricultural labor a more noble activity than artisanship or commerce. What Nankai no Sanjin did in effect, then, was to reverse the positions of peasants and townsmen in the status order, relegating peasants to the last, and implicitly lowest, rung on the social ladder. Whether this reshuffling is a mark of the author's own class pride is unclear, for we know little of Nankai no Sanjin's identity or social status. What is certain, however, is that in according the peasantry so undistinguished a place in his topography, Nankai no Sanjin shared a common bias of the shudō text, which, when it paid attention to peasants at all, was apt to portray them in a rather uncomplimentary light.

The peasantry constitutes a far less conspicuous presence in the shudō text than its absolute numbers might lead one to expect. Saikaku, for example, sets none of the stories in his *Great Mirror* in a specifically peasant milieu, instead focusing upon the two main components of the urban population, samurai and townsmen. Peasants emerge occasionally on the sidelines of the narrative, yet with little elaboration of character or suggestion of individuality. For the most part, they are a faceless mass, evoked primarily to demonstrate the extent of shudō's cultural sway: "Even . . . farm boys slaving in the fields," writes Saikaku in one story, "yearned to sacrifice their lives for the sake of male love." The "even" is, of course, significant, suggesting an essential disharmony between the drudgery of peasant life and the refinement proper to the "way";

similarly, the "sons of merchants sweating over their scales" and "salt makers' sons burnt black on the beaches" who appear together with the toiling farm boy in this passage provide an implicit contrast with such shudō ideals as lack of greed and fairness of complexion. (Yoshida, we may note, recommended that youths lightly powder their faces.) The *Lucid Mirror*, too, is largely silent on the subject of the peasantry, except to say that peasant *wakashu*, along with their townsman counterparts and youths loved by priests, must be taught the virtue of filial piety.

It should not be imagined, however, that the peasant's low profile in the shudō text necessarily reflects a lower incidence of male-male erotic relations among that segment of the population—a matter about which we have little knowledge, and, given the paucity of reliable data, perhaps never shall. Instead, it is an understandable consequence of the shudō text's fundamentally urban perspective. In the eyes of city dwellers, peasants were an unsophisticated and culturally impoverished lot: it is no coincidence that several terms for "boor," including *yabo* and *denbu*, incorporate characters literally meaning "field." Because peasants lived outside the metropolitan centers of shudō culture, they were imagined to have little appreciation for its intricacies, like the country bumpkin whom Nankai no Sanjin portrays as becoming involved in an argument at a kabuki theater because he refuses to believe that a skilled actor of feminine roles is in fact male. Together with the rustic samurai, the figure of the peasant yokel commonly appears in shudō texts as an object of condescension and humor, providing a conventional foil to the presumed urbanity of townsmen and their city-dwelling samurai neighbors. Yet, just as there were surely peasants who gained a more than rudimentary understanding of urban shudō culture, it is not difficult to imagine that rural communities may have harbored their own forms of male-male erotic culture, the nature of which the metropolitan producers of the shudō text had little interest in and therefore never recorded.

As Nankai no Sanjin's reordering of classes illustrates, the medium of the shudō text permitted authors to fashion a vision of the world different from that prescribed by ruling authorities. While few writers questioned the assumption that society was and should be divided into different status groups, some, including Nankai no Sanjin, sought to relativize the rigidity of the class structure by juxtaposing it against the transcendent and universal qualities of male-male erotic desire. "Even if [the object be] the frail son of a daimyo or other lordly house," asks a townsman character in another of Nankai no Sanjin's stories, "can any love lie beyond reach?" In a society as status-conscious as that of Edo-period Japan, the answer was obviously yes: the humble shop clerk who spoke these words stood little chance of courting, much less winning the hand of, a young aristocrat. Nevertheless, since readers, as well as authors, were more likely to number among the peers of the former than the latter, it was only natural that they should desire to believe that the power of love could overcome class privilege, and would prize a literature that made such liaisons possible. It is this sentiment that I have earlier referred to . . . as a sort of romantic egalitarianism. Yet before we laud it too highly, we should note that the egalitarianism involved was a fundamentally skewed one, operating for the benefit of the nenja rather than the wakashu. Its ideal was not so much a world of equal erotic actors, but one in which youths, and particularly those of good rank (here symbolized by the son of the daimyo, whose "frail" temperament is presumably a sign of his sheltered upbringing), were no less available to men of lower than of higher status. Few would have argued that the street urchin should enjoy equal access to the favors of the daimyo.

THE EGG AND THE SPERM: HOW SCIENCE
HAS CONSTRUCTED A ROMANCE BASED ON
STEREOTYPICAL MALE–FEMALE ROLES

EMILY MARTIN

The theory of the human body is always a part of a world-picture. . . . The theory of the
human body is always a part of a *fantasy*.
—James Hillman, The Myth of Analysis (1972), 220

As an anthropologist, I am intrigued by the possibility that culture shapes how
biological scientists describe what they discover about the natural world. If this
were so, we would be learning about more than the natural world in high
school biology class; we would be learning about cultural beliefs and practices
as if they were part of nature. In the course of my research I realized that the
picture of egg and sperm drawn in popular as well as scientific accounts of
reproductive biology relies on stereotypes central to our cultural definitions of
male and female. The stereotypes imply not only that female biological processes are less worthy than their male counterparts but also that women are less
worthy than men. Part of my goal in writing this article is to shine a bright light
on the gender stereotypes hidden within the scientific language of biology.
Exposed in such a light, I hope they will lose much of their power to harm us.

EGG AND SPERM: A SCIENTIFIC FAIRY TALE

At a fundamental level, all major scientific textbooks depict male and female
reproductive organs as systems for the production of valuable substances, such
as eggs and sperm.[1] In the case of women, the monthly cycle is described as
being designed to produce eggs and prepare a suitable place for them to be
fertilized and grown—all to the end of making babies. But the enthusiasm ends
there. By extolling the female cycle as a productive enterprise, menstruation
must necessarily be viewed as a failure. Medical texts describe menstruation as
the "debris" of the uterine lining, the result of necrosis, or death of tissue. The
descriptions imply that a system has gone awry, making products of no use,

not to specification, unsalable, wasted, scrap. An illustration in a widely used medical text shows menstruation as a chaotic disintegration of form, complementing the many texts that describe it as "ceasing," "dying," "losing," "denuding," "expelling."

Male reproductive physiology is evaluated quite differently. One of the texts that sees menstruation as failed production employs a sort of breathless prose when it describes the maturation of sperm: "The mechanisms which guide the remarkable cellular transformation from spermatid to mature sperm remain uncertain. . . . Perhaps the most amazing characteristic of spermatogenesis is its sheer magnitude: the normal human male may manufacture several hundred million sperm per day." In the classic text *Medical Physiology*, edited by Vernon Mountcastle, the male/female, productive/destructive comparison is more explicit: "Whereas the female *sheds* only a single gamete each month, the seminiferous tubules *produce* hundreds of millions of sperm each day" (emphasis mine).[2] The female author of another text marvels at the length of the microscopic seminiferous tubules, which, if uncoiled and placed end to end, "would span almost one-third of a mile!" She writes, "In an adult male these structures produce millions of sperm cells each day." Later she asks, "How is this feat accomplished?" None of these texts expresses such intense enthusiasm for any female processes. It is surely no accident that the "remarkable" process of making sperm involves precisely what, in the medical view, menstruation does not: production of something deemed valuable.

One could argue that menstruation and spermatogenesis are not analogous processes and, therefore, should not be expected to elicit the same kind of response. The proper female analogy to spermatogenesis, biologically, is ovulation. Yet ovulation does not merit enthusiasm in these texts either. Textbook descriptions stress that all of the ovarian follicles containing ova are already present at birth. Far from being *produced*, as sperm are, they merely sit on the shelf, slowly degenerating and aging like overstocked inventory: "At birth, normal human ovaries contain an estimated one million follicles [each], and no new ones appear after birth. Thus, in marked contrast to the male, the newborn female already has all the germ cells she will ever have. Only a few, perhaps 400, are destined to reach full maturity during her active productive life. All the others degenerate at some point in their development so that few, if any, remain by the time she reaches menopause at approximately 50 years of age." Note the "marked contrast" that this description sets up between male and female: the male, who continuously produces fresh germ cells, and the female, who has stockpiled germ cells by birth and is faced with their degeneration.

Nor are the female organs spared such vivid descriptions. One scientist writes in a newspaper article that a woman's ovaries become old and worn out from ripening eggs every month, even though the woman herself is still relatively young: "When you look through a laparoscope . . . at an ovary that has been through hundreds of cycles, even in a superbly healthy American female, you see a scarred, battered organ."[3]

To avoid the negative connotations that some people associate with the female reproductive system, scientists could begin to describe male and female processes as homologous. They might credit females with "producing" mature ova one at a time, as they're needed each mouth, and describe males as having to face problems of degenerating germ cells. This degeneration would occur throughout life among spermatogonia, the undifferentiated germ cells in the testes that are the long-lived, dormant precursors of sperm.

But the texts have an almost dogged insistence on casting female processes in a negative light. The texts celebrate sperm production because it is continuous from puberty to senescence, while they portray egg production as inferior because it is finished at birth. This makes the female seem unproductive, but some texts will also insist that it is she who is wasteful.[4] In a section heading for *Molecular Biology of the Cell*, a best-selling text, we are told that "Oogenesis is wasteful." The text goes on to emphasize that of the seven million oogonia, or egg germ cells, in the female embryo, most degenerate in the ovary. Of those that do go on to become oocytes, or eggs, many also degenerate, so that at birth only two million eggs remain in the ovaries. Degeneration continues throughout a woman's life: by puberty 300,000 eggs remain, and only a few are present by menopause. "During the 40 or so years of a woman's reproductive life, only 400 to 500 eggs will have been released," the authors write. "All the rest will have degenerated. It is still a mystery why so many eggs are formed only to die in the ovaries."

The real mystery is why the male's vast production of sperm is not seen as wasteful.[5] Assuming that a man "produces" 100 million (10^8) sperm per day (a conservative estimate) during an average reproductive life of sixty years, he would produce well over two trillion sperm in his lifetime. Assuming that a woman "ripens" one egg per lunar month, or thirteen per year, over the course of her forty-year reproductive life, she would total five hundred eggs in her lifetime. But the word "waste" implies an excess, too much produced. Assuming two or three offspring, for every baby a woman produces, she wastes only around two hundred eggs. For every baby a man produces, he wastes more than one trillion (10^{12}) sperm.

How is it that positive images are denied to the bodies of women? A look at language—in this case, scientific language—provides the first clue. Take the egg and the sperm.[6] It is remarkable how "femininely" the egg behaves and how "masculinely" the sperm.[7] The egg is seen as large and passive.[8] It does not *move or journey*, but passively "is transported," "is swept," or even "drifts" along the fallopian tube. In utter contrast, sperm are small, "streamlined," and invariably active. They "deliver" their genes to the egg, "activate the developmental program of the egg," and have a "velocity" that is often remarked upon. Their tails are "strong" and efficiently powered. Together with the forces of ejaculation, they can "propel the semen into the deepest recesses of the vagina." For this they need "energy," "fuel," so that with a "whiplashlike motion and strong lurches" they can "burrow through the egg coat" and "penetrate" it.[9] . . .

The more common picture—egg as damsel in distress, shielded only by her sacred garments; sperm as heroic warrior to the rescue—cannot be proved to be dictated by the biology of these events. While the "facts" of biology may not *always* be constructed in cultural terms, I would argue that in this case they are. The degree of metaphorical content in these descriptions, the extent to which differences between egg and sperm are emphasized, and the parallels between cultural stereotypes of male and female behavior and the character of egg and sperm all point to this conclusion.

NEW RESEARCH, OLD IMAGERY

As new understandings of egg and sperm emerge, textbook gender imagery is being revised. But the new research, far from escaping the stereotypical representations of egg and sperm, simply replicates elements of textbook gender imagery in a different form. The persistence of this imagery calls to mind what Ludwik Fleck termed "the self-contained" nature of scientific thought. As he described it, "The interaction between what is already known, what remains to be learned, and those who are to apprehend it, go [*sic*] to ensure harmony within the system. But at the same time they also preserve the harmony of illusions, which is quite secure within the confines of a given thought style."[10] We need to understand the way in which the cultural content in scientific descriptions changes as biological discoveries unfold, and whether that cultural content is solidly entrenched or easily changed.

In all of the texts quoted above, sperm are described as penetrating the egg, and specific substances on a sperm's head are described as binding to the egg. Recently, this description of events was rewritten in a biophysics lab at Johns Hopkins University—transforming the egg from the passive to the active party.

Prior to this research, it was thought that the zona, the inner vestments of the egg, formed an impenetrable barrier. Sperm overcame the barrier by mechanically burrowing through, thrashing their tails and slowly working their way along. Later research showed that the sperm released digestive enzymes that chemically broke down the zona; thus, scientists presumed that the sperm used mechanical *and* chemical means to get through to the egg.

In this recent investigation, the researchers began to ask questions about the mechanical force of the sperm's tail. (The lab's goal was to develop a contraceptive that worked topically on sperm.) They discovered, to their great surprise, that the forward thrust of sperm is extremely weak, which contradicts the assumption that sperm are forceful penetrators.[11] Rather than thrusting forward, the sperm's head was now seen to move mostly back and forth. The sideways motion of the sperm's tail makes the head move sideways with a force that is ten times stronger than its forward movement. So even if the overall force of the sperm were strong enough to mechanically break the zona, most of its force would be directed sideways rather than forward. In fact, its strongest tendency, by tenfold, is to escape by attempting to pry itself off the egg. Sperm, then, must be exceptionally efficient at *escaping* from any cell surface they contact. And the surface of the egg must be designed to trap the sperm and prevent their escape. Otherwise, few if any sperm would reach the egg.

The researchers at Johns Hopkins concluded that the sperm and egg stick together because of adhesive molecules on the surfaces of each. The egg traps the sperm and adheres to it so tightly that the sperm's head is forced to lie flat against the surface of the zona, a little bit, they told me, "like Br'er Rabbit getting more and more stuck to tar baby the more he wriggles." The trapped sperm continues to wiggle ineffectually side to side. The mechanical force of its tail is so weak that a sperm cannot break even one chemical bond. This is where the digestive enzymes released by the sperm come in. If they start to soften the zona just at the tip of the sperm and the sides remain stuck, then the weak, flailing sperm can get oriented in the right direction and make it through the zona—provided that its bonds to the zona dissolve as it moves in.

Although this new version of the saga of the egg and the sperm broke through cultural expectations, the researchers who made the discovery continued to write papers and abstracts as if the sperm were the active party who attacks, binds, penetrates, and enters the egg. The only difference was that sperm were now seen as performing these actions weakly.[12] Not until August 1987, more than three years after the findings described above, did these researchers re-conceptualize the process to give the egg a more active role. They began to describe the zona as an aggressive sperm catcher, covered with adhesive molecules that can capture a sperm with a single bond and clasp it to the zona's surface.[13] In the words of their published account: "The innermost vestment, the *zona pellucida*, is a glycoprotein shell, which captures and tethers the sperm before they penetrate it. . . . The sperm is captured at the initial contact between the sperm tip and the *zona*. . . . Since the thrust [of the sperm] is much smaller than the force needed to break a single affinity bond, the first bond made upon the tip-first meeting of the sperm and *zona* can result in the capture of the sperm."

Experiments in another lab reveal similar patterns of data interpretation. Gerald Schatten and Helen Schatten set out to show that, contrary to conventional wisdom, the "egg is not merely a large, yolk-filled sphere into which the sperm burrows to endow new life. Rather, recent research suggests the almost heretical view that sperm and egg are mutually active partners." This sounds like a departure from the stereotypical textbook view, but further reading reveals Schatten and Schatten's conformity to the aggressive-sperm metaphor. They describe how "the sperm and egg first touch when, from the tip of the sperm's triangular head, a long, thin filament shoots out and harpoons the egg." Then we learn that "remarkably, the harpoon is not so much fired as assembled at great speed, molecule by molecule, from a pool of protein stored in a specialized region called the acrosome. The filament may grow as much as twenty times longer than the sperm head itself before its tip reaches the egg and sticks." Why not call this "making a bridge" or "throwing out a line" rather than firing a harpoon? Harpoons pierce prey and injure or kill them, while this filament only sticks. And why not focus, as the Hopkins lab did, on the stickiness of the egg, rather than the stickiness of the sperm?[14] Later in the article, the Schattens replicate the common view of the sperm's perilous journey into the warm darkness of the vagina, this time for the purpose of explaining its journey into the egg itself: "[The sperm] still has an arduous journey ahead. It must penetrate farther into the egg's huge sphere of cytoplasm and somehow locate the nucleus, so that the two cells' chromosomes can fuse. The sperm dives down into the cytoplasm, its tail beating. But it is soon interrupted by the sudden and swift migration of the egg nucleus, which rushes toward the sperm with a velocity triple that of the movement of chromosomes during cell division, crossing the entire egg in about a minute." . . .

SOCIAL IMPLICATIONS: THINKING BEYOND

All . . . of these revisionist accounts of egg and sperm cannot seem to escape the hierarchical imagery of older accounts. Even though each new account gives the egg a larger and more active role, taken together they bring into play another cultural stereotype: woman as a dangerous and aggressive threat. In the Johns Hopkins lab's revised model, the egg ends up as the female aggressor who "captures and tethers" the sperm with her sticky zona, rather like a spider lying in wait in her web. The Schatten lab has the egg's nucleus "interrupt" the sperm's dive with a "sudden and swift" rush by which she "clasps the sperm and guides its nucleus to the center." Wassarman's description of the surface of the egg "covered with thousands of plasma membrane-bound projections, called microvilli" that reach out and clasp the sperm adds to the spiderlike imagery.

These images grant the egg an active role but at the cost of appearing disturbingly aggressive. Images of woman as dangerous and aggressive, the femme fatale who victimizes men, are widespread in Western literature and culture. More specific is the connection of spider imagery with the idea of an engulfing, devouring mother. New data did not lead scientists to eliminate gender stereotypes in their descriptions of egg and sperm. Instead, scientists simply began to describe egg and sperm in different, but no less damaging, terms.

Can we envision a less stereotypical view? Biology itself provides another model that could be applied to the egg and the sperm. The cybernetic model—with its feedback loops, flexible adaptation to change, coordination of the parts within a whole, evolution over time, and changing response to the environment—is common in genetics, endocrinology, and ecology and has a growing influence in medicine in general.[15] This model has the potential to shift our imagery from the negative, in which the female reproductive system is castigated both for not producing eggs after birth and for producing (and thus wasting) too many eggs overall, to something more positive. The female reproductive system could be seen as responding to the environment (pregnancy or menopause), adjusting to monthly changes (menstruation), and flexibly changing from reproductivity after puberty to nonreproductivity later in life. The sperm and egg's interaction could also be described in cybernetic terms. J. F. Hartman's research in reproductive biology demonstrated fifteen years ago that if an egg is killed by being pricked with a needle, live sperm cannot get through the zona. Clearly, this evidence shows that the egg and sperm *do* interact on more mutual terms, making biology's refusal to portray them that way all the more disturbing.

We would do well to be aware, however, that cybernetic imagery is hardly neutral. In the past, cybernetic models have played an important part in the imposition of social control. These models inherently provide a way of thinking about a "field" of interacting components. Once the field can be seen, it can become the object of new forms of knowledge, which in turn can allow new forms of social control to be exerted over the components of the field. During the 1950s, for example, medicine began to recognize the psychosocial *environment* of the patient: the patient's family and its psychodynamics. Professions such as social work began to focus on this new environment, and the resulting knowledge became one way to further control the patient. Patients began to be seen not as isolated, individual bodies, but as psychosocial entities located in an "ecological" system: management of "the patient's psychology was a new entrée to patient control."

The models that biologists use to describe their data can have important social effects. During the nineteenth century, the social and natural sciences strongly influenced each other: the social ideas of Malthus about how to avoid the natural increase of the poor inspired Darwin's *Origin of Species*. Once the *Origin* stood as a description of the natural world, complete with competition and market struggles, it could be reimported into social science as social Darwinism, in order to justify the social order of the time. What we are seeing now is similar: the importation of cultural ideas about passive females and heroic males into the "personalities" of gametes. This amounts to the "implanting of social imagery on representations of nature so as to lay a firm basis for reimporting exactly that same imagery as natural explanations of social phenomena."[16]

Further research would show us exactly what social effects are being wrought from the biological imagery of egg and sperm. At the very least, the imagery keeps alive some of the hoariest old stereotypes about weak damsels in distress and their strong male rescuers. That these stereotypes are now being written in at the level of the *cell* constitutes a powerful move to make them seem so natural as to be beyond alteration.

The stereotypical imagery might also encourage people to imagine that what results from the interaction of egg and sperm—a fertilized egg—is the result of deliberate "human" action at the cellular level. Whatever the intentions of the human couple, in this microscopic "culture" a cellular "bride" (or femme fatale) and a cellular "groom" (her victim) make a cellular baby. Rosalind Petchesky points out that through visual representations such as sonograms, we are given "*images* of younger and younger, and tinier and tinier, fetuses being 'saved.' " This leads to "the point of visibility being 'pushed back' *indefinitely*."[17] Endowing egg and sperm with intentional action, a key aspect of personhood in our culture, lays the foundation for the point of viability being pushed back to the moment of fertilization. This will likely lead to greater acceptance of technological developments and new forms of scrutiny and manipulation, for the benefit of these inner "persons": court-ordered restrictions on a pregnant woman's activities in order to protect her fetus, fetal surgery, amniocentesis, and rescinding of abortion rights, to name but a few examples.

Even if we succeed in substituting more egalitarian, interactive metaphors to describe the activities of egg and sperm, and manage to avoid the pitfalls of cybernetic models, we would still be guilty of endowing cellular entities with personhood. More crucial, then, than what *kinds* of personalities we bestow on cells is the very fact that we are doing it at all. This process could ultimately have the most disturbing social consequences.

One clear feminist challenge is to wake up sleeping metaphors in science, particularly those involved in descriptions of the egg and the sperm. Although the literary convention is to call such metaphors "dead," they are not so much dead as sleeping, hidden within the scientific content of texts—and all the more powerful for it. Waking up such metaphors, by becoming aware of when we are projecting cultural imagery onto what we study, will improve our ability to investigate and understand nature. Waking up such metaphors, by becoming aware of their implications, will rob them of their power to naturalize our social conventions about gender.

NOTES

1. The textbooks I consulted are the main ones used in classes for undergraduate premedical students or medical students (or those held on reserve in the library for these classes) during the past few years at Johns Hopkins University, These texts are widely used at other universities in the country as well. [Bibliographic references to most of these standard medical textbooks have been deleted for the sake of space. Eds.]

2. Vernon B. Mountcastle, *Medical Physiology,* 14th ed. (London: Mosby, 1980), 2:1624.

3. Melvin Konner, "Childbearing and Age," *New York Times Magazine* (December 27, 1987), 22–23, esp. 22.

4. I have found but one exception to the opinion that the female is wasteful: "Smallpox being the nasty disease it is, one might expect nature to have designed antibody molecules with combining sites that specifically recognize the epitopes on smallpox virus. Nature differs from technology, however: it thinks nothing of wastefulness. (For example, rather than improving the chance that a spermatozoon will meet an egg cell, nature finds it easier to produce millions of spermatozoa.)" (Niels Kaj Jerne, "The Immune System," *Scientific American* 229, no. 1 (July 1973): 53). Thanks to a *Signs* reviewer for bringing this reference to my attention.

5. In her essay "Have Only Men Evolved?" (in *Discovering Reality: Feminist Perspectives on Epistemology, Metaphysics, Methodology, and Philosophy of Science,* ed. Sandra Harding and Merrill B. Hintikka [Dordrecht: Reidel, 1983], 45–69, esp. 60–61), Ruth

Hubbard points out that sociobiologists have said the female invests more energy than the male in the production of her large gametes, claiming that this explains why the female provides parental care. Hubbard questions whether it "really takes more 'energy' to generate the one or relatively few eggs than the large excess of sperms required to achieve fertilization." For further critique of how the greater size of eggs is interpreted in sociobiology, see Donna Haraway, "Investment Strategies for the Evolving Portfolio of Primate Females," in *Body/Politics*, ed. Mary Jacobus, Evelyn Fox Keller, and Sally Shuttleworth (New York: Routledge, 1990), 155–56.

6. The sources I used for this article provide compelling information on interactions among sperm. Lack of space prevents me from taking up this theme here, but the elements include competition, hierarchy, and sacrifice. For a newspaper report, see Malcolm W. Browne, "Some Thoughts on Self-Sacrifice," *New York Times* (July 5, 1988), C6. For a literary rendition, see John Barth, "Night-Sea Journey," in his *Lost in the Funhouse* (Garden City, N.Y.: Doubleday, 1968), 3–13.

7. See Carol Delaney, "The Meaning of Paternity and the Virgin Birth Debate," *Man* 21, no. 3 (September 1986): 494–513. She discusses the difference between this scientific view that women contribute genetic material to the fetus and the claim of long-standing Western folk theories that the origin and identity of the fetus comes from the male, as in the metaphor of planting a seed in soil.

8. For a suggested direct link between human behavior and purportedly passive eggs and active sperm, see Erik H. Erikson, "Inner and Outer Space: Reflections on Womanhood," *Daedalus* 93, no. 2 (Spring 1964): 582–606, esp. 591.

9. All biology texts quoted above use the word "penetrate."

10. Ludwik Fleck, *Genesis and Development of a Scientific Fact*, ed. Thaddeus J. Trenn and Robert K. Merton (Chicago: University of Chicago Press, 1979), 38.

11. Far less is known about the physiology of sperm than comparable female substances, which some feminists claim is no accident. Greater scientific scrutiny of female reproduction has long enabled the burden of birth control to be placed on women. In this case, the researchers' discovery did not depend on development of any new technology. The experiments made use of glass pipettes, a manometer, and a simple microscope, all of which have been available for more than one hundred years.

12. Jay Baltz and Richard A. Cone, "What Force Is Needed to Tether a Sperm?" (abstract for Society for the Study of Reproduction, 1985), and "Flagellar Torque on the Head Determines the Force Needed to Tether a Sperm" (abstract for Biophysical Society, 1986).

13. Jay M. Baltz, David F. Katz, and Richard A. Cone, "The Mechanics of the Sperm-Egg Interaction at the Zona Pellucida," *Biophysical Journal* 54, no. 4 (October 1988): 643–54. Lab members were somewhat familiar with work on metaphors in the biology of female reproduction. Richard Cone, who runs the lab, is my husband, and he talked with them about my earlier research on the subject from time to time. Even though my current research focuses on biological imagery and I heard about the lab's work from my husband every day, I myself did not recognize the role of imagery in the sperm research until many weeks after the period of research and writing I describe. Therefore, I assume that any awareness the lab members may have had about how underlying metaphor might be guiding this particular research was fairly inchoate.

14. Surprisingly, in an article intended for a general audience, the authors do not point out that these are sea urchin sperm and note that human sperm do not shoot out filaments at all.

15. William Ray Arney and Richard Bergen, *Medicine and the Management of Living* (Chicago: University of Chicago Press, 1984), 68.

16. David Harvey, personal communication, November 1989.

17. Rosalind Petchesky, "Fetal Images: The Power of Visual Culture in the Politics of Reproduction," *Feminist Studies* 13, no. 2 (Summer 1987): 263–92, esp. 272.

WE ALWAYS MAKE LOVE WITH WORLDS

GILLES DELEUZE AND FÉLIX GUATTARI

We use the term *Libido* to designate the specific energy of desiring-machines; and the transformations of this energy—*Numen* and *Voluptas*—are never desexualizations or sublimations. This terminology indeed seems extremely arbitrary. Considering the two ways in which the desiring-machines must be viewed, what they have to do with a properly sexual energy is not immediately clear: either they are assigned to the molecular order that is their own, or they are assigned to the molar order where they form the organic or social machines, and invest organic or social surroundings. It is in fact difficult to present sexual energy as directly cosmic and infra-atomic, and at the same time as directly sociohistorical. It would be futile to say that love has to do with proteins and society. This would amount to reviving yet once more the old attempts at liquidating Freudianism, by substituting for the libido a vague cosmic energy capable of all of the metamorphoses, or a kind of socialized energy capable of all the investments. Or would we do better to review Reich's final attempt, involving a "biogenesis" that not without justification is qualified as a schizoparanoiac mode of reasoning? It will be remembered that Reich concluded in favor of an infra-atomic cosmic energy—the orgone—generative of an electrical flux and carrying submicroscopic particles, the bions. This energy produced differences in potential or intensities distributed on the body considered from a molecular viewpoint, and was associated with a mechanics of fluids in this same body considered from a molar viewpoint. What defined the libido as sexuality was therefore the association of the two modes of operation, mechanical and electrical, in a sequence with two poles, molar and molecular (mechanical tension, electrical charge, electrical discharge, mechanical relaxation). Reich thought he had thus overcome the alternative between mechanism and vitalism, since these functions, mechanical and electrical, existed in matter in general, but were combined in a particular sequence within the living. And above all he upheld the basic psychoanalytic truth, the supreme disavowal of which he was able to denounce in Freud: the independence of sexuality with regard to reproduction, the subordination of progressive or regressive reproduction to sexuality as a cycle.

If the details of Reich's final theory are taken into consideration, we admit that its simultaneously schizophrenic and paranoiac nature is no obstacle where we are concerned—on the contrary. We admit that any comparison of sexuality with cosmic phenomena such as "electrical storms," "the blue color of the sky and the blue-gray of atmospheric haze," the blue of the orgone, "St. Elmo's fire, and the bluish formations [of] sunspot activity," fluids and flows, matter and particles, in the end appear to us more adequate than the reduction of sexuality to the pitiful little familialist secret. We think that Lawrence and Miller have a more accurate evaluation of sexuality than Freud, even from the viewpoint of the famous scientificity. It is not the neurotic stretched out on the couch who speaks to us of love, of its force and its despair, but the mute stroll of the schizo, Lenz's outing in the mountains and under the stars, the immobile voyage in intensities on the body without organs. As to the whole of Reichian theory, it possesses the incomparable advantage of showing the double pole of the libido, as a molecular formation on the submicroscopic scale, and as an investment of the molar formations on the scale of social and organic aggregates. All that is missing is the confirmations of common sense: why, in what sense is this sexuality?

Cynicism has said, or claimed to have said, everything there is to say about love: that it is a matter of a copulation of social and organic machines on a large scale (at bottom, love is in the organs; at bottom, love is a matter of economic determinations, money). But what is properly cynical is to claim a scandal where there is none to be found, and to pass for bold while lacking boldness. Better the delirium of common sense than its platitude. For the prime evidence points to the fact that desire does not take as its object persons or things, but the entire surroundings that it traverses, the vibrations and flows of every sort to which it is joined, introducing therein breaks and captures—an always nomadic and migrant desire, characterized first of all by its "gigantism": no one has shown this more clearly than Charles Fourier. In a word, the social as well as biological surroundings are the object of unconscious investments that are necessarily desiring or libidinal, in contrast with the preconscious investments of need or of interest. The libido as sexual energy is the direct investment of masses, of large aggregates, and of social and organic fields. We have difficulty understanding what principles psychoanalysis uses to support its conception of desire, when it maintains that the libido must be desexualized or even sublimated in order to proceed to the social investments, and inversely that the libido only resexualizes these investments during the course of pathological regression. Unless the assumption of such a conception is still familialism—that is, an assumption holding that sexuality operates only in the family, and must be transformed in order to invest larger aggregates.

The truth is that sexuality is everywhere: the way a bureaucrat fondles his records, a judge administers justice, a businessman causes money to circulate; the way the bourgeoisie fucks the proletariat; and so on. And there is no need to resort to metaphors, any more than for the libido to go by way of metamorphoses. Hitler got the fascists sexually aroused. Flags, nations, armies, banks get a lot of people aroused. A revolutionary machine is nothing if it does not acquire at least as much force as these coercive machines have for producing breaks and mobilizing flows. It is not through a desexualizing extension that the libido invests the large aggregates. On the contrary, it is through a restriction, a blockage, and a reduction that the libido is made to repress its flows in order to contain them in the narrow cells of the type "couple," "family," "person," "objects." And doubtless such a blockage is necessarily justified: the libido does not come to consciousness except in relation to a given body, a given person that it takes as object. But our "object choice" itself refers to a conjunction of flows of life and of society that this body and this person intercept, receive, and transmit, always within a biological, social, and historical field where we are equally immersed or with which we communicate. The persons to whom our loves are dedicated, including the parental persons, intervene only as points of connection, of disjunction, of conjunction of flows whose libidinal tenor of a properly unconscious investment they translate. Thus no matter how well grounded the love blockage is, it curiously changes its function, depending on whether it engages desire in the Oedipal impasses of the couple and the family in the service of the repressive machines, or whether on the contrary it condenses a free energy capable of fueling a revolutionary machine. (Here again, everything has already been said by Fourier, when he shows the two contrary directions of the "captivation" or the "mechanization" of the passions.) But we always make love with worlds. And our love addresses itself to this libidinal property of our lover, to either close himself off or open up to more spacious worlds, to masses and large aggregates. There is always something statistical in our loves, and something belonging to the laws of large numbers. And isn't it in this way that we must understand the famous formula of Marx?—the relationship between man and woman is "the direct, natural, and necessary relation of person to person." That is, the relationship between the two sexes (man and woman) is only the measure of the relationship of sexuality in general, insofar as it invests large aggregates (man and man)?

Whence what came to be called the species determination of the sexuality of the two sexes. And must it not also be said that the phallus is not one sex, but sexuality in its entirety, which is to say the sign of the large aggregate invested by the libido, whence the two sexes necessarily derive, both in their separation (the two homosexual series of man and man, woman and woman) and in their statistical relations within this aggregate?

But Marx says something even more mysterious: that the true difference is not the difference between the two sexes, but the difference between the human sex and the "nonhuman" sex. It is clearly not a question of animals, nor of animal sexuality. Something quite different is involved. If sexuality is the unconscious investment of the large molar aggregates, it is because on its other side sexuality is identical with the interplay of the molecular elements that constitute these aggregates under determinate conditions. The dwarfism of desire as a correlate to its gigantism. Sexuality and the desiring-machines are one and the same inasmuch as these machines are present and operating in the social machines, in their field, their formation, their functioning. Desiring-machines are the nonhuman sex, the molecular machinic elements, their arrangements and their syntheses, without which there would be neither a human sex specifically determined in the large aggregates, nor a human sexuality capable of investing these aggregates. In a few sentences Marx, who is nonetheless so miserly and reticent where sexuality is concerned, exploded something that will hold Freud and all of psychoanalysis forever captive: *the anthropomorphic representation of sex!*

What we call anthropomorphic representation is just as much the idea that there are two sexes as the idea that there is only one. We know how Freudianism is permeated by this bizarre notion that there is finally only one sex, the masculine, in relation to which the woman, the feminine, is defined as a lack, an absence. It could be thought at first that such a hypothesis founds the omnipotence of a male homosexuality. Yet this is not at all the case, what is founded here is rather the statistical aggregate of intersexual loves. For if the woman is defined as a lack in relation to the man, the man in his turn lacks what is lacking in the woman, simply in another fashion: the idea of a single sex necessarily leads to the erection of a phallus as an object on high, which distributes lack as two nonsuperimposable sides and makes the two sexes communicate in a common absence—*castration.* Women, as psychoanalysts or psychoanalyzed, can then rejoice in showing man the way, and in recuperating equality in difference. Whence the irresistibly comical nature of the formulas according to which one gains access to desire through castration. But the idea that there are two sexes, after all, is no better. This time, like Melanie Klein, one attempts to define the female sex by means of positive characteristics, even if they be terrifying. At least in this way one avoids phallocentrism, if not anthropomorphism. But this time, far from founding the communication between the two sexes, one founds instead their separation into two homosexual series that remain statistical. And one does not by any means escape castration. It is simply that castration, instead of being the principle of sex conceived as the masculine sex (the great castrated soaring Phallus), becomes the result of sex conceived as the feminine sex (the little hidden absorbed penis). We maintain therefore that *castration is the basis for the anthropomorphic and molar representation of sexuality*. Castration is the universal belief that brings together and disperses both men and women under the yoke of one and the same illusion of consciousness, and makes them adore this yoke. Every attempt to determine the nonhuman nature of sex—for example, "the Great Other" in Lacan—while conserving myth and castration, is defeated from the start. And what does Jean-Francois Lyotard mean, in his commentary—so profound, nevertheless—on Marx's text, when he sees the opening of the nonhuman as having to be "the entry of the subject into desire through castration"? Long live castration, so that desire may be strong? Only fantasies are truly desired? What a perverse, human, all-too-human idea! An idea originating in bad conscience, and not in the unconscious. Anthropomorphic molar representation culminates in the very thing that founds it, the ideology of lack. The molecular unconscious, on the contrary, knows nothing of castration, because partial objects lack nothing and form free multiplicities as such; because the multiple breaks never cease producing flows, instead of repressing them, cutting them at a single stroke—the only break capable of exhausting them; because the syntheses constitute local and nonspecific connections, inclusive disjunctions, nomadic conjunctions: everywhere a microscopic transsexuality, resulting in the woman containing as many men as the man, and the man as many women, all capable of entering—men with women, women with men—into relations of production of desire that overturn the statistical order of the sexes. Making love is not just becoming as one, or even two, but becoming as a hundred thousand.

Desiring-machines or the nonhuman sex: not one or even two sexes, but *n* sexes. Schizoanalysis is the variable analysis of the *n* sexes in a subject, beyond the anthropomorphic representation that society imposes on this subject, and with which it represents its own sexuality. The schizoanalytic slogan of the desiring-revolution will be first of all: to each its own sexes.

PART VII

Bodies at the Margin, or

Attending to Distress and Difference

Bodies at the Margin, or

Attending to Distress and Difference

INTRODUCTION

Contemporary biomedicine unites biology and clinical practice in a powerful institutional and epistemological complex that has been in existence for only two centuries. Underpinning biomedical knowledge and practice is the assumption that nature is constituted by a set of laws entirely independent of history and the social order. The emergence of the disciplines of physiology and biochemistry in the mid-nineteenth century, on the heels of gross anatomy, permitted the hardening of certain ideas that had been in the air for many years in pre-Enlightenment Europe. For the first time, the human body began to be widely understood as neither a creation of God nor a product of humoral forces, but as an autonomous, universal structure—fully part of a newly secularized Nature. By extension, the normal and the pathological were delineated and systematized in the service of a clinical practice that sought to examine the workings of the body alone. In other words, biomedicine made it plausible to assess and control individual disease and health without reference to a sick person, a social milieu, or lived experience. At the same time, the emergent social sciences concerned themselves with the remainder, persons and societies, entities associated with mind more than with matter. Significant human variation became a subject for social anthropology, the science of a dualistically conceived "Man."

The existence of many different forms of biomedicine in the world today, and the flourishing of medical pluralism, often in association with emerging nationalisms, is indisputable.[1] Despite these regional and local variations, it is fair to argue that biomedical explanations about disease causation focus commonly on efficient or proximate causes, that is, on specific identifiable factors located inside the body—pathogens, toxins, lesions. These are "things" that, it is believed, produce detectable, systematically diagnosable changes. Most recently, molecular genetics has vastly expanded the universe of causes, devoting enormous amounts of funding to research that attempts to pinpoint genetic predispositions and the interactive functioning among genes, proteins, and other molecules. Such an approach tends to reinforce the idea that diseases are entities in themselves, affecting all bodies similarly, knowable in isolation, and in theory without moral, social, or political significance (but current attention to genetic diversity may temper this trend).

The common assumptions that states of health and illness are confined to individual bodies, that illness and disease are best managed by medical specialists, that an absence of measurable disease in effect signifies that nothing is amiss, and that the preservation of health is primarily the responsibility of individuals are products of the biomedically inflected times in which we live. These ideas have political ramifications, but they could not have taken root without a tacit agreement about the existence of a universal material body whose condition is determined by nature alone.

Over the past thirty years a corpus of social science literature, and increasingly articles published by medical professionals, have countered the dominant viewpoint of the biomedical sciences. Commenting both on the situation in wealthy countries and in locations where poverty is widespread, these researchers have shown how human distress of all kinds has increasingly been medicalized and individualized. Sometimes when social problems are recast as biomedical concerns, laypeople can be empowered to add expert knowledge and economic resources to the toolkit they can use to overcome suffering and stigma or to improve their lives. But where the biomedical gaze prevails, the systematic social inequalities and political forces implicated in complex social and bodily syndromes tend to be ignored or deliberately set to one side.[2] Life cycle transitions, for example, are now globally medicalized, commencing in Europe and North America at the beginning of the twentieth century with pregnancy and birth and later including adolescence, death and dying, and—toward the end of the twentieth century—midlife. One aspect of this process of medicalization is the formulation of new syndromes that reframe social behavior and embodied experience as biological disorder. Ranging from hysteria in the late nineteenth century to, more recently, attention deficit and hyperactivity disorder (ADHD), posttraumatic stress disorder, fetal alcohol syndrome, and the deficiencies associated with menopause and andropause, social scientists have shown to what extent social processes are silenced by the imposition of medical categories (themselves disputed by involved professionals). The analysis of medicalizing processes highlights the way in which politics—notably the interests of pharmaceutical companies and in some cases the medical profession—has permitted conditions and syndromes such as these to flourish and become widely understood by the public as incontestable biological facts. On the other hand, Gulf War syndrome, black lung, and similar conditions for which compensation is at issue are destined to be disputed endlessly, it seems.

Paradoxically, with some significant exceptions, much research about the social construction of disease has paid little or no heed to the subjective experience of distress and suffering, and the human body itself is left black-boxed and unattended to. The focus of such research is often limited to analyses of the creation of professional discourse and the management of individuals who seek out help from health care professionals. The essays that follow bring another dimension to our understanding of the body in distress by emphasizing the relationship among discourses (subjective, popular, and professional), embodiment, and lived experience. Through this approach the embodied individual is reinserted into social and political contexts, the contribution of which to the physical reality of disease and distress then becomes evident.

Barbara Duden is an influential historian of European medicine who, reflecting explicitly on the challenges faced by the historian of embodiment, presents us with a moment in biomedicine prior to the biological reductions discussed above. Drawing on case records, doctors' memoirs and journals—especially those of a Dr. Johann Storch—and professional medical debates of the eighteenth century, Duden reflects on how she was required to revise her expectations as it dawned on her that neither the women she was trying to understand nor their doctors experienced embodiment in ways she had previously taken for granted. She turned to a process of reconstructing Storch's terms and ordering criteria, which she organized as "meaning clusters." It was necessary to do this more by inference than by direct reading, since Storch was seldom explicit about the nature of the bodies he had learned to treat.

Here we include one of the meaning clusters she discovered as she read texts for the body implicit in them, finding a body that hid within itself transforming flows of humoral and other substances: "The inside was a porous place, a place of metamorphosis: fluids changed in the body, they transformed their materiality, form, color, consistency, and place of exit, and yet apparently they remained essentially alike." Storch did have ideas of the body that diverged from those of his patients in some respects, but they all concurred in attributing agency and intentionality to bodily changes. They thought in terms of *what the body wants*, reading the direction of its transformations as evidence of goals that were normally hidden from view and outside of conscious experience. Perhaps even for Duden's contemporary readers this is a way of understanding bodily experience that is not entirely unfamiliar, despite a modernist tendency to produce a body-consciousness steeped in a mechanistic anatomical model.

To attribute "wants" to the body itself feels strange only for a moment, and perhaps only for those whose bodies have given little trouble and can easily be thought of as automatically and without intention generating "normal" physiological effects. For the rest of us, it is not really strange to think (still in a very Cartesian fashion) that "my body is trying to tell me something."

Interestingly, formal knowledge of the body at the time Storch was practicing was not a very good guide to either patients' or doctors' experiences and interpretations. The learned doctor whose cases Duden reads here was aware of an anatomically derived medicine, one that moderns would find more recognizable and legitimate, but he did not find it possible to mobilize its principles in his clinical practice. When we apply this insight of Duden's to historical methods, it becomes clear that reading learned treatises to understand ordinary practice has pitfalls, many of which continue today to trouble the relation between academic knowledge and allegedly empirical practice.

While doing research in rural southern Italy, Mariella Pandolfi, who is trained both in psychoanalysis and anthropology, found that in and of itself neither listening like a therapist nor gathering information from informants through a crafted verbal exchange like an anthropologist was sufficient to the task of portraying the complexity of lived bodily experience. The intertextual position of her two roles enabled Pandolfi to capture what had previously gone unvoiced, and she begins to understand how village women, located at the margins of their society, create an identity of belonging through storytelling. Moreover, when narrating the "troubles" of their bodies, women insert the emotional events in their lives into both national history and village history. In so doing they describe a metaphorical body that transcends material boundaries and the usual parameters of feeling well or ill. Pain and illness are not the central features of this narrated body, Pandolfi finds, although they most certainly are present—rather, narration is about a form of shared suffering experienced exclusively by village women through the generations, and its telling brings about renewal and affirmations of identity.

When describing their subjective feelings, these rural Italian women used characteristic language to convey graphically how body parts become blocked or paralyzed, or alternatively how they shrink or expand, depending upon the particular event or circumstances. These images are not dissimilar to those glossed under a ubiquitous folk diagnostic category found particularly in the Mediterranean region, in Latin America, and in isolated parts of the English-speaking world such as Newfoundland and the Appalachians. Known as *nervos*, *nervoso*, *nevra* [*sic*], or nerves, the origins of this condition are almost certainly related to Greek humoral pathology but are not recognized by biomedicine. When people with disorders such as these visit doctors, a common experience is that their distress is either dismissed as being merely psycho-somatic, or else they are diagnosed with depression.[3] So-called anomalous bodily complaints are transformed and resorted into the categories belonging to the body proper.

In her essay, Nancy Scheper-Hughes argues that in the northeastern part of Brazil *nervos* has become the primary idiom though which hunger is ex-pressed. Conversion of hunger into *nervos* ensures that systemic problems of poverty and political-economic oppression become cases of individual sick-ness for which no social blame can be allotted. The bodies of men and women so afflicted can be understood, Scheper-Hughes suggests, as both metaphors and metonyms for the sociopolitical system at large and as the media through which their weakness, literal and structural, is simultaneously expressed. The ongoing battle between the strong and the weak that infuses everyday life results in a "social illness" that speaks to the glaring contradictions in Brazilian *Nordestino* society. These contradictions are incorporated so profoundly into "reality" that the poor more often than not participate in this hegemony, shouldering the moral blame for their physical condition. Lining up to see a doctor is the only reasonable response to this situation, and local doctors in turn medicate their patients, knowing full well that *nervos* is in large part a product of politically induced distress but having little beyond medicines to offer.

The literature of medical anthropology is replete with examples such as this one. The circumstances are not always as extreme as in northeastern Brazil, but again and again where violence is present, whether symbolic, structural, or literal, the effects are evident in depressed and broken bodies. In their work in China on depression and neurasthenia Arthur and Joan Kleinman take up a related theme, but from an entirely different vantage point. They are at pains to emphasize the reality of psychobiological processes, but above all they are concerned with the "cultural construction of the depression-neurasthenia-chronic pain complex" as an illustration of the relationship between suffering individuals and society. They set out by reminding their readers that soma-tization—the expression of distress in the form of physical symptoms ranging from pain to dizziness and weakness—is manifest everywhere in the world, including in the industrialized West. Moreover, complaints about chronic pain and other somatized distresses are very often related directly to working condi-tions and are rarely relieved by medication, though they can be affected by improvement in the work environment.

The Kleinmans, by situating the practice of psychiatry in China in the recent historical and political context of that society, are able to show why Chinese psychiatrists and their patients are more comfortable with classifying symptoms that might well be understood as depression in North America or Europe as neurasthenia—a label connoting physical debility that lacks the negative moral load associated with depression and for which medication can easily be prescribed. Even though Chinese psychiatrists are today encouraged to recast neurasthenia as depression, at the time of the Kleinmans' research they remained reluctant to do so. This essay shows clearly how certain physical symptoms and associated subjective feelings are selectively highlighted, while others that may be equally real are effaced or transposed during the processes of diagnosis and healing and in the creating of public and private narratives about the causes and meaning of distress.

The marginal bodies under discussion in Alice Domurat Dreger's essay are those of people born intersexed. Dreger creates a brief, moving genealogy of the changed relationship of intersexed people with the medical profession over the course of the nineteenth century and to the present time. She points out that today the birth of an intersexed infant is regarded as a "social emergency," and the central goal of the medical profession is to bring about "normalization" by means of hormones or surgical interventions. When she started researching this subject, Dreger quickly noticed how much more "human" intersexuals looked in early to mid-nineteenth-century medical pictures than they do in contemporary medical texts. Formerly the identity of the patient was not hidden by the use of facial masks and pseudonyms, as is rigorously required today, and Dreger notes that many intersexed people were at once medical cases and circuslike exhibits. Some actively cooperated with the medical profession in what both parties appear to have understood as a tacit exchange of goods and services. The medical men enhanced their reputations as experts in unusual anatomies (and incidentally, as Dreger puts it, "enjoyed some free voyeurism"), while their patients obtained expert opinions, occasional treatment, and official written testimony to their strangeness, certifications that could then be used as added incentive to attract the paying public to "freak" shows.

Of course, Dreger is not suggesting that we should return to such spectacularizing practices; it is against this historical backdrop, however, that she is able to highlight a late nineteenth-century shift in medical attitudes about bodies deemed of medical interest. She sees a movement away from a more or less egalitarian encounter to a problematic assumption that unlimited access to *all* unusual bodies via a one-way gaze is a medical right and ultimately for the good of humanity as a whole. One of these services to humanity, the one routinely provided for intersexed people, is the "normalization of congenitally unusual anatomies," with the ultimate objective of preventing suffering. Dreger notes that doctors are often shocked to hear that many intersexuals are highly ambivalent about the idea of normalization and that they especially do not relish being minutely examined, anonymized, and displayed for purposes of medical education, either alive or after death as parts in glass jars. Dreger moves beyond the obvious ethical issues to remind her readers that identity is significantly grounded in large part in the experience of one's anatomy. And she reminds us that, just as the intersexed are not "freaks," those with disabilities, short stature, obesity, and an infinite number of other less common bodily characteristics are persistently and systematically subject to marginalization. Such marginalization is far from trivial; the advent of prenatal genetic testing followed by selective abortion makes it possible to banish some "natural" human variation altogether, in a high-tech speed-up of human evolution. This new, laissez-faire "utopian eugenics" is already in action—the result of decisions that individuals and families make after undergoing genetic testing or screening accompanied by genetic counseling.[4]

The authors of the essays in part VII highlight a tension between "the body's insistence on meaning" and the normalizing processes of contemporary biomedicine.[5] Each essay is illustrative of an approach that has gained enormous appeal over the past decade or so, one in which an appreciation of the material body itself, embodiment, subjective experience, history, memory, culturally informed moral judgments, and politics local and global are indispensable to the analysis. In putting body politics at the heart of a broadly anthropological agenda, the dualities of mind and body, individual and society, nature and culture are to a large extent obliterated. These studies reveal an irony associated with medicalization: even as medicine has become increasingly effective in the global relief of physical suffering, its normalizing practices frequently function to increase dis-ease and perpetuate discrimination.

NOTES

1. Nichter and Lock 2002; Selin ed. 2003.
2. See, for example, Nguyen and Peschard, 2004.
3. Lock 1990.
4. Kitcher 1996.
5. Kirmayer 1992:3.

ADDITIONAL READINGS

Canguilhem, Georges. 1989. *The Normal and the Pathological.* New York: Zone Books.

Cohen, Laurence. 1998. *No Aging in India: Alzheimer's, the Bad Family, and Other Modern Things.* Berkeley: University of California Press.

Dreger, Alice Domurat. 2004. *One of Us: Conjoined Twins and the Future of Normal.* Cambridge: Harvard University Press.

Kohrman, Matthew. 2005. *Bodies of Difference: Experiences of Disability and Institutional Advocacy in the Making of Modern China.* Berkeley: University of California Press.

Lock, Margaret. 1993. *Encounters with Aging: Mythologies of Menopause in Japan and North America.* Berkeley: University of California Press.

Young, Allan. 1995. *The Harmony of Illusions: Inventing Posttraumatic Stress Disorder.* Princeton: Princeton University Press.

THE WOMAN BENEATH THE SKIN: A DOCTOR'S
PATIENTS IN EIGHTEENTH-CENTURY GERMANY

BARBARA DUDEN

THE PERCEPTION OF THE BODY

I happened upon [Johann] Storch's *Diseases of Women* by accident, at a time when my own work on the perception of parturition in the eighteenth century was already well advanced. Reading Storch, I soon realized that my own assumptions, and what I considered self-evident, could not do justice to his text. At first I read and collected cases from Storch's compendium with incomprehension. Only through a long and arduous process of reflecting on the compiled registers of all the cases did I begin to sense the importance of Storch's terms and ordering criteria. For time and again I caught myself ordering "things" according to my own anatomico-physiological grid, or according to criteria that were familiar to me from the history of eighteenth-century science. The temptation was great to interpret Storch's notions through [Georg Ernst] Stahl's doctrines, and thus to dissolve them within a doctrine to which the doctors around Halle supposedly adhered. The more deeply I read my way into the source, the more impossible it seemed to accomplish such a reductionism. The most important elements in Storch's ideas about the body were those he had in common with his women patients; thus they were not influenced by Stahl, vitalism, and Halle.

As time went by it seemed less and less meaningful to me to order my information according to "things," such as the frequency of eclampsia or ruptured perinea, or according to social status and the meager social indicators. It proved impossible to trace these stories back to "real" illnesses, since many of the "diseases," such as the insatiable need of a nun from Eichsfeld to be butchered, make little sense within modern medical concepts. Where was the dividing line between the real and the unreal? Was there even such a dividing line in Eisenach?

Gradually the outlines of a perception that the doctor and at least most of the women had in common emerged, a perception of the "body" foreign to me. The doctor's description of this "body" became the central focus of my attention. And at that point, my own certainties became an obstacle. I found myself unable to infer directly from the source to a "real corporeality," since the latter was always only implicitly hinted at.

I therefore decided to create a construct to help my reflections. I called it "orientational patterns of perception guiding the doctor's practice." Why this complex label? What the source describes and wants to describe for young *practici* are exemplary practical procedures. In each case the doctor described his perception of and prescriptions for a woman. In doing so he invariably referred in each case to her "body," even though he rarely ever actually saw it, and if he did then it was, except for face and hands, through the multilayered clothing of the time.

Yet Storch's reflections were guided by a conception of the body that was accessible to me only insofar as it informed his practice. If I wanted to adhere faithfully and closely to my source, I could hazard only to say: "I know the notions orienting the doctor's practice"—without knowing to what extent this statement contained the totality of the body he experienced.

In order to depict this intuitively grasped body conception, I had no choice but to probe it for describable elements, to break it down into meaning clusters. I could also have spoken of meaning bundles, focal points, or image elements. In contrast to my own objectified criteria, with the help of which everything or nothing in the text could be ordered without revealing a contemporary meaning, many of these focal points proved heuristically very productive. In nearly all cases they could also be closely related to conceptual patterns, themes, and motifs that were familiar to me from the scholarly literature.

A dialectical dialogue between contemporary sources, "themes" discussed in the secondary literature, and my own material allowed me to reduce the number of these orientational notions to a dozen. At the same time, the fact that I could relate these meaning clusters to each other created a much more precise, clearer, and more solid picture of the conception mirrored in the sources, which initially I could grasp only intuitively.

What is directly mirrored in the source are the doctor's conceptions. The women's conceptions of their own bodies are mirrored only indirectly through Storch's report. In my commentary on the different orientational notions I try to address the differences that emerged between the doctor's practice-guiding conceptions and the notions of how personal suffering was embodied, which are reflected in the women's complaints. A few very clear differences appeared time and again, like the one between the doctor who tries to entice the stagnant blood to open up, and the women who want to expel it. I found no reason to suspect contemporary gender differences behind these divergent attitudes. Storch's wish to "cleanse" the women with borax, instead of "promoting" them with saffron as they themselves wanted, shows that Stahl's theory influenced not only Storch's thinking but also his concrete perception of the body. I was surprised to discover how close the "body" of Storch was to that of the women when it came to actual practice. Storch's diaries have often struck me as a twenty-five-year struggle to force his own body concept, which was repeatedly confirmed as he listened to the women's complaints, into the scheme of Stahl's categories.

METAMORPHOSIS WITHIN THE BODY

The body in Eisenach is opaque. It is a place of hidden activities. As long as a person was alive, his body could not be opened, his inside could not be deciphered, could not be seen. People could speculate about its inside only with the help of signs that appeared on the body or emanated from it. All this was self-evident before our anatomical certainties became common knowledge.

Vesalius' anatomical illustrations and Harvey's experiments on the living body were known to the educated practitioner of the early eighteenth century —even if Storch did not know about the latter. But this knowledge, gained by examining dead or dying bodies, was transferred very slowly into the conceptual world of the practitioner. In practice the dead body did not yet cast its shadow on the living body. The dissecting knife could reveal the deadly devastation of an organ, and the anatomist could explain the damage as the result of a disease, but this did not explain the way a disease affected the living body. When Storch occasionally "opened" a deceased patient, whom he had never "inspected" or touched while she was alive, he was investigating her body, not the ailments through which he had known her during her lifetime.

Anatomy, the body dissected into its component parts, did not correspond to the composition of the living body. Even if by Storch's time the dead body had for decades been depicted as geometricized interior space, the living body remained a realm into which physiological theories and general concepts were projected, in which a pre-Cartesian worldview was refracted and reflected. As long as modern etiology and the modern visualization of the body interior had not prevailed, the only limits to the imagined processes inside were those drawn by the culture's conception of the world. Despite the geometricization of the Baroque, the disciplining of the dancer, the fencer, or the soldier when loading his musket, the inner body remained a world apart. Thus, in Eisenach at least, the body of a living person could bring forth whatever seemed plausible to the prevailing notions of the time. What seems physiologically fixed to a twentieth-century person could appear within very different explanatory frameworks in the cosmology of an early eighteenth-century doctor.

In early eighteenth-century Eisenach the inside of the body was a sphere of surprising changeability. The possible transformations within this sphere of incessant metamorphosis seem unlimited. Storch, who could not "view" the inside any more than could his contemporaries, speculated on rules about inner processes, which he inferred above all from the body's emanations. No body could have been descriptively grasped in its hidden interior as a universal and universally valid entity, for only the diversity of individual stories could be recorded. These stories resisted any attempt at generalization, since they appeared only as experienced processes. The inside could be grasped only as the place of an experienced but invisible flowing. Thus the external flows were aids to understanding the inside. The doctor and the women interpreted them as signs of inner movements. The doctor recognized a similarity, an analogy, or a difference between fluids that issued forth; he claimed that this allowed him an insight into what was going on within.

> The wife of a princely footman, twenty years of age, lay sick after first childbirth with "cold sweat," "heat," and a feverish rash. Over the course of several days she suffered repeatedly from loose bowels and complained about the "drying up of the milk." She had frequent diarrhea "that looked whitish like milk"; later the diarrhea came out "white, like curdled cheese." (6:606, case 28 1)

In this case Storch thought that the discharge, "white, like curdled cheese," "originated in the regurgitated milk." To substantiate his theory he listed a number of other authors who had observed that women's milk could take irregular paths, *vias extraordinarias* (608). Apparently milk could pass from the breasts to the stomach and there be excreted as white milk. It was also documented that milk could move inside to other locations, again without changing its substance. Storch cites several examples from the literature: milk was repelled from the breasts "but issued forth from the mouth with the spittle"; a young girl had plasters placed onto her swollen breasts, whereupon "the menses broke out, which in color, smell, and taste resembled milk" (3:223, case 32). The discharge from the "genital members" could be viewed as milk. We are also told about an incision for bleeding from which "pure milk" had been drawn, and of the discharge of milk instead of urine (2:323). The transformation was also documented in a different sequence: a nursing woman took elderberry juice as a purgative, whereupon her child "fell into a sweat, which in color and taste was like the elderberry juice" (7:132). What shows itself on the body's surface can be observed and examined, and it speaks of an inner process through "smell, color, and taste," through its "likeness" with analogous matter.

Storch's conception of the inner metamorphosis allowed him to assume that milk was not a substance that could be engendered only at *one* location inside the body and be excreted from only one body opening (the breasts). Milk could emerge from various openings and could resemble other excretions. Since milk appeared "on the outside," it must have strayed along "erroneous paths" on the inside.

Let us look at another case:

> The young wife of a tanner, who had developed an "oozing sore" on one of her legs while still unmarried, was pregnant. During the pregnancy this oozing on her legs disappeared. After the birth she ran a fever in childbed and developed an inflammation in her breast. When "the latter subsided, she said she got a red and itchy rash" and a "stopped-up belly." Storch gave her prescriptions against these ailments, whereupon she had a "copious opening." She discharged "sedes" in good quantity, whereupon the rash "dried up." From this evacuation she got well. (6:9, case 2)

The doctor tried to explain to himself the course of the illness: the feces, the feverish rash, and the inflamed breast were external and internal forms of some "humoral matter, which prior to the pregnancy had issued at the thigh." From the multiplicity of complaints he deduced an inner causal principle: it was *one* form of matter which in the body turned both inward, from the outer leg to the breast (breast inflammation), and outward, expressing itself as the rash. Eventually it was expelled as excrement. It was obviously bad and impure matter, which could move about in the body and had to be gotten rid of. The doctor's gaze turned materials that were quite different into one and the same matter, which he interpreted on the basis of a belief in a continuous inner transformation.

The women's complaints also reflect the notion that all matter, though ultimately the same, is involved in constant transformations. The women said that bad breath appeared when their freckles disappeared (2:434); that sweat smelled of urine (3:223); that loose bowels appeared periodically if the menses failed to flow (8:196, case 47); and that the stopped-up monthly blood exited through bloody sputum—as a chambermaid from the country reported (2:329, case 86). The inside was obviously a place of metamorphosis. How else could freckles transform themselves into bad breath? A woman who felt she was too vehemently assaulted by her husband during marital *congressus* complained of "wind" in her womb. The following day she reported that the "wind had entirely gone out through her ears" (8:363, case 87). Another woman, sick in childbed, sent word that she was having "a strange sensation, as if her breathing and speech went out through her ears" (6:349, case 243). The wind in the womb could rise upward, speech could exit through the ears. In this world the fluid inside the body could apparently assume different forms, yet remain always the same substance. The inside was a porous place, a place of metamorphosis: fluids changed in the body, they transformed their materiality, form, color, consistency, and place of exit, and yet apparently they remained essentially alike.

THE MEANING OF AN INNER PROCESS

On May 17, 1722, a young girl, the daughter of an Eisenach burgher, went for an hour-long walk with her sweetheart, danced at a village dance, and drank some perry; later she went home and complained of a headache. Exhausted, she went to bed early. The next morning the doctor had to be summoned very early. He found her "struck by a fit of senselessness." The girl could not swallow drugs, so Storch had a blister-inducing plaster applied to her calf. But all efforts were in vain. Two days later, on the evening of May 19, the girl died. (2: 127, case 21)

The doctor thinks, "searches," tries to explain the case. Not until eight days later did another girl reveal to him "the incident about the perry-drinking and dancing, and that she had been accompanied and entertained by a sweetheart." Now Storch was certain that he should have insisted, against her parents wishes, that a bleeding at the feet be tried. This, however, had not been done. This is how Storch pieced together the case: the menses rose to the girl's head, they had turned upward from down below. In addition, the blood had been agitated by her amorous feelings, and the perry and dancing had done their part to intensify the inner movement, the maelstrom. It also seemed significant that "her menses had previously been only pale in color."

In order to understand the doctor's thinking and to avoid dismissing his interpretation of the causality in this case as "nonsense," it is necessary to unravel and isolate individual linkages connecting explanatory structures. Nothing was certain, nothing was given as pure factuality. But the ambiguity of each "thing" can be inferred only from its specific meaning, from the causal and material ties that link a specific thing to other phenomena.

What kind of linkages were these? Our source makes many connections between bodily phenomena and the outside world, between bodily phenomena and events, between bodily phenomena and occurrences past and present. These connections give meaning to the various experiences. Each bodily phenomena [sic] can point in three directions: it "comes from something," it is introduced with a "because," that is, it has a cause; it "leads to something."—in the source described as "wherefore," that is, it has a result; it has an inner meaning, that is, it has a goal, an orientation which in each case was fought out as a change for the "better" or the "worse." A pain, an external flow, was not simply "there," it was a sign *for* something. The body and individual parts of the body "want" something. In the case of the girl, the headache was a sign of a hidden cause: the agitated blood, caused by perry and love. The upsurge had an effect: the turning of the monthly blood to the head. And it had a goal, for nature was misled by love into letting too much blood surge upward. These three orientations are implicit in almost all of Storch's stories: the linkage of each phenomenon to an event that preceded it in time—the external and internal *cause*; the linkage to the present and the future—the *effect*; and the inner, unexpressed *meaning*—what the body wants. They are ordering elements of his language and form the core of his self-conception as a doctor. After all, it was the ability to interpret signs correctly that distinguished him from the pure empiricists, who considered only superficial phenomena and did not know how to get at the hidden causes. The ability to recognize the "characteristics of diseases" was the secret of true medical learnedness, for the "signs of sickness" are "those clues whereby one can unravel and lay bare an obscure matter, or . . . they are the visible manifestations that reveal a hidden and unknown matter." Interpreting the signs of disease was further complicated by the fact that the doctor and the women did not share the same assumptions. Although the women thought in terms of these orientations—cause, effect, meaning—and expressed this in their complaints, they sometimes did so differently from the doctor. They had their personal signs and their own experience for interpreting things, and a different understanding of what the body needed. A phenomenon that the doctor tried to explain in complicated deductions might be a personal peculiarity that the women could interpret from personal experience. A lady had recurring pustules on her cheek, and she knew that they announced the menses; another woman linked a periodic toothache to her monthlies (2:295, case 70). With another woman, her friends and family knew exactly that her cough and "hoarseness" indicated the "occasion to congratulate her on being pregnant" (3:82, case 7).

The women had their own tradition for understanding the signs; a myriad of phenomena could point to pregnancy, imminent death, the onset of the menses, or an impending consumption. It is here, in the prognostic interpretation of personal signs and in understanding what the "signs of the women" mean, that the doctor and the women diverged most strongly. Their categories of meaning could be far apart. The interpretation of the invisible intentionality may be the locus of the cultural discontinuity that separated the women and the doctor. Here, and perhaps only here, the element that separated Storch as an academically trained doctor from his women—his physiological, philosophical frame of interpretation, the remnants of Aristotelian training—created a gap and a conflict between him and the women. For Storch the body had the ability, even in a malady, to aim at something good: healing. A sign that was bad in itself could thus indicate a good as well as a bad prognosis: vomiting, spitting of blood could be seen as nature's remedies for ridding herself of the filth in the upper body parts. The women speak less of the body's healing power. The doctor seeks to fathom the cause, the nature, the orientation of the inner body, in order to support it. Hence he will "lure" the blood, show it the proper paths from the outside. The women, on the other hand, interpreted the same signs as a call to force open the stopped-up body from the inside. They want to "expel" the blood, to thoroughly "cleanse" the inside. A woman who suffered from rheumatism in her arms cooked up some warm wine with saffron in order to drive the matter out other arms (8:137, case 36). The doctor saw this as something dangerous: it was better to carefully lead "errant nature" to the proper issues. The signs of the body spoke to the women more of an obstinate and stubborn interior threatened by stoppage. To the doctor they revealed an untimely or erroneous effort of the interior to rid itself of a burden.

MEMORY WITHIN THE BODY: WOMEN'S NARRATIVE

AND IDENTITY IN A SOUTHERN ITALIAN VILLAGE

MARIELLA PANDOLFI

ANALOGY AS METHOD

[Jorge Luis] Borges' manual of imaginary zoology describes the difficult existence of the A Bao A Qu. An animal without precise shape or color, the A Bao A Qu lives at the foot of the Tower of Victory and waits. It waits for visitors to come and give it color and form. But only if the new arrival is spiritually superior will the A Bao A Qu begin to take on shape and splendor and begin to climb the stairs of the tower. The A Bao A Qu's pursuit of perfection is totally dependent on others. Borges tells us nothing about the "splendid prince" or, rather, the extraordinary visitor, or whether he really is charismatic. That is not the question. The A Bao A Qu vegetates in lethargy until someone offers him a form that will define him. But as soon as the meeting occurs, the moment the A Bao A Qu achieves form and color and reaches the head of the stairs, he tumbles to the bottom and begins awaiting the next visitor. Identity here is defined in two different ways: one, everyday life consists in waiting; two, identity is delineated through a human relationship perceived as an extraordinary event.

Experience is never an isolated text, and identity is fundamentally defined through relationship. Plato said much the same thing in asserting that man was punished by being divided in half and must find that other half in order to achieve identity. Freud covered the same ground in his writings on narcissism, as did Lacan with his mirror stage.

What caught my interest in Borges' fable was not the moment of the fabulous encounter but the waiting, the lethargy of formless existence. How does the A Bao A Qu perceive himself as a being without form, how does he survive from one day to the next? Of course the encounter with the other is important, but the daily reality of the A Bao A Qu's life is the waiting, his experience of waiting for something out of the ordinary to happen.

There is a reason for mentioning Borges' evocative story in connection with the fieldwork I have been conducting in a Southern Italian village for the past five years. It seems to be a particularly good metaphoric description of daily life and the experience of women, which I came to understand through my ambiguous roles first as a therapist and then as an ethnographer, gradually shifting from "therapeutic" attention to observing the way suffering is expressed. Elsewhere I have spoken of the space between clinical and ethnographic work as an "in between" methodological space, a space where emotional experience expressed in a clinical setting meets the emotional experience that emerges in ethnographic work. I realized that the particular way "feeling bad or ill" is expressed is not just individual but is actually the genre the women use in describing themselves. The grounding of each individual experience was a language shared by all women regardless of age or class, a language that I would call both ancestral and symptomatologic. My initial role as a listening therapist did not allow sufficient room to gather all the valences of a symptomatologic discourse that increasingly proved to be ancestral: the clinical listening stance certainly helped to highlight this particular way of describing one's own "being," but the psychoanalytic grid alone was too reductive for a diachronic interpretation of the male–female opposition which, in the course of centuries, had organized and periodically reconfirmed a society through life cycle rituals. Listening and therapeutic space could not contain the "excess meaning," but the particular way the women have of perceiving and describing themselves, their history, and the history of the village would probably not have emerged in an exclusively anthropological approach either.

The "intertextual" position of my two roles seems to have been a particularly congenial locus for grasping emotional developments that neither the theory of drives nor that of cultural domestication alone could have fully comprehended. There may be a risk of being trapped in the excess meaning and ending up wondering who is the real A Bao A Qu, the ethnographer-clinician or the village women. Is the awaited event, the event that will give form, the meeting of informant and informed, or is it the flow of events of village life that the women experience as something outside the ordinary course of daily life?

INDIVIDUAL EXPERIENCE AND EXTRAORDINARY EVENT

The ideas of lived experience, of opening to the world, of presence in the world are essentially phenomenological notions, and especially [Wilhelm] Dilthey's notion of *Erlebnis* as "bringing life close to life" takes on extremely pregnant meaning in my interpretation. More recently a new anthropological perspective focused attention on this notion. Of course, as we shall see, the sense of experience changes according to whether the individual approaches life from the center (men) or from the margin (women). Special attention to women informants (women who speak of the darker, unvoiced aspects of their lives, of suffering that is sometimes barely whispered) is in line with the anthropological debate that has given new respect and dignity to matters of everyday life and theoretical status to what women have to say. Following [Ellen] Corin (1985), what turns out in anthropological discourse to be an extremely vital issue is the central role occupied in society by what is seemingly on the outskirts, the borderland, the margin of social exchange. It is along these lines that many writers have been rethinking the centrality of knowledge developed from particular informants (women), who by virtue of the fact that they are not at the center of the social scene can choose special, non-normative itineraries to bring out usage, knowledge, and representations established by the group's official tradition. And it is precisely because they perceive life as something felt rather than accomplished that life is anticipated and recounted as an extraordinary event. In this line the notion of the "extraordinary event" means that everyday existence can only be remembered and anticipated through events that interrupt the ordinary course of events. Existence can only be perceived and narrated to the extent that it breaks the rhythm of daily existence, which is felt to be like an overwhelming wave; past history can only be narrated through events that do not meet the expectations and the hopes of daily life. The notion of the extraordinary event also involves the subjective dimension: the way a woman experiences and expresses her historical context and her position as a historical subject. How are her desires, feelings, and conscious and unconscious emotional dynamics articulated? There is a strong relationship between "being ill," the sense of daily life as a wave, and the narration of daily life with a series of extraordinary events; there is a strong connection between two forms of expression in the narration: the narration of "outside," and the narration of "inside." In one sense the narration is always the narration of suffering, in the other the narration establishes a new identity.

NARRATIVES

For the village women, in fact, identity is established through storytelling. The central point is that the dimension in which the women project themselves is that of expectation or realization of the event, real or metaphoric, past or present. Excluded from the central masculine world of "doing" and "making," and excluded from the sense of this making and doing, women wait and often reconstruct their "own" stories or histories. For example one woman said to me:

"What I would like is another time, another history, another body. Instead, a woman is a gypsy for life. That is her fate. First she is a stranger in her husband's house."

Women rebuild life experience through narrative, through the discourse and the memory of suffering interwoven with the past and the present. But if they want to talk about the village, or about family events, past and present, they begin by narrating the troubles in their own bodies. They describe a constellation of symptoms using terms borrowed from the interpenetration of official and traditional medicine, and delineate a new narrative grammar that transposes the historical extraordinary event into the body's extraordinary event.

"I was sweating and vomiting because my grandson had left me. I raised him like a son. It was the same sweat and the same vomit that I had when my father and my brother went to Australia in 1946."

The narrative of symbolic physiology expands beyond all limits of space and time to take in the whole hostile and uncontrollable world. "My history is written in my body," is how a fifty-year-old woman expressed this; and a forty-two-year-old woman said, "The ailments of the body and soul never let a woman sleep." Generally narration seems to take three forms:

> narration of the historic events upon which the life of San Marco is based; narration of personal and family stories, either real or impregnated with traditional beliefs; and narration through one's own body; narration through real—or, more often, imagined—illness, feelings of distress, or profound suffering. This third form envelops the other two and delineates a new minimal identity.

Here I am particularly concerned with the third form, in which the collective and personal history are transformed into the narration of bodily experience. The symptomatological narratives are expressed in three different dimensions.

The first dimension is the dimension of *severed time* (the history of the village). The memory of foreign domination, times of earthquake, the times of emigrations. The narrative involves the external parts of a parcellized body: legs, back, arms, hands, feet. Legs are paralyzed by fear, the back is bowed as it was in the past; hands and feet are blocked when people emigrate; and eyes go blind in order not to see the ruin of a family and a home. The second dimension, the dimension of *filled space*, refers to social change, to women's roles, involving the internal part of the body. Organs expand, shrink, or change place inside the body:

"When I have to come home from university on the weekend and return to my father's house, my liver shifts places with my heart, and my brain is on fire."

"When my brother asks me about my sexual life my heart gets lost."

The third dimension is the dimension of *blocked movement* and refers to emotions, subjective feelings, the nocturnal plane of sentiments. This dimension involves body fluids, changes in the state of liquids, and changes in the rhythm of the blood.

"The evil of women is envy; the evil of men is strength, it steals your honor. Evil dries the blood, fear of others makes the blood clot."

The narratives that outline the new bodily identity show that it is an emotional event which establishes the operating rhythm of the body's organs and fluids. "Fear kills my veins, it closes my brain." The experience or recollection of a strong emotion immediately shifts the narrative from the normative or commonsense plane to an inner, subjective plane that is hidden from the outer world and sealed within every woman's body. This "minimal" identity, phenomenologically understood, is reawakened and built out of narration in the presence of an emotional event. So the emotional event is the high point of this identity, but it has a dual aspect. The emotion is the event that generates the narrative, but the event is linked to a historical event and the history of the family. An ongoing exploration of one's own body turns up the signs and traces of the history of every woman and the signs and traces of the history of the village. The development traverses a line from emotion to symptom to trace. And it is in the parcelled narrated body that history is petrified, the history of the person and the village. At the same time, a new locus of subjective identity is created, the body, a sore body that hurts and expresses the discomfort of living.

THE SPIRAL OF SUFFERING

What was fundamental was to give meaning to what I have called "the spiral of suffering," which takes the form of a choice of what is certainly not a linear therapeutic itinerary and the form of narrative. Speaking about oneself by way of the inner parts of one's body, fragmenting one's perception by naming aching, "ill" parts means that the village women can reconstruct for themselves what might be called a minimal identity, an identity that is not yet social or socialized, nor is it merely idiosyncratic. What in the past was beyond control and too painful to be remembered except in fragments becomes inscribed in the narrative of a body, phenomenologically understood, the only guarantee of actually being in the world. The suffering, metaphorical body is parcelled out and narrated in an idiom that is different than that of psychosomatic manifestations, although it may express symptoms arising from actual physical pathology.

For the women of San Marco, giving a narrative account of themselves by way of the body means narrating symptoms and pathology that may be real or may be the fruit of hypochondriacal tendencies. More importantly, it means transcending the boundary between feeling well and feeling ill, which is to say that the illness is not the central feature. It may exist or it may not exist; it may be something feared or something actually experienced; it may be in an acute phase, or the person may be on the way to recovery. The narrative seems full of symptoms and pathology, but it should rather be understood as the narrative of a body permeated by age-old suffering. Examples of symptomatologic discourse:

"My legs feel heavy, my heart is bursting, I never sleep, my liver has gone to my brain."

"My hands are shrivelling up, my eyes hurt, my heart has seen things that made it shrink so small it no longer beats."

"My blood has dried up, my blood has withdrawn and I no longer breathe as my blood has turned to water and I've lost the strength of my arms."

There is no grammar of illness here, no other order, albeit pathological: it is not a matter of suffering an illness, it is the story of how each woman suffers history, a sense of suffering transmitted from generation to generation and one that is experienced exclusively by women. These traces are, as it were, an inscription of history in the body itself, the idea of history being subsumed in the inscription of a symbolic itinerary in the woman's body, and the body becomes a sort of memorial to one's personal history and the history of the village as well. A woman tells her own story and the history of the village by way of the body's internal parts, parts that give pain, fall ill, or seem out of order. The course of a person's place in history is narrated and recalled by events that break up the flow of everyday history and become inscribed forever in one's body. What might be interpreted as a symptom or a constellation of symptoms, marking the course of an illness, actually turns out to be a trace or a congeries of traces signaling unusual or disturbing events. The area between illness and emotion is not at all clear or well defined in my work, nor is it in the women's narrative. There may at times be actual illness; at other times it may be hidden or expressed together with other bodily metaphors. Being well or ill is not something that depends on physical health, having a good doctor, or taking the right medicine. Feeling well or not depends on whether life spares us or bows us, and you are bowed and brought lower not so much by misfortune or death, which are considered almost natural aspects, as by events that alter and disrupt social relations (emigration in the past, rigid relations between the families of husband and wife, gossip, family quarrels, etc.).

The data that have been collected show that the anthropology of medicine must identify emotional features, passions, and affects that Western medicine often medicalizes. Recent anthropological debate has fostered efforts to remove "suffocating masks" (Boyer 1989) and has led to extremely interesting epistemological revisions. Room has certainly been made for more complex thought even in terms of the anthropology of illness, suffering, and the world of emotional experience via a rhetorical approach. Most of all the interpretative, or post-interpretative debate in anthropology has helped me move more surely in the area of the women's narrative.

Rather, as I have already remarked, these symptoms are the traces of an interrupted, broken identity, events that are the marks of interrupted personal or shared history. They are aggregated fragments of bodies and events. The body has always represented the first and the ultimate beachhead in defining and outlining a person's subjectivity. Phenomenology from Husserl to Merleau-Ponty and all of phenomenological psychopathology, has always demonstrated this fact. It is hardly worth repeating, but in line with [Karl] Jaspers, [Ludwig] Binswanger, and others, a subject's relationship with a symptom or illness always has a precise valence. Hypochondriacal manifestations or obsessive concern with parts of the body are ultimately a person's way of finding anchorage in a world threatening to collapse. The risk of fragmentation is preceded by proliferation of symptoms. A person who has already lost the world tries to use symptoms, traces of the person's body, to find other traces of a history in the world, a relation with other people who are already lost. But in the spiral of suffering narrated by the women of San Marco, symptomatologic discourse does not fragment; rather it reconstructs a view of world and being in the world.

KNOTS

I should like to conclude with a story that subsumes the forms of narration I have been describing and the dialogic experience in a global view of world. In the following interpretation social roles, the perception of danger, the sense of suffering and the experience of illness all act simultaneously and in like manner.

Donatella is a witch. At least that's what the villagers call her. She lives in a smaller village near San Marco. After meeting her for the first time, I suddenly suffered a violent attack of stomach cramps. I went to the doctor as soon as I got back to the village. I thought the cramps were due to my own ambiguous attitude. Donatella had frightened me. I had wanted to ask her something very intimate because everybody told me her powers were extraordinary. But I was there to work. And my ambivalent attitude to her probably had its effects. Two years later I went back and, to put to rest my unease about our previous meeting, I asked her point-blank:

"Why did you make me suffer? I got terribly sick right after our last meeting."

"Naturally," Donatella said, "I dissolved all the blood."

"What do you mean?"

"Your stomach was full of knots, knots of blood."

"But why?"

"Why? Because you were afraid of people. You didn't know what was happening to you. And so your stomach, like your whole body, was no longer protected. The evil forces could go in and out through your stomach whenever they wanted to. It's always like that when one is ripped by fear or when pain or mourning rob you of the soul or force to live: all the evil forces get inside you. In your case you suffered a lot. But the others were only able to turn your blood into knots. Another time, you'll not recover."

I wanted to know why the body was the object, and why the stomach.

"Have you had any children?" she asked. "No? Well, then you've got space in there to put everything in . . ."

"And men?"

"You know men, men do, they don't feel, and if you make an effort you put it on the outside. You see it's all a question of seeing whether your body receives the external forces suddenly when something happens to you or when you do something. If you do something, if you decide to do something your body closes up and the outside forces can't come in any more. When the body is closed up it gives you all the force you need to move and to act. Men always keep their bodies closed up."

NANCY SCHEPER-HUGHES

Nervos, nervoso, or *doença de nervos* is a large and expansive folk diagnostic category of distress. It is, along with such related conditions as *fraqueza* (weakness) and *loucura* (madness), seething with meanings (some of them contradictory) that have to be unraveled and decoded for what the terms reveal as well as conceal. In fact, nervos is a common complaint among poor and marginalized people in many parts of the world, but especially in the Mediterranean and in Latin America. The phenomenon has been the subject of extensive inquiries by anthropologists, who have tended (as with the analysis of hunger) toward symbolic and psychological explanations. Nervos has generally been understood as a flexible folk idiom of distress having its probable origins in Greek humoral pathology. Often nervos is described as the somatization of emotional stress originating in domestic or work relations. Gender conflicts, status deprivation, marital tensions and suppressed rage have been suspected in the etymology of *nervos* (or *nervios, nevra* [*sic*], or "bad nerves," depending on locality). In all, nervos is a broad folk syndrome (hardly culturally specific) under which can sometimes fall other common folk afflictions such as *pasmo* (nervous paralysis), or *susto* (magical fright), *mau olhado* (evil eye), and "falling out" syndrome among poor black Americans.

What all of these ills have in common is a core set of symptoms. All are "wasting" sicknesses, gravely debilitating, sometimes chronic, that leave the victim weak, shaky, dizzy and disoriented, tired and confused, sad and depressed, and alternately elated or enraged. It is curious that in the vast and for the most part uninspiring literature on *nervos,* there is no mention of the correspondence between the symptoms of *nervos* and the physiological effects of hunger. I would not want to make the mistake of simply equating the two (conceptually and symbolically, at least, nervos and *fome* are quite distinct in the minds of the people of the Alto [do Cruzeiro]) or suggest that in stripping away the cultural layers that surround a diagnosis of *nervos,* one will always find the primary, existential, subjective experience of hunger, the *delírio de fome*, at its base. Nonetheless, it does not seem likely that the situation I am describing here is completely unique to Northeast Brazil.

On the Alto do Cruzeiro today *nervos* has become the primary idiom through which both hunger and hunger anxiety (as well as many other ills and afflictions) are expressed. People are more likely today to describe their misery in terms of *nervos* than in terms of hunger. They will say, "I couldn't sleep all night, and I woke up crying and shaking with *nervos*" before they will say, "I went to bed hungry, and then I woke up shaking, nervous, and angry," although the latter is often implied in the former. Sleeping disorders are not surprising in a population raised from early childhood with the mandate to go to bed early when they are hungry. People on the Alto sleep off hunger the way we tend to sleep off a bad drunk.

Closely related to *nervos* is the idiom of *fraqueza*; a person who "suffers from nerves" is understood to be both sick and weak, lacking in strength, stamina, and resistance. And weakness has physical, social, and moral dimensions. Tired, overworked, and chronically malnourished squatters see themselves and their children as innately sick and weak, constitutionally nervous, and in need of medications and doctoring.

But this was not always so. There was a time, even at the start of the politically repressive years of the mid-1960s, when the people of the Alto spoke freely of fainting from hunger. Today one hears of people fainting from "weakness" or nerves, a presumed personal deficiency. There was a time not long ago when people of the Alto understood nervousness (and rage) as a primary symptom of hunger, as the *delírio de fome*. Today hunger (like racism) is a disallowed discourse in the shantytowns of Bom Jesus da Mata, and the rage and the dangerous madness of hunger have been metaphorized. "It doesn't help [*não adianta*] to complain of hunger," offers Manoel. Consequently, today the only "madness" of hunger is the delirium that allows hungry people to see in their wasted and tremulous limbs a chronic feebleness of body and mind.

The transition from a popular discourse on hunger to one on sickness is subtle but essential in the perception of the body and its needs. A hungry body needs food. A sick and "nervous" body needs medications. A hungry body exists as a potent critique of the society in which it exists. A sick body implicates no one. Such is the special privilege of sickness as a *neutral* social role, its exemptive status. In sickness there is (ideally) no blame, no guilt, no responsibility. Sickness falls into the moral category of bad things that "just happen" to people. Not only the sick person but society and its "sickening" social relations are gotten off the hook. Although the abuses of the sickness exemption by "malingering" patients are well known to clinicians as well as to medical sociologists, here I wish to explore a "malingering" social system. . . .

EMBODIED LIVES, SOMATIC CULTURE

How are we to make sense of *nervos*? Are the *Nordestino* cane cutters suffering, in addition to everything else, from a kind of metaphorical delirium that clouds and obscures their vision? Is false consciousness sufficiently explanatory? Or can we best understand *nervos* as an alternative form of embodiment, or body praxis?

Embodiment concerns the ways that people come to "inhabit" their bodies so that these become in every sense of the term "habituated." This is a play on Marcel Mauss's original meaning of "habitus" (a term later appropriated by Pierre Bourdieu) by which Mauss meant all the acquired habits and somatic tactics that represent the "cultural arts" of using and being in the body and in the world. From the phenomenological perspective, all the mundane activities of working, eating, grooming, resting and sleeping, having sex, and getting sick and getting well are forms of body praxis and expressive of dynamic social, cultural, and political relations. It is easy to overlook the simple observation that people who live by and through their bodies in manual and wage labor—who live by their wits and by their guts—inhabit those bodies and experience them in ways very different from our own. I am suggesting that the structure of individual and collective sentiments down to the feel of one's body is a function of one's position and role in the technical and productive order. . . .

Among the agricultural wage laborers living on the hillside shantytown of Alto do Cruzeiro, who sell their labor for as little as one dollar a day, socioeconomic and political contradictions often take shape in the "natural" contradictions of sick and afflicted bodies. In addition to the expectable epidemics of parasitic and other infectious diseases, there are the more unpredictable explosions of chaotic and unruly symptoms, whose causes do not readily materialize under the microscope. I am referring to symptoms like those associated with *nervos*, the trembling, fainting, seizures, and paralysis of limbs, symptoms that disrespect and breech mind and body, the individual and social bodies. In the exchange of meanings between the body personal and the social body, the nervous-hungry, nervous-weak body of the cane cutter offers itself both as metaphor and metonym for the sociopolitical system and for the weak position of the rural worker in the current economic order. In "lying down" on the job, in refusing to return to the work that has overdetermined most of their child and adult lives, the workers are employing a body language that can be seen as a form of surrender and as a language of defeat. But one can also see a drama of mockery and refusal. For if the folk ailment *nervos* attacks the legs, it leaves the arms and hands unparalyzed and free for less physically ruinous work, such as cutting hair. And so young men suffering from nervous paralysis can and do press their legitimate claims as "sick men" on their political bosses and patrons to find them alternative, "sitting down" work. In this context *nervos* may be seen as a version of the work slowdown or sickout, the so-called Italian strike.

But *nervos* is an expansive and polysemic folk concept. Women, too, suffer from nervos, both the *nervos de trabalhar muito* (the overwork nerves from which male cane cutters also suffer) and the *nervos de sofrer muito* (sufferers' nerves). Sufferers' nerves attack those who have endured a recent, especially a violent, shock or tragedy. Widows and the mothers of husbands and sons who have been murdered in violent altercations in the shantytown or abducted and "disappeared" by the active local death squads . . . are especially prone to the mute, enraged, white-knuckled shaking of sufferers' nerves. In these instances Taussig's (1989a) notion of the "nervous system" as a generative metaphor linking the tensions of the anatomical nervous system with the chaos and irritability of an unstable social system is useful. And so one could read the current nervousness of the people of the Alto—expressed in an epidemic of *nervoso*—as a collective and embodied response to the nervous political system just now emerging after nearly a quarter century of repressive military rule but with many vestiges of the authoritarian police state still in place. On the Alto do Cruzeiro the military presence is most often felt in the late-night knock on the door, followed by the scuffle and abduction of a loved one—father, husband, or adolescent son.

The "epidemic" of sufferers' nerves, sustos, and pasmos signifies a general state of alarm, of panic. It is a way of expressing the state of things when one must move back and forth between an acceptance of the given situation as "normal," "expectable," and routine—as "normal" and predictable as one's hunger—and a partial awareness of the real "state of emergency" into which the community has been plunged. And so the rural workers and *moradores* of the Alto are thrown from time to time into a state of disequilibrium, nervous agitation, shock, crisis, *nervos*, especially following incidents of violence and police brutality in the shantytown. To raise one's voice in active political protest is impossible and wildly dangerous. To be totally silenced, however, is intolerable. One is a man or a woman, after all. Into "impossible" situations such as these, the nervous, shaking, agitated, angry body may be enlisted to keep alive the perception that a real "state of emergency" exists. In this instance nervous sickness "publicizes" the danger, the fright, the "abnormality of the normal." Black Elena, who has lost both her husband and eldest son to the local death squads, has been struck mute. She *cannot* speak. But she sits outside her hut near the top of the Cruzeiro, dressed in white, and she shakes and trembles and raises her clenched fists in a paroxysm of anger nerves. Who can reduce this complex, somatic, and political idiom to an insipid discourse on patient somatization?

THE BODY AS BATTLEGROUND: THE MADNESS OF NERVOS

But there still remain the "negative" expressions of this somatic culture in the tendency of these same exploited and exhausted workers to blame their situation, their daily problems of basic survival, on bodies (their own) that have seemingly collapsed, given way on them. Insofar as they describe the body in terms of its immediate "use" value, they call it "good and strong" or "worthless." A man slaps at his wasted limbs (as though they were detachable appendages from the self) and says that they are now completely "useless." A woman pulls at her breasts or a man clutches his genitals and declares them "finished," "used up," "sucked dry." They describe organs that are "full of water" or "full of pus" and others that are *apodrecendo por dentro*, "rotting away from within." "Here," says Dona Irene, "put your ear to my belly. Can you hear that nasty army of critters, those amoebas, chomping away at my liver-loaf?"

In the folk system nervos may be understood as the zero point from which radiates a set of core conceptual oppositions: those between *força/fraqueza* (strength/weakness), *corpo/cabeça* (body/head, mind, morality), and *ricos/pobres* (rich/poor). Underlying and uniting these core oppositions is a single, unifying metaphor that gives shape and meaning to people's day-to-day realities.

It is the driving and compelling image of "life as a *luta*," as a series of uphill "struggles" along the *caminho*, the "path" of life. One cannot escape this generative metaphor; it crops up everywhere as an all-purpose explanation of the meaning of human existence. The *Nordestino* metaphor of the *luta* portrays life as a veritable battleground between strong and weak, powerful and powerless, young and old, male and female, and, above all, rich and poor. The *luta* requires strength, intelligence, cunning, courage, and know-how (*jeito*). But these physical, psychological, and moral qualities are seen as inequitably distributed, thereby putting the poor, the young, and the female in a relatively disadvantaged and "disgraced" position, making them particularly vulnerable to sickness, suffering, and death. Above all, it is *força*, an elusive, almost animistic constellation of strength, grace, beauty, and power, that triumphs. The folk concept of *força* is similar to what Max Weber meant by charisma. *Força* is the ultimate *jeito*, the real "knack" for survival. The rich and males have *força*, the poor and females have *fraqueza*.

These perceived class and gender differences emerge at birth. Alto women comment on the natural beauty of the infants of the rich, born fat, strong, fair, unblemished, pure, whereas their own infants are born weak, skinny, ugly, already blemished with marks and spots. Some poor infants are born weak and "wasted" before their lives have even begun, and they are labeled with the folk pediatric disorder *gasto* (spent), a quality of incurable *nervoso infantil*. Similarly, adolescent girls are prone to sickness at puberty, a time when the *força de mulher*—the female principle, sexual heat, and vitality—comes rushing from the girl's loins in her *regras*, her periodic menses, the "rules," the discipline of life. The softer among the girls sicken at this time, and some even die.

The rich fare better over all and at all stages of life, just as men fare better than women. The rich are "exempted" from the struggle that is life and appear to lead enchanted lives. Their days and nights are given to erotic pleasures (*sacanagem*) and to indulgence in rich and fatty foods; yet rarely do their bodies show the telltale signs of moral dissipation and wretched excess: bad blood and wasted livers. The poor, who can hardly afford to *brincar* (have fun, also used with reference to sex play) at all, are like "walking corpses" with their *sangue ruim, sangue fraco, sangue sujo* (bad, weak, dirty blood); their ruined and wasted livers (*fígado estragado*); and their dirty and pus-filled skin eruptions, leprosy, yaws, and syphilis. These illnesses come from "inside," and they are not sent from God but come from man, the wages of extravagance, sin, and wretched excess. The body reflects the interior moral life; it is a template for the soul and the spirit.

Within this ethno-anatomical system there are key sites that serve as conduits and filters for the body, trapping the many impurities that can attack the body from without and weaken it. The liver, the blood, and mother's milk are three such filters, and the very negative evaluation of this organ and these fluids by the people of the Alto reveals a profoundly damaged body image. The filter metaphor is particularly appropriate, however, to people accustomed to worrying about their contaminated water supply and who, in clearing the porous candle that traps filth and slime from their own water supply, often wonder aloud whether their own body "filters" may not be just as filthy.

One falls sick with tuberculosis, venereal disease, leprosy, liver disease, and heart disease because of the way one has lived: an agitated, nervous life given to excess. Bad blood or sick blood is the result of bad living, and people with these nervous diseases are said to be *estragado*, "wasted" by drugs, alcohol, or sex. If unchecked, these afflictions brought on by dissipation and excess lead to *loucura*, the most acute and dangerous form of *nervos*.

Dona Célia, once a powerful and feared old *mãe de santos* (a priestess in the Afro-Brazilian possession religion, Xangô), fell sick after Easter in 1987. Within a few months her already lean body became even more wasted, an *esqueleto* (skeleton), she commented sadly, and she lacked the strength to pull herself out of her hammock. A stay at the local hospital resolved nothing, and she was discharged without a diagnosis or any treatment beyond intravenous *soro* (sugar, salt, potassium, water). "So many ways of being sick," mused Célia, "and yet only one treatment for all the *pobres* [the poor]." Her illness, she said, was nervos. Her nerves were frayed and jumpy and brought on wild flutterings in her chest, so that her heart seemed like a wild, caged bird beating its wings to escape. There were other symptoms as well, but it was an infernal itching that was driving her mad.

When I visited her, Célia was straddling her tattered old hammock, busily casting a spell to bring about the return from São Paulo of an errant husband who had abandoned his young wife, leaving her both very lonely and very pregnant. I waited respectfully until the long incantation was completed and the candle at her feet was almost extinguished.

"That will 'burn' his ears all right," Célia reassured the tearful young client with a roguish smile on her face. The Franciscan sister, Juliana, passing by the open door, shook her head and said disapprovingly, "Can a reunion brought about by magic be worth anything?" "Oh, it's worth something, Sister," replied Célia. "I work with the spirit messengers of the saints, not with the devil!"

"How are you doing, *comadre* Célia?" I inquired.

"Poorly, *comadre*," she replied. "I no longer sleep, and the vexation [*vex-ame*] in my chest never leaves me. I can't eat and every day I grow weaker. I have a terrible *frieza* [coldness] in my head, and it's difficult for me to concentrate. I can't even remember my spells, I'm becoming so forgetful. But it's the strange itch, the *coceira esquisita*, that I can't stand. It gives me such agony, I fear that I am going to lose my mind."

Célia's neighbors were divided on the diagnosis. Most accepted that Célia's illness was nervos, but they disagreed on its origins, whether it came *por dentro* or *por fora* (from inside Célia or from outside) and whether it was a "natural" disease that came from God or an evil disease that came from man (through witchcraft). Those who were friendly to the old woman said that Célia was simply "wasted" from years of hard fieldwork. In other words, hers was simply a case of *nervos de trabalhar muito*. But those who were fearful of the old woman, resented her, or accused her of witchcraft dismissed nervos as secondary to her "true" illness: *lepra* (leprosy) resulting from her "sick" and "dirty" blood, the wages of the old sorceress' extravagance. They pointed to Célia's many moral infractions: her ritualized use of marijuana and other drugs in the practice of Xangô, her casting of spells both for good and evil, her many lovers over the years—in short, her generally independent and irreverent attitude toward the dominant Catholic mores of the community.

I stood helplessly by as Célia gradually began to slip away, daily growing more thin and haggard from her ordeal. It was painful to see a once strong and powerfully built woman so physically reduced and humbled. Although I was able to reassure Célia that she was suffering from a bad case of scabies, not from the dreaded *lepra*, I could do nothing to alleviate her nervous symptoms: her weakness, her melancholy, the *agonia* in her heart, and her adamant refusal to eat the small bits of food offered to her by her loyal friends and her few compassionate neighbors. Everything filled her with "nausea," she said. It was no use; she would never eat again.

As a going away present I brought Célia a hand-carved black *figa* (a wooden fetish, in the shape of a clenched fist with a thumb clasped between the fore and middle fingers, used to ward off evil) that I had purchased in Bahia, where Afro-Brazilian religion is practiced with greater acceptance and with more openness than in rural Pernambuco. Célia was so weak that she could barely speak, but she grabbed onto the holy object with a passion that startled me. After implanting a forceful kiss on the *figa*, with it she made a sweeping sign of the cross over her own withered body, and then she blessed me with it as well. I have been blessed many times in my life as a Catholic, but never did I feel as protected and enclosed as in that moment, or as humble.

Less than a week later (but after I had already left Bom Jesus), a few friends gathered to carry Célia in a municipal coffin to her pauper's grave in the local cemetery. There would be no marker and no inscription to honor the remains of the devout sorceress, so I could not visit the grave on my return. Célia's sullen and blasphemous daughter, Ninha, cursed her dead mother and tossed her magical apparatus in the place where pigs forage and garbage is burned on the Alto do Cruzeiro. "She'll pay for that," said Nita Maravilhosa, Nita the Marvelous, who was the old sorceress's apprentice on the Alto do Cruzeiro. . . .

Nervos is a social illness. It speaks to the ruptures, fault lines, and glaring social contradictions in *Nordestino* society. It is a commentary on the precarious conditions of Alto life. *Doença de nervos* announces a general crisis or general collapse of the body as well as a disorganization of social relations.

What, after all, does it mean to say, as did Sebastiana, "My sickness is real, my life," my nervous, agitated, threatened life? *Fraqueza* is as much a state of social as of individual "weakness," for the people of the Alto are accus to referring to their home, work, food, or marketplace (as well as their own bodies) as *fraco*. The metaphor of the *luta* and its accompanying moral economy of the body, expressed through the idioms of nervousness and weakness, are a microcosm of the moral economy of the plantation society in which strength, force, and power always win. *Nervos* and *fraqueza* are poignant reminders of the miserable conditions of Alto life, where individuals must often compete for precious little.

Rather than a torrent of indiscriminate sensations and symptoms, nervos is a somewhat inchoate, oblique, but nonetheless critical reflection by the poor on their bodies and on the work that has sapped their force and their vitality, leaving them dizzy, unbalanced, and, as it were, without "a leg to stand on." But nervos is also the "double," the second and "social" illness that has gathered around the primary experience of chronic hunger, a hunger that has made them irritable, depressed, angry, and tired and has paralyzed them so that they sense their legs giving way beneath the weight of their affliction.

On the one hand, *nervos* speaks to a profound sort of mind/body alienation, a collective delusion such that the sick-poor of the Alto can, like Seu Manoel, fall into a mood of self-blaming that is painful to witness, angrily calling himself a worthless *rato de mato* (forest rat) who is *inutilizado*, "useless," a zero. On the other hand, the discourse on *nervos* speaks obliquely to the structural "weaknesses" of the social, economic, and moral order. The idiom of nervos also allows hungry, irritable, and angry *Nordestinos* a "safe" way to express and register their anger and discontent. The recent history of the persecution of the Peasant Leagues and the rural labor movement in Pernambuco has impressed on rural workers the political reality in which they live. If it is dangerous to engage in political protest, and if it is . . . pointless to *reclamar com Deus*, to "complain to, or argue with, God" (and it would seem so), hungry and frustrated people are left with the possibility of transforming angry and nervous hunger into an illness, covertly expressing their disallowed feelings and sensations through the idiom of *nervos*, now cast as a "mental" problem. When they do so, the health care system, the pharmaceutical industry, commerce, and the political machinery of the community are fully prepared to back them up in their unhappy and anything but free "choice" of symptoms.

SOMATIZATION: THE INTERCONNECTIONS IN CHINESE SOCIETY AMONG CULTURE, DEPRESSIVE EXPERIENCES, AND THE MEANINGS OF PAIN

ARTHUR KLEINMAN AND JOAN KLEINMAN

The theme we will develop is that the disorder depression, though diagnosable worldwide with standard criteria based on psychobiological dysfunction that appears to be universal, can be looked at in another way as fundamentally a relationship between an individual and society. Depressive illness discloses how an individual relates to society, but it also suggests how society affects individuals: their interaction, behavior and even their cognitive, affective, and physiological processes. Indeed the study of depression in society shows us the sociosomatic reticulum (the symbolic bridge) that connects individuals to each other and to their life world. Depression is a profoundly social affect and disorder, certain of whose sources and consequences are structures and relations in the social world. Social structures and relations are deeply affective, thus embodied in the individual and his disorders.

This dialectical relationship between depression (or for that matter any disorder) and society is mediated by the meanings and legitimacies that symptoms take on in local systems of power. [In the longer work excerpted here,] we will interpret chronic pain in the lives of patients we studied in the People's Republic of China through the meanings and legitimacies this "disvalued experience" holds for them, their families, their work relationships, and their physicians. Inasmuch as these individuals construed the experience as neurasthenia, we will briefly return to the controversy regarding neurasthenia and depression in China and the West. We will reexamine this controversy, which we dealt with at great length in an earlier publication, to analyze the cultural construction of the depression-neurasthenia-chronic pain connection as a complement to our interpretation of its social production. That will also illumine the process "whereby any culture externalizes its social categories onto nature, and then turns to nature in order to validate its social norms as natural" (Taussig 1980). Hence depression, neurasthenia, and chronic pain open a window on Chinese psychiatry and society; they do the same for Western psychiatry and the culture from which it takes its origin.

To this connection, and the dialectic between symptom and society it expresses, we will apply the somewhat less unwieldy, yet admittedly still inelegant term, *somatization*. We define somatization as the expression of personal and social distress in an idiom of bodily complaints and medical help seeking. Somatization is, we believe, the appropriate focus for interdisciplinary collaboration between anthropology and psychiatry. If the psychiatric investigation of depression cross-culturally is to become more discriminating in its analysis of depression's social sources and cultural variation, it must draw on interpretive theory and methodology from anthropology. Similarly, if anthropological analyses are to account for the psychobiological processes in depression (and other disorders), and not to do so is to distort the nature of their cross-cultural subject matter, then anthropology too must proceed with an integrative (culture nature) framework for analysis and comparison, albeit one with different emphases than psychiatry. This chapter reviews our attempt to forge such an anthropological psychiatry and psychiatric anthropology.

PERSONAL ORIENTATIONS

We have developed this interdisciplinary approach in Chinese society over the past ten years, first in Taiwan and among Chinese-Americans, and since 1980 in the People's Republic. The studies that we discuss now were all conducted at the Hunan Medical College, one of China's leading centers of psychiatry. In 1980 we spent five months in the Department of Psychiatry at the Hunan Medical College interviewing one hundred patients who carried the diagnosis of neurasthenia as outpatients there. We were intrigued by the fact that neurasthenia—a diagnosis invented in North America and popular in the first third of this century throughout the West but now no longer officially sanctioned or used in the United States and also much less frequently encountered in Western Europe—was the most common psychiatric outpatient diagnosis for neurotic disorders in China, whereas depressive disorder—the most frequent psychiatric outpatient diagnosis in the West and indeed one of the leading diagnoses in primary care in the United States—was, at the time of our study, hardly diagnosed at all in China. We wondered if our Chinese psychiatric colleagues labeled neurasthenia what American and Western European psychiatrists labeled depression. Our earlier research in Taiwan and interviews in various outpatient clinics throughout China in 1978 suggested that this might well be the case. . . . If so, what could such findings inform us about the universal and culture-particular aspects of depression?

But we were also interested in two other questions: What could we learn from differences in the diagnostic approaches of psychiatrists in China and North America about the different varieties of clinical constructions of social reality in the two societies? What would a comparison of the antecedents and consequents of depression in each society, furthermore, tell us about those societies? We were interested, for example, in peering through the window of depression and neurasthenia at everyday life in China and observing its problems and the indigenous approaches that deal with them. . . .

SOMATIZATION IN NORTH AMERICA, THE UNITED KINGDOM, AND THE NON-WESTERN WORLD

In many societies in the non-Westem world somatization has been shown to be the predominant expression of mental illness. For example, high rates of somatization in depressive disorder have been found in clinic-based studies in Saudi Arabia, Iraq, West Africa, India, the Sudan and the Philippines, Taiwan, Hong Kong, and in matched comparisons more for Peruvian than for North American depressed patients and for East Africans more than Londoners. Marsella, in a major review of the cross-cultural literature, finds the somatic expression of depression to have higher prevalence generally in non-Westem societies (Marsella 1979) . . .

What tends to get lost in such studies, however, is, first, that somatization is also very common in the West and, second, that somatization is not limited to depression and other psychiatric disorders. Indeed, it may not always represent pathology or even maladaptation. . . . Somatization appears to have had an even higher prevalence rate in the West prior to the emergence of an increasingly psychological idiom of distress in the Victorian middle class. This psychologizing process has been related to the cultural transformation shaped by modernism, in which a deeper interiorization of the self has been culturally constituted as the now dominant Western ethnopsychology. It may be that this psychological idiom, one of Western culture's most powerful self-images, is the personal concomitant of the societal process of *rationalization* that Max Weber saw as modernism's leading edge, "the process by which explicit, abstract, intellectually calculable rules and procedures are increasingly substituted for sentiment, tradition and rule of thumb in all spheres of human activity" (Wrong 1976:247). That is to say, "affect" as currently conceived and even experienced among the middle class in the West is shaped as "deep" psychological experience and rationalized into discretely labeled emotions (depression, anxiety, anger) that previously were regarded and felt as principally bodily experiences. As bodily experience, "feeling" was expressed and interpreted more subtly, indirectly, globally, and above all somatically. Both psychologization and somatization, then, are cultural constructions of psychobiological processes: the former the creation of the Western mode of modernization that now influences the elites of non-Western societies; the latter the creation of more traditional cultural orientations worldwide, including that of the more rural, the poorer and the less educated in the West. From the cross-cultural perspective, it is not somatization in China and the West but psychologization in the West that appears unusual and requires explanation. . . .

CHINA RESEARCH: 1980

[In our 1980 research in the Department of Psychiatry of the Hunan Medical College, details of which are deleted here,] patients actively rejected the psychiatric label "depression" and reaffirmed the organic medical label "neurasthenia." Not only did the traditional Chinese cultural factors already noted seem to foster this behavior, so did newer political considerations. During the Great Proletarian Cultural Revolution, all mental illness, including, most notably, depression, had been called into question by the Maoists as wrong political thinking. This penumbra of meaning still affects the term *depression*, which also connotes withdrawal and passivity, behaviors that in China's often passionate context of aroused political energy seem suspiciously like disaffiliation and alienation. Such connotations spelled disaster during the Cultural Revolution, and even in the pragmatic political atmosphere of present-day, post-Maoist China these are not public attributions with which patients and families wish to be associated. Indeed, neurasthenia is a much safer, and more readily accessible and widely shared, public idiom of frustration and demoralization, to which Chinese continue to resort in great numbers. Though most demoralization relates to local issues in work and family, for some this is a dual discourse (overt physical complaints, covert political ones) to express dissatisfaction with the broader political situation. Because these insights are not unknown to Chinese psychiatrists, it is even more difficult for them to transform neurasthenia into depression and examine certain of the wider social structural sources of this form of human misery. Our analysis suggests that in spite of the undoubtedly great international professional pressures on Chinese psychiatry to recast neurasthenia as depression, neurasthenia and depression have a much more complex relationship, and the former, not the latter, is (at least at present) a more socially suitable and culturally approved diagnostic category in Chinese society. . . .

CODA: DEPRESSIVE AFFECT

Affect (feeling) is integral to human nature. But even though we may all "feel" the same psychobiologically produced patterns of autonomic arousal and neuroendocrine dysregulation, our unique biography and interpersonal context and the particular collective representations of our culture lead to divergent social productions and cultural constructions of specific affects. Even when each of us feels depressed, the perception, interpretation, and labeling (the construction) of the experience are distinctive, and in that sense what we actually "feel" is different. There may well be an obdurately human core to the psychobiological and social experience of loss and of lowered self-esteem that is terribly similar; nonetheless, to experience depression principally as headaches or as existential despair is not to experience the same feeling, even if "headaches" radiate symbolic meanings of frustration and unhappiness, and if despair causes headaches.

Depression and other affects can also be viewed as particular moral stances in an ideologically constructed behavioral field. There is a long Chinese tradition of paradigmatic exemplars of moral behavior whose ethical stances are simultaneously conveyed as emotion and as political statement. Perhaps it is this tradition that makes depression so potentially dangerous an affect in China: it points to the social sources of human misery that have not been altered by Communist revolution and the building of a new socialist state. Much the same can be said of our own society, though we are socialized to view affect as a natural ("gut") component of the self, not an interpersonal response or manipulation or a moral act with political connotation. In our society, however, the commercialization of human feeling ("the managed heart" of television and the movies and sports and politics) is a visible sign of the capitalist construction of affect, which has its surface (and probably equally superficial) counterparts in the socialist construction of enthusiasm, selflessness, and the other "red" emotions. Intriguingly, during the Gilded Age of late nineteenth-century America, when Beard was popularizing the term, neurasthenia sounded a note of moral criticism of an age of "wear and tear" in which the body's supplies could not keep up with society's demands, and when to be neurasthenic was to withdraw from the frenetic race for material success that characterized this great age of business and replace competition with rest, contemplation, and the sensitivities of an earlier age. The cross-cultural parallels are striking.

But affect as moral position in a social field of behavior is not affect as private experience. To make either pole of this dialectic, as relativists and materialists do, stand for affect en tout is to mistake its thoroughly interactionist nature. The outward movement from private feeling to public meaning that transforms affect in one direction has as its reciprocal the opposite inward cultural transformation that organizes meaning into feeling as personal experience.

We can regard universal psychobiological and social processes (loss, powerlessness, failure) as providing the substrate with which cultural norms react to create affect as a public and private form of experience. This cultural reaction is constrained both by the universal substrate and the culturally and personally particular reagent. The product, meaningful affective experience, is constituted out of both, but each has been changed. It is as unavailing to interpret only collective representation and personal significance in depression or headache as it is to explain them away either by neurophysiological processes or by social universals of the human condition. A disembodied affect is equally artifactual as an affectless body. Clearly, if psychiatry, anthropology, and anthropological psychiatry are to advance the cultural study of depression and other emotions, they must study both sides of the reaction.

Depression as an affect in China, the United States, and other societies for which we possess adequate clinical ethnographies appears to have something to do with loss of crucial social relationships, withdrawal from established social structural positions, and undermining of the cultural norms guiding the self. It is an emotion that poses a threat to social arrangements and symbolic meanings, and not just via suicide. Demoralization, despondency, hopelessness, withdrawal, loss of interest in the social environment are asocial. They call basic norms and relations and institutions into question, undoing the ties of the symbolic reticulum that connect person to society; thus these asocial emotions underline with poignancy and pathos the problems of bafflement and suffering and social order that Weber saw as so fundamental to the social enterprise. In egocentric societies such as our own, perhaps this affect is not nearly as threatening as it would be in sociocentric societies such as China. Indeed, perhaps the expression of personal alienation and existential despair is intrinsic to the egocentric community as a liminal state that discloses the limits of the possible and establishes the border for solipsism and narcissism. But in sociocentric China, it would be both more threatening to cultural norms and less useful for social control to sanction egocentric depression as a liminal state. Rather somatized affect—feeling as pain and bodily, not psychic, suffering—may be the appropriate liminal state.

Here pain is an opportunity to reintegrate the sick person into the social support group and to reaffirm (in the Durkheimian sense) the norms of solidarity and social control. Depressive affect is socially and culturally unsanctioned and therefore suppressed. Somatization is sanctioned and expressed, and carries both cultural cachet and social efficacy. Depressive affect is unacceptable in China because it *means* stigmatized mental illness, breakdown of social harmony—in modern terms political alienation, in traditional terms display of excessively negative feeling harmful to health. It does not signify what it *means* in middle-class white American society—the heroic romance of the lonely individual testing his existential condition by being obdurately solitary, the equity of each independent person naked before his just god (see Burton's *Anatomy of Melancholy*), the immortality of the personal soul, the narcissistic conception of man's ultimate, egocentric rights, the bitter disillusionment with sentimentality at not "making it" in the marketplace.

How we see culture entering into this picture is as the *systematized relations* between feeling, self-concept, interpersonal communication, practical action, ideology, and relationships of power. These systematized relations are distinctive in different societies. This makes for a difference in more than content. There are differences in the very structure of the links between affect, self, and social reality which are produced and reproduced distinctively in different social worlds. Culture always particularizes. The cross-cultural continuities, and there are many, in emotion come from the constraining influences of the separate elements in the cultural system: shared psychobiological processes, universal aspects of social relations, the limited variations in the politics of power.

REFERENCES

Marsella, A. 1979. "Depressive Experience and Disorder Across Cultures." In *Handbook of Cross-Cultural Psychology*, vol. 8, edited by H. Triandis and J. Draguns. Boston: Allyn and Bacon.

Taussig, M. 1980. *The Devil and Commodity Fetishism in South America*. Chapel Hill: University of North Carolina Press.

Wrong, D. H. 1976. *Skeptical Sociology*. New York: Columbia University Press.

This past April I woke up suddenly from a frightening yet laughable little dream in which I couldn't breathe because Governor John Engler, dressed in a power suit, was sitting on my naked chest. My dreams have always been this transparent; they would bore a psychotherapist. I knew immediately when I awoke what this dream was about. The next day I had an appointment with a professional photographer who was going to take a picture of me, bare except for my wedding ring on my left ring finger and a hospital bracelet around my right wrist. After he developed the picture in black and white—assuming I didn't chicken out—he would use PhotoShop to make three changes: impose a stark measurement grid behind me, black out my eyes with a rectangular band, and blur what my mate, Aron, calls "the naughty bits."

This picture would then be used for an anthology I was editing about the medical treatment of people born intersexed—the kind of people who used to be called hermaphrodites. I wanted to use this picture to make a point about the difference it makes whether people (including doctors and medical students) see intersexed people primarily the way medical books show them, or the way intersexed people see themselves. The volume, *Intersex in the Age of Ethics*, includes autobiographies of living intersexed people, and accompanying many of the autobiographies are photos of the authors looking like "normal" people. They are shown with their pets and their lovers, clothed and smiling, with clear, focused eyes—very much *not* blacked out.

The chief aim of the inclusion of a textbook-style picture was to contrast clearly these two kinds of images. The picture illuminated the paradox of the masking of patients: making patients anonymous by using pseudonyms (or no names) and by shielding their faces is great for protecting their privacy, but it is also terrible for the way in which it immediately dehumanizes them. Contributing a photo of myself in the medical textbook style also showed how anyone, even a non-intersexed person like me, could look rather pathologic if photographed this way.

I learned from contriving this "medical" photo of myself that the intersex activist Cheryl Chase was absolutely right when she told me the only thing the black band over the eyes accomplishes is saving the viewer from having the subject stare back. Even with my blackened eyes and blurred parts, those who know me can recognize me in that picture. This being the case, the decision to do this photo shoot was not an easy one, as indicated by the stressful dream in which Governor Engler embodied my university and my choice to have the "naughty bits" blurred, something you would never see in medical texts about intersex, since the whole point of those photos is to show the sexual anatomy. Yet the decision to do the photo addressed the lament, chiming in my consciousness, that I had heard time and again from intersexed people about their medical "exhibitions." These people were talking about the general problem of medical textbooks showing intersexed people not just as different but as tragically deformed. But they also spoke of specific personal experiences. They themselves, as children and adolescents, had been repeatedly subjected to physical and visual examinations by medical students, residents, and attending physicians. Although it was certainly not the medical professionals' intentions, these "exhibitions" had left the subjects feeling freakish and violated—"like insects tacked to a board for study."

This outcome is painfully ironic, since the central goal of the medical treatment of intersex is to help intersexed people feel normal and happy. Protocols for treating intersex children are founded upon the belief that ambiguous sexual anatomy constitutes "a *social* emergency." Although ambiguous genitals may signal an underlying metabolic disorder, they themselves are not diseased; they just look different—sometimes very different. Specialist clinicians often use "normalizing" surgical and hormonal treatments to try to make intersexed children's anatomy look non-ambiguous, because they understand that social responses to ambiguity can cause the intersexed person (and those around him or her) confusion and distress. So, paradoxically, in the effort to train students and residents how to alleviate the freakish feelings sometimes associated with unusual anatomies, people with unusual anatomies are examined, presented, and represented in ways that make them feel freakish.

I thought this phenomenon was specific to intersex until one of my undergraduate students figured out that I study the biomedical treatment of unusual anatomies and came seeking sympathy for what she had had to experience as the result of being born with a facial deformity. She confessed to me, without my prompting, that she would need another operation soon and that she could handle the surgery, but dreaded the inevitable endless line of white coats come to look at her face. We talked about her jumble of emotions, the gratitude and anger she felt for her surgeon as she was obligated to sit mute while he instructed his students where he would cut her. I asked her how she dealt with her troubles, and she told me that, whenever she complained, people assured her what a rich life she was living. She said, laughing under her tears, she would pretty happily trade her heavy cream life for one of skimmed milk.

I first came to the topic of intersex nine years ago, when, as a graduate student in History and Philosophy of Science, I was groping around for a dissertation topic. My advisor, the historian of embryology Fred Churchill, knew I was interested in the interplay of gender and science in the nineteenth century, and he suggested I look at hermaphrodites since at that point very few historians or philosophers had. I thought I would look at the history of embryology and evolutionary theory—Darwin's barnacles and all that—but Fred kept telling me to look at the medical journals. I couldn't figure out why he would suggest this route; I had never heard of human hermaphrodites, and figured that if they existed, they must be very rare. When, after two years, I finally decided to follow his suggestion, I immediately discovered the enormous medical literature on the phenomenon. My apartment quickly flooded with textual and photographic visions of people who so recently had been invisible to me.

I kept close by a letter from Fred, typed on university stationery, explaining the legitimacy of my research, just in case I ever got raided or had my bags searched on the way home from a research trip. But in spite of this precaution, only very gradually did events convince me that I needed to think more critically about the medical representation of intersexed people. First, I had an article on the topic of the Victorian biomedical treatment of hermaphroditism accepted to *Victorian Studies*, and the journal's editors wanted to put two of the pictures that accompanied my article on the volume's cover. These were sketches and photographs from 1890s British medical journals. But the editors discovered that, if they did this, they might have to wrap in brown paper those copies going to Canada: a new anti-pornography regulation could otherwise cause all sorts of problems. Being immersed in a medical culture (and being married to a fourth-year medical student), it took me a few months of protesting "but these are just medical pictures!" before I realized the issue was more complicated than that. I had trouble coming to terms with the fact that medical pictures in a different context might well count as pornographic. Gynecological exams were a lot easier for me before I realized this.

When the *Victorian Studies* article came out, I received a few emails from people who told me that they found my work interesting because they themselves had been born intersexed. I should have known that this would happen;

my research led me to believe there would be a lot of people out there who had the same conditions with which late Victorian doctors struggled. But the medical literature had kept hermaphrodites as essentially voiceless subjects, uneducated about their conditions. I had gotten used to them being that way.

Nevertheless, it would have been obnoxious to write about the history of intersex without listening to what intersexed people thought about my ideas, so I started to have conversations with intersexuals—but only very reluctantly. It wasn't that I was afraid of any kind of "p.c. policing" tactics, because most of them didn't want to police my work, but to engage with it. I am sure instead my reluctance came from my all-too-personal encounter with the basic threat of unusual anatomies: they force the question of what and who is normal, and the much tougher question of why we should prefer the "normal." Is it good to be "the normal one"?

We like to ground particular identities (woman, straight, adult, educated, authoritative) in particular anatomies (female, not-too-butch, physically mature, dull haircut and glasses, able-bodied and articulate), and the fact was that these people messed up in the anatomy/identity rules and dichotomies and hierarchies I enjoyed keeping stable. I didn't mind writing about them as "other," but when they started to step up off the page and to the phone, that made me nervous. No longer would it just be me writing historical representations of medical representations of them. Indeed, for the first time I had to face full force the fact that I was—like the doctors whose history I was studying—creating yet another representation of them as "other." And were they "other"?

I met, electronically, an intersexed person who had appeared on a daytime talk show to talk about her anatomy and experiences. Offline I pooh-poohed this tacky exhibitionism, and online I told her I was sorry she felt she had to go that route. She surprised me by informing me that she had enjoyed the experience and believed it was important work for her to do. She had much more control over her representation on a talk show than she did in the medical arena—for one thing, she got to keep her clothes on around all these strangers—and rather than making her feel freakish or ashamed the way the medical profession (accidentally) had, the talk show host and audience validated her.

By that point in my research, I had been consulting present-day medical books for my work, but only to learn the current biological understanding of the various kinds of intersex. Finally I started to read those present-day texts with a critical eye, and when I did, one of the first things that struck me was how much more human intersexuals looked in the nineteenth-century pictures, especially in the early and mid-nineteenth-century ones. In those pictures, we see the whole body, face and all. Often the subject stands in a classical pose, looking almost proud. The person's identity is not hidden with pseudonyms or facial masks. The focus is still on the whole person, not just the parts.

These early nineteenth-century pictures show the influence of contemporary showy public exhibitions of and by the people with unusual anatomies, and indeed hint at the very blurry distinctions that existed then between medical and public displays. An 1815 sketch of the hermaphrodite Marie-Madeleine Lefort shows her at the Paris Academy of Medicine. She appears to be on a stage, with curtains behind her. Her hair is up in a turban as if to make her seem even more theatrical and more exotic than she already is, with a combination of a well-groomed moustache, plump breasts, a recognizable vulva, and a sizable phallus.

At the very time my ears began to fill with living intersexuals' complaints about the "raw deal" they were getting from medicine, I began to uncover evidence of a time when the "deals" made between medical professionals and people with unusual anatomies were much more overt and perhaps more evenly rewarding. For example, in the 1830s, the hermaphroditic Gottlieb Göttlich earned fame and fortune by exhibiting himself at medical schools and to the lay public across Europe and the British Isles. When Göttlich was born on 6 March 1798, in the Saxon village of Nieder Leuba, those present at the birth presumed Göttlich female, and so the child was baptized and raised Marie Rosine Göttlich. Like dozens (and perhaps hundreds) of hermaphrodites after her, Marie Rosine's sex came under suspicion when an apparent double hernia drew medical attention and the suspicion that the "herniated" organs of this "woman" were in fact descended testicles. In November of 1832, Professor Friedrich Tiedemann at the University of Heidelberg examined Göttlich and declared Marie Rosine was "evidently a man, with genitals of uncommon conformation. She will dress herself, therefore, in men's clothes, and adopt the name of Gottlieb."

Göttlich made the most of this sex-change idea, engineering for himself a very fruitful change of identity. With written testimonies to his masculinity in hand from the well-known Tiedemann of Heidelberg and Johann Blumenbach of Göttingen, Göttlich donned male clothing, obtained a passport that indicated his new sex, and went on the road, something a person with a female passport and little source of income would have had a much harder time doing. During his 1830s career as a traveling hermaphrodite catering to the curiosity of medical men and laypeople, Göttlich made his way all over the European continent and British Isles. At medical schools, he stripped down, posed with a proud expression, and was examined by scores of men of science and medicine who came to see this curious case and render their often conflicting opinions as to the "true sex." Importantly for my point here, in exchange for letting them examine and publish reports about him, the medical and scientific men gave Göttlich certificates which testified that his case was of deep interest to the medical man, the naturalist, the phrenologist, and the physiologist. These were used in turn to generate still more interest and profit. In fact, when a corrective operation was offered him Göttlich "declined all surgical aid." He remained "averse to a proposal of this kind, since it would at once deprive him of his . . . easy and profitable mode of subsistence."

Exhibitors like Göttlich frequently set up, in the early and mid–nineteenth century, these sorts of tacit exchange of goods and services with medical men. Those with peculiar anatomies let intensely curious medical and scientific men examine them, and the biomedical men not only enjoyed some free voyeurism —it is clear voyeurism was part of the attraction—but they also published their accounts of the unusual anatomies and thereby increased medical knowledge of the conditions, occasional medical treatment, and written expert testimonies to their strangeness. Exhibitors like Göttlich used these testimonies in advertisements and penny pamphlets to drum up ever more business.

This concept of fair exchange comes through loud and clear in the report of one of the medical men, William H. Pancoast, M.D., who attended the Philadelphia autopsy of the Bunker twins. When Chang and Eng Bunker, the famous "Siamese Twins," died on 17 January 1874, in Mount Airy, North Carolina, their widows initially decided to keep the brothers' conjoined remains in a cool cellar and allow any curious soul a glance for twenty-five cents. (They probably did this to help support the twenty-one children Chang and Eng left behind,) but on 1 February a contingent of medical men from Philadelphia showed up to claim the remains for "science and humanity." As Pancoast remarked in his autopsy report to the College of Physicians in Philadelphia:

> To advance their own interests [Chang and Eng] frequently consulted medical men in different parts of America and Europe, as to the safety of a surgical operation to divide the band and release them from their peculiar connection; [but] these consultations [with medical men] were mainly used to excite the curiosity of the public, as it is believed by those who knew them well, that they never, except once, seriously contemplated such an operation.

In exchange for medical men's long-running help in "excit[ing] the curiosity of the public," Pancoast "held [it] to be a duty to science and humanity, that the family of the deceased [Bunker twins] should [now] permit an autopsy. [For] the twins had availed themselves most freely of the services of our profession in both hemispheres, and it was considered by many but as a proper and necessary return" that the family should turn over corpses, free of charge, so the medical men could finally satisfy their curiosity about the details of the Bunkers' internal anatomy.

Compared to the Bunkers', the biographies of the conjoined sisters Millie and Christina are much more troubling because the sisters, born to an enslaved African-American woman in 1851, were sold, bought, kidnapped, and repeatedly exhibited by people who did not have their best interests in mind. It does appear that they finally wound up with a caretaker who fairly shared the profits of exhibition with them, but that alleged happy ending to their biographies might in fact have been concocted by yet another inhumane handler for publicity purposes. I am interested in Millie and Christina's biographies here chiefly because they again exemplify the way in which a chance to examine a peculiar anatomy was traded for a salable expert testimony. When young children, Millie-Christina were brought to New Orleans, for a "command performance,"

> in obedience to a request from the medical faculty of that city, asking that she be brought there for scientific examination. Rooms were taken and every preparation made for the contemplated examination, after which she was to be placed on public exhibition. . . . The examination . . . at length took place and proved most satisfactory, every physician in attendance concurring in pronouncing her Nature's greatest wonder. *Being endorsed by the medical faculty*, she was now put on public exhibition.

Millie-Christina repeatedly and without charge performed their songs and dances privately for doctors, and the amazed medical men in turn handed them usable endorsements. Millie-Christina's penny pamphlets sold to the general public included "certificates of eminent medical men," and so it was appropriate that one stanza of Millie-Christina's theme song chimed:

> Two heads, four arms, four feet,
> All in one perfect body meet,
> I am most wonderfully made
> All scientific men have said.

Indeed, exhibitors used the automatic respectability of the testimony of medical men to include in their penny pamphlets intimate details of the sexual anatomy of those exhibited, discourse which would have otherwise have been considered quite lewd. For example, because it was presented in the form of a straightforward quotation from two medical doctors, a report of Millie-Christina's sexual anatomy could be included in a pamphlet sold to the public. It was safe and acceptable for the lay public to read that Millie-Christina had "separate bladders, but one common vagina, one uterus to be recognized, and one perfect anus," as long as this information came from the lips of a medical doctor. Doctors thereby gentrified and legitimated a performance that might otherwise be simply distasteful. Medicine was consciously used to keep the charge of pornography at bay—in the very way I used it, unconsciously, to keep myself from having to see my own subject as anything akin to pornography.

Robert Bogdan has noted in his studies of freak shows that many so-called freaks of the nineteenth century earned quite a lot of money exhibiting themselves and were thereby able to lead financially secure lives, and the trade in medical testimonies obviously helped in this realm. The Bunkers used their substantial income from exhibitions to buy farms and slaves; as long as the farms were profitable (i.e., before the Emancipation of enslaved people), they chose not to exhibit themselves for money. According to the tales told about them, Millie and Christina eventually used the money they earned to buy the plantation on which their mother had been enslaved. Today, by contrast, although unusual anatomies are often displayed in medical circles, on television, and in the popular press, the profit of exhibition is not typically earned chiefly or directly by the exhibited, and "respectable" people are obligated, as I felt I was, to find public exhibition via venues like talk shows pathetic and distasteful. "These things are *medical problems!*" we cry. "They belong in the confines of the clinic."

Now, I am not looking to suggest that we return to the "freak show" era, nor to suggest that doctors "help" their patients by providing them with testimonies of how odd they look. We do not know whether most nineteenth-century exhibitors would have chosen this means of profit if an alternative had been available, or whether many would choose it today if it were a real option. I am hardly romantic about the great age of exhibitions. But I do find it remarkable, given where we are today, to recognize that there was a time when the doctors were quite publicly thrilled to get an audience with these patients, when they would, in very public prose, celebrate them as extraordinary, bizarre, amazing—when they would recognize these people as authorities of a unique and strangely attractive experience.

A critical shift happened subtly in the nineteenth century, as medical professionals became more prestigious and more aligned with science. Earlier physicians were quite willing to exchange concrete and enthusiastic testimonies for access to particular interesting bodies (and the stories of the experiences that came with those bodies). By contrast, later physicians offered instead the much more abstract value of "the good of humanity" in exchange for ready and unlimited access to *all* unusual bodies. We see over the course of the nineteenth century the fading of the idea that the medical or scientific man should have to actually give something immediate in exchange for access to interesting anatomies.

Today, biomedical professionals tend to feel a primary *right* to seeing and using and owning unusual anatomies—whether those anatomies be ancient skeletal remains, extraordinary genes, or patients deemed "facinomas." The right of access is claimed not because the professionals have given that particular unusual person something in return, but because, in a very abstract and universalized sense, science and medicine serve all of humanity.

One of the many "services to all humanity" that biomedical science seeks to perform is the prevention and "normalization" of congenitally unusual anatomies. This is why biomedicine gets free and easy access to those unusual anatomies. But think of the irony. People with unusual anatomies (profoundly short, intersexed, and so on) hear medical professionals saying: "We get to see you, examine you, and display you at will, because we're trying hard to fix and prevent people like you." Of course, what medicine is really trying to prevent and alleviate is the *suffering* of these kinds of people, but when the fact is—as it is for all of us—that one's identity is very grounded in the experience of one's anatomy, the elimination of the anatomical experience at some level equates to the elimination of the self.

How shocked are some doctors to hear that intersexuals might not want to be displayed so that they and others like them can be "normalized," to hear people with profound short stature say the three inches gained from human growth hormone injections might not make up for the constant negative attention to whether they measure up. How shocked are some scientists to be told that Native Americans might not be interested in hearing what "really" happened to their ancestors' bodies. And I find myself in the middle, deeply uncomfortable, not just questioning the "standards" of sex, of height, even of digit numbers, but now questioning the standards of science and medicine. What has happened here? Don't we all agree on the search for truth, the pursuit of beauty? Don't we know what is pornography and exploitation, and what is science and medicine? Don't we know what counts as the search for a fact, and what counts as a mythical quest, and what counts as the copping of a cheap thrill?

In the end, watching my own partner go through medical school and then a residency, I've come to think that the display of patients is not simply about educating physicians with regard to the medical conditions. As in the nineteenth century, a given physician's examination and display of a patient is as much about establishing that physician's place within the institutions of medicine as it is about establishing the patient's disease within the nosologies of medicine. The display of the patient is therefore necessarily about constructing the patient as the unauthoritative, needy "other."

If I am correct about this complex motivation for medical display, then however well intentioned medical professionals are—and I do believe the vast majority are very well intentioned—the practice of displaying voiceless, virtually identity-less patients to large numbers of other medical professionals will not change simply by the recognition that it may harm the patient.

The word *monster* shares the root with *demonstrate*, the *monstrous* is that which portends. The ancient reaction to "monsters" was to kill the messenger. The more recent reaction has been to paint the anatomically unusual person as unfortunate, in need of paternalistic care. In this unique age of ours, between the age of killing the messenger and preventing the messenger, in the age in which the messengers are starting to speak, to object, to engage, to be heard, I wonder what the reaction will be. I know that I have found it terrribly difficult to convince some that people with unusual anatomies not only *do* have a voice, but that they *should* have a voice, a say in the deal, the right to criticize and perhaps even dictate how they will be displayed, how the knowledge and technologies that come from the study of their bodies will be used.

I suppose for some people, what I am suggesting is the equivalent of a freak show, a show in which medicine itself becomes the subject on display to be prodded and questioned. And when one is desperately sick and frightened, in need of medical rescue, the idea of stripping medicine bare and examining *it* seems obscene indeed. It seems to add only to the pain medicine is supposed to kill. I know that the one time I lay in a hospital bed, wondering what was wrong with me, I would have happily accepted an all-powerful, all-seeing version of medicine, no matter how little of my flesh that hospital gown covered.

It was on the occasion of my first visit to the Mütter Museum [of medical history in Philadelphia], when I was just out of graduate school, that I had the revelation that jarring bodies get put in jars—contained physically and conceptually—in an attempt to contain all the confusion and pain and even thrill they cause.

During my second visit, after having talked to a lot more people living with unusual anatomies, it dawned on me how strange it would be, if I were a giant or a dwarf, a conjoined twin or a hermaphrodite, to see my kind of people displayed so that my kind of people might dwindle in number. I had a jolting daydream of a time when the "freaks" would come to repatriate their own ancestors, tired of the deal getting more and more unbalanced.

And it was during my last visit, the first time I managed to visit with Aron (now a chief resident) and his very different professional vision, that I came to understand the intense and almost universal attraction to these anatomies, these bodies that loosened all the boundaries and set us all in motion. As we both stared into the cabinets, I was not just seeing a historical artifact, and he was not just seeing a genetic anomaly, I understood finally how that attraction to their gravity has led to so many power struggles over them. I wondered to myself if the universal attraction might lead to some kind of peace. But to recognize the subject as who we might be—even who we might *want* to be—is to dissolve all the glass that separates us and let the monster out of the jar.

PART VIII

Capitalist Production, or Accounting

the Commodification of Bodily Life

Capitalist Production, or Accounting the
Commodification of Bodily Life

INTRODUCTION

The body, assisted by deliberately crafted artifacts—from hand tools to computers, air conditioners to satellites—is a principal means by which both the necessities of life and a surplus of valued goods are secured. In the process of transforming natural resources into a built world that can support collective human life, the requirements and capacities of bodies provide a quiet but continuous demand and inspiration. The human body is a fulcrum around which an economy of everyday life is enacted and communities, polities, and civilizations are formed.

Karl Marx is famous for showing how the very notion of labor and the form that the production and distribution of commodities takes were fundamentally and permanently changed as a result of industrialization and the emergence of the capitalist mode of production. His argument about the alienation experienced in the late nineteenth century, as people began to lose control of their working lives and the products of their labor, is as relevant today as it was in the nineteenth century. The flexible workforce required for the functioning of the global neoliberal economy, intimately linked with instability and exploitation, has its roots in the experience of workers who were forced off the land and into factories during the industrial revolution of the eighteenth and nineteenth centuries.

Hence the oddly familiar politics to be found in the first essay in this group, published in 1967 by the British historian E. P. Thompson. The main theme of the article—the way in which industrial capitalism transformed not only work-discipline, but also the workers' experience of time and the very rhythm of their lives, thus fundamentally altering entire communities forever—makes an excellent introduction to the lived experience of capitalist alienation and the commodification of embodied labor.

Thompson contrasts the experience of time in preindustrial "task-oriented" economies based on herding, farming, fishing, cottage industry, and small-scale manufacturing with the way in which time came to be managed in the twentieth-century factory. The transition from a preindustrial economy to one dominated by industry was not smooth, as Thompson notes, and involved many rebellions on the part of the workers as well as the evolution of a lively popular culture of dissent. He highlights the variation and irregularity of working lives associated with preindustrial society, the mixing of tasks and types of labor, the fit with the season, the weather, and so on. Thompson also notes, in a section we had to delete from this long essay, that "agricultural improvers" in the eighteenth century complained bitterly about how laborers wasted time at fairs, at the weekly market, and, of course, in the alehouses, and how these improvers made it their business to discipline the workers. Thompson picks out one form of discipline that was indispensable to successful industrialization, namely, time measurement as a means of labor exploitation. He describes the way in which an obsession about the prudent use of time— encouraged by the increasing availability of better and cheaper timepieces— quickly spread among the wealthy and educated classes in England. Not only factories but also domestic life and schools were inundated by rules for the appropriate way to pass the day and for morally correct behavior. One strength of Thompson's article is to show how these attitudes were mirrored in and promoted by the beliefs and practices espoused in the Methodist sects of the early nineteenth century, whose congregations were made up almost entirely of laborers and their families. Thompson argues that evidence of the internalization of these disciplinary practices is clear in the regularity, the "methodical paying-out of energy" on the part of working people, and by the loss of a capacity to relax in the old uninhibited ways.

The setting for the ethnography carried out by Aihwa Ong, on which her essay in this collection is based, is late twentieth-century Malaysia. Despite the vast difference in time and space location, it complements Thompson's essay in several striking ways. Ong documents how, as part of the restructuring of the political economy of Malaysia in the 1970s, a massive movement of people of all ages from rural to urban areas took place. This migration caused not only physical disruption but extreme psychological stress, made manifest in the factories she studied as bouts of spirit possession, particularly among young women. Ong documents how in factories operated by foreign corporations such outbursts were medicalized and labeled as hysteria. Affected women were taken off the shop floor and sedated while managers lamented the consequent slowing down of production. Through an elucidation of local knowledge about spirit possession and what is thought of as appropriate gendered behavior, Ong concludes that the women were responding to sensations of a loss of control over their bodies, fear of becoming polluted as a result of unhygienic factory conditions, especially the unavoidable use of Western toilets, and exposure to the indecorous social and gender relations that were customary in the factories. The women were overcome by mounting fears about moral disorder and bodily pollution as a result of their enforced loss of decorum.

There was, of course, no opportunity for open, self-conscious dissent because the livelihood of workers and their families was entirely dependent upon employment provided by the factories; even the corporeal resistance exhibited by the young women and described so graphically by Ong was not tolerated by the managerial classes—production goals inevitably come before the interests of the workers. In eighteenth- and nineteenth-century England, workers whose behavior or bodies failed to measure up to the required discipline were simply thrown out. In the multinational factories of today, dissenters are sometimes medically managed in the hope that order can be restored, and intruding spirits (an unanticipated product of the global commodification of labor) are banished in judicious deployments of "local culture."

A large corpus of anthropological literature shows that when, for political reasons, open dissent is virtually impossible, the body very frequently becomes a site for the inarticulate expression of resistance. In modern times, a biomedicine focused on bodies alone has increasingly become a hegemonic resource that can be drawn on to discipline unruly bodies in the workforce (see also part VII). This disciplining of bodies for capitalist production is not confined to the workforce and to child and slave labor, however—bodily discipline willingly embraced among the middle classes worldwide as part of aesthetic and health-related practices is deeply embedded in consumer and capitalist culture.

Marx showed how commodities appear as things-in-themselves, fetishes with the mysterious power to take on intrinsic value. The popular perception of the commodity forgets the work of production through which both use-value and exchange value are made. The remaining essays in this section focus on commodities as they are consumed, rather than on their production through labor. Brad Weiss in his phenomenological ethnography of the Haya in northwest Tanzania emphasizes the way in which commodities, when they circulate into new locations, "transform experience and in the process are themselves transformed." Weiss shows how one assumption common among theoreticians of modernization, that incorporation of societies into a capitalistic economy will bring about a "rationalization of cultural values, is by no means always borne out. Weiss also disputes the opposite assumption, articulated less often, that because local values become increasingly irrelevant, alienation is likely to result as modernization takes place. Rather than showing us East Africans alienated by global capitalist forces, he describes the symbolic and practical means through which Haya produce and reproduce embodied relations to things and each other: "lived worlds."

Weiss's essay is about an illness recognized among the Haya only two or three years before he wrote his essay in the early 1990s. The condition is limited to infants and young children whose erupting teeth are sometimes said to be made of plastic or to be like nylon. The children, if left untreated, get fevers, are unable to consume food, and suffer severe wasting. The only treatment is removal of the affected teeth. Weiss elaborates on the significance of teething in Haya society and its close association with the important ceremony of naming the child, both of which actions signify the first signs of independence from mother and the movement of the child into a wider sphere where the agency of men is dominant. He also shows how food consumption is central to the socialization of children and then makes links between the appearance of the illness of plastic teeth and that of the "wasting" disease (HIV/AIDS) among the Haya.

The final link in this fine piece of ethnographic work shows how plastic objects of many kinds have come to be important commodities in everyday life among the Haya. Plastic is known for particular qualities, not all bad, but among which "coldness" has especially negative connotations. Moreover, these goods enter into circulation in the Haya community as fully realized objects ready for use, unlike more traditional locally made goods. Plastic embodies a new form of political economy—one in which production is entirely divorced from local spatiotemporal social relations.

Weiss concludes that plastic teeth are a vivid image of the degree to which the bodies of the Haya have been infected with the power of the larger transformations they are experiencing. Plastic teeth have appeared as part of a nexus of changes—the collapse of the coffee market, the HIV/AIDS epidemic, land fragmentation, and currency devaluation—and they represent an attempt to reconfigure local values as Haya try to remake themselves in the face of these onslaughts.

The essays by Matthew Schmidt and Lisa Jean Moore and by Margaret Lock both deal with the commodification of body products, sperm and genetic material, respectively. Only with the development of biomedical technologies that facilitate the extraction, preparation, preservation, and reuse of substances like these (as well as human organs, tissue, embryos, stem cells, and so on) has the direct commodification of body parts become widely feasible. In the process embodiment has grown more hybrid, more cyborg, and more subject to modification in response to diverse desires.

The essay by Schmidt and Moore is about semen banking and the marketing of donor semen. Semen banks, as the authors show, are part of the contemporary medical-industrial complex; their existence highlights the promotional strategies used to construct public understanding of banked semen—"technosemen"—as technologically superior to natural semen. The result, they argue, contributes to the maintenance of hierarchies among men. Constructions of embodied difference among donors are inscribed in semen bank catalogues through the judicious use of metaphorical language, thus enabling a blurring of genotypic and phenotypic attributes. Race/ethnic origin is without exception the category that is listed first in the order of sperm attributes. Schmidt and Moore demonstrate how semen banks anthropomorphize the semen they sell, thus endowing technosemen with cyborg characteristics. Promotional materials depict their product to be superior to the "dirty" and "unpredictable" semen produced "naturally," even going so far as to suggest that the natural variety may well result in defective children. This discourse of risk, in which technosemen fares unnaturally well, is highly profitable for semen banks. The authors close with the question raised in connection with almost all forms of reproductive technologies: will they become the wave of the future, and if so with what impact on society at large? The commodification of eggs and sperm and the popularity of new reproductive technologies provide an excellent lens through which to view the late twentieth-century transformation of close human relations and the very idea of the family through recent technological innovations.

Lock's article is concerned with "bioprospecting," that is, the procurement of material known as "biologicals" from people labeled by scientists as indigenous or isolated. Such people, it is presumed, have unusual variations of the genes common to all humans, variations that may be of use in developing new drugs or in understanding prehistoric migrations of peoples. When blood is taken from targeted populations and converted into "immortalized cell lines," the meanings and values associated with blood are entirely transformed. Local moral orders in these processes are often put at risk. There are those for whom blood has a semi-sacred quality, represents genealogical ties, and is a substance that should not be bought or sold. However, once it is commodified and transformed into a cell line, it becomes attractive to venture capital. Some cell lines are potentially worth millions of dollars. Lock documents how the corporeal fetishism associated with the commodification of body parts, involving the reification of cells, genes, organs, and so on, obscures the dramatic transformation in value of such "products." Using the specific example of the Human Genome Diversity Project, a project that has been described by some indigenous peoples as a form of neocolonialism, Lock describes the various forms of resistance to the project exhibited by targeted populations as well as its partial acceptance by others. Examples of this kind are proliferating rapidly in all parts of the globe, resulting in legal and political challenges to new forms of "ownership" of body parts. One can only wonder what Karl Marx would have said about the commodification of human body parts; clearly his fundamental theses of alienation, reification, and commodity fetishism remain robust.

ADDITIONAL READINGS

Campbell, Colin. 1987. *The Romantic Ethic and the Spirit of Modern Consumerism*. New York: Basil Blackwell.
Foster, Robert. 2002. *Materializing the Nation: Commodities, Consumption, and Media in Papua New Guinea*. Bloomington: Indiana University Press.
Lowe, Donald. 1995. *The Body in Late-Capitalist USA*. Durham: Duke University Press.
Weiss, Brad. 2003. *Sacred Trees, Bitter Harvests: Globalizing Coffee in Northwest Tanzania*. Portsmouth, N.H.: Heinemann.

TIME, WORK-DISCIPLINE, AND

INDUSTRIAL CAPITALISM

E. P. THOMPSON

Tess . . . started on her way up the dark and crooked lane or street not made for hasty progress; a street laid out before inches of land had value, and when one-handed clocks sufficiently subdivided the day.
—Thomas Hardy

I

It is a commonplace that the years between 1300 and 1650 saw within the intellectual culture of western Europe important changes in the apprehension of time. In the *Canterbury Tales* the cock still figures in his immemorial role as nature's timepiece: Chauntecleer—

> Caste up his eyen to the brighte sonne,
> That in the signe of Taurus hadde yronne
> Twenty degrees and oon, and somwhat moore,
> He knew by kynde, and by noon oother loore
> That it was pryme, and crew with blisful stevene. . . .

But although "By nature knew he ech ascensioun / Of the equynoxial in thilke toun," the contrast between "nature's" time and clock time is pointed in the image—

> Wel sikerer was his crowyng in his logge
> Than is a clokke, or an abbey orlogge.

This is a very early clock: Chaucer (unlike Chauntecleer) was a Londoner, and was aware of the times of Court, of urban organization, and of that "merchant's time" which Jacques Le Goff, in a suggestive article in *Annales*, has opposed to the time of the medieval church.

I do not wish to argue how far the change was due to the spread of clocks from the fourteenth century onwards, how far this was itself a symptom of a new Puritan discipline and bourgeois exactitude. However we see it, the change is certainly there. The clock steps on to the Elizabethan stage, turning Faustus's last soliloquy into a dialogue with time: "the stars move still, time runs, the clock will strike." Sidereal time, which has been present since literature began, has now moved at one step from the heavens into the home. Mortality and love are both felt to be more poignant as the "Snayly motion of the mooving hand" crosses the dial. When the watch is worn about the neck it lies in proximity to the less regular beating of the heart. The conventional Elizabethan images of time as a devourer, a defacer, a bloody tyrant, a scytheman, are old enough, but there is a new immediacy and insistence.

As the seventeenth century moves on the image of clock-work extends, until, with Newton, it has engrossed the universe. And by the middle of the eighteenth century (if we are to trust Sterne) the clock had penetrated to more intimate levels. For Tristram Shandy's father—"one of the most regular men in everything he did . . . that ever lived"—"had made it a rule for many years of his life,—on the first Sunday night of every month . . . to wind up a large house-clock, which we had standing on the back-stairs head." "He had likewise gradually brought some other little family concernments to the same period," and this enabled Tristram to date his conception very exactly. It also provoked *The Clockmaker's Outcry against the Author.*

> The directions I had for making several clocks for the country are counter-manded; because no modest lady now dares to mention a word about winding-up a clock, without exposing herself to the sly leers and jokes of the family . . . Nay, the common expression of street-walkers is, "Sir, will you have your clock wound up?"

Virtuous matrons (the "clockmaker" complained) are consigning their clocks to lumber rooms as "exciting to acts of carnality."

However, this gross impressionism is unlikely to advance the present enquiry: how far, and in what ways, did this shift in time-sense affect labor discipline and how far did it influence the inward apprehension of time of working people ? If the transition to mature industrial society entailed a severe restructuring of working habits—new disciplines, new incentives, and a new human nature upon which these incentives could bite effectively—how far is this related to changes in the inward notation of time?

It is well known that among primitive peoples the measurement of time is commonly related to familiar processes in the cycle of work or of domestic chores. Evans-Pritchard has analyzed the time-sense of the Nuer:

> The daily timepiece is the cattle clock, the round of pastoral tasks, and the time of day and the passage of time through a day are to a Nuer primarily the succession of these tasks and their relation to one another.

Among the Nandi an occupational definition of time evolved covering not only each hour, but half hours of the day—at 5–30 in the morning the oxen have gone to the grazing-ground, at 6 the sheep have been unfastened, at 6–30 the sun has grown, at 7 it has become warm, at 7–30 the goats have gone to the grazing-ground, etc.—an uncommonly well-regulated economy. In a similar way terms evolve for the measurement of time intervals. In Madagascar time might be measured by "a rice-cooking" (about half an hour) or "the frying of a locust" (a moment). The Cross River natives were reported as saying "the man died in less than the time in which maize is not yet completely roasted" (less than fifteen minutes).[1]

It is not difficult to find examples of this nearer to us in cultural time. Thus in seventeenth-century Chile time was often measured in "credos": an earthquake was described in 1647 as lasting for the period of two credos; while the cooking-time of an egg could be judged by an Ave Maria said aloud. In Burma in recent times monks rose at daybreak "when there is light enough to see the veins in the hand." The *Oxford English Dictionary* gives us English examples— "pater noster wyle," "miserere whyle" (1450) and (in the *New English Dictionary* but not the *Oxford English Dictionary*) "pissing while"—a somewhat arbitrary measurement.

Pierre Bourdieu has explored more closely the attitudes towards time of the Kabyle peasant (in Algeria) in recent years: "An attitude of submission and of nonchalant indifference to the passage of time which no one dreams of mastering, using up, or saving . . . Haste is seen as a lack of decorum combined with diabolical ambition." The clock is sometimes known as "the devil's mill"; there are no precise meal-times; "the notion of an exact appointment is unknown; they agree only to meet 'at the next market.' " A popular song runs: "It is useless to pursue the world. No one will ever overtake it."[2]

Synge, in his well-observed account of the Aran Islands, gives us a classic example:

While I am walking with Michael someone often comes to me to ask the time of day. Few of the people, however, are sufficiently used to modern time to understand in more than a vague way the convention of the hours and when I tell them what o'clock it is by my watch they are not satisfied, and ask how long is left them before the twilight. . . .

The general knowledge of time on the island depends, curiously enough, upon the direction of the wind. Nearly all the cottages are built . . . with two doors opposite each other, the more sheltered of which lies open all day to give light to the interior. If the wind is northerly the south door is opened, and the shadow of the door-post moving across the kitchen floor indicates the hour, as soon, however, as the wind changes to the south the other door is opened, and the people, who never think of putting up a primitive dial, are at a loss. . . .

When the wind is from the north the old woman manages my meals with fair regularity, but on the other days she often makes my tea at three o'clock instead of six. . . ."

Such a disregard for clock time could of course only be possible in a crofting and fishing community whose framework of marketing and administration is minimal, and in which the day's tasks (which might vary from fishing to farming, building, mending of nets, thatching, making a cradle or a coffin) seem to disclose themselves, by the logic of need, before the crofter's eyes.[3] But his account will serve to emphasize the essential conditioning in differing notations of time provided by different work situations and their relation to "natural" rhythms. Clearly hunters must employ certain hours of the night to set their snares. Fishing and seafaring people must integrate their lives with the tides. A petition from Sunderland in 1800 includes the words "considering that this is a seaport in which many people are obliged to be up at all hours of the night to attend the tides and their affairs upon the river." The operative phrase is "attend the tides": the patterning of social time in the seaport follows *upon* the rhythms of the sea; and this appears to be natural and comprehensible to fishermen or seamen: the compulsion is nature's own. . . .

The notation of time which arises in such contexts has been described as task-orientation. It is perhaps the most effective orientation in peasant societies, and it remains important in village and domestic industries. It has by no means lost all relevance in rural parts of Britain today. Three points may be proposed about task-orientation. First, there is a sense in which it is more humanly comprehensible than timed labor. The peasant or laborer appears to attend upon what is an observed necessity. Second, a community in which task-orientation is common appears to show least demarcation between "work" and "life." Social intercourse and labor are intermingled—the working-day lengthens or contracts according to the task—and there is no great sense of conflict between labor and "passing the time of day." Third, to men accustomed to labor timed by the clock, this attitude to labor appears to be wasteful and lacking in urgency.[4]

Such a clear distinction supposes, of course, the independent peasant or craftsman as referent. But the question of task-orientation becomes greatly more complex at the point where labor is employed. The entire family economy of the small farmer may be task-orientated; but within it there may be a division of labor, and allocation of roles, and the discipline of an employer-employed relationship between the farmer and his children. Even here time is beginning to become money, the employer's money. As soon as actual hands are employed the shift from task-orientation to timed labor is marked. . . .

Attention to time in labor depends in large degree upon the need for the synchronization of labor. But insofar as manufacturing industry remained conducted upon a domestic or small workshop scale, without intricate sub-division of processes, the degree of synchronization demanded was slight, and task-orientation was still prevalent. The putting-out system demanded much fetching, carrying, waiting for materials. Bad weather could disrupt not only agriculture, building and transport, but also weaving, where the finished pieces had to be stretched on the tenters to dry. As we get closer to each task, we are surprised to find the multiplicity of subsidiary tasks which the same worker or family group must do in one cottage or workshop. Even in larger workshops men sometimes continued to work at distinct tasks at their own benches or looms, and—except where the fear of the embezzlement of materials imposed stricter supervision—could show some flexibility in coming and going.

Hence we get the characteristic irregularity of labor patterns before the coming of large-scale machine-powered industry. Within the general demands of the week's or fortnight's tasks—the piece of cloth, so many nails or pairs of shoes—the working day might be lengthened or shortened. Moreover, in the early development of manufacturing industry, and of mining, many mixed occupations survived: Cornish tinners who also took a hand in the pilchard fishing; Northern lead-miners who were also smallholders; the village craftsmen who turned their hands to various jobs, in building, carting, joining; the domestic workers who left their work for the harvest; the Pennine small-farmer/weaver. . . .

It is true that the transition to mature industrial society demands analysis in sociological as well as economic terms. Concepts such as "time-preference" and the "backward sloping labor supply curve" are, too often, cumbersome attempts to find economic terms to describe sociological problems. But, equally, the attempt to provide simple models for one single, supposedly neutral, technologically determined process known as "industrialization" (so popular today among well-established sociological circles in the United States) is also suspect. It is not only that the highly developed and technically alert manufacturing industries (and the way of life supported by them) of France or England in the eighteenth century can only by semantic torture be described as "pre-industrial." (And such a description opens the door to endless false analogies between societies at greatly differing economic levels.) It is also that there has never been any single type of "the transition." The stress of the transition falls upon the whole culture: resistance to change and assent to change arise from the whole culture. And this culture includes the systems of power, property-relations, religious institutions, etc., inattention to which merely flattens phenomena and trivializes analysis. Above all, the transition is not to "industrialism" *tout court* but to industrial capitalism or (in the twentieth century) to alternative systems whose features are still indistinct. What we are examining here are not only changes in manufacturing technique which demand greater synchronization of labor and a greater exactitude in time-routines in *any* society; but also these changes as they were lived through in the society of nascent industrial capitalism. We are concerned simultaneously with time-sense in its technological conditioning, and with time-measurement as a means of labor exploitation.

There are reasons why the transition was peculiarly protracted and fraught with conflict in England: among those which are often noted, England's was the first industrial revolution, and there were no Cadillacs, steel mills, or television sets to serve as demonstrations as to the object of the operation. Moreover, the preliminaries to the industrial revolution were so long that, in the manufacturing districts in the early eighteenth century, a vigorous and licensed popular culture had evolved, which the propagandists of discipline regarded with dismay. Josiah Tucker, the dean of Gloucester, declared in 1745 that "the *lower* class of people" were utterly degenerated. Foreigners (he sermonized) found "the *common people* of our *populous cities* to be the most *abandoned*, and *licentious* wretches on earth":

Such brutality and insolence, such debauchery and extravagance, such idleness, irreligion, cursing and swearing, and contempt of all rule and authority . . . Our people are *drunk with the cup of liberty*.[5]

The irregular labor rhythms described in the previous section help us to understand the severity of mercantilist doctrines as to the necessity for holding down wages as a preventative against idleness, and it would seem to be not until the second half of the eighteenth century that "normal" capitalist wage incentives begin to become widely effective. The confrontations over discipline have already been examined by others. My intention here is to touch upon several points which concern time-discipline more particularly. The first is found in the extraordinary Law Book of the Crowley Iron Works. Here, at the very birth of the large-scale unit in manufacturing industry, the old autocrat Crowley, found it necessary to design an entire civil and penal code, running to more than 100,000 words, to govern and regulate his refractory labor-force. The preambles to Orders Number 40 (the Warden at the Mill) and 103 (Monitor) strike the prevailing note of morally-righteous invigilation. From Order 40:

I having by sundry people working by the day with the connivence of the clerks been horribly cheated and paid for much more time than in good conscience I ought and such hath been the baseness & treachery of sundry clerks that they have concealed the sloath & negligence of those paid by the day. . . .

And from Order 103:

Some have pretended a sort of right to loyter, thinking by their readiness and ability to do sufficient in less time than others. Others have been so foolish to think bare attendance without being imployed in business is sufficient. . . . Others so impudent as to glory in their villany and upbrade others for their diligence. . . .

To the end that sloath and villany should be detected and the just and diligent rewarded, I have thought meet to create an account of time by a Monitor, and do order and it is hereby ordered and declared from 5 to 8 and from 7 to 10 is fifteen hours, out of which take 1–1/2 for breakfast, dinner, etc. There will then be thirteen hours and a half neat service. . . .

This service must be calculated "after all deductions for being at taverns, alehouses, coffee houses, breakfast, dinner, playing, sleeping, smoaking, singing, reading of news history, quarelling, contention, disputes or anything foreign to my business, any way loitering."

The Monitor and Warden of the Mill were ordered to keep for each day employee a time-sheet, entered to the minute, with "Come" and "Run." In the Monitor's Order, verse 31 (a later addition), declares:

And whereas I have been informed that sundry clerks have been so unjust as to reckon by clocks going the fastest and the bell ringing before the hour for their going from business, and clocks going too slow and the bell ringing after the hour for their coming to business, and those two black traitors Fowell and Skellerne have knowingly allowed the same, it is therefore ordered that no person upon the account doth reckon by any other clock, bell, watch or dyall but the Monitor's, which clock is never to be altered but by the clock-keeper. . . .

The Warden of the Mill was ordered to keep the watch "so locked up that it may not be in the power of any person to alter the same." His duties also were defined in verse 8:

Every morning at 5 a clock the Warden is to ring the bell for beginning to work, at eight a clock for breakfast, at half an hour after for work again, at twelve a clock for dinner, at one to work and at eight to ring for leaving work and all to be lock'd up.

His book of the account of time was to be delivered in every Tuesday with the following affidavit:

This account of time is done without favour or affection, ill-will or hatred, & do really believe the persons above mentioned have worked in the service of John Crowley Esq the hours above charged.

We are entering here, already in 1700, the familiar landscape of disciplined industrial capitalism, with the time-sheet, the time-keeper, the informers and the fines. Some seventy years later the same discipline was to be imposed in the early cotton mills (although the machinery itself was a powerful supplement to the time-keeper). Lacking the aid of machinery to regulate the pace of work on the pot-bank, that supposedly formidable disciplinarian Josiah Wedgwood was reduced to enforcing discipline upon the potters in surprisingly muted terms. The duties of the Clerk of the Manufactory were:

To be at the works the first in the morning, & settle the people to their business as they come in,—to encourage those who come regularly to their time, letting them know that their regularity is properly noticed, & distinguishing them by repeated marks of approbation, from the less orderly part of the workpeople, by presents or other marks suitable to their ages, &c

Those who come later than the hour appointed should be noticed, and if after repeated marks of disapprobation they do not come in due time, an account of the time they are deficient in should be taken, and so much of their wages stopt as the time comes to if they work by wages, and if they work by the piece they should after frequent notice be sent back to breakfast-time.

These regulations were later tightened somewhat:

Any of the workmen forceing their way through the Lodge after the time alow'd by the Master forfeits 2/-d.

and McKendrick has shown how Wedgwood wrestled with the problem at Etruria and introduced the first recorded system of clocking-in. But it would seem that once the strong presence of Josiah himself was withdrawn the incorrigible potters returned to many of their older ways.

It is too easy, however, to see this only as a matter of factory or workshop discipline, and we may glance briefly at the attempt to impose "time-thrift" in the domestic manufacturing districts, and its impingement upon social and domestic life. Almost all that the masters *wished* to see imposed may be found in the bounds of a single pamphlet, the Rev. J. Clayton's *Friendly Advice to the Poor*, "written and published at the Request of the late and present Officers of the Town of Manchester" in 1755. "If the *sluggard hides his hands* in his bosom, rather than applies them to work; if he spends his Time in Sauntring, impairs his Constitution by Laziness, and dulls his Spirit by Indolence . . ." then he can expect only poverty as his reward. The laborer must not loiter idly in the market-place or waste time in marketing. Clayton complains that "the Churches and Streets [are] crowded with Numbers of Spectators" at weddings and funerals, "who in spight of the Miseries of their Starving Condition . . . make no Scruple of wasting the best Hours in the Day, for the sake of gazing. . . ." The tea-table is "this shameful devourer of Time and Money." So also are wakes and holidays and the annual feasts of friendly societies. So also is "that slothful spending the Morning in Bed":

The necessity of early rising would reduce the poor to a necessity of going to Bed betime; and thereby prevent the Danger of Midnight revels.

Early rising would also "introduce an exact Regularity into their Families, a wonderful Order into their Oeconomy."

The catalogue is familiar, and might equally well be taken from Baxter in the previous century. If we can trust Bamford's *Early Days*, Clayton failed to make many converts from their old way of life among the weavers. Nevertheless, the long dawn chorus of moralists is prelude to the quite sharp attack upon popular customs, sports, and holidays which was made in the last years of the eighteenth century and the first years of the nineteenth.

One other non-industrial institution lay to hand which might be used to inculcate "time-thrift": the school. Clayton complained that the streets of Manchester were full of "idle ragged children; who are not only losing their Time, but learning habits of gaming," etc. He praised charity schools as teaching Industry, Frugality, Order and Regularity: "the Scholars here are obliged to rise betimes and to observe Hours with great Punctuality."[7] William Temple, when advocating, in 1770, that poor children be sent at the age of four to workhouses where they should be employed in manufactures and given two hours' schooling a day, was explicit about the socializing influence of the process:

> There is considerable use in their being, somehow or other, constantly employed at least twelve hours a day, whether they earn their living or not, for by these means, we hope that the rising generation will be so habituated to constant employment that it would at length prove agreeable and entertaining to them. . . .

Powell, in 1772, also saw education as a training in the "habit of industry"; by the time the child reached six or seven it should become "habituated, not to say naturalized to Labor and Fatigue." The Rev. William Turner, writing from Newcastle in 1786, recommended Raikes' schools as "a spectacle of order and regularity," and quoted a manufacturer of hemp and flax in Gloucester as affirming that the schools had effected an extraordinary change: "they are . . . become more tractable and obedient, and less quarrelsome and revengeful." Exhortations to punctuality and regularity are written into the rules of all the early schools:

> Every scholar must be in the school-room on Sundays, at nine o'clock in the morning, and at half-past one in the afternoon, or she shall lose her place the next Sunday, and walk last.

Once within the school gates, the child entered the new universe of disciplined time. At the Methodist Sunday Schools in York the teachers were fined for unpunctuality. The first rule to be learned by the scholars was:

> I am to be present at the School . . . a few minutes before half-past nine o'clock. . . .

Once in attendance, they were under military rule:

> The Superintendent shall again ring,—when, on a motion of his hand, the whole School rise at once from their seats;—on a second motion, the Scholars turn;—on a third, slowly and silently move to the place appointed to repeat their lessons,—he then pronounces the word "Begin." . . .

The onslaught, from so many directions, upon the people's old working habits was not, of course, uncontested. In the first stage, we find simple resistance. But, in the next stage, as the new time-discipline is imposed, so the workers begin to fight, not against time, but about it. The evidence here is not wholly clear. But in the better-organized artisan trades, especially in London, there is no doubt that hours were progressively shortened in the eighteenth century as combination advanced. Lipson cites the case of the London tailors whose hours were shortened in 1721, and again in 1768: on both occasions the mid-day intervals allowed for dinner and drinking were also shortened—the day was compressed. By the end of the eighteenth century there is some evidence that some favored trades had gained something like a ten-hour day.

Such a situation could only persist in exceptional trades and in a favorable labor market. A reference in a pamphlet of 1827 to "the English system of working from 6 o'clock in the morning to 6 in the evening" may be a more reliable indication as to the general expectation as to hours of the mechanic and artisan outside London in the 1820s. In the dishonorable trades and outwork industries hours (when work was available) were probably moving the other way.

It was exactly in those industries—the textile mills and the engineering workshops—where the new time-discipline was most rigorously imposed that the contest over time became most intense. At first some of the worst masters attempted to expropriate the workers of all knowledge of time. "I worked at Mr. Braid's mill," declared one witness:

> There we worked as long as we could see in summer time, and I could not say at what hour it was that we stopped. There was nobody but the master and the master's son who had a watch, and we did not know the time. There was one man who had a watch . . . It was taken from him and given into the master's custody because he had told the men the time of day.

A Dundee witness offers much the same evidence:

> . . . in reality there were no regular hours: masters and managers did with us as they liked. The clocks at the factories were often put forward in the

morning and back at night, and instead of being instruments for the measurement of time, they were used as cloaks for cheatery and oppression.
Though this was known amongst the hands, all were afraid to speak, and a
workman then was afraid to carry a watch, as it was no uncommon event to
dismiss any one who presumed to know too much about the science of
horology.

Petty devices were used to shorten the dinner hour and to lengthen the day.
"Every manufacturer wants to be a gentleman at once," said a witness before
Sadler's Committee:

> and they want to nip every corner that they can, so that the bell will ring to
> leave off when it is half a minute past time, and they will have them in about
> two minutes before time . . . If the clock is as it used to be, the minute hand
> is at the weight, so that as soon as it passes the point of gravity, it drops three
> minutes all at once, so that it leaves them only twenty-seven minutes,
> instead of thirty.

A strike-placard of about the same period from Todmorden put it more bluntly:
"if that piece of dirty suet, 'old Robertshaw's engine-tenter,' do not mind his
own business, and let ours alone, we will shortly ask him how long it is since he
received a gill of ale for running 10 minutes over time." The first generation of
factory workers were taught by their masters the importance of time; the
second generation formed their short-time committees in the ten-hour movement; the third generation struck for overtime or time-and-a-half. They had
accepted the categories of their employers and learned to fight back within
them. They had learned their lesson that time is money, only too well.

IV

We have seen, so far, something of the external pressures which enforced this
discipline. But what of the internalization of this discipline? How far was it
imposed, how far assumed? We should, perhaps, turn the problem around
once again, and place it within the evolution of the Puritan ethic. One cannot
claim that there was anything radically new in the preaching of industry or in
the moral critique of idleness. But there is perhaps a new insistence, a firmer
accent, as those moralists who had accepted this new discipline for themselves
enjoined it upon the working people. Long before the pocket watch had come
within the reach of the artisan, Baxter and his fellows were offering to each
man his own interior moral time-piece. Thus Baxter, in his *Christian Directory*,

plays many variations on the theme of Redeeming the Time: "use every minute of it as a most precious thing, and spend it wholly in the way of duty." The imagery of time as currency is strongly marked, but Baxter would seem to have an audience of merchants and of tradesmen in his mind's eye:

> Remember how gainful the Redeeming of Time is . . . in Merchandize, or any trading; in husbandry or any gaining course, we use to say of a man that hath grown rich by it, that he hath made use of his Time.

Oliver Heywood, in *Youth's Monitor* (1689), is addressing the same audience:

> Observe exchange-time, look to your markets; there are some special seasons, that will favour you in expediting your business with facility and success; there are nicks of time, in which, if your actions fall, they may set you forward apace: seasons of doing or receiving good last not always; the fair continues not all the year . . .

The moral rhetoric passes swiftly between two poles. On the one hand, apostrophes to the brevity of the mortal span, when placed beside the certainty of Judgement. Thus Heywood's *Meetness for Heaven* (1690):

> Time lasts not, but floats away apace; but what is everlasting depends upon it. In this world we either win or lose eternal felicity. The great weight of eternity hangs on the small and brittle thread of life . . . This is our working day, our market time . . . o Sirs, sleep now, and awake in hell, whence there is no redemption.

Or, from *Youth's Monitor* again: time "is too precious a commodity to be undervalued . . . This is the golden chain on which hangs a massy eternity; the loss of time is unsufferable, because irrecoverable." Or from Baxter's *Directory*:

> O where are the brains of those men, and of what metal are their hardened hearts made, that can idle and play away that Time, that little Time, that only Time, which is given them for the everlasting saving of their souls.

On the other hand, we have the bluntest and most mundane admonitions on the husbandry of time. Thus Baxter, in *The Poor Man's Family Book* advises: "Let the time of your Sleep be so much only as health requireth; For precious time is not to be wasted in unnecessary sluggishness": "quickly dress you"; "and follow your labors with constant diligence." Both traditions were extended, by way of Law's *Serious Call*, to John Wesley. The very name of "the Methodists" emphasizes this husbandry of time. In Wesley also we have these two extremes—the jabbing at the nerve of mortality, the practical homily. It was the first (and not hell-fire terrors) which sometimes gave an hysterical edge to his sermons, and brought converts to a sudden sense of sin. He also continues the time-as-currency imagery, but less explicitly as merchant or market-time:

> See that ye walk circumspectly, says the Apostle . . . redeeming the time; saving all the time you can for the best purposes, buying up every fleeting moment out of the hands of sin and Satan, out of the hands of sloth, ease, pleasure, worldly business . . .

Wesley, who never spared himself, and until the age of eighty rose every day at 4 A.M. (he ordered that the boys at Kingswood School must do the same), published in 1786 as a tract his sermon on *The Duty and Advantage of Early Rising*: "By *soaking* . . . so long between warm sheets, the flesh is as it were parboiled, and becomes soft and flabby. The nerves, in the mean time, are quite unstrung." This reminds us of the voice of Isaac Watts' Sluggard. Wherever Watts looked in nature, the "busy little bee" or the sun rising at his "proper hour," he read the same lesson for unregenerate man. Alongside the Methodists, the Evangelicals took up the theme. Hannah More contributed her own imperishable lines on "Early Rising":

> Thou silent murderer. Sloth, no more
> My mind imprison'd keep,
> Nor let me waste another hour
> With thee, thou felon Sleep.

In one of her tracts. *The Two Wealthy Farmers*, she succeeds in bringing the imagery of time-as-currency into the labor-market:

> When I call in my laborers on a Saturday night to pay them, it often brings to my mind the great and general day of account, when I, and you, and all of us, shall be called to our grand and awful reckoning . . . When I see that one of my men has failed of the wages he should have received, because he has been idling at a fair; another has lost a day by a drinking-bout . . . I cannot help saying to myself, Night is come; Saturday night is come. No repentance or diligence on the part of these poor men can now make a bad week's work good. This week is gone into eternity.

Long before the time of Hannah More, however, the theme of the zealous husbandry of time had ceased to be particular to the Puritan, Wesleyan, or Evangelical traditions. It was Benjamin Franklin, who had a life-long technical interest in clocks and who numbered among his acquaintances John White-hurst of Derby, the inventor of the "tell-tale" clock, who gave to it its most unambiguous secular expression:

> Since our Time is reduced to a Standard, and the Bullion of the Day minted out into Hours, the Industrious know how to employ every Piece of Time to a real Advantage in their different Professions: And he that is prodigal of his Hours, is, in effect, a Squanderer of Money. I remember a notable Woman, who was fully sensible of the intrinsic Value of *Time*. Her Husband was a Shoemaker, and an excellent Craftsman, but never minded how the Minutes passed. In vain did she inculcate to him. *That Time is Money.* He had too much Wit to apprehend her, and it prov'd his Ruin. When at the Alehouse among his idle Companions, if one remark'd that the Clock struck Eleven, *What is that,* says he, *among us all?* If she sent him Word by the Boy, that it had struck Twelve; *Tell her to be easy, it can never be more.* If, that it had struck One, *Bid her be comforted, for it can never be less.*

The reminiscence comes directly out of London (one suspects) where Franklin worked as a printer in the 1730s—but never, he reassures us in his *Autobiography*, following the example of his fellow-workers in keeping Saint Monday. It is, in some sense, appropriate that the ideologist who provided Weber with his central text in illustration of the capitalist ethic should come, not from that Old World, but from the New—the world which was to invent the time-recorder, was to pioneer time-and-motion study, and was to reach its apogee with Henry Ford.

v

In all these ways—by the division of labor; the supervision of labor; fines; bells and clocks; money incentives; preachings and schoolings; the suppression of fairs and sports—new labor habits were formed, and a new time-discipline was imposed. It sometimes took several generations (as in the Potteries), and we may doubt how far it was ever fully accomplished: irregular labor rhythms were perpetuated (and even institutionalized) into the present century, notably in London and in the great ports.

Throughout the nineteenth century the propaganda of time-thrift continued to be directed at the working people, the rhetoric becoming more debased, the apostrophes to eternity becoming more shop-soiled, the homilies more mean and banal. . . . But how far did this propaganda really succeed?

How far are we entitled to speak of any radical restructuring of man's social nature and working habits? I have given elsewhere some reasons for supposing that this discipline was indeed internalized, and that we may see in the Methodist sects of the early nineteenth century a figuration of the psychic crisis entailed. Just as the new time-sense of the merchants and gentry in the Renaissance appears to find one expression in the heightened awareness of mortality, so, one might argue, the extension of this sense to the working people during the industrial revolution (together with the hazard and high mortality of the time) helps to explain the obsessive emphasis upon death in sermons and tracts whose consumers were among the working-class. Or (from a positive stand-point) one may note that as the industrial revolution proceeds, wage incentives and expanding consumer drives—the palpable rewards for the productive consumption of time and the evidence of new "predictive" attitudes to the future—are evidently effective. By the 1830s and 1840s it was commonly observed that the English industrial worker was marked off from his fellow Irish worker, not by a greater capacity for hard work, but by his regularity, his methodical paying-out of energy, and perhaps also by a repression, not of enjoyments, but of the capacity to relax in the old, uninhibited ways. . . .

Mature industrial societies of all varieties are marked by time-thrift and by a clear demarcation between "work" and "life." But, having taken the problem so far, we may be permitted to moralize a little, in the eighteenth-century manner, ourselves. The point at issue is not that of the "standard-of-living." If the theorists of growth wish us to say so, then we may agree that the older popular culture was in many ways otiose, intellectually vacant, devoid of quickening, and plain bloody poor. Without time-discipline we could not have the insistent energies of industrial man; and whether this discipline comes in the forms of Methodism, or of Stalinism, or of nationalism, it will come to the developing world.

What needs to be said is not that one way of life is better than the other, but that this is a place of the most far-reaching conflict; that the historical record is not a simple one of neutral and inevitable technological change, but is also one of exploitation and of resistance to exploitation; and that values stand to be lost as well as gained.

NOTES

1. E. E. Evans-Pritchard, *The Nuer* (Oxford: Oxford University Press, 1940), 100–104.
2. Pierre Bourdieu, "The Attitude of the Algerian Peasant Toward Time," in *Mediterranean Countrymen*, ed. Julian Pitt-Rivers (Paris: Mouton, 1963), 55–72.

3. J. M. Synge, *Plays, Poems, and Prose* (London: Dent, 1941), 115–16.

4. Henri Lefebvre, *Critique de la Vie Quotidienne* (Paris: Arche, 1958), ii, 52–56.

5. Josiah Tucker, *Six Sermons* (Bristol: S. Farley, 1772), 70–71.

6. M. W. Flinn, ed., *The Law Book of the Crowley Ironworks* (Durham: Andrews, 1957).

7. J. Clayton, *Friendly Advice to the Poor* (Manchester, 1755).

THE PRODUCTION OF POSSESSION: SPIRITS AND THE MULTINATIONAL CORPORATION IN MALAYSIA

AIHWA ONG

The sanitized environments maintained by multinational corporations in Malaysian "free trade zones" are not immune to sudden spirit attacks on young female workers. Ordinarily quiescent, Malay factory women who are seized by vengeful spirits explode into demonic screaming and rage on the shop floor. Management responses to such unnerving episodes include isolating the possessed workers, pumping them with Valium, and sending them home. Yet a Singapore doctor notes that "a local medicine man can do more good than tranquilizers." Whatever healing technique used, the cure is never certain, for the Malays consider spirit possession an illness that afflicts the soul (*jiwa*). This paper will explore how the reconstitution of illness, bodies, and consciousness is involved in the deployment of healing practices in multinational factories....

I believe that the most appropriate way to deal with spirit visitations in multinational factories is to consider them as part of a "complex negotiation of reality" by an emergent female industrial workforce.[1] Hailing from peasant villages, these workers can be viewed as neophytes in a double sense: as young female adults and as members of a nascent proletariat.

ECONOMIC DEVELOPMENT AND A MEDICAL MONOLOGUE ON MADNESS

As recently as the 1960s, most Malays in Peninsular Malaysia lived in rural *kampung* (villages), engaged in cash cropping or fishing. In 1969, spontaneous outbreaks of racial rioting gave expression to deep-seated resentment over the distribution of power and wealth in this multiethnic society. The Malay-dominated government responded to this crisis by introducing a New Economic Policy intended to "restructure" the political economy. From the early 1970s onward, agricultural and industrialization programs induced the large-scale influx of young rural Malay men and women to enter urban schools and manufacturing plants set up by multinational corporations.

Before the current wave of industrial employment for young single women, spirit possession was mainly manifested by married women, given the particular stresses of being wives, mothers, widows, and divorcees. With urbanization and industrialization, spirit possession became overnight the affliction of young, unmarried women placed in modern organizations, drawing the attention of the press and the scholarly community.

In 1971, seventeen cases of "epidemic hysteria" among schoolgirls were reported, coinciding with the implementation of government policy. This dramatic increase, from twelve cases reported for the entire decade of the 1960s, required an official response. Teoh, a professor of psychology, declared that "epidemic hysteria was not caused by offended spirits but by interpersonal tensions within the school or hostel." Teoh and his colleagues investigated a series of spirit incidents in a rural Selangor school, which they attributed to conflicts between the headmaster and female students. The investigators charged that in interpreting the events as "spirit possession" rather than the symptoms of local conflict, the *bomoh* (spirit healer) by "this devious path . . . avoided infringing on the taboos and sensitivities of the local community." Teoh had found it necessary to intervene by giving the headmaster psychotherapeutic counseling. Thus, spirit incidents in schools occasioned the introduction of a cosmopolitan therapeutic approach whereby rural Malays were "told to accept the . . . change from their old superstitious beliefs to contemporary scientific knowledge."

This dismissal of Malay interpretation of spirit events by Western-trained professionals became routine with the large-scale participation of Malays in capitalist industries. Throughout the 1970s, free-trade zones were established to encourage investments by Japanese, American, and European corporations for setting up plants for offshore production. In seeking to cut costs further, these corporations sought young, unmarried women as a source of cheap and easily controlled labor. This selective labor demand, largely met by kampung society, produced in a single decade a Malay female industrial labor force of over forty-seven thousand. Malay female migrants also crossed the Causeway in the thousands to work in multinational factories based in Singapore.

In a 1978 paper entitled "How to Handle Hysterical Factory Workers" in Singapore, Dr. P. K. Chew complained that "this psychological aberration interrupts production, and can create hazards due to inattention to machinery and careless behaviour." He classified "mass hysteria" incidents according to "frightened" and "seizure" categories, and recommended that incidents of either type should be handled "like an epidemic disease of bacteriological origin." In a Ministry of Labor survey of "epidemic hysteria" incidents in

Singapore-based factories between 1973 and 1978, W. H. Phoon also focused on symptoms ranging from "hysterical seizures" and "trance states" to "frightened spells." The biomedical approach called for the use of sedatives, "isolation" of "infectious" cases, "immunization" of those susceptible to the "disease," and keeping the public informed about the measures taken. Both writers, in looking for an explanation for the outbreak of "epidemic/mass hysteria" among Malay women workers, maintained that "the preference of belief in spirits and low educational level of the workers are obviously key factors." An anthropological study of spirit incidents in a Malacca shoe factory revealed that managers perceived the "real" causes of possession outbreaks to be physical (under-nourishment) and psychological (superstitious beliefs).[2]

These papers on spirit possession episodes in modern organizations adopt the assumptions of medical science, which describe illnesses independent of their local meanings and values. "Mass hysteria" is attributed to the personal failings of the afflicted, and native explanations are denigrated as "super-stitious beliefs" from a worldview out of keeping with the modern setting and pace of social change. "A monologue of reason about madness" was thereby introduced into Malaysian society, coinciding with a shift of focus from the afflicted to their chaotic effects on modern institutions.[3] We will need to recover the Malays' worldview in order to understand their responses to social situations produced by industrialization.

SPIRIT BELIEFS AND WOMEN IN MALAY CULTURE

Spirit beliefs in rural Malay society, overlaid but existing within Islam, are part of the indigenous worldview woven from strands of animistic cosmology and Javanese, Hindu, and Muslim cultures. In Peninsular Malaysia, the super-natural belief system varies according to the historical and local interactions between folk beliefs and Islamic teachings. Local traditions provide conceptual coherence about causation and well-being to village Malays. Through the centuries, the office of the *bomoh*, or practitioner of folk medicine, has been the major means by which these old traditions of causation, illness, and health have been transmitted. In fulfilling the pragmatic and immediate needs of everyday life, the beliefs and practices are often recast in "Islamic" terms.

I am mainly concerned here with the folk model in Sungai Jawa (a pseudo-nym), a village based in Kuala Langat district, rural Selangor, where I con-ducted fieldwork in 1979–80. Since the 1960s, the widespread introduction of Western medical practices and an intensified revitalization of Islam have made spirit beliefs publicly inadmissible. Nevertheless, spirit beliefs and practices are still very much in evidence. Villagers believe that all beings have spiritual essence (*semangati*) but, unlike humans, spirits (*ihantu*) are disembodied beings capable of violating the boundaries between the material and supernatural worlds: invisible beings unbounded by human rules, spirits come to represent transgressions of moral boundaries, which are socially defined in the concentric spaces of homestead, village, and jungle. This scheme roughly coincides with Malay concepts of emotional proximity and distance, and the related dimensions of reduced moral responsibility as one moves from the interior space of household, to the intermediate zone of relatives, and on to the external world of strangers.

The two main classes of spirits recognized by Malays reflect this interior-exterior social/spatial divide: spirits associated with human beings, and the "free" disembodied forms. In Sungai Jawa, *toyol* are the most common familiar spirits, who steal in order to enrich their masters. Accusations of breeding toyol provide the occasion for expressing resentment against economically successful villagers. Birth demons are former human females who died in childbirth and, as *pontianak*, threaten newly born infants and their mothers. Thus, spirit beliefs reflect everyday anxieties about the management of social relations in village society.

It is free spirits that are responsible for attacking people who unknowingly step out of the Malay social order. Free spirits are usually associated with special objects or sites (*keramat*) marking the boundary between human and natural spaces. These include (1) the burial grounds of aboriginal and animal spirits, (2) strangely shaped rocks, hills, or trees associated with highly revered ancestral figures (*datuk*), and (3) animals like were-tigers. As the gatekeepers of social boundaries, spirits guard against human transgressions into amoral spaces. Such accidents require the mystical qualities of the bomoh to readjust spirit relations with the human world.

From Islam, Malays have inherited the belief that men are more endowed with *akai* (reason) than women, who are overly influenced by *hawa nafsu* (human lust). A susceptibility to imbalances in the four humoral elements renders women spiritually weaker than men. Women's hawa nafsu nature is believed to make them especially vulnerable to *latah* (episodes during which the victim breaks out into obscene language and compulsive, imitative behavior) and to spirit attacks (spontaneous episodes in which the afflicted one screams, hyperventilates, or falls down in a trance or a raging fit). However, it is Malay spirit beliefs that explain the transgressions whereby women (more likely than men) become possessed by spirits (*kena hantu*). Their spiritual frailty, polluting bodies, and erotic nature make them especially likely to transgress moral space, and therefore permeable by spirits.

Mary Douglas has noted that taboos operate to control threats to social boundaries.[4] In Malay society, women are hedged in by conventions that keep them out of social roles and spaces dominated by men. Although men are also vulnerable to spirit attacks, women's spiritual, bodily, and social selves are especially offensive to sacred spaces, which they trespass at the risk of inviting spirit attacks. . . .

Until recently, unmarried daughters, most hedged in by village conventions, seem to have been well protected from spirit attack. Nubile girls take special care over the disposal of their cut nails, fallen hair, and menstrual rags, since such materials may fall into ill-wishers' hands and be used for black magic. Menstrual blood is considered dirty and polluting,[5] and the substance most likely to offend keramat spirits. This concern over bodily boundaries is linked to notions about the vulnerable identity and status of young unmarried women. It also operates to keep pubescent girls close to the homestead and on well-marked village paths. In Sungai Jawa, a schoolgirl who urinated on an ant-hill off the beaten track became possessed by a "male" spirit. Scheper-Hughes and Lock remark that when the social norms of small, conservative peasant communities are breached, we would expect to see a "concern with the penetration and violation of bodily exits, entrances and boundaries."[6] Thus, one suspects that when young Malay women break with village traditions, they may come under increased spirit attacks as well as experience an intensified social and bodily vigilance.

Since the early 1970s, when young peasant women began to leave the kampung and enter the unknown worlds of urban boarding schools and foreign factories, the incidence of spirit possession seems to have become more common among them than among married women. I maintain that like other cultural forms, spirit possession incidents may acquire new meanings and speak to new experiences in changing arenas of social relations and boundary definitions.

In kampung society, spirit attacks on married women seem to be associated with their containment in prescribed domestic roles, whereas in modern organizations, spirit victims are young, unmarried women engaged in hitherto alien and male activities. This transition from kampung to urban-industrial contexts has cast village girls into an intermediate status that they find unsettling and fraught with danger to themselves and to Malay culture.

In the 1970s, newspaper reports on the sudden spate of "mass hysteria" among young Malay women in schools and factories interpreted the causes in terms of "superstitious beliefs," "examination tension," "the stresses of urban living," and less frequently, "mounting pressures" which induced "worries" among female operators in multinational factories.

Multinational factories based in free-trade zones were the favored sites of spirit visitations. An American factory in Sungai Way experienced a large-scale incident in 1978, which involved some 120 operators engaged in assembly work requiring the use of microscopes. The factory had to be shut down for three days, and a bomoh was hired to slaughter a goat on the premises. The American director wondered how he was to explain to corporate headquarters that "8,000 hours of production were lost because someone saw a ghost."[7] A Japanese factory based in Pontian, Kelantan, also experienced a spirit attack on 21 workers in 1980. As they were being taken to ambulances, some victims screamed, "I will kill you! Let me go!" In Penang, another American factory was disrupted for three consecutive days after 15 women became afflicted by spirit possession. The victims screamed in fury and put up a terrific struggle against restraining male supervisors, shouting, "Go away." The afflicted were snatched off the shop floor and given injections of sedatives. Hundreds of frightened female workers were also sent home. A factory personnel officer told reporters:

> "Some girls started sobbing and screaming hysterically and when it seemed like spreading, the other workers in the production line were immediately ushered out. . . . It is a common belief among workers that the factory is 'dirty' and supposed to be haunted by a *datuk.*"

Though brief, these reports reveal that spirit possession, believed to be caused by defilement, held the victims in a grip of rage against factory supervisors. Furthermore, the disruptions caused by spirit incidents seem a form of retaliation against the factory supervisors. In what follows, I will draw upon my field research to discuss the complex issues involved in possession imagery and management discourse on spirit incidents in Japanese-owned factories based in Kuala Langat.

The political economy of Islam is set up and orchestrated around the silence of inferiors.—Fatna A. Sabbah, *Woman in the Muslim Unconscious*

Young, unmarried women in Malay society are expected to be shy, obedient, and deferential, to be observed and not heard. In spirit possession episodes, they speak in other voices that refuse to be silenced. Since the afflicted claim amnesia once they have recovered, we are presented with the task of deciphering covert messages embedded in possession incidents.

Spirit visitations in modern factories with sizable numbers of young Malay female workers engender devil images, which dramatically reveal the contradictions between Malay and scientific ways of apprehending the human condition. I. M. Lewis has suggested that in traditionally gender-stratified societies, women's spirit possession episodes are a "thinly disguised protest against the dominant sex."[8] In Malay society, what is being negotiated in possession incidents and their aftermath are complex issues dealing with the violation of different moral boundaries, of which gender oppression is but one dimension. What seems clear is that spirit possession provides a traditional way of rebelling against authority without punishment, since victims are not blamed for their predicament. However, the imagery of spirit possession in modern settings is a rebellion against transgressions of indigenous boundaries governing proper human relations and moral justice.

For Malays, the places occupied by evil spirits are nonhuman territories like swamps/jungles and bodies of water. These amoral domains were kept distant from women's bodies by ideological and physical spatial regulations. The construction of modern buildings, often without regard for Malay concern about moral space, displaces spirits, which take up residence in the toilet tank. Thus, most village women express a horror of the Western-style toilet, which they would avoid if they could. It is the place where their usually discreet disposal of bodily waste is disturbed. Besides their fear of spirits residing in the water tank, an unaccustomed body posture is required to use the toilet, in their hurry to depart, unflushed toilets and soiled sanitary napkins, thrown helter-skelter, offend spirits who may attack them.

A few days after the spirit attacks in the Penang-based American factory, I interviewed some of the workers. Without prompting, factory women pointed out that the production floor and canteen areas were "very clean" but factory toilets were "filthy" (*kotor*). A datuk haunted the toilet, and workers, in their haste to leave, dropped their soiled pads anywhere. In Ackerman and Lee's case study, Malay factory workers believed that they had disturbed the spirits dwelling in a water tank and on factory grounds. Furthermore, the spirits were believed to possess women who had violated moral codes, thereby becoming "unclean." This connection between disturbing spirits and lack of sexual purity is also hinted at in Teoh and his colleagues' account of the school incidents mentioned above. The headmaster had given students instructions in how to wear sanitary napkins, an incident which helped precipitate a series of spirit attacks said to be caused by the "filthy" school toilets and the girls' disposal of soiled pads in a swamp adjacent to the school grounds.

In the Penang factory incident, a worker remembered that a piercing scream from one corner of the shop floor was quickly followed by cries from other benches as women fought against spirits trying to possess them. The incidents had been sparked by datuk visions, sometimes headless, gesticulating angrily at the operators. Even after the bomoh had been sent for, workers had to be accompanied to the toilet by foremen for fear of being attacked by spirits in the stalls.

In Kuala Langat, my fieldwork elicited similar imagery from the workers in two Japanese factories (code-named ENI and EJI) based in the local free-trade zone. In their drive for attaining high production targets, foremen (both Malay and non-Malay) were very zealous in enforcing regulations that confined workers to the workbench. Operators had to ask for permission to go to the toilet, and were sometimes questioned intrusively about their "female problems." Menstruation was seen by management as deserving no consideration even in a workplace where 85–90 percent of the workforce was female. In the EJI plant, foremen sometimes followed workers to the locker room, terrorizing them with their spying. One operator became possessed after screaming that she saw a "hairy leg" when she went to the toilet. A worker from another factory reported:

> Workers saw "things" appear when they went to the toilet. Once, when a woman entered the toilet she saw a tall figure licking sanitary napkins ("Modess" supplied in the cabinet). It had a long tongue, and those sanitary pads . . . cannot be used anymore.

As Taussig remarks, the "language" emanating from our bodies expresses the significance of social dis-ease.[9] The above lurid imagery speaks of the women's loss of control over their bodies as well as their lack of control over social relations in the factory. Furthermore, the image of body alienation also reveals intense guilt (and repressed desire), and the felt need to be on guard against violation by the male management staff who, in the form of fearsome predators, may suddenly materialize anywhere in the factory.

Even the prayer room (*surau*) provided on factory premises for the Muslim workforce, was not safe from spirit harassment. A woman told me of her aunt's fright in the surau at the EJI factory.

> She was in the middle of praying when she fainted because she said . . . her head suddenly spun and something pounced on her from behind.

As mentioned above, spirit attacks also occurred when women were at the workbench, usually during the "graveyard" shift. An ENI factory operator described one incident which took place in May 1979.

> It was the afternoon shift, at about nine o'clock. All was quiet. Suddenly, (the victim) started sobbing, laughed and then shrieked. She flailed at the machine . . . she was violent, she fought as the foreman and technician pulled her away. Altogether, three operators were afflicted. . . . The supervisor and foremen took them to the clinic and told the driver to take them home. . . .
>
> She did not know what had happened . . . she saw a hantu, a were-tiger. Only she saw it, and she started screaming. . . . The foremen would not let us talk with her for fear of recurrence. . . . People say that the workplace is haunted by the hantu who dwells below. . . . Well, this used to be all jungle, it was a burial ground before the factory was built. The devil disturbs those who have a weak constitution.

Spirit possession episodes then were triggered by black apparitions, which materialized in "liminal" spaces such as toilets, the locker room and the prayer room, places where workers sought refuge from harsh work discipline. These were also rooms periodically checked by male supervisors determined to bring workers back to the workbench. The microscope, which after hours of use becomes an instrument of torture, sometimes disclosed spirits lurking within. Other workers pointed to the effect of the steady hum and the factory pollutants, which permanently disturbed graveyard spirits. Unleashed, these vengeful beings were seen to threaten women for transgressing into the zone between the human and nonhuman world, as well as modern spaces formerly the domain of men. By intruding into hitherto forbidden spaces, Malay women workers experienced anxieties about inviting punishment.

Fatna Sabbah observes that "the invasion by women of economic spaces such as factories and offices . . . is often experienced as erotic aggression in the Muslim context."[10] In Malay culture, men and women in public contact must define the situation in nonsexual terms. It is particularly incumbent upon young women to conduct themselves with circumspection and to diffuse sexual tension. However, the modern factory is an arena constituted by a sexual division of labor and constant male surveillance of nubile women in a close, daily context. In Kuala Langat, young factory women felt themselves placed in a situation in which they unintentionally violated taboos defining social and bodily boundaries. The shop floor culture was also charged with the dangers of sexual harassment by male management staff as part of workaday relations. To combat spirit attacks, the Malay factory women felt a greater need for spiritual vigilance in the factory surroundings. Thus the victim in the ENI factory incident was said to be

> possessed, maybe because she was spiritually weak. She was not spiritually vigilant, so that when she saw the hantu she was instantly afraid and screamed. Usually, the hantu likes people who are spiritually weak, yes. . . . one should guard against being easily startled, afraid.

As Foucault observes, people subjected to the "micro-techniques" of power are induced to regulate themselves.[11] The fear of spirit possession thus created self-regulation on the part of workers, thereby contributing to the intensification of corporate and self-control on the shop floor. Thus, as factory workers, Malay women became alienated not only from the products of their labor but also experienced new forms of psychic alienation. Their intrusion into economic spaces outside the home and village was experienced as moral disorder, symbolized by filth and dangerous sexuality. Some workers called for increased "discipline," others for Islamic classes on factory premises to regulate interactions (including dating) between male and female workers. Thus, spirit imagery gave symbolic configuration to the workers' fear and protest over social conditions in the factories. However, these inchoate signs of moral and social chaos were routinely recast by management into an idiom of sickness.

THE WORKER AS PATIENT

. . . Struggles over the meanings of health are part of workers' social critique of work discipline, and of managers' attempts to extend control over the workforce. The management use of workers as "instruments of labor" is paralleled by another set of ideologies, which regards women's bodies as the site of control where gender politics, health, and educational practices intersect.

In the Japanese factories based in Malaysia, management ideology constructs the female body in terms of its biological functionality for, and its anarchic disruption of, production. These ideologies operate to fix women workers in subordinate positions in systems of domination that proliferate in high-tech industries. A Malaysian investment brochure advertises "the oriental girl," for example, as "qualified *by nature and inheritance* to contribute to the efficiency of a bench assembly production line." This biological rationale for the commodification of women's bodies is a part of a pervasive discourse reconceptualizing women for high-tech production requirements. Japanese managers in the free-trade zone talk about the "eyesight," "manual dexterity," and "patience" of young women to perform tedious micro-assembly jobs. An engineer put the female nature-technology relationship in a new light: "Our work is designed for females." Within international capitalism, this notion of women's bodies renders them analogous to the status of the computer chips they make. Computer chips, like "oriental girls," are identical, whether produced in Malaysia, Taiwan, or Sri Lanka. For multinational corporations, women are units of much cheap labor power repackaged under the "nimble fingers" label.

The abstract mode of scientific discourse also separates "normal" from "abnormal" workers, that is, those who do not perform according to factory requirements; in the EJI factory, the Malay personnel manager using the biomedical model to locate the sources of spirit possession among workers noted that the first spirit attack occurred five months after the factory began operation in 1976. Thereafter,

> we had our counter-measure. I think this is a method of how you give initial education to the workers, how you take care of the medical welfare of the workers. The worker who is weak, comes in without breakfast, lacking sleep, then she will see ghosts!

In the factory environment, "spirit attacks" (kena hantu) was often used interchangeably with "mass hysteria," a term adopted from English language press reports on such incidents. In the manager's view, "hysteria" was a symptom of physical adjustment as the women workers "move from home idleness to factory discipline." This explanation also found favor with some members of the workforce. Scientific terms like "penyakit hysteria" (hysteria sickness), and physiological preconditions formulated by the management, became more acceptable to some workers. One woman remarked,

> They say they saw hantu, but I don't know. . . . I believe that maybe they . . . when they come to work, they did not fill their stomachs, they were not full so that they felt hungry. But they were not brave enough to say so.

A male technician used more complex concepts, but remained doubtful.

> I think that this [is caused by] a feeling of "complex"—that maybe "inferiority complex" is pressing them down—their spirit, so that this can be called an illness of the spirit, "conflict jiwa," "emotional conflict." Sometimes they see an old man, in black shrouds, they say, in their microscopes, they say. . . . I myself don't know how. They see hantu in different places. . . . Some time ago an "emergency" incident like this occurred in a boarding school. The victim fainted. Then she became very strong. . . . It required ten or twenty persons to handle her.

In corporate discourse, physical "facts" that contributed to spirit possession were isolated, while psychological notions were used as explanation and as a technique of manipulation. In ENI factory, a bomoh was hired to produce the illusion of exorcism, lulling the workers into a false sense of security. The personnel manager claimed that unlike managers in other Japanese firms who operated on the "basis of feelings," his "psychological approach" helped to prevent recurrent spirit visitations.

> You cannot dispel kampung beliefs. Now and then we call the bomoh to come, every six months or so, to pray, walk around. Then we take pictures of the bomoh in the factory and hang up the pictures. Somehow, the workers seeing these pictures feel safe, [seeing] that the place has been exorcised.

Similarly, whenever a new section of the factory was constructed, the bomoh was sent for to sprinkle holy water, thereby assuring workers that the place was rid of ghosts. Regular bomoh visits and their photographic images were different ways of defining a social reality, which simultaneously acknowledged and manipulated the workers' fear of spirits.

Medical personnel were also involved in the narrow definition of the causes of spirit incidents on the shop floor. A factory nurse periodically toured the shop floor to offer coffee to tired or drowsy workers. Workers had to work eight-hour shifts six days a week—morning, 6:30 A.M. to 2:30 P.M.; afternoon, 2:30 P.M. to 10:30 P.M.; or night, 10:30 P.M. to 6:30A.M.—which divided up the 24-hour daily operation of the factories. They were permitted two ten-minute breaks and a half-hour for a meal. Most workers had to change to a different shift every two weeks. This regime allowed little time for workers to recover from their exhaustion between shifts. In addition, overtime was frequently imposed. The shifts also worked against the human, and especially, female cycle; many freshly recruited workers regularly missed their sleep, meals, and menstrual cycles.

Thus, although management pointed to physiological problems as causing spirit attacks, they seldom acknowledged deeper scientific evidence of health hazards in microchip assembly plants. These include the rapid deterioration of eyesight caused by the prolonged use of microscopes in bonding processes. General exposure to strong solvents, acids, and fumes induced headaches, nausea, dizziness, and skin irritation in workers. More toxic substances used for cleaning purposes exposed workers to lead poisoning, kidney failure, and breast cancer. Other materials used in the fabrication of computer chips have been linked to female workers' painful menstruation, their inability to conceive, and repeated miscarriages. Within the plants, unhappy-looking workers were urged to talk over their problems with the "industrial relations assistant." Complaints of "pain in the chest" were interpreted to mean emotional distress, and the worker was ushered into the clinic for medication in order to maintain discipline and a relentless work schedule.

In the EJI factory, the shop floor supervisor admitted, "I think that hysteria is related to the job in some cases." He explained that workers in the microscope sections were usually the ones to kena hantu, and thought that perhaps they should not begin work doing those tasks. However, he quickly offered other interpretations that had little to do with work conditions: There was one victim whose broken engagement had incurred her mother's wrath; at work she cried and talked to herself, saying, "I am not to be blamed, not me!" Another worker, seized by possession, screamed, "Send me home, send me home!" Apparently, she indicated, her mother had taken all her earnings. Again, through such psychological readings, the causes of spirit attacks produced in the factories were displaced onto workers and their families.

In corporate discourse, both the biomedical and psychological interpretations of spirit possession defined the affliction as an attribute of individuals rather than stemming from the general social situation. Scientific concepts, pharmaceutical treatment, and behavioral intervention all identified and separated recalcitrant workers from "normal" ones; disruptive workers became patients. According to Parsons, the cosmopolitan medical approach tolerates illness as sanctioned social deviance; however, patients have the duty to get well.[12] This attitude implies that those who do not get well cannot be rewarded with "the privileges of being sick." In the ENI factory, the playing out of this logic provided the rationale for dismissing workers who had had two previous experiences of spirit attacks, on the grounds of "security." This policy drew protests from village elders, for whom spirits in the factory were the cause of their daughters' insecurity. The manager agreed verbally with them, but pointed out that these "hysterical, mental types" might hurt themselves when they flailed against the machines, risking electrocution. By appearing to agree with native theory, the management reinterpreted spirit possession as a symbol of flawed character and culture. The sick role was reconceptualized as internally produced by outmoded thought and behavior not adequately adjusted to the demands of factory discipline. The worker patient could have no claim on management sympathy but would have to bear responsibility for her own cultural deficiency. A woman in ENI talked sadly about her friend, the victim of spirits and corporate policy.

> At the time the management wanted to throw her out, to end her work, she cried. She did ask to be reinstated, but she has had three [episodes] already. . . . I think that whether it was right or not [to expel her] depends [on the circumstances], because she has already worked here for a long time; now that she has been thrown out she does not know what she can do, you know.

The non-recognition of social obligations to workers lies at the center of differences in worldview between Malay workers and the foreign management. By treating the signs and symptoms of disease as "things-in-themselves,"[13] the biomedical model freed managers from any moral debt owed the workers. Furthermore, corporate adoption of spirit idiom stigmatized spirit victims, thereby ruling out any serious consideration of their needs. Afflicted and "normal" workers alike were made to see that spirit possession was nothing but confusion and delusion, which should be abandoned in a rational worldview.

THE WORK OF CULTURE: HYGIENE AND DISPOSSESSION

Modern factories transplanted to the Third World are involved in the work of producing exchange as well as symbolic values. Medicine, as a branch of cosmopolitan science, has attained a place in schemes for effecting desired social change in indigenous cultures. While native statements about bizarre events are rejected as irrational, the conceptions of positivist science acquire a quasi-religious flavor. In the process, the native "work of culture," which transforms motives and affects into "publicly accepted sets of meanings and symbols,"[14] is being undermined by an authoritative discourse that suppresses lived experiences apprehended through the worldview of indigenous peoples.

To what extent can the bomoh's work of culture convert the rage and distress of possessed women in Malaysia into socially shared meanings? As discussed above, the spirit imagery speaks of danger and violation as young Malay women intrude into hitherto forbidden spirit or male domains. Their participation as an industrial force is subconsciously perceived by themselves and their families as a threat to the ordering of Malay culture. Second, their employment as production workers places them directly in the control of male strangers who monitor their every move. These social relations, brought about in the process of industrial capitalism, are experienced as a moral disorder in which workers are alienated from their bodies, the products of their work, and their own culture. The spirit idiom is therefore a language of protest against these changing social circumstances. A male technician evaluated the stresses they were under.

> There is a lot of discipline. . . . but when there is too much discipline . . . it is not good. Because of this the operators, with their small wages, will always contest. They often break the machines in ways that are not apparent. . . . Sometimes, they damage the products.

Such Luddite actions in stalling production reverse momentarily the arrangement whereby work regimentation controls the human body. However, the workers' resistance is not limited to the technical problem of work organization, but addresses the violation of moral codes. A young woman explained her sense of having been "tricked" into an intolerable work arrangement.

> For instance, . . . sometimes . . . they want us to raise production. This is what we sometimes challenge. The workers want fair treatment, as for instance, in relation to wages and other matters. We feel that in this situation there are many [issues] to dispute over with the management. . . . with our wages so low we feel as though we have been tricked or forced.

She demands "justice, because sometimes they exhaust us very much as if they do not think that we too are human beings!"

Spirit possession episodes may be taken as expressions both of fear and of resistance against the multiple violations of moral boundaries in the modern factory. They are acts of rebellion, symbolizing what cannot be spoken directly, calling for a renegotiation of obligations between the management and workers. However, technocrats have turned a deaf ear to such protests, to this moral indictment of their woeful cultural judgments about the dispossessed. By choosing to view possession episodes narrowly as sickness caused by physiological and psychological maladjustment, the management also manipulates the bomoh to serve the interests of the factory rather than express the needs of the workers.

Both Japanese factories in Kuala Langat have commenced operations in a spate of spirit possession incidents. A year after operations began in the EJI factory, a well-known bomoh and his retinue were invited to the factory surau, where they read prayers over a basin of "pure water." Those who had been visited by the devil drank from it and washed their faces, a ritual which made them immune to future spirit attacks. The bomoh pronounced the hantu controlling the factory site "very kind"; he merely showed himself but did not disturb people. A month after the ritual, the spirit attacks resumed, but involving smaller numbers of women (one or two) in each incident. The manager claimed that after the exorcist rites, spirit attacks occurred only once a month.

In an interview, an eyewitness reported what happened after a spirit incident erupted:

> The work section was not shut down, we had to continue working. Whenever it happened, the other workers felt frightened. They were not allowed to look because [the management] feared contagion. They would not permit us to leave. When an incident broke out, we had to move away. . . . At ten o'clock they called the bomoh to come . . . because he knew that the hantu had already entered the woman's body. He came in and scattered rice flour water all over the area where the incident broke out. He recited prayers over holy water. He sprinkled rice flour water on places touched by the hantu. . . . The bomoh chanted incantations [*jampi jampi*] chasing the hantu away. He then gave some medicine to the afflicted. . . . He also entered the clinic [to pronounce] jampi jampi.

The primary role of the bomoh hired by corporate management was to ritually cleanse the prayer room, shop floor, and even the factory clinic. After appeasing the spirits, he ritually healed the victims, who were viewed as not responsible for their affliction. However, his work did not extend to curing them after they had been given sedatives and sent home. Instead, through his exorcism and incantations, the bomoh expressed the Malay understanding of these disturbing events, perhaps impressing the other workers that the factory had been purged of spirits. However, he failed to convince the management about the need to create a moral space, in Malay terms, on factory premises. Management did not respond to spirit incidents by reconsidering social relationships on the shop floor; instead, they sought to eliminate the afflicted from the work scene. As the ENI factory nurse, an Indian woman, remarked, "It is an experience working with the Japanese. They do not consult women. To tell you the truth, they don't care about the problem except that it goes away."

This avoidance of the moral challenge was noted by workers in the way management handled the kenduri, the ritual feast that resolved a dispute by bringing the opposing sides together in an agreement over future cooperation. In the American factory incident in Penang, a bomoh was sent for, but worker demands for a feast were ignored. At the EJI factory, cleansing rituals were brought to a close by a feast of saffron rice and chicken curry. This was served to factory managers and officers, but not a single worker (or victim) was invited. This distortion of the Malay rite of commensality did not fail to impress on workers the management rejection of moral responsibility to personal needs—*muafakat*. Women workers remained haunted by their fear of negotiating the liminal spaces between female and male worlds, old and new morality, when mutual obligations between the afflicted and the bomoh, workers and the management, had not been fulfilled.

The work of the bomoh was further thwarted by the medicalization of the afflicted. Spirit possession incidents in factories made visible the conflicted women who did not fit the corporate image of "normal" workers. By standing apart from the workaday routine, possessed workers inadvertently exposed themselves to the cold ministrations of modern medicine, rather than the increased social support they sought. Other workers, terrified of being attacked and by the threat of expulsion, kept up a watchful vigilance. This induced self-regulation was reinforced by the scientific gaze of supervisors and nurses, which further enervated the recalcitrant and frustrated those who resisted. A worker observed,

> [The possessed] don't remember their experiences. Maybe the hantu is still working on their madness, maybe because their experiences have not been stilled, or maybe their souls are not really disturbed. They say there are evil spirits in that place [that is, factory].

In fact, spirit victims maintained a disturbed silence after their "recovery." Neither their families, friends, the bomoh, nor I could get them to talk about their experiences.

Spirit possession episodes in different societies have been labeled "mass psychogenic illness" or "epidemic hysteria" in psychological discourse. Different altered states of consciousness, which variously spring from indigenous understanding of social situations, are reinterpreted in cosmopolitan terms considered universally applicable. In multinational factories located overseas, this ethno-therapeutic model is widely applied and made to seem objective and rational.[15] However, we have seen that such scientific knowledge and practices can display a definite prejudice against the people they are intended to restore to well-being in particular cultural contexts. The reinterpretation of spirit possession may therefore be seen as a shift of locus of patriarchal authority from the bomoh, sanctioned by indigenous religious beliefs, toward professionals sanctioned by scientific training.

In Third World contexts, cosmopolitan medical concepts and drugs often have an anesthetizing effect, which erases the authentic experiences of the sick. More frequently, the proliferation of positivist scientific meanings also produces a fragmentation of the body, a shattering of social obligations, and a separation of individuals from their own culture. Gramsci has defined hegemony as a form of ideological domination based on the consent of the dominated, a consent that is secured through the diffusion of the worldview of the dominant class.[16] In Malaysia, medicine has become part of hegemonic discourse, constructing a "modern" outlook by clearing away the nightmarish visions of Malay workers. However, as a technique of both concealment and control, it operates in a more sinister way than native beliefs in demons. Malay factory women may gradually become dispossessed of spirits and their own culture, but they remain profoundly dis-eased in the "brave new workplace."

NOTES

1. Vincent Crapanzano, Introduction, in Vincent Crapanzano and Vivian Garrison, eds., *Case Studies in Spirit Possession*, 1–10 (New York: John Wiley, 1977), 16.

2. Susan Ackerman and Raymond Lee, "Communication and Cognitive Pluralism in a Spirit Possession Event in Malaysia" *American Ethnologist* 8, no. 4 (1981): 789–99, at 796.

3. Michel Foucault, *Madness and Civilization: A History of Insanity in the Age of Reason*, trans. R. Howard (New York: Pantheon, 1965), xi.

4. Mary Douglas, *Purity and Danger: An Analysis of Pollution and Taboo* (Harmondsworth: Penguin, 1966).

5. Carol Laderman, *Wives and Midwives: Childbirth and Nutrition in Rural Malaysia* (Berkeley: University of California Press, 1983), 74.

6. Nancy Scheper-Hughes and Margaret Lock, "The Mindful Body: A Prolegomenon to Future Work in Medical Anthropology," *Medical Anthropology Quarterly* 1, no. 1 (1987): 1–36, at 19.

7. Linda Lim, "Women Workers in Multinational Corporations: The Case of the Electronics Industry in Malaysia and Singapore" (Ann Arbor: Michigan Occasional Papers in Women's Studies, no. 9, 1978), 33.

8. Ioan M. Lewis, *Ecstatic Religion: An Anthropological Study of Spirit Possession and Shamanism* (Harmondsworth, England: Penguin, 1971), 31.

9. Michael Taussig, *The Devil and Commodity Fetishism in South America* (Chapel Hill: University of North Carolina Press, 1980).

10. Fatna Sabbah, *Women in the Muslim Unconscious* (New York: Pergamon Press, 1984), 17.

11. Michel Foucault, *Discipline and Punish: The Birth of the Prison*, trans. Alan Sheridan (New York: Vintage, 1979).

12. Talcott Parsons, "Illness and the Role of the Physician: A Sociological Perspective," in *Readings from Talcott Parsons*, ed. Peter Hamilton (New York: Tavistock, 1985), 146, 149.

13. Taussig, *The Devil and Commodity Fetishism*, 1.

14. Gananath Obeyesekere, "Depression, Buddhism and the Work of Culture in Sri Lanka," in *Culture and Depression*, ed. Arthur Kleinman and Byron Good (Berkeley: University of California Press), 147.

15. Catherine Lutz, "Depression and the Translation of Emotional Worlds," in *Culture and Depression*, ed. Arthur Kleinman and Byron Good (Berkeley: University of California Press1985), 63–100.

16. Antonio Gramsci, *Selections from the Prison Notebooks*, trans. Quentin Hoare and Geoffrey Nowell Smith (New York: International Publishing, 1971).

PLASTIC TEETH EXTRACTION: THE ICONOGRAPHY

OF HAYA GASTRO-SEXUAL AFFLICTION

BRAD WEISS

The rapid and devastating spread of AIDS in Africa is well known today, and the Haya living in the Kagera Region of Tanzania are all too familiar with its effects. But a considerable number of them are equally concerned with yet another new, deadly disease, one whose symptoms and prognosis both relate to and may lend insight into Haya understandings of the AIDS epidemic. The victims of this disease are infants and children who suffer severe disorders brought on by the growth of what are described as plastic teeth and who will die if they do not have these teeth removed.

In this article I explore the contemporary appearance of this dental affliction by tracing the symbolic valences of its symptoms. I then relate the symbolism and symptomatology to a wide range of sociocultural transformations in Kagera. Such an analysis grounds these particular symptoms, symbols, and transformations in Haya understandings of their bodies and their significance to social action and the world. Mindful of Michael Jackson's objections to studies in which the body "is dismembered so that the symbolic value of its various parts in indigenous discourse can be enumerated" (Jackson 1989:124), I would, nevertheless, argue that the experience of disease and therapy frequently entails a condensation of meanings in discrete forms. Indeed, it seems more fruitful to insist that segmentation and totalization presuppose each other; wholes are always and necessarily a relation of parts and can only be understood in terms of the significance that derives from their dynamic organization. Further, Haya notions of totality, and especially of bodily wholeness, entail an understanding of this relation between parts. Physical disorder and well-being can be assessed through attention to various discrete symptoms or conditions that can be interpreted as expressions of more general states of being. While the body may allow us to remember, then, we often do so by dismembering procedures. The processes through which singular parts come to take on general and diffused meanings thus demand our analytic attention.

I do not intend, however, to break down the practical unity of the body in order to assess its conceptual organization. Rather, I will examine Haya processes of embodiment by "thinking through" teeth and other bodily and symbolic forms. In this way I hope to capture the dynamics of an encompassing order of sociocultural practices in and through which bodily experience is constituted, as well as to address the symbolic processes through which particular, concrete, and occasionally discrete (even dismembered?) phenomena become the objectified forms or iconic expressions of these practices.

An assessment of the body's ability to remember will also alert us to other culturally significant forms of memory. Such forms of affliction as plastic teeth clearly not only concern the dynamics of local sociocultural orders as they orient the body; they also give evidence of the historical nature of these orders as they experience rapid and sweeping transformation. The juxtaposition of the body with tokens of this social transformation, such as plastic and other commodity forms, affords us a privileged position from which to assess the particular Haya experience of history and to recognize that experience as part of global processes.

Theories of historical transformation, especially those that focus on the economic forces of markets and money, tend to rely on teleological notions of change. They presume that the systematic incorporation of local societies into wider economic relations must necessarily result in either a "rationalization" of cultural values or a pervasive sense of "alienation" as local values become increasingly irrelevant to metropolitan projects (Taussig 1980; Wallerstein 1974). Moreover, commodity production and commodities themselves come to embody the "disenchantment" of symbolically nuanced orders of practice and experience (Bourdieu 1979; Weiner and Schneider 1989:13). From the perspectives of these theorists, a uniform, systematic incorporation can only lead to the effacement of meaningful practice. A challenge to such theorizing is, therefore, posed by local accounts and actions which tell us that those factors most essential to any process of social transformation are the experiential and meaningful aspects through which it is realized. Instead of emphasizing the alienating effects of commodity flows, these accounts remind us that commodities flow into situations in which they transform experience and in the process are themselves transformed.

This challenge to social scientific paradigms, thus, suggests another form of re-membering, of properly situating local orders in encompassing relations. Analytic attention to such enigmatic forms of experience as affliction may allow us to get at the ways in which the "world system" becomes what it is through its actualization in *particular*, local forms.

In what follows, I begin by examining Haya constructions of bodily process specifically and dynamic forms of growth more generally. I then explore the ways in which disease and disorder, especially as they are related to the particular forms of affliction presented by plastic teeth, are articulated with these processes. The grounding of sociocultural meanings in a concrete order of social relations and practices ultimately allows us to consider the material, historical transformations of commoditization as aspects of a creative, semantic reconfiguration of everyday life. It is the experience of such material and semantic transformations that is embodied in these forms of affliction.

SYMPTOMS, SIGNS, AND THERAPY

Plastic teeth (in Swahili, *meno ya plasta*, from the English "plastic"; or in Haya, *ebiino*, literally, "bad teeth") are said to be a new illness, one that has arrived in Kagera only in the last two or three years. But its effects are already pervasive; according to most of my Haya friends, all children are now born with or are susceptible to developing plastic teeth. These teeth are said to be made of plastic, to be *nko nylon* (like nylon).

Children who develop plastic teeth present a range of symptoms. The most commonly cited are diarrhea, vomiting, and fever, as well as a refusal to nurse and wasting. The proper treatment in such cases is to have the plastic teeth removed; in most cases, two to four canines or incisors are extracted by a local dentist. Children up to the age of three or four (according to my data and press reports), but typically younger, are thought to be susceptible to the growth of plastic teeth. It should also be pointed out that plastic teeth are often removed before the child, in many cases a neonate, has begun to teethe. In such cases the teeth are removed from below the surface of the child's gums.

Different accounts of the origins of plastic teeth are offered. All Haya that I spoke with agreed that the disease had come to the region from somewhere else: some held that it came from Uganda, while others claimed that milk from the cattle brought by Rwandan migrants was responsible. While there is a widespread familiarity with the symptoms described, certain people, usually women whose own children have had their teeth removed, are sought out in order to make a preliminary diagnosis. The women with whom I talked about diagnosing plastic teeth said that the gums of the afflicted child were often runny with water and "soft."

Once a specialist confirms the diagnosis of plastic teeth, they must be removed. This is always the task of a specialist (although locals claimed that anyone could learn how to remove plastic teeth simply by watching dentists, in much the same way that local women had learned to diagnose the disease through their children's experiences). The procedure, according to those who actually witnessed it, is usually performed with a sharp implement; a hypodermic needle or sharpened bicycle spoke was most often cited as the tool used by the extractor. While the biomedical community holds that the symptoms of what are said to be plastic teeth are simply typical responses to teething, Haya plastic teeth extractors follow a different symptomatology. After the teeth are removed, the dentist provides evidence of the disease, according to witnesses, by showing the child's parent how the newly extracted tooth "plays" (*kyarina*), moving back and forth in a rocking motion. This movement (caused by the enervation of the tooth bud itself, according to biomedical accounts) is said to indicate the presence of a worm (*kijoka*) in the tooth. Thus, the plastic tooth might be said to have a life of its own, one that threatens the life of the child.

It is important to point out that, in addition to sharing with me their widespread knowledge about the symptoms and treatment of plastic teeth, people always pointedly mentioned the cost of the therapy. Always included in their characterization of the dentist's techniques was the fact that he charged 200 shillings, or approximately a dollar per tooth—a significant cost in a region where the median annual income from the sale of coffee totaled approximately $30 in 1988–89. Moreover, the dentist was reported to have amassed a significant amount of money because of his practice, and this certainly seemed to enhance his reputation. The special concern with the specific costs of therapy, as well as the enigmatic appearance of plastic in the human body, suggests that the force of commodities and commoditized images is an important dimension of this phenomenon.

TEETHING, TIMING, AND NAMING

Plastic teeth may have arrived only recently, but the Haya have long been concerned with their children's teeth. Legends about Kanyamaishwa (The Beast), a hunter who helped to found the kingdom of Kiziba, indicate that as a child he was left to die in the forest because he had cut an upper tooth before any of his lower teeth. Even today, the Haya are careful to make certain that their children cut a lower tooth first. Failure to do so is said to be particularly dangerous to the child's mother's brother. Such dangerous teething is called *amahano* (strange or odious action or condition), a term that is also used for incestuous relations. For a child to be born with visible teeth is equally dangerous. The verb for teething in both Swahili and Haya is best translated as "to sprout" (Swahili *kuota*; Haya *oku-mela*). I would suggest that the process of "sprouting" in both human and vegetable growth entails more than the appearance of what lies hidden beneath the surface. For the Haya, such growth is both transformative and creative; that is, teeth (to use this example) do not simply emerge from the gum but are produced by the emergence. This point is relevant to the Haya understanding of plastic teeth as well. The mere fact that infants who have not yet begun to teethe actually have teeth that can be removed from their gums is evidence of ill health. As one friend who had seen his newborn cousin's plastic teeth extracted remarked, "What kind of a person is born with teeth? Maybe a lion, or a calf—but a human being?"

The particular forms of disorder manifest in the ill-timed growth of teeth are suggestive of the qualities that all teeth and the teething process signify for the Haya. The beastly name of the early exile and the explicit association of improperly developed infant teeth with animals reveal the wild and fundamentally inhuman character of this inversion. Further, this brute, animal-like character is related to the subversion of reproductive and sexual relations. Not only is this disorder categorized as *amahano*, just as incest is, but it is seen to be an active threat to affinity in the danger it poses to the mother's brother. Beattie (1960:146) points out that the Nyoro, like the Haya, do not permit a married man to eat in the presence of his mother-in-law. Such an act (which the Haya explicitly associate with illicit sexuality between these in-laws) would be a form of *amahano*. Here, the improper use of teeth (that is, eating of this sort) is linked to dangerous sexuality in the context of affinal relationships. These associations suggest that the successes of affinal attachments are realized, in the most immediate context of experience, through the bodily development of these attachments' most significant product, the body of the child. Teething, then, can be seen to model the domestication of sexuality to which affinity aspires, insofar as its proper development entails a control over the direct manifestation and visible presence of forces that are nonetheless integral to the construction of social, cultural, and biological activity.

The concrete social ordering of productive and reproductive practices suggested by these connections to sexuality, and demonstrated in the category of *amahano*, is further indicated by certain associations of teeth and teething with the sequential unfolding and symbolic qualities of other Haya processes. For example, the verb "to bite" (*okunena*) is related to the verb "to cut" (*okushaia*) in Haya usage. Thus, to say "I am in pain," one can say "*Nanenwa*" (literally, "I am bitten") or "*Nashasha*" (which uses the causative—*okushasha*—of the verb *okushaia*). Both of these usages suggest that pain is characterized by the force of sharp, incisive penetration. Further, the verb *okutema* (to slash), typically used to describe the action of cutting through the banana stem when the stalk is harvested, can be used to describe eating with great gusto and pleasure. The significance of the link between biting and cutting can be seen in the cultural formulation of Haya cultivation techniques. Haya men are solely responsible for the preparation of the plantain and banana plants (*engemu*) that provide the staple food crop. The principal men's activity in this cultivation is o*kushalila* (the benefactive form of okushaia) *engemu,* pruning the plantain plant, which consists of cutting the desiccated leaves and outer husk of the plant and stem to be laid at the base of the plant as mulch. Women do contribute to the growth of plantain and bananas, not by cutting or altering the actual plants themselves, but by weeding and mulching the ground of the farm. An important contrast in these forms of work is noted by the Haya themselves when they point out that "men work above, and women work below."

By elaborating on these practices I do not simply mean to imply that there is an array of symbolic associations or metaphorical connections between teething and cultivating. Rather, I am asserting that there is a systematic patterning to the sequential form of Haya (re)productive processes, and that the meaning embedded in these processes is realized in teething as well as other activities. Control of the sequential development of productive practices is managed by movement in space and time from women's to men's agency. Thus, teeth, which "sprout" like plants, should properly go from below to above; that is, lower teeth should come in before upper. The shift from below to above embodies a passage from a female to a male position. Further, harvesting the banana stalk (okutema) "finishes" the life of the growing plant, just as devouring food (okutema) "finishes" the "life" of the cooked bananas. In both cases, markedly female activities (weeding and mulching the farm and cooking the harvested plantain) create the potential that *subsequent* male activities transform and appropriate. It is the male position as the competition to the developmental process that is undermined by appearing before female potential can be properly generated. From this perspective, then, we can better understand why the premature growth of an upper tooth poses a particular threat of amahano to the child's mother's brother.

The measure of control realized in these processes of productivity and socialization and focused on teething is further demonstrated in Haya associations of teeth and names. Most Haya today will not give a child a "clan name" (*eibala ly'oluganda*) until the youngster has cut a first tooth. Prior to cutting a tooth a child can be called *ekibumba* (from *okubumba,* "to be clogged"), because, as the Haya say, "He can't pronounce any word until he has cut a tooth." Thus speech, or the culturally marked capacity for it, is the requisite to being named.

A child is given a clan name by a senior agnate, preferably its father's father. Naming, then, firmly grounds the identity of the infant in the generational order of clan relations. Moreover, this transformation in identity is constructed in terms of a contrast between agnates and affines. Children are named and recognized with gifts from and parties among their co-residential agnates. The gifts include, most importantly, money, as well as certain amulets that are tied to the child's body. Subsequently, children are taken to their mother's father's home for a second round of gift giving. In this way naming (which is facilitated by the presence of teeth) establishes social identity, not as an essential property of the person but as a relational dimension of the self. Having a name enables a child to be known as a distinct person, and this knowledge and distinction are situated in and realized as a practical relation between patrilineal clan members and their affines.

The sociocultural transformations produced by naming are also embodied in important ways. Children who have not yet cut teeth are said to be totally dependent on their mothers. A mother may not leave her child or be divorced by her husband until the child has its first tooth. It is as though the first tooth and the naming process it indexes sever the immediate intimacy of the bodily attachment of mother and child. One of the gifts commonly given to a named child is an amulet, usually in the form of a bored coin (*ekyapa*) tied with a string around the infant's waist, which ensures that the child will not have any trouble suckling. Moreover, the string used to tie the amulet serves as a means of measuring the growth of the child. The expanding or shrinking belly of the child can be demonstrated by the relative tautness of the string, and this measure assures parents that their children are getting fat. This gift, then, suggests that the immediate intimacy characteristic of mother-child relations becomes problematic after teeth "sprout" and names are given.

TEETH, SPEECH, AND AGENCY

It is also interesting to note that, in a sense, the toothless child is thought to be so thoroughly dependent on its mother as to lack an independent existence. Children without teeth are thought to be immune from attack by snakes. Snakes are animals which, for the Haya, epitomize the surreptitious threat of penetration by dangerous natural forces. Yet when a snake discovers that a child has no teeth, it will simply play with him or her, as such a child poses no threat. In effect, the lack of teeth demonstrates not only the child's inability to act for itself but, reciprocally, the inability of other agents to act on the child. The child without teeth is in a condition of dependence that has a dual or reflexively structured orientation to the wider world. On the one hand, this dependence is characterized by intimate associations between the body of the mother and that of the child; on the other hand, the child is thoroughly disengaged from action in the world. The child can truly be said to be "clogged," as it lacks the capacity either to act or to be acted upon.

These forms of embodiment, then, indicate that the processes of teething and naming effect a passage from the bodily identification of the child with the mother to the socially achieved personhood of one who has a clan name. The transition from undifferentiated corporeal existence to socially inscribed body and person is simultaneously a transition from mother to father, from the immediate intimacy of birth to the generational hierarchies of clanship. What is objectified in the child's teeth, and realized through these social transitions, is a situation of reciprocal engagement in the world. By opening the body, the tooth requires that the child (in practical terms) grasp the world, "take up" its "basic significance" (Merleau-Ponty 1962:102) as an entity external to his or her self (for example, the child suckles from a mother now marked as a distinct person); at the same time, by grasping the world the child comes, in effect, to be grasped by that world and its inscribed orientations and potentials. Thus, in suckling, the child becomes bound by the string which measures weight gain—but which also measures the child's orientation to the sociocultural world that evaluates weight gain. Through their "rootedness" in the processes in which this dialectical form of engagement is generated, teeth are produced as icons of agency.

Speech, itself made possible by the appearance of teeth, epitomizes this engaged situation. To begin with, distinct forms of speech and conversation are identified with distinctions between men and women. Men's verbal exchanges, for example, facilitate a range of activities and experiences, from courtroom disputes to funerary mourning. Language learning, associated with the process of teething, has a gendered dimension: the Haya claim that every child's first word is *Tata* (Father), never *Mawe* (Mother). As Dauer notes, men's speech (by their own evaluation) should properly reflect the capacity for self-control in interpersonal relations. Men's relatively greater ability to be restrained yet forthright (*okwekomya*; literally, "to make one's self bound") contrasts with women's speech styles, which are more subtle and deferential. Women "pass behind" (*okulabya enyuma*), using euphemism and even slips of the tongue in order to avoid confrontation (Dauer 1984:276–85). The point of this contrast is not that men control speech more than women do, but that men's speech is more openly assertive, while women's speech is typically effective through indirection. The distinction between these modes of agency is critical to the significance of teeth, for teething marks a point of differentiation that allows for the assertion (and ascription) of an individuated identity. The infant who teethes might, therefore, be said to pass from a condition of being "clogged" and to acquire an incipient capacity for "restraint."

For the Haya, speech is fundamental to the constitution and transformation of these differentiated forms of personhood. This capacity for transformation, and the meaning of these differentiated identities, can be demonstrated in the particular speech styles of mature men, as well as in the use of certain types of speech over the course of the life cycle. As part of a marriage ceremony, the newly married son will address his senior agnates. He carries a spear and adopts a threatening stance toward his elders while at the same time his speech extols the virtues of their clan and of his position in it as a worthy heir to its fame. This form of "boasting" (*okwebuga*; literally, "to move" or "to push one's self"), then, serves both to establish the mature independence of a married man (as the threatening boast suggests) and to express the necessary subordination of a man to the structure of clan authority with which he identifies. In the same way, naming signifies the independence of the child from its mother, by means of its dependence on and identification with agnatic authority. At these critical moments of production and reproduction, speech coordinates the dialectical process of identity construction in which separate, independent personhood is asserted as an aspect of collective identification.

The significance of speech lies in its articulation of the processes of teething, naming, and speech acquisition. The capacity for speech is the requisite for being named. By being named, the person submits to the authority and subjectivity of the namer. In the Haya case, the structure of authority is embedded in the collective forms of clanship and generation. At the same time, however, the child must possess the capacity for speech (that is, a tooth) in order to receive a name. Thus, the power of the subjective authority to which the child submits as object is realized in the constitution of the child's subjectivity. The objective submission of the child implies the potential for subjective action. The coordination of these potentials demonstrates that the child's self/world, subject/object orientation is reflexively constructed and infused with the relations and meanings of power that produce it.

Considered as an embodied socialization process, Haya teething reveals a concern with the control of well-being that is not restricted to the infant. Haya actions are intended to prevent the penetration of animal forces suggested by upper teeth and to effect a passage from the maternal to the paternal. The child moves from mere bodily being to clan identity and from disengaged closure to opened agency. Through such actions Haya achieve a measure of control over the body while producing the capacities and orientations of that body. How-ever, these empowering transformations are clearly more than mere transi-tions between binary and rather generic positions. Critical to the efficacy of these transformations is the fact that the body is oriented to the world (includ-ing other bodies, other subjects, and a meaningful order of space and time) and at the same time realizes the orientation in actions which reconstruct that world. That is, bodily action and actions on the body (such as teething, nam-ing, speaking) go on in such a lived world and in so doing produce the world in which they "go on" (cf. Munn 1983, 1986). I have characterized this situation at the conjuncture of the subject/object relationship as the engagement of the body and the world, and I suggest that the body is produced, through its various forms of engagement, as an objectification of meaning, power, and the like.

THE HAYA MORAL GASTRONOMY

Given the significance of teething as a socialization process, and of teeth as icons of the forms of agency this process produces, we can go on to address some of the specific ways in which plastic teeth are believed to afflict children. As we have seen, plastic teeth are primarily associated with symptoms relating to food consumption; an afflicted child is unable to retain food and refuses to eat. The body is reduced to a position of absolute isolation and inaction with respect to food. Such children can neither process food inside their bodies nor appropriate food from others. In terms of the symbolic configuration of this condition, the child afflicted by plastic teeth presents a rather horrific negative projection of the newborn infant's idyllic situation. As I have shown, the child who has yet to cut a tooth is in a condition of undifferentiated bodily being, disengaged from the world yet intimately connected to its mother's body. In contrast, the newborn afflicted with plastic teeth is disengaged from both the world and the bodily connections necessary to sustain total dependence. A child should move from a position of total dependence to one of increasingly active separation. But plastic teeth produce a severance so total that the af-flicted child is reduced to a state of absolute dependence on an inactive body. It is appropriate, then, that teeth, which condense the meanings of this com-plex transformation, should be the pathological cause of these gastrological inversions.

Food, for the Haya, is a total social fact. Through its consumption, pro-duction, and exchange, its preparation, presentation, and anticipation, the Haya imbue their food with the moral order of their lives. "Eating and drinking," they say, "are our wealth!" (Swahili *utajiri*; Haya *eitunga*). Of course, the proper control over food and its manifold potentials is never unproblematic. One's relation to food, for example, often indexes or provides a privileged and condensed expression of the experience of illness. The proverb "The parent gets sick, but doesn't vomit" (that is, "No matter how much trouble your children are, you can never totally divest yourself of them") indicates the complex ways in which the experience of food is deployed in social commentary. This saying suggests that vomiting is a condition that expresses one's own alienation from, and is explicitly contrasted to, the well-being embedded in proper food consumption. Upset with his daughter for failing to return from her grandfather's village, my neighbor told me, "There she stays and enjoys herself [*yalya obushemera*; literally, "eating happiness"], and we at home, we're puking!"

The problematic nature of people's relation to food lies not simply in the complications of illness or the threat of food shortage but, as the example of vomiting suggests, in the range of often contradictory forces and meanings that food-related practices entail. In some African idioms, eating and being eaten are the quintessential acts of domination and appropriation, associated especially with political hierarchy and competition as well as with sorcery and affliction. For the Haya, feeding others is a critical form of asserting authority, and those who eat are thereby subordinated. Eating is a form of appropriation and evaluation in which an entire order of empowerment is, literally, consumed.

In the Haya context, moreover, the satisfaction or dissatisfaction produced by how people eat is intrinsically connected to their moral condition. The wealthy person, say the Haya, always eats well. This, however, is not as transparent a claim as it might seem. The wealthy person eats well *not* because he eats a lot or even because he eats particularly desirable food. Indeed, I was often told that such a person only ate a very small amount of food, yet he still got full and fat. The wealthy person eats well because "he has no worries," and his eating well, regardless of the quantity, allows his food to satisfy him. Conversely, a person who has many "worries" begins to thin out, although not because the "worries" stem from a lack of food—the troubled person can simply eat and eat, but he will never get full. Moreover, a rich person is especially susceptible to growing thin when his money runs out, because his losses will give him a lot of "worries." Eating well is essential to well-being, and at the same time well-being determines how well one eats.

In the symbolic construction of this evaluative process, the Haya person realizes and demonstrates his condition through his relation to food consumption. The body signifies this condition, both by getting fat or thin and by experiencing and displaying hunger and the consequent ability to be satiated. From this perspective, the contrast in subjective experience between well-being and disease (in Haya terms, between having no worries and having many) is embodied in the subject's relation to and control over food. The person who has no worries eats little, yet is always satisfied. This condition implies a certain capacity to distance one's self, which suggests a degree of independence from the immediate context of consumption. Yet such a person appropriates the values of consumption and displays this appropriation through both the subjective bodily experience of being full and the appearance of a fat belly. Those who live well and without worries demonstrate their command over the productive dimensions of consumption by means of their control over themselves. In contrast, troubled persons experience and demonstrate an absolute inability to separate the self from eating. No amount of food satisfies them, yet they continue to waste away. Subjectivity, in this condition, is so reduced to the immediate demands of the body that it becomes the object of that body.

Plastic teeth replicate this troubled condition but in an even more extreme form. The person who is plagued by "worries" grows thinner, yet continues to eat and be hungry. Subjective bodily experience continues to orient the person toward an appropriation of the object world, through which that subjectivity seeks to effect control over that world and itself (that is, the hungry person tries to eat well in order to get fat). What is particularly devastating about children with plastic teeth is that they lack both the ability to appropriate food effectively (as diarrhea, vomiting, and wasting suggest) and the subjective experience to orient them to food. The child's refusal to eat or suckle signifies a condition of total detachment from food-related activities. Yet participation in these activities is the very means by which the subjective capacity that is at risk can demonstrate its viability. Plastic teeth are more than mere "worries"; they sever people from their world, from themselves, and from the collective actions through which human beings process the world and so become human. Plastic teeth are indeed a fatal disease.

SEXUALITY AND SOCIAL DISEASES

There is certainly nothing novel in suggesting that food-related practices and experiences are redolent of sexual associations. Seitel has noted that Haya tales are replete with "metaphorical descriptions of sex that employ the vehicle of . . . food" (Seitel 1977:203). Moreover, eating well is seen to be inextricably linked to sexual capacity and experience. When Haya men marry they usually seclude (*okumwalika*; literally, "to bring inside") their wives for a period of at least several months. During this time the bride is supposed to be fattened and, ideally, will become pregnant. The seclusion and fattening give a husband sexual access to his wife and give a wife the "heat" and "strength" for sexual activity. The physiologies of reproduction and alimentation are also directly linked. The Haya point out that a child develops in its mother's womb because it is fed with the food from its father's farm. Further, intercourse is necessary throughout pregnancy in order to "feed" (*okulisa*) and "grow" (*okukuza*) the fetus; and nonuterine siblings are called "stomach of the father" (*eibunda lya tata*), as the stomach is the origin of both men's and women's sexual secreta (*amanyare*).

The associations between feeding, fattening, and reproduction are not exclusively sexual in nature. They clearly link up with general concerns with the form and organization of productive processes, as has been discussed in relation to questions of growth, cultivation, and socialization. They suggest that the fattening achieved by feeding a bride is essential to the productivity of social relations and the success of affinal attachments. Sexuality is one aspect of the affinal experience that is signified by bodily growth, just as the growth of the child signifies another, related dimension of affinity. The productivity and value of sexuality, childbirth, and successful affinal relations are literally embodied in alimentary processes. Yet these are the very processes that are threatened by the appearance of plastic teeth.

I point to the connections between affinity, sexuality, and alimentary experience in order to further suggest (somewhat tentatively) certain relations between plastic teeth and AIDS. To begin with, there are some formal similarities in Haya characterizations of the two diseases. Like plastic teeth, AIDS is said to have originated in Uganda and to have come to Kagera through the sexual contacts of rich Haya businessmen. More significant, both plastic teeth and AIDS are seen to have their most deleterious effects in the physical wasting of the body. Victims of "Slim," as the appellation suggests, are always recognized by their weight loss (and gossip abounds as neighbors remark on—or anticipate—their fellows' fluctuations). The concern with weight loss, as my assessment would suggest, is clearly motivated by more than the simple observation of AIDS symptoms. Those who suffer from AIDS, like those who have plastic teeth, are clearly isolated from the processes through which personal well-being is achieved, and it is this isolation and diminished well-being that are signified by such drastic slimming.

Affinity and sexuality have become increasingly fractious and uncertain aspects of the Haya's lives. Indeed, recurring epidemics of venereal disease and a continuous increase in the number of women from Kagera who practice prostitution in urban East Africa have been abiding interests of social scientists, doctors, and missionaries. This shifting history of disease has also seen the emergence of local medical practitioners like *watu wa sindano* (Swahili for "people of the needle"), whose injections of chloroquine and penicillin are sought to cure malaria and to prevent the infertility of syphilis. There is, then, this historical connection between sexuality, reproduction, and the practices that address their complications, practices which articulate biomedical categories of disease (the word "malaria," for example, has entered the Haya vernacular) in the Haya context of bodiliness and (re)production.

The fact that "rich businessmen" are said to have brought AIDS with them from Uganda may also be relevant to the appearance of plastic teeth. The spread of AIDS that has made sexuality so intransigent is often linked explicitly to the use of money, not simply in the profession of prostitution but in the everyday acts of people (especially women) whose desire for money is said to be insatiable. In this respect, it is important to note the connection to Rwandan migrants that many of my neighbors cited in their etiology of plastic teeth. Rwandans who have come to Kagera in recent years are thought of as a kind of "homeless" population who live largely as unpaid tenants on the farms of absentee Haya landowners. It seems possible that these migrants' dependence on the productivity of land which they do not and cannot control is a powerful image of the intransigence of the local economy in relation to the wider world. The "product" generated in this kind of tenuous situation—that is, the milk from Rwandan cattle—thus carries with it the instability of the conditions under which it is created. It does not seem accidental, given the local experience of an unregulated and unstable movement of money, persons, and objects in a complex economy, that a highly visible commodity form—plastic—has infected children, just as the pervasive flow of commodities has undermined adult sexuality.

A man whose own son had had several plastic teeth removed said to me once, "Plastic teeth are a kind of children's 'Slim'; but fortunately there are doctors to take care of children. We adults don't have any medicines. We just die." This claim clearly reveals a far from subtle and certainly widespread resentment of the medical alternatives for a community whose abject suffering is ever increasing. But, by pointing to the contrast between the therapeutic courses for these diseases, it seems to suggest that the Haya are seeking to redress the ruptures in their world. While AIDS may spread unabated, plastic teeth can be removed. The fact that plastic teeth is a disease of very young children requires that parents demonstrate concrete control over their children in pursuing treatment. The child/parent relationship provides an occasion for the real exercise of power that is otherwise unobtainable. Moreover, the symbolic configuration of these diseases indicates that plastic teeth may be a product of an AIDS-infected world. For the children who are born with plastic teeth suffer from an inability to engage with the objects and processes of the *world* through which agency and identity are realized and signified. And they are the offspring of parents who face the same inability to engage in the stable and enduring attachments with *each other* that enable them to constitute themselves. Is it too far-fetched to suggest that parents in a world of sexual instability and fracture give birth to children who cannot sustain their bodily integrity?

PLASTIC

Thus far I have focused on the significance of teeth and plastic teeth as expressive icons of Haya socialization processes and as symptoms of a disintegration of personal, bodily capacity in the related fields of alimentation and sexuality. I want to turn now to the question of why these teeth are said to be plastic.

Haya use the term *plasta*, from the English "plastic," much as we do to describe a substance. However, in everyday usage *plasta* refers to a plastic bucket with a 20-liter capacity. So, for example, a person might tell you to "use two plastics of water" when making beer. Other objects with which the Haya are very familiar—bowls, plates, containers, and the like—are also made of plastic. My point is not that there is something about a 20-liter bucket that makes it a privileged vehicle for the "idea" of plastic. What I do mean to suggest is that the general substance, plastic, is understood by the Haya in practical terms of commodity forms and media that enter into their lived experience.

These commodity forms are, as I have said, very much enmeshed in everyday life. The Haya commonly note that plastic objects are very useful because they are more durable than other products. Moreover, food and beer are thought to keep longer when stored in plastic. These, I suggest, are not simply utilitarian observations. Rather, given the importance of the concrete, sequential ordering of productive processes (especially ones relating to the consumption of such "valuables" as food and beer), these statements reveal a concern with plastic's capacity to transform and arrest the temporal processes of cooking, preserving, and decaying.

Plastic's ability to alter the temporal ordering of these processes is specifically linked, in Haya commentaries, to the thermal properties of plastic objects. The durability of plastic containers, as well as the durability they impart to the foods they contain, is seen to derive from the fact that plastic keeps things cold. Thus, for example, buckets and jerry cans may be used to store beer for several days; but the beer cannot be *produced* in plastic containers. Proper banana beer (*olubisi*) is always made by men using large dugout canoes. The production process entails extracting banana juices and then storing them in the covered canoe, which must rest on a bed of the residue of desiccated bananas (*enkamelo*, "that which has been squeezed"). The residue is essential to the fermentation process, as it was described to me, because it transmits the necessary heat to the beer. Similarly, drinking gourds, which, particularly for men, are strongly identified with their owners, gradually develop a reddish patina that many of my informants attributed to the heat that is generated by the owner's continuously holding and touching the gourd.

These examples indicate that heat is essential to local notions of ongoing transformation and development, and they further explicitly locate the source of this heat in the bodily connections that people establish with particular objects (the heat of the *enkamelo* is generated by the men who squeeze it, and the warmth that transforms a gourd's appearance comes from its being held). In both respects, these examples recall the significance of heat that is derived from feeding, fattening, and sexual contact. The coldness of plastic, therefore, would suggest the introduction of a kind of fixity to these warm, ongoing processes, and one that stands in an uneasy relation to the forms of bodily connection that are the source of this heat. It should also be noted that other commodity forms are experienced as particularly cold. Houses with iron roofs and cement floors, for example, are said to be colder than thatched beehive huts (*omushonge*), which are rapidly disappearing. These concrete symbolic properties—durability and preservation, as well as coldness—are expressions of the ways in which plastic is experienced within a cultural order as a powerful substance that possesses the capacity to transform as well as threaten those carefully constructed processes of growth, production, and consumption of which it has become an intimate part.

But plastic goods are also clearly recognized as commodities. This means not only that they can be marked as foreign objects, but also that they embody an order of circulating media, especially money, that stands in uneasy relation to such productive processes as eating and sexual activity. The tensions of this relation are well illustrated by the gifts and amulets provided when a Haya child is teething. As we have seen, money is given to the child at this time; indeed, a coin is tied to the child's waist. Here, the first tooth of the child indexes the development of the child and demonstrates its capacity to engage in social interaction as a defined, named person. The gift of money demonstrates the child's new identity—but the meaning of money is mediated by the form and timing of the gift, which carefully controls its circulation. Plastic, and especially plastic teeth, are inversions of this relation between persons and objects. While the child who has cut a tooth demonstrates the capacity to be independent and thus to control and participate in the circulation of objects, plastic teeth possess a life of their own that subverts this very form of independent agency and personhood and thus controls the child they afflict.

What seems especially important about plastic goods is the way in which the substantial and material form of plastic objects—that is, their very plasticity—is closely linked to their form of social circulation, the commodity form. There is clearly a "ready-made" quality to the plastic items available in Kagera. The fact that bowls and sandals, buckets and combs are all seen to be made of the same substance yet assume such thoroughly distinct forms further suggests the dynamic interconvertibility of plastic itself. Plastic goods are, therefore, presented to Haya consumers as *fully realized* objects whose distinctive commodity forms bear no recognizable relation to the common *content* of plastic or to the production process through which such mutability is achieved. In this respect plastic is, again, an inversion of teeth (and other bodily forms). Teeth, as has been shown, are properly created by a carefully mediated sequence of events and actions. Their particular qualities and symbolic configuration are built up in the course of everyday practice, as well as of ritual actions, so that their specific properties are linked to their concrete form. Plastic is meaningful in such a context because it intrinsically condenses or obscures its production process and reveals itself in nonunique forms. Plastic commodities are, then, evidence of the dynamic ability of the contemporary political economy to assume all manner of transposable forms.

I am not suggesting that commoditization exists or can be assessed as a process unto itself. Commoditization is a totalizing process that transforms all productive practice, just as it transforms the potential of all the objects of this practice. For the Haya, commoditization, with its easy convertibility of objects, alters the spatiotemporal connections between persons and things that are generated through work, cultivation, inheritance, affinity, and other generative activities. To cite perhaps the most important instance of this process, land alienation results in a severance of the attachments between clansmen and their clan farms, attachments that can only be realized through the transmission of the land from clansmen to fellow clansmen. Those who sell their clan lands "just eat" their money, but they "can never get full." The symbolic and material potential of this kind of commoditization is often represented in consciousness in the concrete forms of the media through which it operates. Plastic is just such a medium and, I would argue, has the same disruptive potential. Used in the daily course of domestic affairs, plastic is a part of every person's intimate experience. At the same time, plastic carries the symbolic weight of commoditized practices that render transient and dislocate forms of well-being that should characterize such experience. Thus, far from being the predictable scourge of meaning in social relations, material forces and commodity forms like plastic often engender such dense symbolic representations and practices. These forces and forms are always simultaneously destructive and creative, alienating and intimate. Plastic teeth, then, are a vivid image of the degree to which Haya experience and the social body have been imbued, indeed infected, with the power of this historical transformation.

CONCLUSION

I have discussed the significance of an affliction by situating its signifying process in a concrete order of cultural practices and historical transformations. Plastic teeth have emerged under specific conditions and are caught up in a nexus of relations with other diseases, forces, and objects. The imagery of these wider powers is tangibly objectified in the body, but their effects can be traced through every aspect of the condition in which the Haya find themselves today. In a time of increasing land fragmentation, a collapsing coffee market, incomprehensible currency devaluations, and growing rates of HIV seropositivity, the Haya experience an increasingly eclipsed capacity to control the forces that give their lives meaning. Perhaps we should not be surprised to find that through engaging in such a fragmented and fragmenting world the Haya are producing themselves in its image.

REFERENCES

Beattie, John. 1960. "On the Nyoro Concept of Mahano." *African Studies* 19: 145–150.
Bourdieu, Pierre. 1979. "The Disenchantment of the World." In *Algeria 1960*, 1–94. Cambridge: Cambridge University Press.
Dauer, Sheila. 1984. "Haya Greetings." Ph.D. dissertation, University of Pennsylvania.
Jackson, Michael. 1989. *Paths Toward a Clearing*. Bloomington: Indiana University Press.
Merleau-Ponty, Maurice. 1962. *The Phenomenology of Perception*. London: Routledge and Paul.
Munn, Nancy. 1983. "Gawan Kula: Spatiotemoral Control and the Symbolism of Influ-

ence." In *The Kula: New Perspectives on Massim Exchange*, ed. J. Leach and E. Leach, 277–308. Cambridge: Cambridge University Press.

———. 1986. *The Fame of Gawa*. London: Cambridge University Press.

Seitel, Peter. 1977. "Blocking the Wind: A Haya Folktale and Interpretation." *Western Folklore* 36: 189–207.

Taussig, Michael. 1980. *The Devil and Commodity Fetishism in South America*. Chapel Hill: University of North Carolina Press.

Wallerstein, Immanuel. 1974. *The Modern World System*. New York: Academic Press.

Weiner, Annette, and Jane Schneider. 1989. *Cloth and Human Experience*. Washington: Smithsonian Institution Press.

CONSTRUCTING A "GOOD CATCH," PICKING A WINNER: THE DEVELOPMENT OF TECHNOSEMEN AND THE DECONSTRUCTION OF THE MONOLITHIC MALE

MATTHEW SCHMIDT AND LISA JEAN MOORE

Donor insemination (DI), the attempt to impregnate a woman with semen from a donor, can be a relatively simple procedure:

> You suck the semen into a needleless hypodermic syringe (some women use an eye dropper or a turkey baster), gently insert the syringe into your vagina while lying flat on your back with your rear up on a pillow, and empty the syringe into your vagina to deposit the semen as close to your cervix as possible. (*Boston Women's Health Book Collective* 1992:387; see also Federation of Feminist Women's Health Centers 1981)A

Information on how women can inseminate themselves is easily accessible. The necessary technology is available in a kitchen. The ease, cost, and accessibility of this procedure create the potential for radical redefinitions of reproductive processes and the social relationships surrounding those processes. A traditional American homemaker's turkey baster can be transformed into a means of independence from patrilineal kinship relations, and new definitions of families can emerge.

Despite the simplicity of this procedure, the semen banking industry has become enormously profitable over the past twenty years. Many semen banks have become subsidiaries of diversified medical services corporations. As fledgling members of the United States medical-industrial complex, semen banks are now diversifying to offer newer and more elaborate reproductive services and technologies. Industry expansion has largely been predicated upon the use of marketing strategies that influence the ways in which potential consumers perceive the processes of reproduction. These marketing strategies shape emerging discourses over the meaning of reproduction in our society. They portray gender, technology, and the medical community in ways that advance the interests of capital accumulation and may have a substantial impact upon how we choose to procreate. In this chapter we examine the ways in which semen banks use discursive strategies in promotional materials and donor catalogues to construct the semen they sell as technologically superior to "natural" semen. These strategies reify differences among semen donors and contribute to the maintenance of hierarchies among men.

In our first section we briefly outline the emergence of semen banking. Second, we situate semen banking within the United States medical-industrial complex. Third, we discuss the importance of representation in reproductive discourses. Fourth, we explore the means by which semen banks inscribe cyborg identity to their products. Finally, we examine the ways in which such inscriptions portray masculinity and procreation.

SPAWNING NEW FRONTIERS: THE HISTORICAL DEVELOPMENT OF ARTIFICIAL INSEMINATION

In 1550, Bartholomeus Eustacius recommended that a husband guide his semen toward his wife's cervix with his finger after intercourse in order to improve the couple's chances of conception. This was the first recorded suggestion in Western medical literature that humans could control their own reproductive capacities through the manipulation of semen. Throughout the sixteenth, seventeenth, and eighteenth centuries various strategies of artificial insemination for fish and livestock were developed with success. The first recorded incidence of mammalian artificial insemination (AI) was in 1742 by Abbe Lazarro Spallanzani. He "injected dog sperm into a female bitch, who sixty-two days later became the mother of 'three little vivacious puppies'" (Finegold 1976). Successful impregnation of women using AI soon followed. By the nineteenth century physicians across Europe and the United States were using this procedure for married couples of the upper and middle classes who were having difficulty reproducing.

The prevalence of donor insemination was low until the early 1960s. Several conditions were important in preventing earlier development of the industry. First, theologians, reproductive scientists, and women concerned with increasing self-control over reproduction were waging war over its morality. Second, demand for the procedure was relatively low. Finally, the technology which would allow for its full exploitation had not yet been developed. Advancements in cryopreservation techniques during the mid–twentieth century allowed for the indefinite preservation of semen. Commercial bovine semen banking developed rapidly after the introduction of lycerol as a cryopreservative in 1949. However, human semen banking took a different route. Between 1954 and 1972 a number of human semen banks were established in the United States, all of which were university based and research oriented.

The first commercial human semen bank opened in 1972. Since then, semen banking has become a $164-million-per-year industry in the United States. The rise of semen banking has been predicated upon several important social conditions: (1) control by medical professionals; (2) advances in reproductive technologies; and (3) the expansion of the medical-industrial complex.

The growth of the semen industry is dialectically embedded within multiple contemporary situations. On the broadest level, the medical-industrial complex has undergone a significant change in the last two decades involving industry growth in competition for patients, increased investor ownership of medical services—referred to as "corporatization"—and the exploding advances of biotechnology. Increasingly complex processes of medicalization have transformed this method of reproduction into an elaborate event, requiring the assistance of multiple actors in the medical-industrial complex. Irving Zola (1990:401) locates the growth of medicalization in four processes:

> first, through the expansion of what in life is deemed relevant to the good practice of medicine; secondly, through the retention of absolute control over certain technical procedures; third, through the retention of absolute access to certain "taboo" areas; and finally, through the expansion of what in medicine is deemed relevant to good practice of life.

Semen banks participate in each of these medicalizing processes. The vast majority of semen banks are owned and/or operated by physicians. The expansion of control by physicians over access to the technical processes of donor insemination and the information relevant to its practice has become evident in that many banks refuse to send semen directly to buyers. They will send samples only to physicians—ostensibly so that the client will have the greatest chance at impregnation given the physician's superior knowledge and abilities. Finally, by selecting certain donors and rejecting others, semen banks judge the physical, social, and psychological attributes that will produce semen that will, in turn, increase the chances of impregnation and produce healthier babies.

These trends are situated within a broad economic context of supply and demand. Semen banks must address issues of demand for their products. One method is to expand their consumer base. Another is to increase the number of services and products sold to each customer. Semen banks attempt to do both in the materials they normally send to prospective consumers. They address the demand of the market by presenting their products in ways familiar to consumers; shopping for donor semen is presented as being similar to shopping for clothing out of a catalogue. In order to increase sales to "hooked" customers, commercial semen banks develop peripheral services and products (semen accessories) and market them as necessary for reducing the risk of producing abnormal or sub-par children. Thus, semen banks attempt to influence consumer demand and increase sales.

Semen banks produce a great deal of written and visual material. These discursive materials are constructed at the intersection of many social worlds, including medicine, genetics, biology, law, ethics, and marketing. Semen bank promotional materials reveal the values and ideologies of those involved in their production. Linda Singer (1993:38) notes that:

> Advertising depends on marketing, which is the science of constructing, dividing, targeting, and mobilizing consumers. Marketing entails the transformation of an audience from one of potential to actual consumer. This is accomplished through a series of interrelated strategies, most saliently, market segmenting and establishing tactical specificity, which involves recognizing, producing, and proliferating differentiated needs in the services of profitability.

But not everyone approves of how semen banks transform browsers into consumers and semen banking into a highly profitable means of capital accumulation.

While many semen banks are being assimilated into corporate structures, others are maintaining their independent, noncorporate, non-university-based status. Some semen banks cater only to married women, others sell mail-order semen to single women and lesbians. Some banks offer a full line of reproductive services including IVF [In Vitro Fertilization], GIFT [Gamete Intra Fallopian Transfer], and genetic counseling; others simply sell semen. This diversity should not be overemphasized; all but one of the banks we sampled were for-profit ventures, 31 percent offered reproductive services in addition to semen, and more than 50 percent would send semen samples only to a physician.

A number of feminist researchers have made important critiques of semen banking practices as battlegrounds for social struggles between men and women. Little attention has been given to how these practices may affect relations of power among men. We argue that discursive practices used by semen banks to sell products construct differences and hierarchies among men and support the perpetuation of hegemonic forms of masculinity.

Organizations specializing in reproductive products and services are often self-protective and suspicious of outsiders requesting access to information. Tabloid accounts of the semen bank industry have further encouraged semen banks to step up their organizational gatekeeping. We have sidestepped the industry's resistance to scrutiny by analyzing preexisting, publicly available materials.

An accurate listing of the industry is impossible because semen banks are not highly regulated by any federal governmental agency. The American Association of Tissue Banks (AATB) is currently the only accreditation agency that compiles a major listing of tissue banks; we selected our sample from its listing. But not all banks belong to the AATB.

Requests for promotional materials were mailed to all forty-six listed AATB agencies. Thirty-five of these semen banks responded (76 percent). We carefully reviewed each set of materials and analyzed them for common themes and variance. From these thirty-five, we chose a purposive sample of seven banks' materials for in-depth analysis using content analysis and grounded theory.

DOWNLOADING A DREAM DADDY: TECHNOSEMEN AND THE CONSTRUCTION OF MALE DIFFERENCES

In this section, we explore two aspects of the semen enterprise. First, semen banks represent their products as being of superior quality. We define this new, improved semen as *technosemen* and investigate its manufacture and representation. Second, we examine two important outcomes of representing technosemen for sale. These are (1) the deconstruction of a monolithic male and the ensuing production of hegemonic masculinities; and (2) the construction of donor semen as cyborg.

TECHNOSEMEN

From a semen bank's promotional materials:

> We believe that the quality and safety of the individual specimen must take precedence over all other considerations, therefore, this program is guided by the following principles:
> SAFETY Systematic, mandatory testing aims to insure pathogen-free specimens;

CARE Patient-oriented communication meets the complex concerns of re-cipients and helps achieve satisfactory donor matching;

SUCCESS The selection process is specifically designed to include only those donors whose specimens exhibit a high likelihood of fertility through frozen/thawed techniques.

We define *technosemen* as the "new and improved" bodily product that semen banks advertise to clients through their informational pamphlets. Tech-nosemen is the result of technologically based semen analysis and manipula-tion. Technological manipulation of semen is carefully presented to potential clients and described in great detail by each bank. For instance, semen analysis includes sperm counts, morphology, motility testing, functional testing (in-cluding the hamster penetration assay), and sperm washing ranging from the swim-up methods to percoll or the two-step simple wash. Sperm counts in-volve taking a small sample of semen from a donor and counting the number of viable sperm to determine the overall amount. Sperm counts may involve conducting morphology (shape assessment) and motility (movement assess-ment) tests as well. Computer assisted semen analysis (CASA)—which uses digital computer imaging devices connected to microscopes—is now available to conduct all of these tests. This technology has spurred the development of a new language for semen analysis; technicians can now measure the velocity, linearity, and wobble of sperm movements. Each of these new measures is represented as being highly correlated with semen fertility. Morphology testing may be done using electron microscopy; the resultant "spermprints" give fer-tility technicians another means of assessing sperm health.

In addition to these diagnostic tests, postcoital functionality tests are con-ducted to determine the fertility of semen samples. The hamster penetration assay uses hamster eggs to determine if semen is capable of penetrating an egg. Semen antibody testing is common. Often infertility is blamed upon immune reactions to semen from either seminal fluid or cervical mucous. Antibody testing is done to determine the reaction of sperm to these fluids.

Donors and semen samples routinely undergo disease and genetic testing. Semen is cultured for sexually transmitted diseases such as gonorrhea, chla-mydia, and gardnerella. Microscopic analysis reveals any white or red blood cells in specimens. Donors are screened for genetic markers to Tay-Sachs disease and sickle-cell anemia. Donors are also screened for blood type, hepa-titis, syphilis, HIV, CMV, and HTLV-I. Semen banks quarantine semen for six months so that donors can be retested for HIV infection.

Donor semen which passes these diagnostic and function tests is eligible for functionality enhancement by a variety of techniques. Sperm washing is sometimes recommended to consumers if physicians have determined that they have "abnormal" or "hostile" cervical environments. Each type of wash method is conducted in a laboratory. The "swim up method" involves centrifuging the semen sample, removing the seminal fluid, and placing the remaining sperm pellet in an artificial insemination medium. After an hour, the most motile and active semen which "swim" to the top of the solution are retained. Percoll washing involves layering the semen with percoll (a solution to "clean" semen) and centrifuging for thirty minutes. In addition to increasing the motility of semen samples, percoll washing reduces the amount of seminal bacteria in each sample and may reduce the number of sperm with genetic defects. After washing, various performance-enhancing media can be added to sperm. Caffeine can be added to increase motility. A fluid marketed as Sperm Select™ is more viscous than cervical mucous and can allow sperm to swim more easily after insemination. Finally, many banks are now offering pre-sex selection services. A variation of the swim-up method of sperm washing can be used to increase the number of either X or Y chromosome-bearing sperm in each sample, thus, theoretically, increasing the chances that resulting children will be either female or male. The effectiveness of this procedure is a matter of great debate in the reproductive sciences.

Semen analysis, disease testing, and manipulation together create and constitute technosemen. Marketing technosemen can increase demand in the semen market by convincing or, better yet, guaranteeing the general public that technosemen is fertile, uncontaminated, and genetically "engineered" for desirable traits. In this age of AIDS, geneticism, and environmental disasters, semen banks can capitalize on the promise these technological procedures offer, thus increasing their revenues. In so doing, semen banks reinforce and bolster public concern about these material threats to the future of the human race. The processes by which technosemen are produced and advertised to consumers are deeply connected with how the "contents" of the semen are represented in promotional materials.

In "The Materiality of Informatics," N. Katherine Hayles (1992) discusses how the body simultaneously produces and is produced by culture. Hayles posits that through a perception feedback loop characteristic of the late twentieth century, the body becomes "dematerialized." In other words, our means of perceiving the body and embodied experiences are dramatically changing through applications of technologies such as those employed in creating virtual reality or Internet chat rooms. Our experiences of familiar bodily responses (sensations and corporeality) become disengaged from our sub-jectivity. One method of bodily dematerialization is the attachment of dis-cursive inscriptions to bodily products, like semen. This transformation of bodily fluids into discourses disciplines the ways in which people act and per-ceive bodies.

Automated laboratory machines turn semen into technosemen. These ma-chines assist in what Hayles might call decreasing the friction of spermatic materiality. Semen becomes inscribed as malleable. Prior to the development of this technology the physiological character of semen was perceived as much less controllable. Once the genetic code was finally cracked, the manipulation of semen was largely viewed as the manipulation of genetic information. Now through the use of new reproductive technologies, semen can be tailored to fit consumer demand. Semen banks utilize promotional materials that articulate, both explicitly and implicitly, inscribed ideologies based on genetic, eugenic, medical, and other bioscientific discourses. These inscriptions portray malle-able semen as highly desirable to consumers.

BODIES OF WRITING: GENDER, RACE, AND THE CONSTRUCTION OF CORPO(REALITY)

In her well-known essay "The Egg and the Sperm: How Science Has Con-structed a Romance Based on Stereotypical Male-Female Roles," Emily Martin (1991) deconstructs reproductive stories about male and female gametes—cells—in scientific textbooks. She suggests they are tropes that reveal cultural beliefs and practices. Martin (1991:500) is concerned about "keeping alive some of the hoariest old stereotypes about weak damsels in distress and their strong male rescuers." Throughout her analysis, we learn how cells become performers acting out heterosexist fantasies/realities of patriarchal culture; for instance, she (1991:499) cites examples of sperm being personified as going on a "perilous journey" and being "survivors" where the egg is seen as "the prize." Martin's (1991:501) work implores us to investigate the sites of such gendered constructions of biological functions: "More crucial, then, than what kinds of personalities we are bestowing on cells is the very fact that we are doing it at all. This process could ultimately have the most disturbing social consequences." What happens in this process of giving personality to biological objects, ac-cording to Martin . . . , is that we naturalize socially, materially, and bodily experienced, but nonetheless constructed, gender inequities.

In Nancy Stepan's (1986) work, analogy and metaphor in science become naturalizing lenses through which social situations are brought into focus. In her words (Stepan 1986:274),

> because a metaphor or an analogy does not directly present a pre-existing nature but instead helps "construct" that nature, the metaphor generates data that conform to it, so that nature is seen via the metaphor and the metaphor becomes part of the logic of science itself.

Stepan implies that those who do science are culpable participants in processual obfuscation of material conditions through linguistic manipulation.

Both Martin's and Stepan's discussions attend to ways in which difference is constructed through reference to the body. Those who participate in the semen enterprise draw deeply on such cultural metaphors of the body to construct spermatic difference and to create the body and body fluids as incontestable (because of their material reality) sites of true difference.

Semen banks engage in a process of difference naturalization. Constructions of embodied differences among donors are inscribed in semen catalogues. Since catalogue listings refer to specific vials of semen, the semen itself is inscribed with these same differences. For example, there are many shades of skin and varied biographies of people we call "African American"; nevertheless the semen becomes "the African American man with a GPA of 3.2 with interests in sports and music." The commodification of race, through the advertisement practices and discursive constructions, reifies differences between men. In using the metaphor of "African American" or, more commonly, "black" to describe the donor, semen banks not only deny differences among variously colored men but also are constructing differences between these groups and other men. This metaphor reifies absolute differences between "the blacks" and "the whites." Within the semen banks' donor catalogue conventions, the metaphors are taken literally (as often occurs in everyday life), supporting Stepan's claim that "they (the metaphors) tend to lose their metaphorical nature and to be taken seriously" (1986:275).

DISCOURSE TEMPLATES

What are the consequences of semen bank promotional materials on corporeality? In one sense, semen banks dematerialize the body in the perception feedback loop. Body parts are turned into discursive statements, labels, and metaphors in the donor catalogue. For instance, one catalog reads:

Donor ID	Race/Ethnic Origin	Hair	Eyes	Ht./Wt.	Blood	Skin
001	Black/Creole	Dk Br	Br	6–1/180	A+	Med

Yrs College	Occupation/Major	Special Interests
1	Theater Arts/Drama	Comedy, Boxing, Guitar

On one hand, the body of the donor becomes encoded into phenotypic, sociogenetic information. On the other hand, this dematerialization is not complete because semen is part of the body. The site of inscription is recast in the semen that represents a concentrated body, paralleling other liquids from concentrate. Semen banks market the chance to rematerialize, reconstitute, and reproduce the body of the donor. This chance to rematerialize is presented to the client through discursive statements about the donor. Extending Hayles's argument, semen banks complexify the process of de/materialization of the body. Semen is simultaneously a part of the body, a potential body, and, as represented in a donor catalogue, a series of codes.

Semen banks inscribe semen using *discourse templates*. These are modes of representing information that have become routinized and appear in specific types of information presentations. Discourse templates organize novel or exotic information in ways that are highly familiar to an audience. Such templates may traditionally have been used to present other types of information. Templates allow for quick selections among many similar choices. Joel Best (1991) has observed that these templates may create hegemonic dominance over definitions of certain social phenomena.

Semen banks usually choose one or two common discourse templates that may be familiar to their clients, such as personal ads or charts. These templates create comfort zones and mitigate the strangeness of the new market. Consider the personal advertisement template. The conventions of personal ads include abbreviations and brief descriptive narratives like those used to meet people in the personals of newspapers. Donors are introduced in paragraphs, without names. Limited information is presented; more detailed information is available for additional cost. For example:

He has wavy dark brown hair and eyes. He's 5′ 10″ tall and weighs 156 lbs. on a medium frame. His complexion is fair to medium. His ancestry is Eastern European. His blood type is A+. He is currently studying law with a GPA of 3.4. Other interests include music, classical and rock. He also enjoys tennis, cars, ice-skating and juggling.

He has black hair, wavy, with dark brown eyes and a fair complexion. He's 5′ 10″ tall and weighs 145 lbs. on a small to medium frame. He is a student majoring in electrical engineering with a GPA of 3.0. His blood type is o positive. His ancestry is Chinese. His interests include racquetball and tennis. He also plays the piano, clarinet, French horn and percussion instruments.

The discourse template of personal ads thus encourages the client to think of semen purchasing as akin to a dating game; will they choose sperm number one, sperm number two, or lucky sperm number three? Compatibility and socially desirable properties such as evidence of upward mobility, intelligence, and social integration are portrayed as important in choosing the right semen. Efforts are clearly made to convince the woman that she is choosing a man rather than sticky, wriggly, little cells. Sperm may be disembodied, but they are vividly personified.

These templates provide a method of quick and easy rational comparison among many different entries and products. The buyer has the opportunity to choose both attractive semen (which through comparison of phenotypic characteristics will produce an aesthetically pleasing child) and successful semen (which through comparison of social and psychological traits will produce a child who is a winner). The banks use donor catalogues as a means of creating reproductive dominance through the construction of male differences.

Three types of information are highlighted by semen banks; they serve as ways to easily segment the market, but they also do much more. Semen banks prioritize differences believed important to the client through the ordering of the characteristics of donors. These characteristics include (1) race/ethnic origin; (2) social characteristics; and (3) characteristics of social and physical power.

(1) Race/ethnic origin is always the first category presented; several personal ad–type lists are sectioned according to race. Race is thereby created as the primary choice. One bank even stores its semen in vials that are color-coded according to the race of the donor—white vials for Caucasian, black vials for African American, yellow vials for Asian American, and red vials for everyone else. Differences are thus visually reified and metaphorically stereotyped through these vials. Semen from donors with mixed racial and ethnic backgrounds is grouped as (the) Other and colored red (which suggests to us the mythic themes of the color red: revolutionaries, communists, Native Americans, redskins, enemies, emergencies). However, it is not clear that any racial group is presented as more desirable by semen banks. Most semen banks market semen from donors with a wide variety of racial and ethnic backgrounds. Anecdotal evidence suggests that most consumers prefer to

buy semen from donors with similar racial characteristics to their own. Thus, presenting any one racial group or groups as more desirable than others could prove costly to the bottom line.

(2) Social characteristics are usually listed toward the end of each entry. Although they are not given as much significance as phenotypic characteristics, they are presented as consequentially creating differences in semen. These differences imply that semen may be qualitatively different because of the donor's personal history. For example, a donor who likes to run, swim, and read may culturally indicate healthier semen than a donor who smokes, rides a motorcycle, and juggles. The "scientifically disproven" Lamarckian assumption of the inheritance of acquired characteristics is both re-created and sustained in these catalogues.

(3) In addition, these categories of differences among men are fundamentally based on strata of both social and physical power. Categories of height, weight, body build, and favorite sport provide consumers with indicators of the donor's health and ability to be physically dominant. Categories of occupation, grade point average, and years of college provide indicators of social survivability and social dominance. Donors who do not rate highly within these categories are not included in these catalogues, nor presumably is their semen stored in banks as sellable inventory.

Thus, these categories tend to reify power differences among and between men. A hegemonic masculinity is created and reinforced. Carrigan, Connell, and Lee (1985:92) have defined hegemonic masculinity as "a question of how particular groups of men inhabit positions of power and wealth, and how they legitimate and reproduce the social relationships that generate their dominance." Using this definition, Mike Donaldson (1993:653) has argued that "the view that hegemonic masculinity is hegemonic, in so far as it succeeds, in relation to women is true, but partial." Donaldson argues that within such a system of masculinity, one of the defining characteristics is control by a very few men over a great many men and women.

While the call has been made in masculinity studies to deconstruct the monolithic male, we have found that this deconstruction and the pursuant construction of differences among men may itself then involve the perpetuation of hegemonic masculinity rather than its demise. Connell (1987) has suggested that masculinity expresses itself in authoritarian ways through fatherhood. Owners and operators of semen banks are overwhelmingly men and take on the role of surrogate fathers in the reproductive process. In so doing, they have the power to determine who may be a donor and who may not, which social and genetic characteristics should be embodied within the semen they sell, and what the social relations of donors will be to the children produced with their donations. Our research suggests that semen banks are indeed wielding this power to prevent certain types of men from participating in semen donation. We are not suggesting that the semen banking industry is solely responsible for this situation. To a certain extent consumer demand limits the ability of banks to offer a wider selection of donors. Joanna Scheib (1994) has found that women purchasing semen in Australia tend to prefer donors with superior health and social skills. However, semen banks reinforce this desire, limit the categories across which women may choose donors, and construct hierarchies from categories previously unknown to consumers, e.g., sperm motility. While semen banks have the potential to radically open up control to women over their own bodies, to assist in the proliferation of heterogeneous family structures, and to provide men with alternative means for participating in the reproductive process, such promises are arguably being significantly compromised by relations of masculinity within the industry such as those described here.

The construction of technosemen has consequences for the understanding of semen ontology and how semen is inscribed as cyborg.

MY DADDY CAN BEAT UP YOUR DADDY!:
THE PROBLEMATICS OF CONSTRUCTING CYBORG ONTOLOGY

We have used our research to discover the political economy of semen banking. We have shown how the material as well as discursive construction of technosemen plays an important part in the political economy of the semen banking industry. Technosemen is not only a product that promises increased fertility to its consumers, it is the material that provides fecundity to a growing young industry. Out of technosemen a new semen ontology has emerged. At once inscribed with human characteristics provided by donors and super-human characteristics provided by technological manipulation, semen becomes technosemen, bodily fluid becomes cyborg.

Semen banks anthropomorphize the semen they sell across many different dimensions. Through discourse templates, semen is inscribed as embodied, socialized, racialized, and gendered. Some banks use other means to highlight the human characteristics of their semen. One bank prints a cartoon of sperm on the front of their informational pamphlet. Sperm in this image are shown with human facial characteristics, smiling, waving hello, and wrapped in scarfs to protect them from the chilling cold of liquid nitrogen. Here sperm are portrayed as emotional, friendly, communicative, and using tools. In one sense, the human body itself becomes a discourse template in an attempt to make the banking experience "user-friendly."

Semen banks simultaneously reinforce certain cyborg characteristics of semen. Semen is portrayed as encoded, disciplined by technology, and superior in potency to unprocessed semen. The construction of this cyborg ontology is an integral part of semen bank marketing. It allows banks to make claims about the potency of their products, while at the same time making claims as to the "naturalness" of new reproductive technologies. These constructions suggest that new reproductive technologies are not unnatural but rather an improvement upon the inherent unpredictability of natural procreation. As Hermann Rohleder, one of the first physicians to write extensively about reproductive technology, claimed, "Artificial fecundation is not an unnatural method because it is scientifically founded" (Rohleder 1934:142).

One consequence of this cyborgization is that semen banks are able to construct a discourse of reproductive risk, capitalizing upon consumers' concerns about the risks of producing defective children. "Natural" unprocessed semen is described as irrational, dirty, and unpredictable. For example, semen banks emphasize that unprocessed semen can produce genetic defects, cause STDs, and fail to get to the egg. What semen banks are selling is the ability to control risk and to harness the agency of semen in order to coerce it to act more rationally. This is more sellable to the consumer. As with the technologies of childbirth, creating and perpetuating a discourse of risk is a marketing strategy used to encourage the consumers to invest in technosemen. This discourse is beginning to proliferate among various medical and scientific worlds. For example, a recent article in the journal *Nature* reports the research of a zoologist from the University of British Columbia who claims that males are the weak link of our species. She finds that sperm are more likely to create genetic "disasters" than eggs. Her research suggests that it would be less perilous for women and their offspring to be able to reproduce without men or to seek out younger mates who produce fewer genetic deformities in their semen. Although semen banks have not (to our knowledge) capitalized on this particular research, claims such as these enhance the semen banks' ability to market their products and services. Semen banks construct technosemen as [being] less risky to consumers, as being able to create better children, taller, smarter, and more musical, than those produced by "natural" semen.

The development of this discourse of reproductive risk means that technosemen is represented as less an option than an imperative. It becomes a necessary element in conception because it is the product of a process which weeds out unwanted semen (and donors) and creates a superfertile substance. Only some semen is good enough to be made cyborg; thus at its very essence technosemen is divisive. Only elite semen samples are allowed to undergo the disciplining processes of new reproductive technologies. Through the construction of semen as cyborg, hierarchies of potency are established across categories of men.

However, creating a new cyborg ontology establishes the potential for a new geometry of social relations that is non-dichotomous and non-hierarchical. By restructuring the ways in which technosemen is created, both discursively and technologically, the politics of the semen enterprise become potentially liberatory. Strategies that semen banks use to reduce the stigma and strangeness of semen banking, e.g., employing discourse templates for product presentation, could be used to open up a diversity of reproductive possibilities for both men and women. Instead of the ranking of semen and men according to characteristics that represent hegemonic forms of social and physical power, all men would be eligible to participate in semen donation. Rather than excluding the semen of men who do not rank highly across categories of power, new technologies would be used to empower the semen of all men, especially those with fertility problems. However, semen banking has not thus far lived up to this potential. Draper (1993) has explored how the construction of risk works to create a hegemony of corporate control, culturally privileges certain actors, and reinforces stratified social relations based upon inequities. Technosemen has had its cyborg ontology constructed within a web of related interests that utilize a discourse of risk to enhance profits, bolster professional medical dominance over reproduction, and maintain hegemonic forms of masculinity.

IMPLICATIONS

We conclude by highlighting three significant issues raised by our analysis. First, semen banking is reconstructing the moral terrain of reproduction. If indeed it becomes the reproductive wave of the future, whose privilege or responsibility will it be to procure, use, and invest in such "healthy" semen? More specifically, what are the implications of this technology and the configuration of the industry for women, men, and couples? Since access to these technologies is still limited by its cost, will women, men, and couples who are unable to gain access to semen bank services come under increasing moral scrutiny for failing to reproduce appropriately?

Second, what was once a not-for-profit, informal, altruistic service largely available through medical school clinics has been transformed into a for-profit medical industry, largely outside of the new managed care medical-industrial complex. What are the implications for the social organization of reproduction? Is privatization of costs the new trend that will be sustained? Will children produced via such methods and their parents be further privileged in terms of access to insurance coverage because of assumed healthier outcomes?

Finally, feminists need to consider and explore other sites where the construction of male difference occurs. As we have demonstrated, even though men are often portrayed as benefiting as a group from the medicalization of reproduction, feminists must develop more sophisticated analyses of late capitalist technologies that diversify embodied genders. There is not one straw man in late capitalism, there are many. "Capital has fallen in love with difference" (Clarke 1995:146). In addition, men are differentiated across strata related to power and dominance by capitalist enterprises as a means of expanding markets and increasing profits. This process has a twofold effect, both contributing to the reproduction of existing gender discourses by advancing claims of superiority about men's phenotypical, biological, and social characteristics and altering the conditions of men's reproductive capacities. As semen banks begin to compete for customers, they may limit semen collection to men who rate more strongly across these types of categories. Thus, they may limit the abilities of large numbers of men to participate in these alternative modes of procreation. How semen banks broker their commodities is both shaped by how we think about men and, in turn, influences and is constitutive of our future understandings of men.

REFERENCES

Best, Joel. 1991. " 'Road Warriors' on the 'Hair Trigger Highways': Cultural Resources and the Media's Construction of the 1987 Freeway Shootings Problem." *Sociological Inquiry* 61, no. 3: 327–45.

Carrigan, Tim, Bob Connell, and John Lee. 1985. "Toward a New Sociology of Masculinity." *Theory and Society* 14, no. 5: 551–604.

Clarke, Adele. 1995. "Modernity, Post-Modernity, and Reproductive Processes 1890–Present, or 'Mommy, Where Do Cyborgs Come From Anyway?' " In *The Cyborg Handbook,* ed. Chris Hables Gray. New York: Routledge.

Connell, Robert W. 1987. *Gender and Power.* Stanford: Stanford University Press.

Donaldson, Mike. 1993. "What Is Hegemonic Masculinity?" *Theory and Society* 22: 643–57.

Draper, E. 1993. "Fetal Exclusion Policies and Gendered Constructions of Suitable Work." *Social Problems* 40, no. 1: 90–107.

Finegold, Wilfred J. 1976. *Artificial Insemination.* 2d ed. Springfield, Ill.: Thomas.

Hayles, N. Katherine. 1992. "The Materiality of Informatics." *Configurations* 1, no. 1: 147–70.

Martin, Emily. 1991. "The Egg and the Sperm: How Science Has Constructed a Romance Based on Stereotypical Male-Female Roles." *Signs* 16, no. 3: 485–501.

Rohleder, Hermann. 1934. *Test Tube Babies: A History of the Artificial Impregnation of Human Beings.* New York: Panurge Press.

Scheib, Joanna. 1994. "Sperm Donor Selection and the Psychology of Female Mate Choice." *Ethology and Sociobiology* 15, no. 3: 113–29.

Singer, Linda. 1993. *Erotic Welfare: Sexual Theory and Politics in the Age of Epidemic.* New York: Routledge.

Stepan, Nancy. 1986. "Race and Gender: The Role of Analogy in Science." *Isis* 77: 261–77.

Zola, Irving K. 1990. "Medicine as an Institution of Social Control." In *The Sociology of Health and Illness: Critical Perspectives*, ed. Peter Conrad and Rochelle Kern, 3d ed. New York: St. Martin's Press.

ALIENATION OF BODY PARTS AND THE

BIOPOLITICS OF IMMORTALIZED CELL LINES

MARGARET LOCK

In 1993 Isidro Acosta, lawyer and President of the Guaymi General Congress of the Guaymi peoples who live in a remote corner of Panama, met with Adrian Otten in Geneva. At the time, Otten was the senior official responsible for negotiations in connection with Trade Related Intellectual Property (TRIPS), destined to become one facet of the international General Agreement on Tariffs and Trade (GATT). Acosta's purpose was to have human genetic material excluded from patenting under this agreement. But his trip to Geneva was to no avail, with the result that by 1997 the U.S. patenting office had received more than 4,000 applications to patent human genes, over 1,500 of which had been ratified. More than 80 percent of these applications had been made by the private sector, and the number of applications accelerates year by year. Increasingly partnerships between bioscience and the market are considered the most efficient means of advancing knowledge, particularly when genetic sequencing is involved.

When Acosta undertook the trip to Europe it was with a sense of urgency because he had learned, only weeks before, that the Secretary of Commerce of the U.S. government had laid patent claim to a "cell line" created out of the blood taken from a twenty-six-year-old Guaymi woman who had leukemia. If granted, this patent would have given the U.S. government exclusive rights to decide who could use this cell line, and at what cost. The Guaymi General Congress had already made a specific request to the U.S. government that the claim be dropped, but Acosta's journey to Geneva was to try to ensure that patenting of human tissues and cells be outlawed completely. As far as the Guaymi are concerned bio-prospecting (gene hunting) is entirely unacceptable when the end product is patented with the result that life itself becomes the private property of pharmaceutical and other companies. This particular claim to patent was ultimately dropped because the "immortalized" cell line did not prove to be as valuable as the prospectors had envisioned. Similar patent claims made about the same time on cell lines procured from several

Hagahai, who reside in New Guinea, and another from an individual in the Solomon Islands, also created a public furor. The Solomon Islanders were from the outset opposed to patenting and to profiteering from the manipulation of human tissue. In contrast, it was reported that the Hagahai willingly agreed to the donation of blood samples, the creation of cell lines, and to their patenting and further, that should money result from a vaccine or any other end product, then all individuals with a claim to Hagahai ethnicity would receive half the profit. The anthropologist Carol Jenkins, who has been doing research among the Hagahai since 1985, acted as the principal mediator in making the monetary arrangement. Given that the Hagahai had only been "discovered" in the 1980s, at which time they were dubbed a "stone-age tribe," it seems unlikely that many individuals understood the full import of what was involved. Of course, this does not justify any paternalistic claims made by outsiders to the effect that the Hagahai, or any other peoples for that matter, may not decide their own fate in connection with the donation of blood and the creation of cell lines. Eventually, as with the Guaymi case, both the Hagahai and Solomon Island patent claims were dropped, once again when it became clear that the cell lines were not, after all, going to be profitable for pharma-business.

These cases make it clear that it is inappropriate to assume that peoples everywhere exhibit a uniform resistance to the bio-prospecting of human, plant, and animal materials. It is also evident that venture capital is the driving force behind the procurement of DNA from indigenous peoples, and that this activity does indeed exhibit the characteristic features of prospecting, replete with uncertain futures as part of its practices.

Procurement of human materials to make immortalized cells lines, of which the above cases are early examples, was fostered by two international agreements made legally binding in 1993 and 1994, agreements that in effect globalize intellectual property laws. The first, the Convention on Biological Diversity, was adopted at the "Earth Summit" in Rio de Janeiro in 1992, and the second, the GATT TRIPS agreement, was signed in June 1994. These two agreements, once legalized, ensured that virtually all signatories to GATT (now the World Trade Organization), with the exception of only a few very small countries, must agree to intellectual property provisions. In effect, what this means is that individuals who donate their own body parts for research purposes do not retain property rights over such materials, nor can they participate in any profit that might result from manipulation of these materials unless special provision is written into the original transfer agreement, as was the case with the Hagahai. The legal practices of North America, Europe, and Japan, in which genes may be patented once technological artifacts are created out of them (cell lines being the prototypical techno/natural hybrid of this sort), have set the global standard. Genetic information obtained from cell lines is transformed through patenting into private property of a particular kind, namely, intellectual property.

What we are currently witnessing with the escalating procurement of human body materials is a globalized commodity fetishism that goes virtually uncontrolled, in which "regimes of value" (Appadurai 1986), those at the site of production and those at the site of consumption, are at a great remove from one another. Human body parts do not have universal value, and once potentially available for conversion into circulating commodities their worth, and more basically the question of whether or not they are alienable in the first place, is open to dispute. Disputes are implicated in the social exchange of virtually all objects at times, notably because "commodities, like persons, have social lives" (Appadurai 1986:3). But the commodification of human cells, tissues, and organs incites particular concern because boundaries usually assumed to be natural and inviolable are inevitably transgressed, raising concerns about "self" and "other," "identity," "genealogies," group continuity, and so on. Disputation is not simply about ownership, property rights, or alienability; it is also constituted in large part out of a profound angst about a perceived violation of the moral order.

CONTESTED COMMODITIES AND ALIENATION OF BODY PARTS

Commodification of human corpses and of body parts procured from the living and the dead has a long history. Human material has value as trophies of war, religious relics, therapeutic materials, medicinals, and anatomical specimens, among other uses. The value attributed to corpses and their parts is transformed with alienation, a transaction in which, historically in the Christian world, at least, the family of the "donor" rarely participated willingly, for tampering with and division of the body made resurrection impossible. Church doctrine made an exception for saints, the dispersion of whose body parts was necessary for the well-being of the Christian community—holy relics have been traded from the early days of Christianity, and phials of clotted blood and specimens of rubbery hearts were deposited in church precincts from where they emanated spiritual power.

The use of human bodies by the medical profession, although closely linked until the seventeenth century with the Church, was more often than not associated with violence. Vivisection of humans and animals by Herophilus in fourth century BC Alexandria earned him a lasting reputation as the "father of scientific anatomy." Later, commencing in thirteenth-century Italy in Church precincts, professional anatomists performed public dissections of the corpses of criminals and vagrants, a practice that continued until the early nineteenth century in anatomy theaters built in many parts of Europe. This practice ensured that the bodies of individuals on the margins of society acquired outstanding medical value, and Ruth Richardson argues that from the seventeenth century in Europe the human corpse, as already was the case for human relics, was bought and sold like any other commodity. Around this time, too, the practice of robbing graves in order to sell corpses for medical dissection became fashionable.

According to Peter Linebaugh an increase in the trade in corpses in early modern Europe reveals a significant change in attitude towards the dead body:

> The corpse becomes a commodity with all the attributes of a property. It could be owned privately. It could be bought and sold. A value not measured by the grace of heaven nor the fires of hell but quantifiably expressed in the magic of the price list that was placed upon the corpse (1975:72).

Linebaugh documents in gruesome detail the public hangings—the "hanging matches"—that took place in England at this time, notably at the gallows at Tyburn outside London. Surgeons and kin of the hanged person would fight over the body, tear it down, half dead, from the gallows, the one desirous of a body to dissect, and the others, the family, hoping to give the body a decent burial in order, in their minds, to ensure resurrection. The "Tyburn crowd," often in their hundreds, became restless with the passing years, and riots were staged to which the military had to be called to restore order. Incomplete hangings were common and "resurrection," as it was dubbed by the spectators, sometimes occurred once the body was cut down, at which time the victim was usually released, unless the anatomists were the first to get hold of the limp body before its restored life became clearly visible. The Anatomy Act, designed to prohibit the sale of dead bodies, was eventually signed in 1831 after extensive debate, and remains the foundation for modern law in the United Kingdom and North America in connection with the procurement of bodies for scientific purposes . . .

I have made this brief diversion into the commodification of corpses and body parts in Europe to emphasize how competing regimes of corporal value are not limited to the globalized economy of modernity. However, there can be no doubt that the dissemination of capitalism through colonialism, and more recently the current extraction of wealth of all kinds by multinational conglomerates from the "developing" world, has exacerbated the situation. Although, with the exception of blood, much of which was procured from Haiti and other "Third World" locations for American destinations, it was not possible to incorporate living biological material into the global market prior to the last quarter of the twentieth century. The necessary expertise simply did not exist. Nevertheless, a foreshadowing of the contested commodification of body parts so evident today was clear in the extraction of human labor from formerly colonized sites.

Michael Taussig, in writing about the effects of capitalist relations of production among indigenous miners in Colombia and Bolivia, argues that different types of fetishism are at work when the commodity value of labor is contested. He draws on the work of Marcel Mauss on modes of exchange in noncapitalist societies, where it is argued that reciprocity is central to premodern exchange. Mauss's thesis was that the "gifts" that are central to so many noncapitalist economies are not themselves of overriding importance, but that the bonds of reciprocity established between the involved persons, donors and recipients, are primary. In effect, individuals give away a modicum of their "essence" with gifts, which must, therefore, be returned in kind. Taussig argues that the practice of the modern market system strives to deny this metaphysics of persons and things reflected in social exchange and to replace the type of fetishism outlined by Mauss with the commodity fetishism of capitalism as interpreted by Marx. This latter type of fetishism, as is well known, derives from the alienation that arises between persons and the things that they produce that are then put into circulation in a monetary economy. Codified in law as well as in everyday practices, such alienation results in the "phenomenology of the commodity as a self-enclosed entity, dominant over its creators, autonomous, and alive with its own power" (Taussig 1980:124).

Taussig's later research into institutionalized terror and associated healing practices elaborates on the many forms of commodification of the colonized body itself, not simply as labor, but as an object for exoticism and prurient desire. He shows how diverging regimes of value attributed to the native body by exploiter and the exploited result in disorder and disruption of the conventional meanings attributed to human bodies by both parties. Gene prospecting brings to light divergent value systems in connection with human blood, making great the potential for exploitation. Because blood is a renewable resource and simple to donate, it is all too easy for scientists and other outsiders to objectify and fetishize human DNA. Haraway has dubbed this "corporeal fetishism," in that reification of cells, genes, or other body parts obscures a heterogeneous set of interactions between human and nonhuman actors (1997:142). Exotic individuals, themselves subject to objectification, whose blood potentially exhibits rare qualities, are vulnerable to exploitation in the name of advances for the benefit of biocapitalism. Little if any thought is given to protection of donor rights and interests, even though blood is so often attributed with sacred qualities and symbolizes the continuity of peoples.

One other compounding feature is that cell lines often contain, as in the Hagahai case, viral material in addition to human protein. So here is secreted a hybrid of self and a parasitic other, permitting Amar Bhat, a representative of the National Institutes of Health to assert, in defense of the Hagahai and Solomon Island patent claims, that: "they [the involved laboratory] cloned only the genes of the two viruses" (1996). That the virus cannot exist outside of human tissue and by definition cannot be an independent entity is conveniently set aside by Bhat and others when they make this kind of claim based on reification upon reification.

INALIENABLE WEALTH

. . . Different from the fetishization of objects that are created by hand or manufactured, the body and body parts constitute life itself, individual and communal—past, present, and future. Even after death and decay the memorialized body signifies reproduction and continuity. For this reason body tissues, organs, and even fluids are, more often than not, regarded as inalienable, except when made use of in specified ritual practices or in carefully circumscribed activity, such as human reproduction. In order for body parts to be made freely available for exchange they must first be conceptualized as thing-like, as detachable from the body without causing irreparable loss or damage to the individual person or generations to follow. The mystical or transcendental essence associated with body fluids, organs, and tissues must be dissipated. This process of reification and fragmentation of body parts, so characteristic of biomedical knowledge and practices, has been criticized repeatedly by social scientists and feminists over the past three decades as a dehumanizing move. The assumption in much of this criticism is that patients and their families participate in this process of objectification unwillingly. But a careful reading of the literature on medicalization reveals that people are sometimes happy to relinquish the dense social, cultural, and mystical associations associated with their bodies, and are comfortable with a discourse reduced to the material. The moralizing discourses associated with bodies and body parts bring with them a burden of responsibility and more often than not, of blame, something that can be side-stepped through objectification.

On the other hand, recent literature that discusses the social effects of organ transplants shows how fetishism of human organs can be deliberately mobilized in order to promote donation: Transplant coordinators bandy about the metaphor of the "gift of life," and families who donate organs talk about how pleased they are that their dead relative is "living on" in someone else. Contradictions are rife because once organs are procured and stitched into recipients, these same recipients can be severely reprimanded, even thought of as pathological, if they attribute "life-saving" organs with animistic qualities passed along from donors . . .

Moral disputes will no doubt be implicated in the manipulation of certain types of human biological materials no matter to what extent efforts are made to transform these materials into autonomous, reified entities.

IMMORTAL CELL LINES

I turn now to the Human Genome Diversity Project (HGDP), conceived nine years ago. The involved scientists apparently had little inkling of the extent and vigor with which this project would be contested by indigenous peoples, who quickly recognized that they were to be made into the subjects of this endeavor. This ongoing dispute, a commodification contest about corporeal fetishism, provides an excellent example of the ambiguities, misunderstandings, and politics involved in the alienation and procurement of body cells and tissues. . . .

The technology to create immortalized cells lines, without which the HGDP could never have been envisioned, was set in motion by the 1912 Nobel Prize–winning eugenicist Alexis Carrel, after whom two of the litter of recently cloned piglets are named. Carrel spent most of the time between 1910 and 1935 working to perfect a technique known as tissue culture—a technique for growing living fragments of tissue outside the body of the organism. The ability to keep body parts such as chicken heart cells functioning *in vitro* permitted developments in transplant technology on the one hand, and on the other led to a genealogy of technological practices culminating in cloning as we know it today. Carrel argued, given the proper conditions, that cells removed from the body and maintained in a culture that is regularly renewed can go on living and dividing and thus be made immortal through human intervention. Today, many thousands of human cell lines are maintained in hundreds of laboratories around the world. Their source is from patients, healthy research subjects, and cadavers. Every one of us is a potential source of these "biologicals" (a term coined in the early 1980s). Practices that create cell lines "make it increasingly difficult to say where the body is bounded in time, space, or form"

(Landecker 1999:221). The most famous of these cell lines is known as HeLa, named after the African American Henrietta Lacks, from whom the virulent cancer cells were obtained. These cells have been cloned and stored in laboratories all round the world so that, in the words of Anne Enright, "there is more of her [Henrietta Lacks] now, in terms of biomass, than there ever was when she was alive" (2000:8).

Some cell lines are patented and others are not, with remarkably different consequences. Those that are patented make it particularly difficult to separate out their use value in terms of the goals of scientific inquiry, progress, and profit, and sight is easily lost of the "gifts" of donors without which no cell lines would exist. In order to procure patent on biologicals it must be shown that through the "process of their production" the "natural" object has been transformed into an "invention." Even though [these are] a human/nonhuman hybrid, a discontinuity between the human source and the biological invention must be established; in other words, reification of the cells as solely a technological creation is integral to patent claims.

Late last year the U.S. Patent and Trademark Office extended the boundaries of what can be patented to include single nucleotide polymorphisms (SNPs)—the smallest unit of genetic variability. SNPs are crucial to understanding genetic diversity. In August 1999 the U.S.-based CuraGen company announced that it had identified 120,000 human SNPs. A spokesperson stated that CuraGen is "aggressive in making patent filings." A rival company based in France, GenSet, pointed out that it is "essential to demonstrate progress to the market" (Hodgson 1999). It is estimated that the profits for involved companies in mapping genetic diversity will be billions of dollars. The promise of the development of new "designer" drugs as a result of these activities, for those individuals who participate in well-funded health care systems, means that control over human genetic diversity has great allure for the Gene Giants (the major pharmaceutical consortiums). The Gene Giants have in turn been accused by political activists, often purporting to represent minority interests, of promoting a particularly pernicious form of neocolonialism.

ACCOUNTING FOR HUMAN GENETIC DIVERSITY

The HGDP represents an early and special case of a plan to investigate human genetic diversity. Advances in molecular genetics over the past two decades have made it theoretically possible to systematically survey variation in the human genome across the entire human population, and to store these samples as immortalized cell lines. A small group of human geneticists and molecular biologists made a proposal in 1991 to set such a project in motion, but this endeavor remains virtually unfunded and unrealized. Given that we are bombarded daily by information in the media about advances in connection with human genetics, this blockage appears remarkable, particularly when one of the claims made by the involved scientists was that the project will "help to combat the widespread popular fear and ignorance of human genetics and will make a significant contribution to the elimination of racism" (HUGO 1993).

One difficulty for the proposed diversity project is that from the outset it has been upstaged by the multibillion dollar Human Genome Project, which has been primarily concerned, not with genetic difference but, on the contrary, with genetic sameness. Only over the past two decades have geneticists pointed out how remarkably similar to one another human beings are with respect to genes. On average, any two people will be identical for about 99 percent of their DNA. This close similarity suggests to the majority of biological anthropologists that we are descended, somewhere between 150,000 and 200,000 years ago—very recently in evolutionary terms—from a common ancestor, or from a small population of "founders." It is this shared genetic heritage that the Human Genome Project is attempting to represent by mapping the human genome, an artifact destined to become a standardized codification for human life. However, as Richard Lewontin (1992:35) warns, "the human DNA sequence will be a mosaic of some hypothetical average person corresponding to no one." We will all become, in effect, deviations from this abstracted norm. . . .

Despite the fact that we know from research on population genetics that "race is only skin deep" (Lewontin et al. 1984), the use of race as a scientific category persists in epidemiological, psychiatric, and public health publications, and racism, it is painfully evident, remains pervasive. It was into this minefield that the HGDP naively stepped, in the misplaced assumption that the "facts" of science would take the day, supposedly when it was finally understood by the public that we all, genetically speaking, share many more similarities than differences.

RESISTANCE FROM THE OBJECTS OF INVESTIGATION

In his book *Marvelous Possessions*, Stephen Greenblatt (1991) explores the way in which, at the time of the "discovery" of the Americas, new and "wonderful" knowledge about the natural world was garnered from native inhabitants, taken by force or in exchange for cloth, and spirited away to be stored in European libraries and archives for posterity. As his closing paragraph poignantly reminds readers, Greenblatt's theme is not simply one of massive physical and intellectual exploitation. He describes a visit to the village church in Tlacochahuaya in the valley of Oaxaca. There he discovered, tucked away from view in a niche, a stone carving of the Mixtec god of death gazing down from the ceiling into the face of the crucified Jesus, whose effigy was prominently displayed in the nave. Greenblatt comments that "the divinities have exchanged this sightless gaze, this perpetual circulation, for more than four hundred years"—a memorial to the contradictory forces of resistance and assimilation at work when predatory Outsiders mix with "primitive" Others.

Circulation of ideas and knowledge has accelerated geometrically since the time of Cortes, and today the technology of electronic communication sets up unlikely juxtapositions between "tradition" and "late modernity." Native-L, a "First Peoples" news net, has been humming over the past few years with commentary and letters of protest in response to the proposed HGDP. On December 21, 1993, Chief Leon Shenandoah and the Onondaga Council of Chiefs sent an e-mail communication to the National Science Foundation in Arlington, Virginia. They demanded to know why the Project had progressed to its fifth meeting (there had in fact been three official workshops by that time) "without discussion or consent of the indigenous nations and peoples it affects." The Chief and his Council found this situation "unconscionable," one that "violates the canons of anthropology and science." This letter followed an account given a month earlier on the same network of the proposed $23 million project in which up to 15,000 human "specimens" would be collected, many from "isolates of historic interest." The key words in this communication had been taken directly from the language in HGDP documents. A concluding epithet: "Didya ever notice how come there ain't no Injuns on STAR TREK?" completed the angry commentary.

This skirmish is a fine example of the way in which knowledge, which formerly might have remained contained within meeting rooms and in the publications and archives of government, academe, and industry, today becomes quickly available for public debate and politicization due largely to global access to sophisticated communication technology. The native gaze, hypersensitive to exploitation (and no wonder), glowers back. The objects of investigation, the sources of the DNA, were, it seems, unthinkingly conceptualized as specimens, as items from our uncivilized past, in the minds of the planners of the HGDP. However, politically astute representatives of the Iroquois and other Indian nations ensured that misplaced nostalgia about exotic others and unexamined racist notions about "premoderns" did not pass undisputed.

The HGDP is a relatively small research project by today's standards, and even though it has been unable to obtain more than a tiny amount of funding, it nevertheless proceeds in an ad hoc way. The project as conceived in 1991 by two geneticists, Allan Wilson (since deceased) and Luca Cavalli-Sforza, is designed, it is claimed, to gain fresh insights into the origins and evolution of humankind, human migration, reproductive patterns, adaptation to various ecological niches, and also into the global distribution and spread of disease. The ultimate goal, pronounced at the first organizational meeting in 1992, is, quite simply, to find out "who we are as a species and how we came to be" (Roberts 1991). The scale of this project, its range through time and space, exhibits remarkable hubris.

In order to undertake this ambitious task, the group printed a request in the journal *Genomics* asking for researchers worldwide to assist them in an ad hoc way by collecting DNA samples from hundreds of "indigenous" populations with a view to creating a database for the benefit of the scientific community. At the first meeting it was agreed that for the Project proper, DNA samples would be collected from between 400 and 500 populations, in addition to European populations (to be handled separately). The plan, as originally proposed, was to take, as a minimum, blood samples from twenty-five individuals in named populations that would then be preserved as immortalized cell lines for future analysis (thus ensuring that there would be no further need to return for more blood at a later date). The aim was also to collect as many extra blood samples as possible from each selected population. In addition, tissue scrapings would be taken from the inside of the cheek, and probably hair root samples would also be collected (Roberts 1991).

For the 722 groups of people who, without consultation, found themselves in 1992 picked out from a preliminary selection of 7,000 to qualify as "genetic isolates," loud alarm bells started to sound. Their blood would be "immortalized" and stored in facilities, mostly in America, to which it appeared anyone who so desired, for a small charge, could gain access for experimental purposes. Early HGDP publications indicated "highest priority was to be given to groups defined as unique, historically vital populations that are in danger of dying out or being assimilated." It should come as no surprise that this kind of language generated hostile responses, nor that the HGDP became known as the "vampire project." But the involved scientists were, it seems, taken by surprise. Among some leaders of the groups who found themselves targeted, the idea was quickly established that although their blood was going to be immortalized, they themselves were to be allowed to continue on the road to extinction.

In 1993 the Rural Advancement Foundation International (RAFI), the Ottawa-based organization that had first alerted the World Council of Indigenous Peoples about the proposed HGDP, urged HGDP organizers to convene a meeting together with indigenous peoples to address ethical and scientific issues associated with the project. The purpose of the meeting would be to ensure that indigenous organizations would be involved at every stage of the planning and execution of the project, to grant them veto powers, and to place the project under United Nations control. The question of who exactly comes under the rubric of "indigenous" was never made clear; it was assumed that this was self-evident, a "factual" category. These suggestions were ignored for several years, and at a conference held in Montréal in late 1996, one session of which was devoted to the HGDP, no representative of "indigenous" peoples was asked to participate. This oversight by the organizers resulted in a public demonstration outside the hotel. The police were called in and, together with hotel management, they denied the demonstrators entry to the hotel even to pay the required fees to attend the conference, and thus to "exercise their democratic right to speak from the floor," as one conference organizer had suggested they might do.

CONSTRUCTING HUMAN BIOLOGICAL POPULATIONS

Not only is the question of representation of the people to be studied in the planning of the HGDP at issue, so too is its scientific merit. The HGDP, as first conceived, committed a "category fallacy," namely, the imposition of one set of data on another set of a different kind: selecting human groups identified on the basis of a shared culture, and assuming that their genetic constitution is also shared, is to conflate time and space inappropriately.

What is more, analysis of gene pools tells us rather little about the history of relatively ephemeral sociopolitical groupings formed and disbanded throughout history. The San peoples of Southern Africa, for example, at the top of the so-called "genetic isolate list," and therefore a pristine example of an uncontaminated population by HGDP standards, embrace three different language groups, suggesting relatively recent formation as a single group. Wilmsen has shown that the San became isolated only in the nineteenth century, and that their isolation is related directly to colonialism. Prior to that time they were fully integrated into complex local trading networks.

The "*eta* of Japan" were also placed on the HGDP list. This word is exceedingly pejorative, and refers to one among several groups of people who were classified legally and institutionally as outside the rigid class system of Tokugawa Japan (1603–1868). *Buraku* is the less inflammatory descriptor now used to refer to descendants of those individuals deliberately created as Other by the centralized, caste-conscious samurai hegemony of premodern Japan. Arguments persist in Japan about the "origins" of *burakumin* (plural) but, as Fowler notes, the "originary 'trap' into which historians commonly fall . . . is that of appealing to history to explain the contemporary '*buraku* problem' rather than scrutinizing the ideology that uses history as an excuse for discrimination" (2000:15). Fowler quotes Hantaka who states, "It was not, 'In the beginning, there were *burakumin*, if anything, it was, 'In the beginning, there was discrimination,' and *burakumin* were its product" (1999:57). Genetically burakumin are as much "Japanese" as is anyone else who makes claims to Japanese ancestry, and singling them out had the potential to feed into the active discrimination that remains so persistent in Japan.

Many Indian nations of North America were placed on the HGDP list on the assumption that they too are genetically "pure," even though there is no agreement as to how many separate migrations took place across the Bering Strait in prehistoric time. Linguistic studies of founder populations show enormous movement and extensive contact between groups once in the Americas. The image of isolated, exotic cultures living close to nature, from which we moderns became separated as a result of migration and then evolved into a "higher" civilization, is one from an era shaped by Spencer's theory of social evolution. We have been very slow, anthropologists included, to recognize that the "people without history" as Eric Wolf has ironically named "isolated" cultural groups (1982), are not frozen in time, artifacts from the past, in terms of either culture or biology.

It is highly misleading to suggest that we can reconstruct the migratory history of a specific group of people, although HGDP organizers continue to indicate that they can do just this. The contested politics of boundary making looms large here. The experience of the Yuchi, who reside in Oklahoma and who were the first indigenous peoples in North America to be directly contacted by a member of the HGDP about project participation, provides clear evidence of the contradictions evinced by this project. At the time they were approached by the scientists, the Yuchi had recently been denied their request made to the Bureau of Indian Affairs to be recognized as an independent political entity. Their hope was that this would assist them in the preservation of their language and culture and give them some autonomous administrative power. When presenting his case, the HGDP representative explained to the Yuchi that they are "a unique Indian tribe," and that for this reason their DNA should be preserved for posterity, an irony that was not lost on the Yuchi as they turned away the scientist in no uncertain terms.

Presentations at a 1999 conference in Montana entitled "Genetic Research and Native Peoples: Colonialism through Biopiracy" indicated that local accounts of genealogies continue to be honored by many, perhaps the majority of native peoples. A major concern was that land settlement claims and other political issues currently being heard in Washington and Ottawa might be thrown out as invalid should culturally defined groups be found not to be genetically "pure." Such fears are no doubt entirely justified. In an era when a discourse of genetic determinism is rapidly gaining the upper hand, it would be a travesty of justice if political disputes were swayed by arguments based on biology rather than on the history of the Americas of the past five hundred years. It is on the basis of the lived experiences of self-defined groups of people sharing a cultural and linguistic heritage that restitution for abuse and continuing discrimination is being claimed. As with burakumin in Japan, a search for genetic "purity" among North American Indian populations has the potential to cause untold damage to the long overdue piecemeal moves being made towards reparation.

No simple oppositions can be made between the knowledge and practices of the more than 5,000 groups of people currently recognized as indigenous, and those inhabiting the so-called developed world. We live in an era of heterogeneity and of global exchange. Nevertheless, it is clear that in those parts of the world where the collection and commodification of local knowledge, plant, and animal materials by agribusiness, pharmaceutical companies, and other interested parties has become common, concern about research into human genetic diversity is most apparent. Most of these same localities were, of course, formerly subject to colonization and decimation. History is repeating itself on a scale previously unimaginable, and hammering out bargains about some share of the possible profit for local peoples may well result in the creation of new economic dependencies, although a case can be made that people should be left to negotiate any kind of settlement they see fit. . . .

THE POLITICS OF HUMAN DIFFERENCE

In effect, the HGDP conceptualizes "exotic" bodies as a scarce resource, the essence of which can be extracted to transcend time and space and join the never-ending circulation of commodities integral to late modernity. Concerns of individuals from whom the cells are to be taken are primarily about a continued indifference on the part of the dominant world order to their condition. The political activist Aroha Te Pareake Mead, Foreign Policy Convenor and Deputy Convenor of the Maori Congress in Aotearoa, has responded to this indifference with insightful barbs. She says that all human genetic research must be viewed in the context of colonial imperialistic history:

> Human genes are being treated by science in the same way that indigenous "artifacts" were gathered by museums; collected, stored, immortalized, reproduced, engineered—all for the sake of humanity and public education, or so we were asked to believe. (1996, 46)

. . . In commenting on some of the difficulties posed by the HGDP, Donna Haraway notes that the majority of targeted peoples clearly do not consider themselves as a "biodiversity resource." The problem is one of "what may count as modern knowledge and who will count as producers of that knowledge" (1997:249). Commodification of body tissues is contested, and the potential for the accumulation of scientific knowledge from the creation of immortalized cell lines, together with the enormous profit incentive associated with it, is weighed against inalienable possessions. Body cells and tissues represent history, genealogy, and even the survival or demise of entire groups of people. But disentangling who speaks for whom, who represents what interests, and what value blood samples have to the involved parties is like walking in a hall of distorting mirrors. . . .

THE POLITICS OF HUMAN DIFFERENCE

How can regulation be monitored and enforced, and at whose expense, particularly when so much research is initiated by the private sector? Who "owns" genetic material? Individuals? Communities or tribal groups? Corporate organizations? Or humankind? Representatives of indigenous groups for the most part exhibit a preference for group ownership, whereas U.S. property law upholds individual ownership provided that body parts are not separated from the body in question. Other people argue that DNA cannot belong to anyone, or alternatively, that it belongs to us all, and yet others claim that ownership through patenting of body tissues and cells is essential if scientific research is to remain competitive. Contracts drawn up in connection with genetic research focus on entitlement, patenting, access, distribution, and uses to which genetic material may be put. In hammering out the terms of agreement of such contracts, radically different ontological perspectives about the human body, and the uses to which body parts may be put, can readily be discerned.

Above all, it is questions of stigmatization, discrimination and eugenics associated with investigations into genetic diversity that are the greatest source of anxiety. In North America it seems highly unlikely that targeted groups will voluntarily cooperate with such research unless individual and group identity is rigorously protected. Further, it must be absolutely certain that ongoing legal negotiations with governments, most of them in connection with land claims, will in no way be jeopardized. The idea that humankind migrated out of Africa many thousands of years ago is anathema to many indigenous peoples of the Americas. This narrative is in conflict with local accounts of events since "time immemorial," and any dislodging of the idea of "distinct" peoples will, it is feared, be used as a lever to reject land settlement claims . . .

With the incremental procurement, commodification, and worldwide circulation of human DNA, tissue, and body parts, this new form of biopolitics is here to stay. The HGDP may well never get off the ground, but research initiated by NIH and the private sector is moving ahead rapidly. The HGDP was designed solely to create scientific knowledge, and profit was not at issue. Even so, the naivete and dissembling exhibited by the involved scientists does not bode well. In contrast to the HGDP, the activities of the Gene Giants are rarely exposed to public scrutiny. Patenting actually promotes secrecy. Thus far neither the UN Human Rights Commission, the WHO, nor UNESCO's bioethics committee have taken positions on human DNA collection. Disputes about the ownership of human biologicals are part of a broader set of dilemmas urgently posed by the practices of biocapitalism. If this behemoth cannot be made to respond to the concerns of the people whom they target, then, aside from corporate profit and perhaps some new drugs on the market, it is probable that humankind may not benefit much at all from gene prospecting. On the contrary, affiliation across diverse groups, so urgently needed in this global era, may be irrevocably damaged.

REFERENCES

Appadurai, Arjun. 1986. "Introduction: Commodities and the Politics of Value." In *The Social Life of Things: Commodities in Cultural Perspective*, ed. Arjun Appadurai, 3–63. Cambridge: Cambridge University Press.

Bhat, Amar. 1996. "The National Institutes of Health and the Papua New Guinea Cell Line." *Cultural Survival Quarterly* 20 (Summer): 29–31.

Enright, Anne. 2000. "What's Left of Henrietta Lacks?" *London Review of Books*, April 13, 8–10.

Fowler, Edward. 2000. "The Buraku in Modern Japanese Literature: Texts and Contexts." *Journal of Japanese Studies* 26, no. 1: 1–39.

Greenblatt, Stephen. 1991. *Marvelous Possessions: The Wonder of the New World*. Chicago: University of Chicago Press.

Guha, Sumit. 1998. "Lower Strata, Older Races and Aboriginal Peoples: Racial Anthropology and Mythical History Past and Present." *Journal of Asian Studies* 57, no. 2: 423–41.

Haraway, Donna. 1997. *Modest_Witness@Second_Millennium: FemaleMan_Meets_Oncomouse.* New York: Routledge.

Hodgson, John. 1999. "CuraGen Lays Down Markers, 120,000 of Them." *Nature Biotechnology* 17: 951.

HUGO (Human Genome Organization). 1993. *The Human Genome Diversity (HGD) Project: Summary Document*, 1–8. London: HUGO Europe.

Landecker, Hannah. 1999. "Between Beneficence and Chattel: The Human Biological in Law and Science." *Science in Context* 12, no. 1: 203–25.

Lewontin, Richard C. 1992. "The Dream of the Human Genome." *New York Review of Books*, May 28.

Linebaugh, Peter. 1975. "The Tyburn Riot: Against the Surgeons." In (eds.) *Albion's Fatal Tree: Crime and Society in Eighteenth-Century England*, ed. Douglas Hay, Peter Linebaugh, John Rule, E. P. Thompson, and Cal Winslow, 65–117. London: Allen Lane.

Mead, Aroha Te Pareake. 1996. "Geneology, Sacredness, and the Commodities Market." *Cultural Survival Quarterly* (Summer): 46–51.

Richardson, Ruth. 1987. *Death, Dissection and the Destitute.* London: Routledge.

Roberts, Leslie. 1991. "Fight Erupts Over DNA Fingerprinting." *Science* 254: 1721–50.

Taussig, Michael T. 1980. *The Devil and Commodity Fetishism in South America.* Chapel Hill: University of North Carolina Press.

——. 1987. *Shamanism, Colonialism, and the Wild Man: A Study in Terror and Healing.* Chicago: University of Chicago Press.

Wilmsen, Edwin. 1989. *Land Filled with Flies: A Political Economy of the Kalahari.* Chicago: University of Chicago Press.

Wolf, Eric. 1982. *Europe and the People Without History.* Berkeley: University of California Press.

PART IX

Knowing Systems, or Tracking the

Bodies of the Biosciences

Knowing Systems, or Tracking the
Bodies of the Biosciences

INTRODUCTION

Michel Foucault was among the first to emphasize that the language of medicine does not describe a preexisting biological reality. He argued that the "medical gaze" and its associated discourses are the result of a rigorous training in how to see, interpret, and constitute what is concealed in the body. In *The Birth of the Clinic*, written in 1963, Foucault points out that it is misplaced to believe, as many historians have, that toward the end of the eighteenth century there was a relatively sudden appearance of a systematic modern medicine grounded in the scientific method. He agrees that a significant transition indeed took place around this time but suggests that not merely scientific "facts" but new codes of knowledge, new objects of perception, and new institutional forms emerged, permitting a change in the "silent configuration in which language finds support" (1975:xi). In his writing Foucault attempts to elicit what it is that sustains practices, discourse, perceptual experience (the gaze), and its objects together in interdependent formations. Foucault's stated desire was to move beyond the structuralist arguments and the atemporality characteristic of mid-twentieth-century French analyses and to seek out instead the historical conditions of possibility for production of knowledge in the human sciences. He concedes that "deep structures" must be laid bare, but with the important proviso that such structures themselves arise in a historical process.

These insights of Foucault are not above criticism; for one thing he has rather little to say about the material body and its agency, but his writing nevertheless forced the consolidation of a critical approach toward medical knowledge that in the intervening years has become widely accepted. During the past quarter century numerous researchers have challenged the universalizing truth claims of biomedicine and focused instead on the epistemology of knowledge production and on the historical and social construction of illness, disease, medical knowledge, and practices. This literature highlights the relevant discursive backgrounds that make such constructions not only possible but plausible. It opens the way for critique of the commonly held assumption that, as a result of the accumulation of knowledge grounded in the basic sciences, we are inexorably making progress toward a full understanding of *the* human body. It also encourages reflection about the way in which values are inevitably embedded in medical knowledge and practices. The most successful of such arguments do not ignore material reality or deny the successes associated with biomedicine in connection with intervention into bodily processes. But they have opened science and medicine to history in a number of suggestive ways.

Foucault's histories focus on France and England and the emergence of modernity in Europe. He paid virtually no attention to the rich anthropological literature available when he was writing. Many local differences in how bodies, deviance, and discipline were and are perceived, represented, and managed thus appeared at first to be beyond the reach of some of his key concepts. In opening non-Western archives and social worlds to critical historical study, however, his successors have expanded the implications of his genealogical method.

In the first essay of this final section, Shigehisa Kuriyama, a historian of medicine, poses certain fundamental questions in order to move beyond the obvious notion that varying medical traditions reflect mere cultural differences. Like some other research on non-Western knowledge traditions, that of Kuriyama argues that we must ask questions that are subtler and more attentive to embodiment. He asks, for example, "*In what sense* do people think differently? How literally are we to construe the phrase, 'They saw things differently'?" In keeping with his larger comparison of the Greek and Chinese traditions, he here focuses on the history of pulse diagnosis (sphygmology). Kuriyama points out that the "language" of the pulse is, of course, a language of tactile sensation, wherever it is put into practice, but that its use provides quite different information in classical China and Greece. Using a remarkable economy of words to draw the reader into two vastly different worlds, Kuriyama shows how the unfolding of an anatomical understanding of the body in Greece (which historically did not take place in China), and the development of the theory of correspondences in Chinese medicine (which did not take place in Greece) demanded, in each instance, the development of radically "new sensibilities." In other words, the realms of bodily experience had to be transformed in light of theory and practices that permitted novel understandings of the material. These understandings in turn led to differing ideas in China and Greece about the locations of the pulse, its functions and meanings.

Kuriyama states that "the habitual dichotomy of biology and culture would have us situate the divergence in culture in different ways of thinking about the unique and universal body." He cautions us against taking this body as a given fact, and concludes: "Just as training is indispensable to wine-tasting and music appreciation, so the apprehension of the body also requires the cultivation of special sensibilities"—the bodies of both doctor and patient are invested with different meanings and senses, above all in their manners of seeing and touching.

The essays by Rayna Rapp and José van Dijck lead the reader into the world of biomedical imaging, with its various highly authoritative ways of constituting both patient and doctor in the medical gaze. Rapp considers ethnographically the application of the routinized use of sonograms in the management of pregnancy. Discourses create objects, but in the contemporary medical world the medical gaze and its associated discourse are more often than not facilitated through the use of machines and technologies. Today's medical perception is, in effect, a hybrid sensorium, combining the active capacities of both experts and machines. Here, though, Rapp argues that the agency of the patient is also crucial. She shows how, particularly in the case of visualizing technologies, not only the medical practitioner, but the patient herself (in this case the pregnant woman), must actively participate in "reading" the sonogram. She expects to see a baby and this is indeed what she sees, but only with the assistance of medical personnel. Drawing on the seminal work of Edward Yoxen, Rapp argues that even though the images produced by sonograms may appear as babies to pregnant women as they undergo an ultrasound test, the "particularity of the object that the women view is deeply embedded in the practices of its scientific representation." What they see is not literally a fetus, but a visual reconstruction of a fetus that is animated for them by the technician. Often the woman sees nothing at first, but the fetus is then actively made into something: at once a living entity and a second patient secreted away inside the female body. Given the political climate surrounding reproductive choice in the United States, Rapp concludes that fetal images index not only women's responsibilities, obligations, and desires, but also the broader bio-politicized, contested terrain of motherhood. Her essay encourages recognition of the way in which moral judgments are integral to the discursive world that informs clinical practice—the space where the real time construction of the fetus takes place. Like Kuriyama, Rapp makes us appreciate that it is not the medical gaze alone that reproduces the objects of medicine, but also exchanges and engagement in clinical encounters between health care practitioners and patients or clients.

The essay included here by José van Dijck is about a technology designed to permit new ways of seeing that extend the limits of previous scientific and popular perceptions of knowing the human body. This project has received publicity far beyond the biomedical world and has to some extent scandalized public opinion. Van Dijck discusses the technology known as plastination recently developed and patented by a controversial German anatomist, Gunther Von Hagens. Von Hagens subjects donated cadavers to a sequence of chemical treatments so that they can then be modeled into sculptures by the anatomist for exhibition in a permanently fixed state. Van Dijck writes, "The resulting anatomical object looks like a conflation of an opened-up mummy, a skinned corpse, and an artistic sculpture." He is interested in why the large traveling exhibition of plastinated cadavers, *Bodyworlds*, has been such a controversial success in such cities as Vienna, Tokyo, and London. Linking *Bodyworlds* to the history of anatomy, van Dijck stresses that Von Hagen's goal is not to imitate nature, as has been the case with all previous anatomical modeling, but rather to modify it so that the result is a mixture of organic material with rather extreme artifice. Even so, the sculptor insists that his creations are accurate representations, the ultimate in realism; they are fixations of the "real," and he advertises them as such, but Van Dijck's reflection on the exhibit leads him to ask, "What is real?" The very fact that Von Hagens was able to patent his technique means that it has been formally recognized as an invention. Van Dijck thus argues that the technique is part of contemporary "technological culture" in which imitation has been replaced by modification, whether it be by means of genetic engineering, plastic surgery, anabolic steroids, or plastination.

The fascination the public shows for the *Bodyworlds* exhibition stems, van Dijck argues, from the ethical debate that forces itself into people's minds as they view the cadavers: Is this science or art? Is it for instruction or for entertainment? Ambiguity has been a key element in anatomical dissection from the days of Andreas Vesalius's public dissections in the fifteenth century to the present.[1] Von Hagen's artifice enables him to tinker radically with conventionally understood nature/culture boundaries, and in doing so he produces mixed responses of excitement and unease in the viewers of his exhibition, as was also the case for medieval audiences who attended public dissections. But what disturbs van Dijck most about plastinated bodies, these "fleshed-out scans," is the appearance they give of defying physical deterioration—of being superior to nature. This normative aesthetic choice is quite different from either the depictions of earlier anatomists or the visual and manual experience of cadaver dissection in medical school. Von Hagen's sculptures are products of a postmodern world in which boundaries formerly thought of as immutable are broken down through technological interventions. In the essays by Rapp and van Dijck a subtle shift in emphasis has taken place, away from a discourse analysis that follows the heritage of Foucault rather closely, with its emphasis on archival discourses, to a focus on the implementation of specific technologies which themselves have a genealogy, resulting in new material configurations with profound social, political, and moral repercussions.

The essay by Charis Thompson turns our attention from a fixed anatomical body to a rapidly transforming relational body. She shows how, with the implementation of the technologies of donor egg in-vitro fertilization (IVF) and gestational surrogacy, working out "who is related to whom and how" becomes problematic, making the category of motherhood highly contestable. Thompson reminds us that contemporary Westerners usually assume that "blood" relations simply reflect "genetic" relationships. Further, the ideal is that biological and social parentage would be one and the same, although this situation is quite obviously breaking down and being legally contested in many places. The ability to make use of IVF technologies involving donor eggs and surrogate mothers means that blood (gestational) and genetic (hereditary) relationships are drawing apart, although the specificities of the technologies and the body parts involved mean that this separation takes form in a number of ways.

Thompson illustrates this process through the experiences of women undergoing procedures that make use of eggs donated by a daughter, a sister, or a friend of the same ethnicity. She uses the concept of opacity to reveal the process whereby determination of who will count as the mother is actually made through resort to ideas about biology or nature, socioeconomic factors (including who is paying for the treatment), legal factors about ownership of gametes, who provided the sperm, what constitutes the nuclear family, and so on. By analogy, Thompson uses the concept of transparency where certain activities, such as egg donation, enable hereditary relatedness, but are then eliminated in configuring the ultimate web of kinship relations. Surrogate mothers are almost without exception excluded from kin relations despite claims they might make on the basis of blood; similarly a woman's daughter is expected to respond to a new baby as a sibling, regardless of her relatedness to the child and its genetic mother. Thompson concludes that we are dealing here with " a flexible ontological space for designating biological motherhood" that has the possibility to challenge genetic essentialism. A "complicated choreography [is] necessary to disambiguate transparent and opaque interventions." If this work is left undone, assisted reproduction can lead to errors and incalculable unhappiness in determining who is related to whom. Clearly, more ethnographic research is called for, particularly among families created through IVF interventions.

The last essay in this section is also concerned with genetics. In this instance Keith Wailoo is concerned with what he deliberately terms "racial identity." He sets out by reminding readers that biomedical sciences, in particular the diagnostic categories they produce, give rise to entirely new categories of persons—the HIV-positive person, for example, is someone not yet clinically diseased, but the act of labeling has enormous social, economic, and political significance. Wailoo's historical account focuses on the labeling of nondiseased "heterozygotes" bringing about "a new template of cultural understanding on which novel interpretations of historically defined identities might be developed." Genealogies of the "racial diseases" discussed in this essay—sickle cell anemia (associated primarily with people of recent African decent), Tay-Sachs disease (associated with Ashkenazi Jews), and cystic fibrosis (associated with Anglo-Americans)—show clearly that these conditions and the meanings associated with them are not simply uncovered through science, but rather are biological reinterpretations of particular group histories, identities, and memories. Recent knowledge obtained by means of molecular genetics supports these reinscriptions of race and ethnicity. The heterozygote, those individuals who carry a single trait for a specific disease, are known in contemporary genetic jargon as carriers. Not themselves sick, two heterozygous individuals bearing similar traits can produce children who are diseased. As Wailoo notes, an ability to expose heterozygotes by means of genetic testing and screening raises "widespread anxieties about social interaction and new methods of surveillance." The notion of risk is drawn on to justify screening for heterozygotes whose marriages and pregnancies can be monitored, and numerous wanted pregnancies have been prevented or terminated on the basis of genetic test results. Wailoo reveals how social policies in connection with these diseases have varied enormously with significant political consequences for individuals and communities. In the cases of Tay-Sachs disease and cystic fibrosis, individuals and communities often seek out and cooperate wholeheartedly with screening, whereas the history of screening for sickle cell disease is a story of heightened stigma and discrimination against African Americans.

Through this comparative discussion Wailoo is able to show graphically how the heterozygote becomes during the course of the twentieth century "the site of community, memory, and the making of bodies." Wailoo concludes that the heterozygote is a "cultural work in progress"—continually subject to being remade in light of emerging scientific knowledge and reinterpreted by involved peoples and policy makers. Physical manifestations of these diseases have not changed, but the earlier language of heredity has been replaced by one associated with the new molecular genetics in which concerns about prediction and control now dominate.

In 1997, Donna Haraway, writing about gene mapping, argued that the gene has in effect become a phantom object because its inescapable heterogeneous relationality has been reduced for the purposes of representation and manipulation to something seemingly fixed and objective. Following Michel Callon, Haraway reminds us, as does Wailoo, that "technoscience's work is a cultural production" in which both human and nonhuman actors are involved.[2]

Much has been written about what it means to claim that we are "mapping" genes when in fact all that is being carried out is sequencing of the amino acids adenine, thymine, cytosine, and guanine. Twenty years ago it was argued, with liberal use of mixed metaphors, that the "map" of the human genome would reveal the "blueprint" for each one of us. The biologist Richard Lewontin was quick to point out that in reality the map would correspond closely to the individual genome of no one and that we are all, in effect, deviants from the master design created in laboratories out of DNA samples collected from a very small number of individuals. But, even more significant, the reductionist assumption behind the mapping of the human genome has proven to be extraordinarily shortsighted. The metaphors of map and territory that accorded so well with a structural approach to bodies from the era of systematic anatomy forward are finally on the wane. Evelyn Fox Keller argues that, with the formal announcement of the almost complete sequencing of the human genome in 2001, twentieth-century genetics reached its apotheosis, and, paradoxically, this sudden awakening was a result of the insights obtained from gene sequencing. A significant shift in the world of molecularization has taken place, from a focus on structure to one on function and from genes per se to gene-gene and gene-protein interactions and to complex biological pathways. The map of the human genome is not, after all, a map at all, but simply a departure point, and no one now believes that information about genes will lead to "the secret of life." Fox Keller argues that structural genomics has given us "the insight we needed to confront our own hubris."[3]

Gene fetishism and the assumptions about embodiment that have accompanied it may eventually decline, but for the moment the majority continue to be captivated, it seems, by the vision that has engaged us so powerfully since the late nineteenth century, that we can best understand our bodies and ourselves through an unveiling of material structures. The move to functional genomics (genes in action) and to proteomics (proteins and their products in action) does not necessarily diminish this reductionistic vision, although it forces us to recognize material complexity and contingency as never before. The haplotype map, designed to document in full the 0.1 to 1 percent genetic difference among humans, is currently occupying numerous geneticists. Along with it, human taxonomies threaten to reinvent race; the possibility of a newly scientized racism, never far below the surface, is once again demanding serious attention. To look elsewhere for the significance of biological life would begin to go beyond the body proper, a task with which the institutions of formal knowledge production in modern society are perhaps just beginning to grapple. An emergent science of a recast body makes the work of historians, sociologists, and anthropologists of the biosciences all the more significant.

NOTES

1. See Good and Good 1993.
2. Haraway 1997.
3. Fox Keller 2000:8.

ADDITIONAL READINGS

Farquhar, Judith. 1994. *Knowing Practice: The Clinical Encounter of Chinese Medicine*. Boulder: Westview Press.

Franklin, Sarah, and Margaret Lock. 2003. *Remaking Life and Death: Toward an Anthropology of the Biosciences*. Santa Fe: School of American Research.

Kuriyama, Shigehisa. 1999. *The Expressiveness of the Body and the Divergence of Greek and Chinese Medicine*. New York: Zone Books.

Moss, Lenny. 2003. *What Genes Can't Do*. Cambridge: MIT Press.

Rapp, Rayna. 1999. *Testing Women, Testing the Fetus: The Social Impact of Amniocentesis in America*. New York: Routledge.

Turney, Jon. 1998. *Frankenstein's Footsteps: Science, Genetics and Popular Culture*. New Haven: Yale University Press.

Zimmermann, Francis. 1987. *The Jungle and the Aroma of Meats: An Ecological Theme in Hindu Medicine*. Berkeley: University of California Press.

PULSE DIAGNOSIS IN THE GREEK AND CHINESE TRADITIONS

SHIGEHISA KURIYAMA

The enigma that motivates this essay is a simple but compelling one. In its most general form, the enigma can be expressed quite succinctly: Why is there a history to medicine? The human body, after all, during the time span with which historians of medicine concern themselves, has no history. We assume that the same biological principles that governed the body in Egypt of the Middle Kingdom, or China of Han times still hold true for the body in contemporary America. The body is timeless and universal. Why, then, do we find such distinctions as Greek medicine and modern medicine, Egyptian medicine and Chinese medicine? How are we to understand the diversity of medical traditions when we believe the body to be just one? The foundation of history is the apprehension of plurality. The question is: whence the plurality?

One could, of course, always respond that the various medical traditions reflect cultural differences, that they arise from "different ways of looking at things," "different modes of thought." But that is merely to restate the problem. For the challenge lies precisely in coming to terms with what we mean by "different ways of thinking." *In what sense* do people think differently? What meaning can be given to the explanation, "They had different conceptions of the body"? How literally are we to construe the phrase, "They saw things differently"?

What I propose to do in this essay is explore this general enigma of medical diversity within the specific context of the history of pulse diagnosis. More particularly, by focusing on one key and concrete aspect of the divergence of classical Greek and Chinese sphygmology, I hope to articulate an example that will illuminate not only the elusive nuances by which medical traditions diverge but also the special demands these nuances make of the individual who seeks to understand them. . . .

Perhaps the first task should be to ask whether Greek and Chinese sphyg-
mologies do indeed represent distinct ways of interpreting the pulse. The
question, as physicians in Enlightenment France and England discovered, is far
from a trivial one. . . .

The language of the pulse is, needless to say, a language of tactile sensation.
But the crucial observation for the Chinese was that in fact translation begins
before the fingers have touched the patient. Already in the selection of a locus
of attention, in the decision as to *where* to feel the pulse, a critical interpretation has taken place.

One discerns at least four major methods of pulse analysis in classical
Chinese medical literature. The first, about which only fragmentary accounts
have survived, involved the examination of twelve different sites of pulsation.
Though the absence of a full exposition makes it impossible to reconstruct all
the details of this method, one thing is certain: it is the locus of pulsation that
determines the content of the sphygmological message. Thus, according to the
vestiges preserved in later works such as the *Shang han lun* and the *Jingui yao
lue* a "floating pulse" discerned at the dorsalis pedis artery suggested excessive
gastric function, whereas the same pulse found in the ulnar artery would be
indicative of a "wind" ailment.

A second theory of interpretation is that outlined in the chapter of the *Su
wen* entitled "Discourse on the three parts and nine signs." This system required comparison of pulses at nine different points of the body—three on the
head, three on the arm, and three on the feet—with each pulse indicating the
function of a different organ or part of the body. For example, the superficial
temporal artery reflected the state of the eyes and ears, the radial artery the
condition of the lungs, and the posterior tibialis artery, kidney function. It
is this particular method of diagnosis that figures most prominently in the
Su wen.

The *Ling Shu*, by contrast, promotes a third and entirely different approach
to structuring the topology of pulsation. Here, the sites of diagnosis are reduced to just two—the carotid artery pulse, associated with the yang principle,
and the radial artery pulse, linked with the yin principle. By comparing the
strength and breadth of pulsation at these two locations it becomes possible to
determine whether the ailment involves a yin or yang deficiency (or excess),
and the degree of any excess.

Finally, in the latter Han period classic the *Nan Jing*, the radial artery pulses alone have become the focus of attention. As is evident from the explicit references to the "twelve pulses" and the "three parts and nine signs" contained in the early sections of this work, the *Nan Jing* synthesizes much of the previous tradition in a particularly elegant and compact form. . . .

There are further complexities in each of the systems, but the main point, I think, is already quite clear: for the Chinese the language of the pulse was governed by a topological grammar. Throughout the involved history of interpretations and reinterpretations of the pulse, there remained one constant assumption, namely, that it is the specificity of diagnostic sites that structures the basic meaning of pulsation. The goal of all early Chinese sphygmological theories was to define a framework for the comparison of the various distinct pulses of the body.

. . . This notion of a topological pulse grammar was quite foreign to the European medical tradition. The concept of differentiated sites of examination does not appear in the extensive pulse writings of Galen, and it achieves brief prominence only in the eighteenth century with the Montpellier physicians Henri Fouquet and Théophile de Bordeu. In part, this obliviousness to local variations in pulse quality might be explained by the fact that Greek sphygmology, for its part, developed foci of interest which the Chinese largely neglected. More specifically, Greek pulse interpretation minutely scrutinized the temporal qualities of the pulse—regularity, rate, and the relative durations of the diastole and systole.

But . . . a deeper, more fundamental factor [underlies] both the Greek neglect of topological variation and the fascination with rhythmicity, and that is the primacy of anatomy. Herophilus, one of the founders of Greek sphygmology, was also one of the first anatomists. For him and his medical descendents, the pulse was first and last the contraction and expansion of arteries, which ultimately all arose from the heart. Such a view not only generated a cardio-centric orientation to pulse-taking, and focused attention on cardiac rhythms, but also made the Chinese site theory appear "phantastical" and "absurd."

To summarize, the Chinese constructed their interpretation of the pulse on the specificity of diagnostic sites. On the other hand, not only was the Greek tradition oblivious to local variations, but the anatomical vision of the pulsing artery actively militated against the conception of a topological grammar. *This*, then, is our example of medical divergence. This exemplifies one way in which Greek and Chinese medicine differ.

But what exactly is that way? What does this example signify? Or more properly, what must we do to make sense of it? As we reflect upon this divergence we are invited along two paths of approach. One challenges us with the enigmatic character of the Chinese perspective. In his discussion of pulse sites physician Li Kao (1180–1251) remarks, "The three fingers are separated by hairbreadths, but the diseases they distinguish are a thousand leagues apart." What kind of conception of the body is it that endows three structurally indistinguishable points, separated by only millimeters, with such decisive and vital significance? What type of thinking, what type of experience would lead people to scrutinize pulse variation in different parts of the body?

Certainly not the experience of anatomy. And this raises the other question. Why is it that the Greek (and consequently our own) understanding of the body came to be so dominated by the results of dissection? How did it come about that the body came to mean the body we can *see*? . . .

THE THEORY OF CORRESPONDENCE

At the heart of Chinese medical thought is the correspondence of the microcosm of the human body and the macrocosm of heaven and earth. "Man is paired with the macrocosm (*t'ien ti*)," writes Li Yen-shih (1628–97). "The human body is a majestic microcosm. There are no principles in the two realms which do not correspond." The passage dates from Ch'ing times, but it expresses views already pervasive in the early Han dynasty and ubiquitous in Chinese medical and philosophical literature. The unity of heaven and man, all expositors agree, is the *idée-clé* of Chinese medicine. As such it should guide us into the enigma of a topological sphygmology. Our first step must be to take note of the etymology of "correspondence" (*ying*). The translation here is a felicitous one in that it captures the dynamic character of the Chinese microcosm-macrocosm theory, the primacy of mutual *response* as the relationship uniting the two realms. Thus the association between man and heaven and earth is not simply that of morphological, numerical, or other metaphorical parallels. It is not just that "the head rises up and is round and resembles the shape of heaven," or that "the body has five viscera that correspond to the number of the Five Agents," or even that "the body has its orifices and veins that resemble rivers and valleys." Similarity is important because, as Tung Chung-shu (179–104 BC) titles chapter 57 of his *Ch'un ch'iu fan lu*, "Things of the same kind activate each other."

> Therefore when the note F is struck in the seven-stringed or twenty-one-stringed lute, the F note in other lutes sound of themselves in response. This is a case of similar things being activated because they are similar in kind.

Their activity takes place in sound and is invisible. Not seeing the form of their activity, people say that they sound of themselves. . . . In reality, it is not that they do so themselves, but there is something that causes them.[1]

The theory of correspondences is thus not an academic exercise in analogical association; it is a serious attempt to analyze the dynamics of change. What is it that propels the universe through its transformations? The forces at work cannot, of course, be directly seen. But they can be apprehended in their regularities, in the patterns of responsiveness. "Heaven possesses yin and yang, and man also possesses yin and yang. When the universe's material force of yin arises, man's material force of yin arises in response." Thus, as Tung Chung-shu explains, "When the sky is dark and it is about to rain, people want to sleep, because the material force of yin is at work. . . . And people who are happy do not want to lie down because the yang of happiness and staying up require each other." Time in this view is apprehended as the interplay of basic forces in a vast network of response, and human physiology is united to the process of universal change.

This is as far as most expositions of correspondence theory go. In order to elucidate the foundations of a topological sphygmology, however, we need to proceed further into much subtler territory. Immediately following his previously cited remarks on man as microcosm, the Ch'ing physician Li Yen-shih continues:

North is *k'an*, the site of water. South is *li*, the site of fire. East is *chen*, the site of wood. West is *tui*, the site of metal. The center is *k'un*, the site of earth. Try facing south and looking at the sites in the two hands. The heart belongs to fire, and resides in the *ts'un* site. This also is in the south. The kidneys belong to water. They reside in the *ch'ih* site. This also is in the north. The liver belongs to wood. It resides in the left. This also is in the east. The lungs belong to metal. They reside in the right. This also is in the west. The spleen belongs to earth. It resides in the *kuan* site. This also is in the center.

The passage exemplifies the recurring challenge presented by Chinese medical texts. On the one hand the text presents a superficial transparency. Why have a site-specific interpretation of the pulse? We are explicitly directed to the correspondence of microcosm and macrocosm. The reasoning is clear enough: the relative dispositions of the diagnostic sites are identical to the positions of the four directions and the center.

. . . It is of course the inference that puzzles. For us the chasm separating topological similarity and functional identity is not easily crossed. It does not seem enough to say that one point is a few millimeters north of another in order to explain the radical polarity in diagnostic significance. Note, however, that the difficulty here does not lie in the analogy between macrocosm and microcosm. Rather, the real difficulty lies in the idea, explicit here, that *places have dynamic propensities. K'an, li, chen, tui,* and *k'un* are hexagrams from the *I Ching (Book of Changes)*; water, fire, wood, metal and earth are of course the five agents. All of these express directions of transformation; all of these remind us (for in the Chinese context this is not a matter of debate) that positions in space are not just abstract points in an isotropic coordinate system but rather gradients of change. Each site in the universe and, by correspondence, each site on the body is engaged in a specific network of responsiveness by virtue of nothing other than its position. Place, not form, determines function. It is apparent that if we could accept *ts'un* and *ch'ih* not as parts of the same artery but rather as independent loci with their own responsive sensibilities, we could feel much more at home with the Chinese construal of the pulse. If we could accept the dynamic character of places—even on a macroscopic level—then much of our puzzlement over a topological sphygmology would be resolved. But can we accept it?

The question confronts us with the real challenge of comparative investigations. What kind of statement is "North is *k'an*, the site of water"? It is a statement, I have suggested, about the dynamic propensities of north. But the question is, what *kind* of statement? The issue of kind is crucial here, because on it hinges the nature of the historian's task: different kinds of statements require different modes of acceptance. We accept a joke by laughing. We accept an order by obeying. We accept a mathematical theorem by assenting to the logic of its proof. How must we accept "North is *k'an*, the site of water"?

What I propose is that we consider this and other such statements about the transformational potential of sites in much the same way that we consider expressions such as "Bordeaux wines of 1935 are distinguished by their chicory bouquet" or "Steinway pianos have a baroque timbre." That is, I urge that they are statements of fact of a special kind. In stating what is the case, they neither call for nor permit critical examination, rejection, or assent. Rather, they define directions of learning: they invite us to develop our sensibility in such a way as to discern the realities described. For the untrained individual the *k'an* proclivities of the north may seem as unreal or at least as meaningless as "chicory bouquet" and "baroque timbre" are for those unversed in wine-tasting or music; but for the cultivated palate, to take just one example, the chicory bouquet is a fact of immediate experience. To accept the statement "North is *k'an*, the site of water," thus, means nothing more nor less than engaging oneself in a process of self-transformation. The dynamic character of places is a fact that one must learn to sense. This is the real challenge of a comparative history of medicine: the inquiry into history is an inquiry into the possible realms of human experience.

How can we learn to sense places? An attempt at a full explanation would require an essay in itself. One way, however, of approaching the specificity of sites on the body is to consider the specificity of sites on the earth. Here, at least, some of the grosser differences are evident even to the non-initiate. A mountain crag, a grassy meadow, a tropical jungle—each clearly has a different "feel" about it; that is to say, each induces a different response in us. Moreover, we know that if, like some animals, we could develop our sensitivity to the earth's magnetic field, we would find that different terrestrial sites do indeed have different energy potentials. In point of fact there is an important tradition of individuals in China who have made a profession precisely out of an educated sensitivity to terrestrial sites. I mean the geomancers.

The parallels between geomancy and medicine are suggestive. In selecting sites and orientations for buildings and graves, the geomancer is called upon to discern the energetic dispositions of each place in much the same way as the physician who selects sites and directions of needle insertion in acupuncture. Along these lines, the names of acupuncture sites are revealing: points are variously labeled seas, ponds, springs, marshes, mountains, valleys, ravines, hills, and bends. The body, in short, is conceived as a highly variegated land-scape. The standard general term for acupuncture points, *hsueh*, is the same technical term applied to geomantic sites.

The sphygmologist who scrutinized the specificity of diagnostic sites thus had a kindred brother in the geomancer who studied the earth. The local "feel" of places and the development of the sensitivity required to discern its fine nuances give us some hint of the topological intuitions underlying medical experience. To penetrate more profoundly into the Chinese strategy of pulse interpretation, however, we need at this point to return to consideration of the correspondence theory in general.

Correspondence—the responsiveness uniting the microcosm of the human body and the macrocosm of heaven and earth—is, as we have noted, the living heart of Chinese medicine. It defines, as it were, the natural order to things. But the paradox of the human condition is that man must learn to be natural. While in one sense the unity of heaven and man represents an immutable reality, it is at the same time a reality that is actualized only through self-cultivation. In his commentary to the *Su wen*, Ma Shih notes, "Man's *ch'i* communicates (*t'ung*) with the *ch'i* of heaven. But only the sage can perfect this heavenly *ch'i* and consolidate the basis of life; the masses lose this communication."

And again: "Follow the order of time and no harm can arise. But only the sage knows how to move the spirit (*ching ch'i*) in his body in accordance with the four seasons."

Just as spring calls forth a burst of green vegetation and winter induces the hibernation of animals, so man's dispositions and activities should respond to the shifting seasons. With each season the body should manifest distinct physiological patterns as evident to the trained physician as the shifting patterns of flora and fauna. But because the average individual, preoccupied with myriad desires, has lost his natural sensitivity, correspondence frequently fails, and the result is sickness, debility, and premature death. Only the sage who has recaptured the ability to sense and respond to the shifting energies of time can be a *chenjen*, that is, truly human.

We thus discover that at the heart of correspondence theory is an essential ambiguity: there is a tension between correspondence as a description of what actually *is* the case, and correspondence as a normative guide for what *should be* the case. The statement "The human body is a microcosm" defines not a state but a process, the path by which man returns to the natural responsiveness that integrates him into the universe of change. It is again a statement of the kind, "Bordeaux wines of 1935 are distinguished by their chicory bouquet." But what it asks us to educate is not simply our palates but our whole body; and what it invites us to respond to is not the nuances of fine wines but the subtle yet pervasive forces of time. It is in this sense that correspondence expresses the Chinese approach to the problem of change: one apprehends change by embodying it.

The consequences of this project of embodiment, for medical theory in general and the understanding of the pulse in particular, are decisive. For the effort to embody change naturally focuses attention on the body known, as it were, from the inside, the body discovered subjectively in the process of cultivating responsive sensitivity. Thus, while accounts of dissection appear already in Han dynasty texts, the principal structures of the Chinese conception of the body, such as meridians, points, and the *tan t'ian*, or cinnabar field, are all objectively invisible. Recall Floyer's characterizations of Chinese anatomy as "phantastical" and "absurd."

But the consequences for sphygmology are even more striking. Close study of a text like the *Nan ching* reveals that pulsation and the movement of blood are understood not in terms of the beating heart—which receives no mention at all—but rather in terms of the somatic function that figures most prominently in the process of self-transformation, namely, respiration. What determines the pulse is not anatomy but the breath. While the *Nan ching* is by no means a text of Taoist yoga, its analysis of inspiration and expiration, and the primacy of the so-called inter-renal pulse (*shen chien chih tung*) all evidence the unmistakable imprint of techniques of breath control on the development of medical theory. The path of embodying change led away from an anatomical, cardio-centric interpretation of the pulse. What paths, then, lead to it? Whence the anatomical approach to the body? Chinese sphygmology suggests there are other paths into the body. We must now ask why the Greeks took the path of dissection.

THE RISE OF GREEK ANATOMY

Few problems in the history of medicine are as important or call for as much profound reflection as the rise of the anatomical conception of the body. By this I am thinking not only of the centrality of dissection in the European medical tradition, the obvious legacy of which is still with us today; nor only of the more general identification of "the body" with the results of the dissected cadaver. I am referring also and above all to the insight that the rise of anatomy provides into that critical elision in Western thinking by which experience and observation become virtually synonymous, and both of these become fused into the conception of scientific rationality.

Why did anatomical dissection come to figure so prominently in the Greek understanding of the body? The question has rarely been asked. Instead, the tendency of most historians has been to invert the question and to focus on obstacles to anatomical progress, thus ascribing an inevitability or at least "naturalness" to dissection as an approach to the body. This view, in turn, has often rested upon the assumption of the compelling utility of anatomical knowledge, particularly in surgery. Galen, however, in his *Anatomical Procedures*, explains that contemporary anatomists have "obviously elaborated with care the part of anatomy that is completely useless to physicians, or which gives little or only occasional help." He continues:

> The most useful part of the science of anatomy lies in just that exact study neglected by professed experts. It would have been better to be ignorant of how many valves there are in each orifice of the heart, or how many vessels minister to it, or how or whence they come, or how the paired cranial nerves reach the brain, than not to know what muscles extend and flex the upper and lower arm and wrist, or thigh, leg, and foot.[2]

Galen's advocacy of a practical anatomy, thus, is the exception that proves the rule. Clearly the professed experts of anatomy were intrigued by matters that extended well beyond immediate utility. While surgical applications no doubt enhanced the appeal of anatomical study, they were evidently not the only nor even the principal reason for fascination with dissection.

But if not surgical utility, then what? What was it about dissection that so appealed to Greek students of the body? In addition to physicians, Galen cites three other types of people interested in anatomy: the man of science (*aner physikos*) who loves knowledge for its own sake, the man who seeks only to demonstrate that Nature does nought in vain, and the one who wants anatomical data for investigating some function, physical or mental. Those familiar with Galen's *On the utility of the parts* and his teleological research into the unity of form and function will see that these three motivations are often but different aspects of a single endeavor. To know the body is to see how Nature designed each part perfectly for its end, that is to say, its function.

The critical role played by teleological concerns in the rise of Greek anatomy is one that has not been sufficiently appreciated. Yet, at least from the time of Plato's *Timaeus* the study of the body's structure was a concern not only of the Greek physician but also, and perhaps more importantly, of the Greek philosopher. Why study anatomy? Certainly not just to see what the body is made up of. In his *Parts of animals*, Aristotle concedes that "it is not possible without considerable disgust to look upon the blood, flesh, bones, blood-vessels, and similar parts of which the human body is constructed."[3] Aristotle defends the study of anatomy, however, and stresses that the purpose of dissection is not the sight of the immediately sensible materials per se, which are repulsive, but rather the contemplation (*theoria*) of the purposive design of Nature. As long as one trains one's eyes somehow to see beyond the matter of which man and animals are composed and to apprehend the whole configuration, the Form as it reflects Nature's ends, then this gruesome enterprise of anatomy can even be characterized as beautiful. Moreover, it recommends itself to the philosopher by its convenience. For the reality that underlies the world of becoming, the divine realm of unchanging Being we all "long to learn about," offers little evidence to our senses. Plants and animals, on the other hand, because we live among them, can be studied more readily. What we must do thus is to scrutinize these, not in their perishable materiality, but in their formal design, as refracted images of the divine.

Underlying the rise of anatomical studies in Greek medicine, then, was the powerful religio-philosophical motivation of teleology. Plato talks of the Demiurge; Aristotle, of the contemplation of the divine; and of course Galen's

On the utility of the parts is nothing less than an epic meditation on the religious awe inspired by human structure. Accompanying teleology, however, and in some sense constituting the conditions that make teleology viable, are deeper assumptions about the nature of knowledge. In dissection one attempts to *know* the divine by *seeing* it. But one must still know *how* to see.

Here we may recall the haunting myth of the Cave in Book Seven of Plato's *Republic*. In this, one of the most memorable passages in Greek philosophy, we get as close as Plato ever gets to a positive description of genuine epistemological experience, the apprehension of true knowledge. Note however—and this is crucial—that it is not a question of ordinary vision. What is seen is not the shadowy quotidian realm of material beings to which ordinary mortals are riveted, but rather the sun, "*which provides the seasons and the years*, which governs everything in the visible world, and is also in some way the cause of those other things (shadows) which he used to see."[4] The apprehension of the Good, of that Being from which time and all the shadowy world of becoming originate, involves an act of contemplation. The myth of the Cave hinges on an inseparable link between visibility and intelligibility.

> If you interpret the upward journey and the contemplation of things above as the upward journey of the soul to the intelligible realm, you will grasp what I surmise since you were keen to hear it. Whether it is true or not only the god knows, but this is how I see it, namely that in the intelligible world the Form of the Good is the last to be seen, and with difficulty; when seen it must be reckoned to be the cause of all that is right and beautiful, to have produced in the visible world both light and the fount of light, while in the intelligible world it is itself that which produces and controls truth and intelligence, and he who is to act intelligently in public or in private must see it.[5]

There are two points to be noted here. First of all, I suggest that we need to reflect more seriously on the fact that the term Form (*eidos*) is etymologically derived from the verb to see (*idein*). I point out this etymology partly to remind ourselves of the often-noted peculiarity of classical Greek whereby a large number of verbs about thinking and knowing are related to the notion of seeing. For example, Bruno Snell, in his etymological reflections on the Homeric concept of *noos*, or "mind," reminds us that the verbal form *noein* means to "acquire a clear mental image of something. Hence the significance of *noos*. It is the mind as the recipient of clear images, or more briefly, the organ of clear images. . . . *Noos* is, as it were, the mental eye which exercises an unclouded vision."[6]

But there is another reason for reflecting on the etymology of Form. It points up an ambiguous tension. Forms are, in contrast to the ceaselessly changing entities of the visible world, the unchanging realities apprehended by the soul. But they are still in some sense seen. This brings me to my second point, and that is the acquired character of seeing. *Seeing is a learned skill.* This is a point of critical importance in appreciating the history of anatomy. In an unthinking moment we may think of dissection as a straightforward ostensive exercise, a pointing out of structures that are there for all to see. But as Galen never tires of repeating, anatomical vision requires a long and difficult apprenticeship. Without proper training and constant practice one sees nothing in a dissection, much less in casual observations of wounds. In order to see, one must paradoxically have seen before. Only the experienced dissector will find more in a dissected cadaver than messy gore, only he will be able to discern the delicate perfection of structures, only he will apprehend the divine.

The rise of anatomy in Greece, thus, like the development of theory of correspondence in China, required the development of new sensibilities, the extension of the realms of human experience. The teleological perfection of the human body would be no more evident to the untrained dissector than the chicory bouquet of Bordeaux 1935 would be to the individual who has only gulped cheap wines. This was the Greek approach to the problem of change: Change had an originator, called variously the Demiurge, or Nature, or God, and his intentions could be known by the dissector of ever-refined vision. In China, change was to be embodied. In the Greek tradition, it was to be seen.

CONCLUSION

At the outset of this essay we raised the enigma of medical diversity. We posed the question, why are there a variety of distinct medical traditions when the human body is just one? Having examined the divergence of Greek and Chinese sphygmology, we can now sketch a way of pursuing this problem.

In what sense do medical traditions diverge? The habitual dichotomy of biology and culture would have us situate the divergence in culture in different ways of thinking about the unique and universal body. But what is the body? What makes it up, and how does it work? Habitually, more out of sheer laziness and inertia rather than thoughtful philosophical conviction, we identify the body with the body known to modern medicine. We take the body as a given fact.

The most important lesson, however, which our cursory review of Greek and Chinese medical thought suggests to us is that the body is not a known given, an immediate fact of experience. Rather, it is an unknown realm that must be approached. It is in the different paths of approach into this unknown realm that we must situate the divergence of medical cultures.

These paths are largely paths of perceptual education. Just as training is indispensable to wine-tasting and music appreciation, so the apprehension of the body also requires the cultivation of special sensibilities. The dissector who learns to discern the fine complexities of anatomical structure has a different conception of the body from the Taoist who learns to control his heartbeat rate. The term "conception" here must be appreciated etymologically: the body is constructed out of educated experiences; and without an intuition of these experiences, much of medical theory appears nonsensical or at least baffling. The study of the comparative history of medicine might thus be imagined as an adventure into that unknown land that is the body. What we have before us are the charts of previous travelers. What we discern of the land depends on the paths taken. The study of history becomes an engagement not only with past thinking but with our own bodies; as we attempt to pursue historically the nuances of human experiential possibility, we become involved in a curious dialectic whereby the experience and awareness of our own bodies become inseparable from the understanding of the past.

NOTES

1. These phrases come from Tung Chung-shu's *Ch'un Ch'iu Fan Lu*, chap. 5. For a translation of this chapter, see Wing-tsit Chan, *A Sourcebook in Chinese Philosophy* (Princeton: Princeton University Press, 1963), 280–82.

2. Charles Singer, trans., *Galen on Anatomical Procedures* (London, New York, Toronto: Oxford University Press, 1956), 34.

3. Aristotle, *Parts of Animals* 645a, trans. A. L. Peck (Loeb Edition).

4. Plato, *The Republic* 516b–c. trans., G. M. A. Grube (Pan Books, 1974).

5. Ibid., 517b–c.

6. Bruno Snell, *The Discovery of Mind in Greek Philosophy and Literature* (New York: Dover, 1982), 13.

REAL-TIME FETUS: THE ROLE OF THE SONOGRAM
IN THE AGE OF MONITORED REPRODUCTION

RAYNA RAPP

With her own eyes, she could now pretend to see reality in the cloudy image derived from her insides. And in this luminescence, her exposed innards throw a shadow over the future. She takes a further step—a giant leap—toward becoming a participant in her own skinning, in the dissolution of the historical frontier between inside and outside.
—Barbara Duden

Ultrasound Fetal Imaging has rapidly gained wide diffusion as a screening device during pregnancy. In the United States it is estimated that more than 50 percent of pregnant women currently undergo at least one sonogram examination during their pregnancies, and in urban areas like New York City, where my research was conducted, informal estimates place the population under surveillance at 90 percent. Some medical authorities have suggested that the technology is being dramatically overused, given the difficulties of assessing its long-term safety effects. The rapid diffusion of ultrasound in pregnancy has raised epidemiological questions on two fronts: consensus studies suggest that the technology is safe in the short run, but only the analysis of much greater longitudinal data will reveal whether or not it has low-level biological effects on fetal auditory or neurological systems. Important questions concerning its cost-effectiveness have also been raised. Yet most obstetrical clinicians consider ultrasound to be the best invention since sliced bread and use its instant visualizations to measure, date, position, and intervene in pregnancies, while "reassuring" their patients that their fetuses are developing in a normal manner.

Wherever the evolving debate on over-routinization may lead, some uses of the technology seem unambiguously acceptable to epidemiologists, obstetrical and radiological clinicians, and pregnant women and their supporters. Most notably, second-trimester sonography is routinely used prior to amniocentesis. This procedure has clear medical utility, enhancing the safety of the amniotic tap and screening for some physical fetal anomalies as well. This essay is drawn from my experience in observing and interviewing women (and sometimes men) about the use of ultrasound accompanying amniocentesis. I want to stress that such amnio-related sonograms constitute a very small percentage of all the sonograms performed in any obstetrical service. . . .

IMAGING TECHNOLOGICAL HISTORY

In 1958 Jerome Lejeune, a French geneticist working at the Hôpital Saint-Louis in Paris, peered through an aged microscope at a sample of smooth muscle tissue taken from three patients with Down syndrome. His cardiologist-colleague Marthe Gauthier had used the then-innovative techniques of tissue culture to treat the sample. The full complement of human chromosomes had only recently been confirmed as forty-six in 1955–56. Lejeune was trying to assess whether his patients with Down syndrome lacked one human chromosome, as some abnormal fruit flies lacked one fruit fly chromosome. Instead, he (or, perhaps more accurately, he and Gauthier) discovered that they had a surfeit of chromosomes: then, and in subsequent studies, tissue samples taken from people with Down syndrome yielded a chromosome count of forty-seven. With great hesitation, Lejeune published his results in 1959. At the same time, a research group at the University of Edinburgh independently arrived at the same findings, confirming the Paris research. As a provost at University College wrote to pioneering English geneticist Lionel Penrose, "It must be one of the most important things that have happened in genetical studies for a long time" (Kevles 1985:248).

A few years later, in 1967, American researchers reported the first detection of a fetal chromosome problem in a sample of amniotic fluid drawn from the womb of a pregnant woman. Amniocentesis—the technique of extracting amniotic fluid transabdominally through a catheter—was first performed and described in Germany in 1882 by a doctor attempting to relieve harmful pressure on the fetus of a pregnant woman. It became an experimental treatment for polyhydramnios (excess fetal fluid that threatened fetal development) but was not widely used until the 1950s, when researchers in Great Britain and the United States discovered they could deploy the same technique to test for maternal-fetal blood group incompatibility and to assess the severity of Rh disease. In critical cases, amniocentesis led to intrauterine transfusion. It also permitted the assessment of fetal lung maturity, so that fetuses with serious disease could be delivered as early, and as safely, as possible.

Once it became possible to isolate and grow fetal cells from amniotic fluid, the examination of their chromosomes quickly followed. Some of the earliest "fishing expeditions" inside women's wombs were undertaken on mothers who had hemophilic sons and wanted to know the sex of their present pregnancies. If the fetus was determined to be chromosomally male, it ran a 50 percent risk of having the disease. By the time researchers in Sweden, Japan, Great Britain, and the United States began experimenting with amniotic prenatal chromosome diagnosis, the technology to invade the uterus and extract its liquids was well known. In the same decades, abortion reform throughout much of the West made the termination of problem pregnancies more easily available; even today, in the midst of concerted efforts in both legislatures and courts to dismantle abortion rights in the United States, surveys report strong public support for legal abortion in the case of serious "fetal deformity." The success of prenatal diagnosis must thus be situated in its social and legal, as well as its medical, contexts.

The evolution of amniocentesis was accompanied by other, closely related technological developments. In the same year that Lejeune observed the karyotype (chromosome picture) of his Down syndrome patients, midwifery professor Ian Donald and his colleagues in Glasgow published an article titled "Investigation of Abdominal Masses by Pulsed Ultrasound," in which they described the adaptation of sonar naval technology to observe fetuses inside their mothers' wombs. While the idea of *sonar* (*so*und *na*vigation and *r*anging) had been patented in England directly after the *Titanic* disaster for the detection of icebergs at sea, it was the French who developed it to detect enemy submarines during World War I. In the decades after the Great War, commercial engineers and physicists in the Soviet Union and later the United States investigated the technology's potential for revealing metallurgical flaws. Initial medical experimental uses focused on the energy generated by sound waves as a rehabilitative therapy. It took decades of clinical experimentation before the image-producing capacities of the technology were fully recognized and methods to harness it developed.

Diagnosis of pregnancy was thus neither the first nor the most obvious use claimed for sonography, which was initially (and wrongly) thought to be beneficial in scanning brain tumors and, later, kidney masses. While the sonogram has certainly proved to have many medical uses (especially in cardiology), its most routinized successes developed in obstetrics. Pulse-echo sonography (ultrasound) works by bouncing sound waves against the fetus, creating an image as the waves return to a cathode. After decades of perfecting transduction (the ability to attach the machine to the exterior of a patient's body so as to image interior soft tissue) and sectoral scanning (the capacity to render a three-dimensional object in two dimensions in regularized segments), it became clear to Donald and his colleagues that sonograms offered the possibility of normalizing representations of the fetus throughout its gestation. After their reports, sonography's diffusion in obstetrics was rapid and dramatic, and physicians hailed it as "totally safe" long before any actual safety studies were conducted.

As many feminists have pointed out, the technology intervened in the doctor-patient relationship dramatically, allowing the physician to bypass pregnant women's self-reports in favor of a "window" on the developing fetus. Additionally, radiologists and obstetricians working together could use sonograms in the developing technology of amniocentesis. Sonography enabled them to picture where the fetus wasn't and the fluid was, rather than groping blindly for a pocket of liquid into which to insert the amnio-bound catheter. As "real-time" sonography became available, doctors were able to observe a moving image of the fetus while sampling its environment. When used in conjunction with sonography, experimental invasive techniques of the womb became safer, and the miscarriage rates attributable to these procedures dropped dramatically.

The technology of prenatal diagnosis continues to evolve at a rapid pace. During the decade in which I have been investigating its social and cultural geography, another intrauterine technology has waxed and waned. Initially hailed as a revolutionary replacement for amniocentesis, cvs (chorionic villus sampling) works on preplacental tissue, allowing diagnoses to be completed within the first trimester. Despite the attraction of ever earlier diagnosis, serious safety objections have been raised. More recently, experiments in early (twelve-week) amniocentesis are moving from the anecdotal to the clinical, although controlled national trials have yet to be completed. Additionally, inexpensive MSAFP (maternal serum alfa fetoprotein) screening has become routinized throughout the United States. When elevated values for this biochemical marker produced by the fetus are found, they may indicate a neural tube problem. But the test itself produces a large number of false positives, leading to increased use of amniocentesis as a backup diagnostic technology. . . .

THE ROLE OF THE SONOGRAM IN THE AGE OF MONITORED REPRODUCTION

Above all else, a close encounter with prenatal testing increases women's worries about the specific health status of the fetuses they carry. Generic pregnancy fears might once have crystallized around the desire to carry a "healthy" baby. For a woman having amniocentesis, there is now a focus on precise conditions;

chromosomes and AFP levels index a panoply of anxieties with newly medicalized names. The specificity and reality of childhood disability become exquisitely focused through prenatal testing, engaging a complex mix of science and superstition as pregnant women and their supporters encounter potential diagnoses:

> Down syndrome, I knew about Down syndrome. What I didn't know about was all that other stuff. There's more to worry about than just Down syndrome, now I know there's other heredity problems. And this spine business [spina bifida], I wasn't exactly acquainted with that. Something more to worry about. (Lacey Smythe, African American secretary, 38)

> I remember thinking, "Oh, my God, it's like a message direct from inside." In the old days, our mothers certainly never knew this, the picture of the inside of their wombs, the small swimming thing. But we do. We're the first ones to follow pregnancy in books, day by day, with photos. We know exactly when the arms bud off, when the little eyes sew shut. And if something goes wrong, we know when that happens, too. They called it "an error in cell division." It feels like the cells could have a car crash, and produce this wreckage, and that's the extra chromosome, that's Down's. (Pat Gordon, white college professor, 37)

> Suddenly, I'm starting to see all these kids with Down syndrome on the street. Who knows if they're really Down's kids, or if I'm imagining it. And now you're asking all these questions, and I'm trying not to think about spina bifida. I never even knew spina bifida was a problem. But after counseling, I do. (Enid Zimmerman, white municipal service planner, 41)

The power of the sonographic imaging which accompanies the test has complex effects, funneling the pregnant woman's consciousness of her fetus into highly focused and routinized channels. But how are these channels constructed through imaging? The gray-and-white blobs of the sonogram do not speak for themselves but must be interpreted. As many sociologists and historians of science and technology have pointed out, the objects of scientific and medical scrutiny must be rendered; they are rarely perceived or manipulated in their "natural" state. It is their marking, scaling, and fixity as measurable, graphable images that enable them to be used for diagnosis, experimentation, or intervention. The power of scientific images may, in large measure, be attributed to their mobile status: they condense and represent an argument about causality that can be moved around and deployed to normalize individual cases and theoretical points of view. Viewed on a television screen or snapped with Polaroid-like cameras, sonograms may appear to pregnant women and their supporters as "babies." But the particularity of the object women view is deeply embedded in the practices of its scientific representation.

> The partial rotation of the beam and the electronic recording of the echoes as spots of light thus "renders" . . . the internal two-dimensional structure of an organ or a limb or a test object in a given plane. The resulting image is certainly not artifactual. It registers features, like the fat-muscle interface, that really exist. Yet it picks out only those features that reflect ultrasound. (Yoxen 1989:292)

But surely pregnant women and their supporters are not thinking about the embedded, reductive, and normalizing aspects of imaging technology as they "meet" their baby on a television monitor for the first time. Such uterine "baby pictures" are resources for intense parental speculation and pleasure, for they make the pregnancy "real" from the inside, weeks before kicks and bulges protrude into the outside world. The real-time fetus is a social fetus, available for public viewing and commentary at a much earlier stage than the moment of quickening, which used to stand for its entry into the world beyond the mother's belly.

Perhaps sonograms also enable fathers and mothers to "share" what was formerly an entirely female experience of early pregnancy, increasing and hastening men's kinship claims. And surely they increase the speed with which fetal development is recognized as a process independent of the mother's embodied consciousness. As one white college teacher commented to me, "It put my pregnancy into fast-forward." She thus neatly aligned sonography with videotapes, that other near-ubiquitous forum for home-viewing. One couple, who disparagingly referred to themselves as "yuppies," brought their own video camera and tape recorder to the sonogram examination because they wanted to capture and domesticate the fetal heart beat. The acceleration of a subjective connection to the pregnancy thus passes through and is augmented by technologies external to the pregnant woman herself.

Of course, modern imaging technologies provide powerful framings for the health and meaning of a pregnancy that appear radically new and individualistic, but they do not hold exclusive rights to the air space in which the image of pregnancy is interpreted. Public commentary on pregnancy has ebbed and flowed with the development of religious discourse, the representational arts, and the history of science and medicine. Current biomedical interpretations pass through other "images that possessed power within their own time and to which other images and ideas clung" (Stafford 1991:xvii). In the process, pregnancy is constantly relocated as an object of speculation, investigation, and intervention.

Contemporary feminists have alerted us to the changing relationship among a pregnant woman, her fetus, and the social world indexed in reproductive medicine; they have also provided ample evidence for older representational politics. Sonograms reinscribe prior debates and interpretations about the meaning of pregnancy which have deep roots in Western history; residues of those discourses shape what we take to be modern notions of sex and its biological embodiment. Pregnancy, for example, figures in the tensions and agreement between medieval theology and natural philosophy. In that period, women's "fleshiness" was associated with the body in body/soul dualities, and her fetus was a cause for speculation about maternal and paternal contributions to God's purpose and perfection. Women's reproductive capacities invoked reflection on divine regulation and causes of oddity. Medieval texts and artifacts evinced enormous curiosity about pregnancy, monstrous births, and the relation of blood to milk, couched as problems of permeability and stability. Because biological sex was thought, for example, to be extremely labile, "the nature and cause of hermaphroditism, as of other embryological anomalies, was much discussed" (Bynum 1989:187).

In a later period, the eighteenth century, "The activity of visibilizing, or incarnating, the invisible became endowed with a special urgency in early modem art and medical experimentation" (Stafford 1991:17). A fascination with conception—with eggs, fetuses, grotesques, and biological monsters—provided the ground for artistic and scientific debates about how matter was formed—all at once, as the preformationists asserted, or in developmental stages, as the epigenicists insisted. Embryology and pregnancy were two fertile fields for the representation of theories of causality. Beginning in the same period and continuing throughout the nineteenth century, the study of anatomy focused on sexual difference and its representations, using models, cadavers, and works of art.

By the turn of the twentieth century, representations of the body (especially the female body) were once again relocated, this time at the flourishing intersection of photography, early cinema, and medical research, where "the body emerged as a (visual) apparatus . . . an embodiment of . . . the techniques of the observer" (Bruno 1992:249). The imbrication of theology, natural philosophy, artistic and media conventions and emergent medicine thus provided verbal, plastic, and artistic representations of women and fetuses long before the advent of the new reproductive technologies.

Women and their fetuses were embedded and represented in social and power-laden discourses long before the present moment; this much is not new. But in prior times, individual fetuses made their presence public only slowly, over a period of months, and that presence was attached to signs—whether miraculous, mundane, or scientific—that passed through the woman's codification. A woman's physical and emotional state might reveal internal signs of pregnancy in hormonally induced swollen breasts, skin changes, energy loss, dizziness, or nausea, all of which were experienced kinesically and holistically. Later a midwife or physician might pick up a fetal heartbeat through a wooden trumpet, a stethoscope, or, more recently, a Doppler machine. But the passage from internal to external signs was slow, and almost all of the cues depended on the pregnant woman's reportage.

Now, however, sonography bypasses women's multifaceted embodiment and consciousness, providing independent medical knowledge of the fetus. Moreover, the technological framework reduces the range of relevant clues for whose interpretation women act as gatekeepers. A technology of exclusively visual signs that renders "a collection of echoes" into a representation of a baby substitutes for prior, embodied states. This reduction also sharpens the focus from a diffuse knowledge of women's embodied experiences to a finely tuned image of the fetus as a separate entity or "patient." This visual representation can then be described by radiologists, obstetricians, and technologists in terms that grant it physical, moral, and subjective personhood.

Indeed, one ethnographic study of sonographers and their pregnant patients powerfully described the code-switching that medical professionals perform. Among medical peers, sonograms are described in the neutral language of science. But when speaking to pregnant women, sonographers attribute motives to fetal activity and presence: a fetus that is hard to visualize is "hiding" or "shy"; an active fetus is described as "swimming," "playing," or even "partying." "Showing the baby" drives its personification. Thus the routinization of a new reproductive technology (or, more properly, a technology whose routinization is most powerfully occurring in the prenatal context: sonograms are also used to visualize the human heart, and abdominal masses, but I doubt whether these uses are personified) provides medical professionals with a "toy" through which they can simultaneously provide a compelling service and stake their claim to authority. The need to both monopolize a new professional turf and popularize its value here contributes to radiologists' and technologists' perhaps unconscious desire to personify the fetus.

Perhaps the most powerful aspect of that personification process is the sexing of the fetus. The technology often (although not always) allows radiologists to visualize fetal sex organs at the mid-trimester examination that precedes amniocentesis. And whether or not the radiologist "can tell," the chromosome analysis always reveals fetal sex. As Barbara Katz Rothman's study pointed out a decade ago, knowledge of fetal sex increases the velocity of a pregnancy; in our culture a sexed fetus is no longer a developmental imaginary but a "little slugger in a Mets uniform or a ballerina in a pink tutu" (Rothman 1986). Lost in the rush to fetal sexing is the slower process by which even a newborn may remain relatively unsexed, or at least episodically sexed, in the experiences of new parents.

Not everyone wants to know the sex of the fetus. In the interviews I conducted, genetic counselors reported that about half their clients would rather retain the mystery. But in my interviews with pregnant women, less than a quarter didn't want to know, and they were almost always those bearing a second or subsequent child. For first-time parents, knowing the sex is a powerful lure. From personal experience during a second pregnancy in which I didn't want to know the sex of the fetus, I know it is difficult for obstetricians to keep their mouths shut once fetal sex has been entered into the charts of their pregnant patients. Genetic counselors often caution those who would rather not know to announce their preferences firmly when they enter the radiology suite. Otherwise, a loquacious medical staffer is likely to point out the sex.

Some of the lure of sexing is based on control of knowledge. To the question "Why do/don't you want to know the sex of your fetus?" many people (and virtually every Jewish person in the sample!) answered, "Because if the doctor [or technologist, geneticist, or clinic secretary] knows, then I should know, too."

> I didn't like the idea that someone knew something about my baby that I didn't know . . . I don't care whether it's a boy or a girl, it really isn't that, it's merely that information exists, and other people have it. (Laura Forman, white theater producer, 35)

> As long as it's known, I feel the parents should know, you know. I mean, we shouldn't be the last to know, it's that kind of a feeling. (Carola Musky, white schoolteacher 39)

For such respondents, once technology exists to provide the information, ignoring it constitutes deprivation. Such a structure of sentiment surely drives the proliferation of knowledge generation and consumption. For others, the need to know is cosmological:

> Just like that, because it's a miracle of science to know what God provides for you, that's why I want to know. (Felicia Bautista, Dominican factory worker, 37)

They tell me it's a boy. After three girls! I still don't believe it. I'll believe it when I see it. I heard from a neighbor they sometimes make mistakes. I'll believe it when I see it. But knowing, that's a gift. (Cynthia Baker, African American homemaker, 40)

For some, fetal sex knowledge genderizes in conventional ways:

Because if it's a girl, you got to be more careful with girls. You can't just let anyone take care of them. (Rafael Trujillo, Puerto Rican unemployed worker, 43)

I want a girl, but my name is dying. If it's a girl, well, we'll just have to plan for a second. (John Freeman, African American computer technologist, 32)

Let's face it, knowing the sex made it go from a fetus to a child. I can't tell you how, but now I feel more protective, it's more real. And because it's a girl, I feel more connected to it, to my mother, to my sisters. Jeremy asked me which sister I want to name it for. I don't know if I want to do that. But the possibility made her more of a baby, a full kid, a living child. (Marise Blanc, white college professor, 35)

Several women from working-class families claimed they wanted sex information for practical reasons:

I figured at this point, financially, instead of buying all those different kinds of clothes, you just buy one specific set. (Angela Carponi, white homemaker, 33)

During the course of my research, I was invited to a baby shower for a pregnant genetic counselor with whom I worked closely. Her colleagues (who had analyzed her fetal chromosomes) had purchased appropriately pink items, but she refused to take them home, saying:

It's gonna cause a war in my family. My mother wants a girl. His mother wants a boy. They'll both be happy at the birth. But if they find out now, they'll tear each other apart.

In some cases the prospective parents may have a difference of opinion. The decision to know or not to know must then be negotiated:

I want to know, but Frank doesn't want to know, he absolutely doesn't want to know, he doesn't want some doc, you know, telling him before he has the real experience, finding it out together, in life, not as information. (Marcia Long, white psychologist, 37)

Like amniocentesis itself, which feeds on age-old pregnancy anxieties, the curiosity and mystery of fetal sexing is now reified and revealed through technology. Old cultural preoccupations with genderizing "who the baby will be" are thus put through the sieve of new technologies of knowledge.

Many women are delighted to claim this new knowledge as their own, using it for old purposes:

> I wanted Frank to get more involved. I didn't say, "Come with me to the first sonogram because it will make you more involved." I think I just said, "I'd really like it if you came with me to this sonogram thing." He's not that affective, he isn't really that connected to the baby yet. But he is connected to me. So I knew the sonogram would get him more connected, through me. . . . And I think it was true because he seemed moved, emotionally moved, and for the first time, after the sonogram, he started talking to people about the baby. (Marcia Long, white psychologist, 37)

> I was frustrated that so many women came alone. I brought my husband, it's important, very important. For good and for bad, it's women's burden, and her husband should know about it, and share it. If there's a miscarriage, he should share the pain. It's his creature, too, and if he sees it on the television, he will know it as his own, love it from the very beginning. He will see God's work, marvel, and share. The family should be united for the test, so that men share the power of God's work and the joy of a new baby. (Juana Martes, Dominican home care attendant, 38)

> I took my husband and my son to see it. I thought he was, you know . . . [RR: Not committed?] Yeah, uncommitted. I thought it would touch home. That's why I did it. I had a feeling. Use a little psychology. My husband, he was amazed by what he saw. (Diana Mendosa, Puerto Rican nurse, 35)

In such stories, a new technology supports an old female strategy of attempting to heighten male involvement with pregnancy.

Many women are also happy to align their descriptions to what technicians and physicians orchestrate:

> It was wonderful, I said, "It's great, can I leave now?" I mean, I didn't want the amnio, I just wanted to see my baby. I saw the spine, the bladder, the orbs for the eyes, the penis, everything, I saw all of it, I loved it. That was very satisfying. Maybe they do it as bribery, so you won't jump off the table. I feel like there's not much discrepancy between the sonogram and what it feels like inside me. (Alicia Williams, African American public relations executive, 36)

It's a creature from the moment of conception. On the TV screen I saw it all, a little head, a beating heart, even fingers and a back bone. It like a baby but indistinct, blurry . . . As it grows, it will get bigger, and more distinct, almost like tuning in the television. It corresponds to what's inside me now. (Guana Martes, Dominican home care attendant, 38)

It looked very alien, like a little space creature. It was clasping and unclasping its hands, and it had its fist under its chin . . . it was moving around so I could see the arms and fingers, which was nice, then it kind of got up on its legs, kind of pushed itself up, and you could see the whole spinal column, and the heart and the eye sockets and the shape of its skull. It's like a halfway baby now, yes, it's a halfway baby, and it's an inside-out feeling. (Marge Steinberg, white social worker, 39)

Like the pregnant woman who used the video analogy, Marge Steinberg was drawing on the fetus-as-voyager imagery that moviegoers and television watchers recognize from films like *2001: A Space Odyssey* and the Right to Life's *Silent Scream,* and, more recently, the Volvo advertisement selling safety to pregnant couples and their "passengers." The ad, which presents an ultrasound fetal image accompanied by the message, "Is Something Inside Telling You to Buy a Volvo?" was withdrawn after public protest over what was widely perceived as capitalizing on a sacred terrain. Sonographic fetal images perform practical and aesthetic service in the world at large, where women get to know them long before they arrive in the obstetrics suite.

Many women also recognize that their viewing is orchestrated and their internal state has been interpreted:

It was nothing, really, it looked like nothing. Then they showed it to me, and made it something. (Ileana Mendez, Ecuadorian-born baby-sitter, 37)

To tell you the truth, it didn't really look like a baby, I couldn't really tell what it was, they had to tell me. (Letty Sharp, white hospital clerk, 36)

You could see at certain points. Towards the end, I couldn't really tell what was what, and then there was the feet. I saw the legs crossed, and then it looked like a little baby, cute. After they told me what to look for, then I knew I was really pregnant. (Sandra MacAlister, African American administrative secretary, 41)

That baby was so active, jumping around, I swear it was mugging for the camera. But I couldn't tell what was what, I know somebody's in there, but I don't know who. (Lauren Smith, Anglo American lawyer, 39)

The "mysterious voyager" image provided by sonograms is compelling, ubiquitous, and unavoidable. When I asked women to describe their internal images of their pregnancies, most of them depended on the stereotypes of fetal space creatures: "Like *2001*," "Just like in *A Child Is Born*, you know, kind of pinkish-creamy," "Floating," and "A little traveler inside me" were common answers. Only a few could imagine other descriptive referents, and these women had luxuriant animals and vegetables blooming in their bellies:

> I could just imagine it like a little fish, you know, the one that jumps a lot, like a sardine, no, not a sardine, it goes uphill. [RR: Do you mean a salmon?] Yeah, that's right, a salmon, that's what I feel, this child goes so low sometimes it jumps like it's going to go through my vagina, that's how it jumps, all alive. (Angela Carponi, white homemaker, 33)

> It's got lumps and bumps, and they're growing, organic, you know, sort of like a cauliflower. (Marcia Long, white psychologist, 37)

For most women, internal images of their pregnancies had been refocused through the lens of sonography, eclipsing any alternative, less standardized, embodied notions of what a fetus felt like. Their internal states were now technologically redescribed. This process of reinscription has a visual history: in 1965, *Life* Magazine published Lennart Nilsson's photographs of autopsied embryos, presented in "living technicolor," as the "Drama of Life Before Birth." Greeted with far more awe and credibility in the United States than in his native Sweden, Nilsson's imagery was once again presented in the rejuvenated *Life* of 1990, where endoscopic photography enabled him to track sperm and egg uniting, as well as fetal development. In the quarter century separating the two photographic essays, the mother had become not only transparent but also a potentially hostile environment for the sperm, eggs, and fetuses she carries from "The First Days of Creation." We cannot argue that women (and their supporters) were re-educated to see themselves and their fetuses as separate entities exclusively by *Life* Magazine's enormously popular articles and Nilsson's (1977) best-selling book, *A Child Is Born*. But these images, which came to permeate sex education, right-to-life literature, and obstetricians' examining rooms, surely contributed to the narrowing of aperture and sharpening of focus on the fetus rather than on pregnant women.

While media presentations of technologically assisted viewing are presented as sources of visual pleasure, actual sonograms may provoke anxiety as well. If the fetus has become "real" through its imaging, as mysterious as an underwater dream and as intimate as a videotape, it has also become vulnerable:

I saw the sonogram of the twins, and I was thrilled. But I really couldn't read it, I didn't know what it meant. They had to interpret it for you, to say, "Here's a heart, these are arms." Afterwards, it made me queasy—they made the babies real for me by telling me what was there. If they hadn't interpreted, it would have just been gray blobs, and now I'm more frightened to get the results of the amnio back. (Daphne McCarle, white college professor, 41)

Because as soon as you see the sonogram, it's very real. They focused on the heart and it was beating, and then you could see the head . . . and the doctor was really terrific, like, there was all this excitement in the room, and she gave me a picture and they're all very positive . . . but you're trying to contain yourself from feeling that way because you know the only reason you're having this test is because you're more likely than the average person to have a problem. So I walked out of there pretty high . . . but I really have been trying to hold back the feeling pending results. (Laura Forman, white theater producer, 35)

With sonography and amniocentesis, one *can* be "just a little pregnant." Laura Forman's comments on self-containment surely echo Rothman's (1986) analysis of the effects of having a "tentative pregnancy." A woman's growing awareness of the fetus she is carrying is here reshaped by her need to maintain a distance from it, "just in case" something wrong should be discovered and she should be confronted with the choice to end or continue the pregnancy. Even as the sonogram personifies the fetus, the amniocentesis puts its situation in question. Simultaneously distanced and substantiated, the pregnancy is suspended in time and status, awaiting a medical judgment of quality control. Personal and social constructions—of maternal sentiment on the one hand and disability beliefs on the other—activate and agitate the fetal imaginary powerfully pictured through these technologies. . . .

REFERENCES

Bruno, Giuliana. 1992. "Spectatorial Embodiments: Anatomies of the Visible and the Female Bodyscape." *Camera Obscura* 28: 239–62.

Bynum, Caroline Walker. 1989. "The Female Body and Religious Practice in the Later Middle Ages." In *Fragments for a History of the Human Body*, part I, ed. Ramona Naddaff, Michel Feher, and Nadia Tazi. New York: Zone Books.

Kevles, Daniel J. 1985. *In the Name of Eugenics: Genetics and the Uses of Human Heredity.* New York: Knopf.

Rothman, Barbara Katz. 1986. *The Tentative Pregnancy: Prenatal Diagnosis and the Future of Motherhood*. New York: W. W. Norton.

Stafford, Barbara Maria. 1991. *Body Criticism: Imaging the Unseen in Enlightenment Art and Medicine*. Cambridge: MIT Press.

Yoxen, Edward (1989). "Seeing with Sound: A Study of the Development of Medical Images." In *The Social Construction of Technological Systems*, ed. Wiebe E. Bijker, Thomas P. Hughes, and Trevor Pinch. Cambridge: MIT Press.

QUIT SNIVELING, CRYO-BABY, WE'LL WORK
OUT WHICH ONE'S YOUR MAMA!

CHARIS THOMPSON

Contemporary Westerners are used to thinking of there being just two biological parents who both donate genetic material, a bilateral or cognatic kinship
pattern inscribed in our understanding of biogenetics. A baby is the product of
the fusion of the mother's and the father's genetic material. Kin are divided
into "blood relations" and non–blood relations, and it is usually assumed that
blood relations simply reflect genetic relationship. The three most familiar
kinds of disruption to this biogenetic understanding of what parents are all
preserve the genetic "natural" basis for Western kinship. In the case of commercial sperm donation, this model is not disrupted but, if anything, is exaggerated and reinforced. The vast majority of sperm banks sort their donors by
"genetic" traits so that clients can pick a donor who is a phenotypic match for a
husband, or pick a Nobel laureate, on the basis of a theory of the genetic
inheritance of these traits. The inseminated woman is taken to be related to the
baby in the usual way. In adoption, the "biological" parents or "birth" mother
are distinguished from the "social" parents, preserving the distinction between
the two. "Blended" families (where children from different marriages cohabit,
or where children live with stepparents or with a single parent after a divorce
or separation) muddy the waters of social parenting but leave intact the understanding of who the biological parents of each child are.

In donor egg in-vitro fertilization, however, the overlapping biological
idioms of blood and genes come apart. The maternal genetic material is contributed by the egg, which is derived from the ovaries of the woman who is the
donor. Nonetheless, the embryo grows in and out of the substance of another
woman's body; the fetus is fed by and takes form from the gestational woman's
blood, oxygen, and placenta. This disrupts the coherence of the natural ground
for bilateral linear descent and creates a schism between the concepts that
seemed to map so perfectly onto one another. If "blood relation" does not
designate the same entities in all circumstances as "genetically related," then

the naturalistic reduction of blood to genes that validated both the science and bilateral system of kinship reckoning does not work in some settings.

Gestational surrogacy means that eggs from one woman are fertilized with her partner's sperm in-vitro (occasionally donor eggs or donor sperm are used in place of the gametes of one partner from the paying patient couple) and then transferred to the uterus of a different woman who gestates the pregnancy, known as a gestational surrogate. The woman from whom the eggs were derived and her partner have custody of the child and are the parents (if donor eggs from a third woman were used, the woman from the paying couple is still the mother, and she adopts the baby at birth, as in conventional surrogacy).

Gestational surrogacy is procedurally identical to donor egg in-vitro fertilization: eggs from one woman are fertilized and gestated in the uterus of another woman. Two things make donor egg in-vitro fertilization and gestational surrogacy different from one another. First, the sperm with which the eggs are fertilized comes from the gestational woman's partner (or a donor standing in for him and picked by that couple) in the case of donor egg IVF. The sperm comes from the partner of the provider of the eggs (or a donor standing in for him) in the case of gestational surrogacy. From a lab perspective there is no difference—sperm collected by masturbation is prepared and added to retrieved eggs, and the embryos are incubated and transferred identically in both cases. The sperm comes from the person standing in the right sociolegal relationship to whichever of the women is designated as the mother-to-be. Who is designated as the mother-to-be is the second difference between the two procedures. Whether the gestational woman or the egg provider woman is the mother-to-be depends on who came into the clinic for treatment for infertility, the various parties' reproductive history, and in the case of private clinics, who is paying. Where additional donors are used or where the egg donor or gestational surrogate is being contracted on a commercial basis, the importance of who is paying for the treatment in deciding who the designated parents are is reinforced.

Gestational surrogacy also separates blood and genes, but whereas donor egg IVF traces motherhood through the blood half of this separation, gestational surrogacy traces it through the genetic half. Studying what happens in areas where the settled ontological hierarchy and distinctions break down should reveal possibilities for other ways of "doing" kinship that configure the mixture of nature and culture differently.

CASES 1 AND 2 IN-VITRO FERTILIZATION WITH DONOR EGG:
GIOVANNA AND PAULA

One afternoon I was in an examination room getting ready for the next patient's ultrasound scan, which had been ordered to see whether her ovaries were clear of cysts after hormonal down-regulation in preparation for a cycle of in-vitro fertilization (IVF). The patient was already in the room, changed and ready for her scan, so we talked as we awaited the physician's arrival. The patient, Giovanna, described herself as an Italian American approaching forty. She explained that she had tried but "failed" IVF before, using her own eggs and her husband's sperm. Egg quantity and quality are negatively affected by maternal age over thirty-five, and rates of implantation of in-vitro embryos are said to be negatively affected by both the age of the woman from whose body the eggs are retrieved and the age of the woman in whose womb the embryos are attempting to implant (Flamigni et al. 1994). Her response to the superovulatory drugs and the doctor's recommendation had persuaded Giovanna to try to get pregnant using eggs from another woman (donor eggs). Almost all clinics report better implantation rates in IVF using donor eggs, which are retrieved from women under thirty-five years old, than those obtained using the patient's eggs if the patient is over thirty-five.

In discussions with psychologists and nurse coordinators at IVF clinics I had learned that there is a typical trajectory for patients, such that treatments that are more expensive or more invasive or, especially, that involve donor sperm or eggs take a while to become acceptable. Frequently patients will start off being adamant that they will stop before a certain phase of treatment—say, before trying IVF or before moving to donor gametes. The passage of time and failure to get pregnant on other protocols often cause that boundary to shift. For women, accepting donor gametes is reportedly easier than for men, which correlates well with the lay belief that women find the idea of adoption easier to accept than their male partners. For women, conventional surrogacy is apparently harder to accept than donor egg or gestational surrogacy because the designated mother-to-be neither provides the eggs nor gestates the fetus. Donor egg IVF and gestational surrogacy are both easier to accept because the first involves gestating the fetus, whereas the last involves the development of an embryo formed from one of her eggs. By the time I was speaking with Giovanna she presented herself as committed about pursuing the option of an IVF pregnancy using donor eggs.

Giovanna said that she had decided to use a donor who was a good friend, rather than an anonymous commercial donor. She described her friend as excited and ready to help. Choosing a friend for a donor seemed to be an important part of reconfiguring the experience of pregnancy: if conception was not to occur inside her body or with her eggs, then it was preferable that she have emotional attachments of friendship to (and could make the corresponding demands on) the woman who was to be her donor. She did not say anything about her husband's relationship to her friend or specifically address his feelings about expanding the pregnancy such that his sperm would fertilize eggs derived from her friend, but she did say that he was supportive of her pursuing this as a treatment option. Choosing a friend as her donor was also important in the ways in which Giovanna was articulating the kinship issues raised by this splitting of the biological basis of descent into a gestation component and a genetic component.

Giovanna explained to me that her friend was also Italian American. She described this shared ethnic classification as being "enough genetic similarity." Further, Giovanna accorded her gestational role a rich biological significance: she said that the baby would grow inside her, nourished by her blood and made out of the very stuff of her body all the way from a two-celled embryo to a fully formed baby. There are several contemporary arenas where the line between nature and nurture has been creeping further and further back in pregnancy: fetal monitoring and fetal surgery, right-to-life political movements, the improved survival rates of significantly premature infants, and a predominantly child-based perspective for discussing the social and ethical dimensions of human reproduction. In these arenas gestation is increasingly assimilated to the care one provides to a child once it is born. Giovanna could have described her procedure as a further retraction of the boundary between nature and culture, bringing culture right back to everything that happens after fertilization. This would have preserved a unitary and grounding function for the natural component of kinship (reduced to genetics). Her use of donor gametes would then have fitted the general trend of de-privatizing pregnancy. Instead Giovanna cast her gestation in biological terms, appealing to blood and shared bodily substance.

Giovanna pried apart the *natural* basis for specifying mother/child relations into separable components. In addition, she complicated the natural status of the genetic component that would be derived from her friend by *socializing* genetics.[1] She said that what mattered to her in genetic inheritance was that the donor share a similar history to her own. She said that because her friend was also Italian American they both came from the same kind of home, and that as they both had Italian mothers, they had grown up with the same cultural influences. So genes were coding for ethnicity, which Giovanna was expressing as a national/natural category of Italian Americanness. This ethnic category itself coded for a sociocultural life history that I imagine Giovanna would say goes far beyond anything that is mandated by the natural destiny of having Italian American genes. Genes figure in Giovanna's kinship reckoning because there is a chain of transactions between the natural and the cultural that not only grounds the cultural in the natural but gives the natural its explanatory power by its links to culturally relevant categories.

When in scientistic mode, it is tempting to think that reductions of higher levels of organization or greater sociality to simpler or more natural or causal underlying factors work without remainder, and in one direction only. Applied here, the scientistic impulse is to assume that our biology underwrites our sociocultural potential and not the other way around, and that biology is sufficient to account for sociocultural reality. Even though there are infinite contingency and variability in the ways things actually turn out, for any actual state of human affairs, biology can give a full account of that state of affairs. Giovanna's separation of biology into shared bodily substance and genetics and her formulation of what matters to her about genetics in the context of her procedure resist both these assumptions. The reduction to genes is only meaningful because it codes back to sociocultural aspects of being Italian American (it is not unidirectional). Likewise, the ethnic category is not just a category that performs a transitional function between nature and culture, but is a category of elision, collecting disparate elements and linking them without any assumption that every one of the sociocultural aspects of having an Italian American mother, for example, needs to map back onto biology.

In a related case, an African American patient (Paula) whom I met at the clinic only once spontaneously offered commentary on the kinship implications of her upcoming procedure. Paula had undergone "premature ovarian failure" and entered menopause in her early thirties before she had had any children. She and her husband had decided to try donor egg IVF, and she was hoping to be able to carry a pregnancy. They had not yet chosen an egg donor, and Paula said that she would first ask her sister and a friend, to see if either of them would be willing to be her egg donor. Paula expressed a strong preference for using a donor from her community. She said, laughing, that using a donor was not as strange as it might at first seem. After all, "It's just like we've been doing all along!" When I asked her what she meant by that, she explained that in African American communities it was not unusual for women to "mother" or "second-mother" their sister's or daughter's or friend's children.

This explanation has several interesting elements. First, it rejects the idea that these procedures are wrong because they are unnatural or because they are exploitative of the donor. If being a donor is just one more way of doing something that is already a prevalent social phenomenon—dividing different aspects of mothering across generations, between friends and between sisters—then it is not a radical departure from existing social practice. In presenting it like this, Paula is normalizing her reproductive options. Rather than being exploitative, using a donor is assimilated to other ways in which women help one another to lead livable lives. But just as much as normalizing her own reproductive plight, Paula's explanation suggests the possibility of legitimizing socially shared motherhood through the naturalization of shared mother-hood. Legitimizing social "deviance" by pointing to its natural basis is a famil-iar strategy. In this case, however, the strategy mobilizes only newly available biological possibilities (pregnancy with one woman providing the egg and another providing the shared bodily substance) and reverses the usual form of the argument where naturalization is legitimizing because it is inevitable and because it precedes social forms.

Both Giovanna and Paula are pursuing new ontologies out of, but different from, their lived experiences of kinship. From the heart of biomedicine they are changing and extending the reference of the word *mother*. In this regard, they are practical metaphysicians. . . .

CASES 5 AND 6 INTERGENERATIONAL DONOR EGG IN-VITRO FERTILIZATION: FLORA; AND INTERGENERATIONAL GESTATIONAL SURROGACY: VANESSA/UTE

These two cases complicate the kinship situations described above by adding an intergenerational element. Case 5 (Flora) involves a case of egg donation where the egg donor is the daughter of the woman who is trying to get pregnant. Case 6 (Vanessa/Ute) describes a gestational surrogacy where the surrogate became pregnant with an embryo formed from the sperm of the husband of the couple trying to have a baby and a donated egg from a daughter from a previous marriage of the wife. In cases like these, the "mother" (as defined from the treatment perspective) is also the grandmother in the famil-iar way of accounting: the donor daughter is also the genetic mother. In case 6 the designated mother is neither the genetic nor the gestational mother, but by reversing the direction of linearity (the daughter passes on genetic material to her mother instead of the other way around), genetic relatedness is still mobilized . . . to override the shared bodily substance of the gestational surro-gate and fetus. The daughters in cases 5 and 6 are both genetically related to the fetuses, but the genetic information has been made available through them in a chain of relatedness that leads back to their mothers; the daughters are biological intermediaries. To avoid incest and intergenerational coercion, they are also made socially invisible. This social invisibility is achieved by the lack of duty toward any offspring conceived with their eggs, and will no doubt be enforced by the lived experience of relating to these children as siblings rather than sons or daughters.

A fifty-one-year-old woman, Flora, came in for treatment. She was perimenopausal, and had five grown children from a previous marriage. She did not fit the typical patient profile of the elite white postponed-childbearing woman. Flora was Mexican, and crossed the border from the affluent suburb of Tijuana where she lived to southern California for her treatment. With five children she already had what many couples would probably consider "too many" children. She had recently been remarried to a man many years her junior who had not yet had children. Flora was quite explicit about the gender, age, and financial relations between herself and her new younger husband, and her desire, as she put it, "to give him a child."

As mentioned in the discussion of Giovanna's case above, conception rates for women over forty are extremely low on assisted reproduction protocols but can be significantly improved by using donor eggs. It was therefore suggested that if Flora wanted any significant chance of getting pregnant, she should find an egg donor. The donor eggs would be inseminated by Flora's husband's sperm, and resulting embryos would either be transferred to Flora's uterus or frozen for use in subsequent cycles. Donor consent covers all eggs retrieved at the time from the donor, so any frozen embryos would also "belong" to Flora and her husband.

Flora read widely in the medical and popular literature and frequently made suggestions about or fine-tuned her own protocol. She also picked her own egg donor: one of her daughters in her early twenties, who herself already had children. The mother's and daughter's cycles were synchronized, and the daughter was given superovulatory drugs to stimulate the simultaneous maturation of several preovulatory follicles. The daughter responded dramatically to the drugs and at the time of egg recovery the physician and embryologist removed sixty-five eggs from her ovaries (ten eggs +/- five is "normal"). The eggs were inseminated with Flora's husband's sperm, forty-five fertilized, and five fresh embryos were transferred to Flora's uterus that cycle. Twenty grade 1 embryos were frozen at the 2PN stage (when two pro nuclei are visible, indicating that fertilization has taken place, but before cell division has begun), and the remaining embryos were frozen at the 3 and 4 cell stage after the fresh embryo transfer. Flora did not get pregnant in the fresh cycle or in the first two frozen cycles, but did in the third.

Flora and her daughter did not seem overly perturbed by the intergenerational confusion of a mother giving birth to her own "grandchildren" and to her daughter's "daughter/sister." Instead they discussed the mother's and daughter's genetic similarity. Like Paula (case 2 above) they also assimilated their case to existing social practice, in this case to the prevalence of generation-skipping parenting (where a grandparent parents a child socially and legally) in communities with which they were familiar. Nonetheless, Flora signaled some ambivalence on the part of the daughter. When I asked who her donor was, she replied, "My daughter." I then asked, "Is she excited?" and Flora replied, "Not exactly, but she doesn't mind doing it." The daughter herself told me when the mother was out of the room that she didn't mind helping them have a single baby, but that the huge number of embryos stored away was unsettling. After all, she said, her mother already had a family; "she doesn't need to start a whole new family—one baby is one thing but . . . !"

The daughter's reluctance to see her mother having "a whole new family" might have been due in part to a distrust or disapproval of her mother's relations to her new husband, or to a reluctance to have the grandmother (Flora) of her children back being a mother of babies again. The daughter's anxiety about the stock of embryos that were frozen, however, seemed to be at least partly an anxiety about the existence of unaccounted-for embryos using her eggs and her stepfather's sperm. Using one of her eggs to help initiate a pregnancy that was clearly tied into a trajectory on which it was Flora's and her husband's child placed Flora as an *opaque intermediary* (placed Flora as significant in the flow of kinship) between the daughter and the husband and protected the daughter from any direct connection to her stepfather. The embryos in the freezer, however, were in limbo. If they were not used to initiate a pregnancy in Flora, then they existed as the conjoined gametes of the daughter and the stepfather. The status of these embryos—even though technically owned by Flora and her husband—did seem to raise anxiety associated with inappropriate kinship.

Case 6 concerns Vanessa, who started up her own surrogacy agency shortly after being herself a commercial gestational surrogate and giving birth to a baby for another couple. I met Vanessa at a local fast-food restaurant which she had suggested as our rendezvous point. She explained over the phone to me that this was the neutral ground where she usually met with her surrogates and recipient couples before drawing up contracts or getting lawyers involved.

Vanessa told me that she had seen a program on television in which she noticed the "joy in the mother's eyes" when the baby was handed over by the surrogate. Vanessa's family was in some financial difficulty at the time as their small-scale manufacturing operation had just closed. Since she had had "no trouble" during pregnancy or birth with her own four children, she thought that she would look into the possibility of being a surrogate herself. She expressed this decision in a religious idiom, as a chance offered by God simultaneously to do good and to make a fresh start. She contacted two agencies and decided to work with the agency that was based closer to her home even though they paid their surrogates $1,500 less than the other program ($12,000 as opposed to $13,500).

Vanessa was introduced to "her" couple in the middle of 1992. She described the couple as a "German woman of about forty" (Ute) and an "Asian man." It was agreed that they would try gestational surrogacy. Vanessa was given Lupron to shut down the production of estrogen and progesterone and thus put a halt to ovulation, so as to synchronize her cycle with Ute's cycle. On the first treatment cycle Ute ovulated before the physician took her to surgery for ovum pickup and they got only one egg at surgery. The one egg was successfully fertilized with Ute's husband's sperm, incubated for two days, and transferred to Vanessa, but Vanessa did not get pregnant.

For the second treatment cycle, Ute and her husband decided to combine gestational surrogacy with donor egg IVF. Ute had an adult daughter from a previous marriage who agreed (as Flora's daughter above had) to be her mother's and stepfather's donor. I asked Vanessa whether the daughter or husband or Ute were worried about the genetic stepfather/daughter relationship. Vanessa said that the only things that were raised were that it was going to be kept secret and that using the daughter's egg was the next best thing to using Ute's eggs because of the genetic similarity.

Vanessa did not get pregnant on the fresh IVF cycle using the daughter's eggs, but there had been sufficient embryos to freeze some for a subsequent attempt. On the frozen cycle, Ute's daughter's and husband's remaining embryos were thawed and transferred to Vanessa's uterus, and Vanessa became pregnant with a singleton. The pregnancy, unlike her four previous ones, was not easy for Vanessa. The difficulty of the last months and of the birth made her determined never to be a surrogate again and, as she described it, propelled her into the decision to start her own surrogacy agency. Under her surrogacy contract she was not allowed to make any of her own medical decisions while pregnant, and she had to consult with the recipient couple and the doctor before taking any medications or changing her agreed-upon routine in any way. The recipient couple and the physicians took over jurisdiction of her body for the twelve months of treatment and pregnancy. Nonetheless, Vanessa described the pregnancy almost wistfully, explaining how exhilarating the intimacy with the couple was and how spoiled she felt. During the pregnancy Ute and her husband took her out, bought her fancy maternity clothes, and so on. Vanessa underlined the importance of the relationship between the surrogate and the recipient couple in all surrogacy arrangements by saying that they are "the couple you're going to be a relative with for a year and a half."

Despite her experience first as a surrogate and subsequently working with other surrogates and recipient couples, Vanessa seemed surprised by the severing of the ties of relationship between herself and the recipient couple after the baby was born and handed over. She said that the couple stopped contacting her and that when she called them to find out how they and the baby were, the couple made excuses and hung up quickly. Other gestational surrogates I have talked to have recounted similar experiences. It seems that Vanessa's relationship to the couple for the year and a half was enacted because she was prosthetically embodying their germ plasm and growing their child, but that, on the contrary, she was at no point related to the recipient couple or the baby in her own right. Designating Ute as the mother meant that Vanessa *had* to be transparent or irrelevant to the baby's kinship. Once the baby was born, Vanessa was in many ways just like any other instrumental intermediary in establishing the pregnancy, such as the embryologist or even the petri dish. The fact that she cared for and had good reasons for continuing to feel connected to the couple and the baby and so did things like make phone calls meant perhaps that she needed some postnatal management to be kept in the background. But the logic of disconnection was the same as for other intermediaries in the recipient couples' reproduction.

MAKING KINSHIP: OPACITY AND TRANSPARENCY

In the cases of Giovanna and Paula, . . . described above, I do not know whether a pregnancy has been or will be established (cases 1 and 2). I try here hypothetically to isolate some of the strategies that were used in the clinical setting in each case for delineating who the mother was for each child. These strategies are not exhaustive and cannot be expected to be invariant in different arenas of the patients' lives either. For example, legal and familial constraints bring their own sources of plasticity and relative invariance which are very powerful in determining kin. But the clinic is one significant site of negotiation of kinship, and it is of particular interest because it articulates between the public and the private and because it illustrates flexibility in biological and scientific practice. I emphasize the mothers' and not the fathers' relatedness because in the cases I have chosen it is in motherhood that the procedures raise a challenge to biological essentialism. For each case I distinguish different significant intermediaries in establishing the pregnancy. I then sort the intermediaries as to whether they are opaque or transparent in the determination of who counts as the mother. I also ask where and under what conditions the ways of designating the mother are liable to break down or be contested.

In recounting the cases above I referred to some of the women as "opaque" in the establishment of a pregnancy. For analytical purposes, I call any stage in the establishment of a pregnancy opaque if it gives rise to relatedness and personhood. A stage does this kinship work if it brings two people into relationship. Here, I am interested in working out who the *mother* is, so an opaque stage in this determination would be a stage that answers who the mother is. As the cases showed, there are many resources for making a stage opaque that are not necessarily well differentiated. Biology and nature are resources; so are a wide range of socioeconomic factors, such as who is paying for treatment, legal factors such as who owns gametes and embryos, and familial factors such as whose partner is providing the sperm and who is projected to have future financial and "nuclear family" responsibility for the child. A stage can be made securely opaque by separating but bringing into coordination the biological and social accounts of the relationship. Depending on the kind of parenting in question, different kinds of coordination are appropriate.

By analogy, I call a stage *transparent* if it enables heredity or relatedness but does not itself thereby get configured in the web of kinship relations. A woman is transparent in conceiving and bearing a child as regards who the mother is if she is an *intermediary* in the pregnancy and childbirth without by her actions becoming the mother or contesting who the mother is. When people—the paradigm opaque nodes of kinship webs—are biologically involved in kinship transmission—the paradigm means of making kinship in modern Western societies—but are transparent to the process, there is a prima facie tension. The cases described above exhibited different strategies for achieving kinship transparence of some people and not others. Breakdown (from the point of view of clinics or designated recipient couples), contestation, and prohibition usually occur when an attempt by at least some of the actors to render one or more stage as transparent is contested or fails.

In the first case, that of Giovanna, the Italian American woman who was planning to undergo a donor egg procedure with the eggs of an Italian American friend of hers, the two potential candidates for motherhood were the donor friend and Giovanna herself. The friend was made appropriately transparent using three strategies. First, the friend's contribution of the eggs was biologically *minimized* by stressing the small percentage of the pregnancy that would be spent at the gamete and embryo stage versus the length of time that the fetus would grow inside Giovanna. Second, genetics were *redeployed*, so that the friend's genes were figured as deriving from a common ethnic gene pool (Italian American) of which Giovanna was a member and so in which she was also represented. Third, the friend was secured as an intermediary in the pregnancy by stressing her other bonds to Giovanna of friendship and mutual obligation. She could be instrumental to the baby because she was very much not purely instrumental to Giovanna.

To further disambiguate who should be considered the mother, Giovanna put forward strategies through which she could assert her own opacity. Giovanna's opacity was to come from her gestation of the baby and provision of the bodily substance and bodily functioning out of which the baby would grow and be given life. Giovanna stressed the significance of the gestational component of reproduction and emphasized the importance of the experiential aspects of being pregnant and giving birth in designating motherhood. Further, Giovanna was married to, and would parent with, the provider of the sperm.

In Paula's case (the African American woman who was going to use an African American friend or one of her sisters as her donor), the strategies of making transparency and opacity were similar to those employed by Giovanna, but there were also interesting differences. The most striking difference was that Paula drew the line between opacity and transparency more tenuously, being content to leave more ambiguity in the designation of motherhood. This tolerance of ambiguity was part of her way of legitimating the procedure itself and simultaneously legitimating the social parenting practices with which she drew analogies.

Like Giovanna, Paula gave an ethnic or racial interpretation to the genes such that by getting genetic African Americanness from her donor the baby would share racial sameness with Paula. There was a difference in how they discussed this ethnic or racial commonalty, however. Whereas Giovanna routed her friend's genetics through cultural sameness and back to herself, Paula seemed not to trace a route from her donor back to herself; just coming from the same racial group and whatever its gene pool might be was enough. The trope that genes code for racial distinctions, inclusions and exclusions, and purity seemed to be more significant here than the equally prevalent trope that genes function as the thing that provides the definitive mark of individuality (the DNA fingerprint) which is passed down cognatically from a mother's and a father's individual contribution. In fact, Paula was the only one of the people whose cases are discussed in this essay who did not bring genetics back to herself as an individual so as to have it function as a resource in designating her as the mother. Giovanna, Flora, and Ute all also mobilized the idea of genetic similarity with their donors, but unlike Paula they all started with the donor as an individual and ended up with the recipient as an individual. Thus, according to my definitions above, Paula's donor did not need to be transparent for race.

Similarly, the logic of transparency versus opacity was elided in Paula's analogy between her procedure and socially shared mothering. She took legitimacy for the procedure from the fact that shared parenting was commonplace among people she identified with racially, and she also commented on the natural confirmation of these social patterns that her upcoming procedure would provide. Socially and biologically her motherhood—including its ambiguity— would be recognizable and legitimate. She would be sufficiently opaque, and her donor sufficiently transparent without either needing to be wholly so. . . .

In Flora's case, where she was using her daughter as her egg donor, a significant strategy in making her daughter transparent was to see the genetic material not as deriving from the daughter and her eggs per se but as being passed on in the daughter's eggs, in a manner that could be retraced to Flora. The daughter was made an intermediary by making her gamete contribution a *detour* to her mother Flora's genetic material, which could no longer be accessed directly from Flora. Flora, perhaps more closely than any of the other donor egg patients whose cases are described here, attempted to recapture in her own claims to motherhood the genetic idiom of linear descent.

Flora's case was complicated by the intergenerational element, however, and her daughter's transparence was threatened from at least two sources. Flora and her daughter discussed the similarity of what they were doing to other intergenerational parenting in which a grandparent can be the social and legal parent. Drawing an analogy with prevalent social practices, as Paula had done, Flora strengthened the legitimacy of the procedure and so stabilized her claim to being the mother of the child. In making this analogy, however, the daughter's transparence was in danger of being compromised, because grandparenting often allows for the "real" parent to reclaim his or her parental jurisdiction. Unlike for Paula, this analogy to social practices did not loosen the designation of who was to count as mother but was meant to disambiguate it even beyond the norm for the social practices with which she was drawing the analogy. Both Flora and her daughter were adamant that there was to be no ambiguity in who was to be the mother—Flora. Flora's marriage to the provider of the sperm and the incestuous implications of reckoning it otherwise reinforced the disanalogy with social grandparent parenting. The other threat to the daughter's transparence came from the unintended stockpiling of embryos formed from the daughter's eggs fertilized with her stepfather's sperm. The frozen embryos in the lab were only tenuously tied to Flora's reproduction, and their quasi-independence and potential to initiate other pregnancies were troubling to the daughter. The large numbers of embryos owned by her mother but created from her eggs faced the daughter with a situation where she no longer had control over her instrumental role. This made it hard for her to be confident of the maintenance of her own transparence.

In Vanessa's pregnancy, there were three potential candidates for sufficient opacity to be designated as the mother. There was Vanessa herself, the commercial gestational surrogate; there was the intended mother who was the wife of the recipient couple; and there was the wife's daughter who was the egg donor. . . . Vanessa was made transparent by assimilating her role in gestating the fetus to the provision of a temporary caring environment. . . . Vanessa's intermediary role was not elicited by the obligations of a prior relationship to the recipient couple. Instead her services were contracted commercially, and Vanessa had no further claim on the child after the birth. This disconnection was underwritten by the assumption of contractual arrangements that both parties agree that recompense is satisfactory despite possible incommensurability of the things being exchanged. Furthermore, contracts assume that the transaction itself is limiting and does not set up any subsequent relationship or further obligation. Vanessa's transparence was threatened when she experienced the sudden severing of relations after the birth of the baby as baffling and troubling. The contractual relation assured the temporary intermediary relation of shared bodily substance with the recipient couple, but insofar as it was uncontested it sustained no further contact.

Ute's daughter, who was acting as the egg donor, was made transparent in two ways. First, the genetic contribution was described as being closely *similar* to her mother's genetic material. This is not quite the same argument as that made by Flora and her daughter, where the daughter became a vehicle or detour through which genetic material originally from Flora had passed; in this case the mother and daughter simply alluded to the similarity of the mother's and daughter's genes. The daughter was further rendered transparent by the recipient family's commitment to keeping her role a secret.

PROGNOSIS

Ute, the recipient mother, was made opaque not by herself capturing either of the predominant biological idioms (genetics as represented by providing the egg, or blood and shared bodily substance as represented by gestation). Instead, the genetic component from the daughter was similar enough to stand in for her genetic contribution, and the blood component was made intermediary in the contracting out of the gestation. Neither natural base was sufficiently strong to overwhelm her claim to be the mother, which she asserted through being married to the person who provided the sperm, through her daughter's compliance with her desires, and through having the buying power to contract Vanessa.

The donor egg procedures seem to offer the potential somewhat to transform biological kinship in the directions indicated, for example, by Giovanna when she codes genetics back to socioeconomic factors and thereby deessentializes genetics, and when she draws on the trope of blood relation and shared bodily substance without genetics. Developments in other contemporary sites in the United States over the same period as this fieldwork, such as decisions in legal custody disputes, have tended to favor genetic essentialism in determining motherhood, although this has not been universal. If an overwhelming predominance of nonclinical cultural contexts come down on the side of genetic essentialism, donor egg procedures might well become assimilated to adoption or artificial insemination with donor sperm. The bid to make gestation in donor egg procedures opaque vis-à-vis biological kinship would have failed. A newer procedure, reports of which began to appear in journals and in the press in 1997, uses the eggs of older patients but "revivifies" them by injecting cytoplasmic material derived from the eggs of younger women into the older eggs before fertilization. This donor procedure preserves the genetic connection between the recipient mother and the fetus, and so brings the genetic and blood idioms back into line. It thereby tightens up again the flexible ontological space for designating biological motherhood that had been opened up by donor egg procedures.

Paula used a more mixed ontology for motherhood than Giovanna. For her it was satisfactory if there was no definitive answer to *which* one was the mother, and she was happy to accept that in some ways there was more than one mother. Likewise, because she raised the possibility of biologically enacting what she described as already prevalent social practices (shared mothering), there was less at stake in having gestation without genetics be biologically opaque. Shared mothering was not presented as necessarily involving a natural kinship rift; indeed it was presented as a practice which preserves racial identity and integrity. Conventional adoption, on the other hand, has historically gone in one direction: African American children being adopted into Caucasian families, disrupting racial identification without disrupting racism. For Paula, having her procedure assimilated to some kinds of social parenting would still entitle her to make appeal to notions of biological sameness.

. . . The likely stability of Flora's and Ute's kinship opacity and claims to motherhood is hard to gauge. Ute's opacity could have been challenged by Vanessa's desire for contact with the child after its birth, set against the purchasing power of the contracting couple. Flora is likely to encounter social censure, just as she did in the clinic, for her desire to bear a child "for" her younger husband. If either Ute or Flora confesses to using her daughter as a donor, she may be condemned for putting her daughter through medical interventions so as to regain her own youth. If either daughter contests the circumstances of her role at a later date, the settlement of who is the designated mother might break down. But it is also possible that parenting with donor eggs for perimenopausal and postmenopausal women will play a part in breaking down some of the more oppressive aspects for women of the "biological clock." If Flora and Ute can maintain their opacity with regard to their claims to motherhood, they will be cases to hold against the elision of women's identities, femininity, youth, and ovulation. If Flora is buying into the cult of youth to keep her husband, as she claims, she may yet be subverting the wider essentialist identification of women's identities with their youthful biologies, as she also maintains. . . .

The mere fact of establishing a pregnancy does not sort out the chains of kinship by itself. As the cases recounted above illustrate, there is a complicated choreography necessary to disambiguate transparent and opaque interventions. If the work is not done (social, biological, economic, legal, familial) to sort out opaque and transparent links, there will be errors and breakdowns in determining who is related to whom and how.

NOTE

1. The ambivalence in feminist theory about the relative virtues of naturalism and social constructivism is entirely appropriate: sometimes important political work is done by socializing what has previously been taken to be natural and deterministic; sometimes the reverse is necessary. What cases like these drawn from infertility medicine show us is that it is not necessary for the theorist to champion one strategy rather than another. Modern medicine gives us many cases where the choreography between the natural and the social is managed flexibly by ordinary people (the practitioners and patients using the technologies in question).

BODYWORLDS: THE ART OF PLASTINATED

CADAVERS

JOSÉ VAN DIJCK

In the 1950s, when synthetic materials had recently been introduced, people used to admire plastic tulips for their realistic quality. Consumers were
charmed by the obvious advantages of these fake flowers: they never withered,
and every tulip looked absolutely perfect. When I buy a bouquet of real tulips
these days, it strikes me how much they resemble plastic ones. By and large, the
famous Dutch tulip is no longer an exclusive product of nature, for its cultivation increasingly depends on treatment with chemical and biotechnological
means. The advantages are obvious: the flowers remain fresh much longer, and
every single tulip meets the requirements of standardized size, shape, and
color. Whereas before, we wanted the artificial object to look like a real one, we
have now entered an era in which we want the real object to look like "perfected nature." We are no longer satisfied with a plastic imitation of an organic
object, yet neither are we satisfied with nature's own imperfect products. So we
tinker with flowers and treat them with chemical and other techniques, until
they meet our aesthetic standards. The contemporary tulip, in other words,
has become an intricate object, an amalgam of organic material, cultural
norms, and technological tooling.

This new preference for the enhancement—instead of imitation—of natural
material also pertains to the human body. Dentists who, in the 1960s, did not
think twice about pulling a patient's teeth and replacing them with a set of
dentures (cheap and low-maintenance), now make every effort to save the
original ivories. They have an extensive collection of tools and plastic materials
at their disposal to perfect our pearly whites, until they resemble the (re-
touched) teeth of fashion models in magazine pictures. In a similar vein, our
physical appearance can be optimized by plastic surgery, anabolic steroids, and
perhaps, in the near future, by genetic therapy. "Natural silicone breasts" is no
longer an oxymoron, but an indication of a reality in which female bodies are
reshaped by cultural norms with the help of advanced technology. The preference for a manipulable body perfectly fits a material, technological culture in

which imitation has been replaced by *modification*. Just like the tulip, the body has become a mixture of organic matter and artifice.

If the living body has become a mix of nature and artifice, it is no great surprise to find this also applying to the dead body. In the past twenty years, Gunther Von Hagens, a German anatomist from Heidelberg, has developed a preservation technique that he has dubbed "plastination." It involves a sequence of chemical treatments of the corpse, which is then modeled into a sculpture by the anatomist's hand and scalpel. The resulting anatomical object looks like a conflation of an opened-up mummy, a skinned corpse, and an artistic sculpture. Von Hagens calls his collection of cadavers "anatomical art," which he defines as "the aesthetic and instructive representation of the inside of the body."[1] After its first public showing in Japan, Von Hagens's remarkable collection *Körperwelten* (Bodyworlds) was exhibited in Mannheim in 1997–98 and in Vienna in 1999. The German exhibition lasted four months and attracted more than a million visitors—an exorbitant figure for what was advertised as a scientific exhibition—and the Vienna event was kept open twenty-four hours a day, seven days a week to accommodate all visitors. Even shows in major art museums devoted to the work of canonized painters seldom receive this much popular attention.

What, then, makes the plastinated bodies from Mannheim so fascinating? Why did Bodyworlds become such a success? Evidently, in our increasingly medicalized society, people's interest in the human body has risen in proportion to their interest in its normally hidden dimension. And yet, other anatomical-pathological museums in Europe have offered inside glimpses of the body as well, without attracting anywhere near the numbers of visitors that Bodyworlds has. A factor contributing to its popularity may well have been the discussion, fanned by the German media, about the ethicality of this exhibition. Newspapers and television shows raised the question whether the display of real human cadavers was indeed legitimate, and if so, for what purposes. Did Bodyworlds serve any scientific goal at all, or was its prime intention the display of artistic objects? Some reporters even suggested that while any other country could, in good consciousness, feature such an exhibition, Germany could not, due to its dubious medical experiments on living and dead bodies during the Nazi era. Undoubtedly, this media attention drew more visitors to the Mannheim exposition, but that still does not fully account for its immense popularity.

The appeal of Bodyworlds, as well as the controversy surrounding the exhibition, can be properly understood only if we approach the phenomenon from a historical perspective. Von Hagens's plastinated cadavers perfectly fit the long-standing scientific tradition of anatomical body production, as much as they prolong artistic conventions of anatomical representation. By the same token, the remarkable exhibition setting can be traced to the cultural roots of public anatomy lessons and the artful display of body parts. From the history of anatomy we have learned that anatomical practices, objects, and representations have always been an intricate mixture of science and art, and a hybrid of medical instruction and popular entertainment. During the Mannheim exhibition, the ethical debate centered primarily on the question whether plastination should be looked upon as either science or art, either instruction or entertainment. What makes Von Hagens's anatomical art controversial, though, is not that it cannot be classified in clearly defined boxes, but that it defies the very categories on which ethical judgments are grounded.

THE HISTORICAL TRADITION OF ANATOMICAL BODIES AND MODELS

Throughout the history of anatomical practice, anatomists have tried to reconcile the contradictory requirements of authenticity and didactical value in the teaching of medical knowledge. On the one hand, the anatomical body should consist of real flesh, so that cutting into a cadaver teaches future doctors the organic complexity of a living human body. Yet working with dead bodies also has a distinct drawback: it is difficult to demonstrate certain aspects of physiology, such as blood circulation or the complex web of muscular tissue. For students to conceptualize anatomical structures, a human cadaver should also be pliable in order to single out particular aspects. Body *models*, shaped and sculpted to show distinct parts or features, have been produced since the early Renaissance, to serve as teaching aids in medical schools. Models have the advantage that certain physiological features can be disproportionately accentuated in order to convey particular anatomical insights. An obvious drawback of body models is that they do not give students a feel for organic texture. From the time of Vesalius to the days of Von Hagens, we see anatomists struggling to combine a preference for authentic bodies with the educational advantages of body models.

During the early Renaissance, watching an anatomist perform a dissection was the only way for future doctors and artists to get a sense of the body's insides. When the Belgian anatomist Andreas Vesalius in Bologna, and his colleagues Jacobus Sylvius in Paris and later Nicolaas Tulp in Amsterdam, performed public dissections, neither students nor onlookers were allowed to touch any body parts. Dead bodies were prone to quick decay, so dissection had to be executed swiftly and expertly. Vesalius (1514–64), while demonstrating the intricacies of a corpse, opened up our gaze, as Jonathan Sawday aptly puts it, to the depths of physiological reality.[2] The naked realism of dead bodies on the dissection table, combined with the public knowledge of their criminal pasts, provided a mesmerizing spectacle for a large audience who paid a substantial fee to attend these anatomy lessons. However, the educational value of these messy performances was minimal to anyone except, perhaps, the anatomist.

The necessity to preserve corpses beyond several days, as well as anatomists' desire to demonstrate particular physiological features, stimulated the invention of better conservation methods. Between the early twelfth and sixteenth centuries, various techniques for embalming or preserving corpses had been experimented with. The Dutch anatomist Frederick Ruysch (1658–1731), successor to the illustrious Tulp, developed unprecedented standards for the preservation and display of bodies. He injected the veins with a mixture of talc, tallow, cinnabar, oil of lavender, and colored pigments, the precise recipe of which he kept secret; as a result, the body would last much longer, sometimes up to a full year, and dissection was less messy due to the replacement of blood by preservative. Yet Ruysch's technique did more than ameliorate the material preconditions for dissection: it allowed for a new kind of anatomical artifact— a work of art, rather than a scientific work object. As Julie V. Hansen observes, Ruysch created "a new aesthetic of anatomy that melded the acts of demonstration and display with the stylistic and emblematic meanings of Vanitas art."[3] Besides performing public dissections, Ruysch built up a collection of body parts such as hands, limbs, or heads, carefully conserving each one in a separate glass jar. To enliven his objects and disguise the brutality of death and dismemberment, he embellished the compartmentalized cadavers with flowers or garments. His favorite displays were little bodies of fetuses or stillborn babies, which he clothed with scarves and embroidered baby-hats, replacing their eyes with glass to make them look like innocent infants. Even though Ruysch was one of the most respected Dutch anatomists in the seventeenth century, he is consistently referred to as an artist who elevated anatomical bodies to the status of sculpture and painting. As Hansen argues: "Under Ruysch's hand, the body was not dead, it was nature revealed, admired as the handiwork of God, the invisible made visible."[4]

Ludmilla Jordanova, in a similar vein, emphasizes that conflicting requirements of authenticity and didactic value emerge repeatedly throughout the history of anatomical artifacts. After the Renaissance, medical education increasingly called for hands-on practice with anatomical bodies. Increased demand and tougher laws regarding obtaining cadavers forced anatomists to look for body substitutes. Even as practical solutions to the shortage of real cadavers led to the creation of "fake bodies" in the seventeenth and eighteenth centuries, these models were equally subjected to the norms of accuracy, durability, and technical flexibility. The development of wax models catered to these educational needs, and had some advantages over real bodies.[5] Beeswax had the unique quality of resembling organic texture, while it was also fully pliable. Parts of the wax model could be taken out to allow the student to look into the organic complexity, or to manipulate individual parts and organs. In the second half of the eighteenth century, Bolognese sculptors like Lelli and Morandi and Florentine masters like Caldani, Fontana, and Piranesi lifted the craft of wax modeling up to the status of art; they began to be commissioned by a royal Maecenas, and their models were bought up by private collectors. From clinical-instructional settings, the wax models moved to private collections and later to museums, where they can still be admired. . . .

The invention of new chemical techniques, particularly the application of formaldehyde in the nineteenth century, allowed anatomists to extend the preservation of cadavers, and enabled students to participate in actual dissections. Dissections were no longer public events, as they had been in the Renaissance, but took place behind the closed doors of the hospital laboratory. Through various modes of public display—most notably, formaldehyde-drenched body parts in glass bottles—we can further trace the typical hybrid requirements of authenticity and pedagogical value. In contrast to the embellished body parts from, for instance, Ruysch's collection, nineteenth-century exhibitions of organs in glass jars show a preference for unadorned, straightforward anatomical parts. Reproductive organs affected by sexually transmitted diseases, or livers degenerated by alcoholism, clearly served a double pedagogical mission: the specimens obviously instructed doctors about the regularities and irregularities of human anatomy, yet their broader aim was to teach ordinary men and women the laws of moral behavior. The primary appeal of such anatomical collections was their focus on the aberrant, especially the monstrous aspect of pathological cases, such as embryos with spina bifida and fetuses with hydrocephalus. Although pathological creatures and "monsters" as objects of spectacle date back at least as far as the European fairs and curiosity cabinets of the Middle Ages, their display in anatomical exhibitions rendered them part of an authoritative medical culture. Monstrosities and deformed fetuses preserved in formaldehyde commanded respect, not only for the relentless power of nature (and, in its wake, the arm of God), but also for medical science, which was capable of dethroning this power. A mixture of authenticity and educational value, of titillation and moralism, characterizes the nineteenth-century specimens still to be found in anatomical museums today.

In the early twentieth century, we may notice a transition from individual organs to erect, full-fledged body models, and from parts in glass bottles to specimens treated with translucent chemicals, such as plastic. One of the most popular anatomical displays in Germany was a model called the "Transparent Man." First exhibited in 1911 at the Hygienic Exhibition in Dresden, it traveled to world exhibitions in Paris, Chicago, New York, and Berlin before ending up in the German Hygiene Museum. The Transparent Man—soon accompanied by a female counterpart—consisted of a real human skeleton stuffed with fake inner organs harvested from various wax models and protected by a thin layer of celluloid. Most remarkable about the model is not just its erect figure, but its pose: arms outstretched, palms open, gaze directed upward, the Transparent Man conveys to the viewer an image of the divine superhuman, looking up only to God. In 1926, the museum's curator, Franz Tschakert, applied a layer of cellon—an early form of plastic—to the body's surface, which gave it a translucent shine. Significantly, the Transparent Man became a symbol for the eugenics movement in the 1930s, when it was shipped to the United States as part of an exhibition called "Eugenics in the New Germany." The vitreous figure seemed designed to overcome corporeality, with its association of uncleanliness. After World War II, replicas of the Transparent Couple surfaced in Moscow, a birthday gift of the East German government to Stalin; here, the models were claimed to signify the victory of science over the imperfection of individuality—the body represented as a well-ordered mechanism perfectly manageable by medicine. In retrospect, the Transparent Man is less an anatomical object intended to popularize anatomical knowledge than an interesting token of German's history, tainted by National Socialist ideology and later by Stalin's communism.

The European tradition in anatomical modeling is pivotal to understanding the popularity of Bodyworlds as well as the controversy surrounding the exhibition. The technique of plastination is supposed to be both a continuation and an enhancement of the centuries-old tradition. Gunther Von Hagens's assertion that "the plastinates are only complete if their authentic representation agrees with their educational function" betrays the same tension between authenticity and the desire to instruct that has defined the manufacture of anatomical bodies and models from the early sixteenth century on.[6] Plastination, according to its inventor, manages to combine the qualities of real bodies with the advantages of body models; in his view, authenticity and didactical value—organic materiality and pedagogical plasticity—are not mutually exclusive features of an anatomical object. His plastination technique is based on a special chemical treatment that renders cadavers pliable while also preventing them from decaying, and keeps the "original" body intact while still accentuating specific physiological details. By carving out the relevant parts and discarding the surrounding tissue, Von Hagens highlights specific features of the body, such as muscles, bones, respiratory functions, or heart functions. For instance, Bodyworlds displayed plastinated sculptures that consist only of bone structure, alongside so-called muscle men—corpses that feature only muscle tissue. In other words, the plastination technique purportedly allows for preservation of the organic material, while the corpse simultaneously functions as a body model.

Put briefly, the plastination technique works as follows: The corpse is first immersed in formaldehyde, so as to stop its decay. Next, the cadaver may be prepared in either of two ways. It can be used for *Scheibeplastinate*, which means that the body is cut into slices sometimes less than a millimeter thick. These slices are covered first with a thin layer of plastic, and subsequently pressed between glass plates. More impressive, however, is the second category of plastinates, the *Ganzkörperplastinate*. In the production of these anatomical objects the whole body is left intact, except that certain parts are removed, as a result of which others become visible or pronounced. After being dipped in formaldehyde, the body is submerged in a warm basin to get rid of the remaining body fat, and fluids are then replaced by acetone. The final phase of the chemical process consists of impregnation under pressure: in a warm basin the acetone is replaced by synthetic resin. The *Ganzkörper* are subsequently put into the desired position. A final treatment with gas or hot air determines the ultimate fixed form—a form that, according to Von Hagens, allows the corpse to be preserved for at least two thousand years.

As opposed to the artificiality of wax or plastic anatomical models—including the Transparent Man, which is a combination of both—the "realness" of the plastinated object is advertised as an important asset. Von Hagens stresses the relevance of the authenticity of the plastinates, thus putting them on a higher plane than body models, which are, after all, *imitations* of bodies. Texts accompanying the Ganzkörper at the exhibition in Mannheim stated explicitly that the bodies were not compilations of various cadavers or partial imitations, but that they were "real" and "intact." But what is "real"? The objects are manipulated with chemicals to such an extent that they can hardly be regarded as "real" bodies. Just as in the case of living bodies that have been altered by plastic surgery or anabolic steroids, it is almost impossible to use the term "authentic" in this branch of anatomy. By constantly foregrounding the realness of his cadavers, Von Hagens downplays the role of chemical modi-fication—yet it is precisely this technique that he has patented. The novelty of this method is not the selection of chemicals used in preparing the bodies, but rather the way in which they are applied. The plastinated cadaver is thus as much an organic artifact as it is the result of technological tooling. Like the engineers of genetically modified corn and wheat, who insist that these are "natural" products, Von Hagens understates the process of chemical manipu-lation. The plastinated sculptures, however, are as much "imitations" of bodies as are body models, and they sometimes look less "real" (more like plastic) than eighteenth-century wax figures.

The educational or moral function that dominated nineteenth-century an-atomical collections returns explicitly in the plastinated organs or body parts. Visitors to Bodyworlds seemed particularly eager to look at the displays of physical defects, both congenital ones and those caused by disease after birth. Tumors of the liver, ulcers, enlargements of the spleen, and specimens of arteriosclerosis show the ruthless destruction of the human body. Plastinates of pathological embryonic growth, such as a fetus without brains and one with hydrocephalus, illustrate what may go wrong during the human reproduction process. Unlike the bottled specimens preserved in formaldehyde from the nineteenth century, the exhibition's catalogue explains, the absence of glass jars and fluids allows a more "authentic" or "unmediated" look at physiological reality. In Von Hagens's plastinates, too, a claimed unmediated realism goes hand in hand with outright moralism. Visitors were confronted with unam-biguous messages about various types of self-induced physical degeneration. Plastinates of tar-covered black lungs were displayed alongside white, per-fectly healthy ones; similarly, a healthy liver and one affected by excessive alcohol consumption were shown side by side. In this way, Bodyworlds should be understood as a direct continuation of the realist-moralist tradition in the art of anatomy. At the same time, the Mannheim exhibition provided a metacommentary on the twentieth-century "nature" of the flesh. Whereas in nineteenth-century displays, natural bodies were shown to be prone to degen-eration, either through God's hand or through immoral behavior, the plasti-nated cadavers celebrate the power of humankind to interfere with life and death. Von Hagens seems to move away from the idea that the body is just an organic object, which people can influence negatively by, for instance, smoking or drinking. In the course of medical history, there have been an increasing number of inventions aimed at countering—if only temporarily—physical de-terioration. One of the plastinates explicitly comments on the influence of technology in medicine: the "Orthopedical Plastinate" is covered from top to bottom with all kinds of internal and external prostheses, ranging from a metal knee and external fixtures for broken bones to a pacemaker and a replacement for a fractured jawbone. This remarkable plastinate not only demonstrates technological progress in medical science, but also entails a statement about the contemporary living body: human beings have become hybrid constructs, amalgams of organic and technological parts—cyborgs, in Donna Haraway's definition.[7] The natural body is no longer a given, as both longevity and quality of life can be manipulated. Technological and chemical aids are promoted as "natural" extensions of the living human body, just as the process of plastination prolongs the durability of the dead body.

The history of material production teaches us that the anatomical body has always been regarded as a hybrid object, one of art as well as science, whereby concerns for authenticity and instruction tended to compete with each other. In some periods, the authenticity standard prevailed; at other times, the instruction criterion was foregrounded. In introducing the method of plastination, Von Hagens claims to have moved beyond the body-or-model dilemma, because his cadavers are both real and modifiable. He has repeatedly stressed the authenticity of his anatomical creations, yet their modified nature is the very reason why his plastinated corpses can be patented. Although in many ways his work is a continuation of age-old traditions in the material production of anatomical objects, he also adds a commentary on the "nature" of the human body: humans are no longer subjected to divine nature, as science and scientists to a large extent control longevity and quality of life. By the same token, we can witness a similar mixture of continued tradition and postmodern commentary in Bodyworlds' recasting of artistic conventions.

ANATOMICAL BODIES AS ARTISTIC REPRESENTATIONS

In one of his famous essays, art historian Erwin Panofsky suggests that the rise of anatomy during the sixteenth and seventeenth centuries cannot be understood in isolation from the renaissance in art; the history of anatomy is deeply embedded in art history.[8] He even argues that in order to determine the scientific value of anatomical art, it should be evaluated from the perspective of the art historian. During the sixteenth century, the accumulated knowledge of the body was represented visually in drawings and engravings produced by anatomists and their craftsmen. Anatomical atlases are still admired for their clear depiction of contemporary anatomical insights, but even more so for their artistic qualities mirroring the conventions of early Renaissance art. The drawings in Vesalius's *De humani corporis fabrica* (1543), for instance, are reminiscent of ancient Greek sculpture with its strong bundles of muscles and round, broad-shouldered torsos. A characteristic aspect of Vesalius's engravings is that the dissected organs are surrounded by a healthy, living body, distracting from the rather repellent look of death; the scientific reality of the image is embellished, aestheticized, so as to make it more pleasing to the eye. Vesalius's skeletons and so-called muscle men are also imprinted with the principles of sculptural tradition; although they refer to dead bodies, they pose as upright, living figures. Classical conventions of Renaissance sculpture and painting determined the formative elements of his anatomical representations.

Panofsky's view that artistic techniques of representation dominate and shape scientific insights is corroborated by Ludmilla Jordanova, who, in a close analysis of eighteenth-century wax models, shows how neoclassical ideas determined the representation of scientific insights in this genre of anatomical objects.[9] These anatomical models provide perfect specimens of bodies that are partially opened up, thus showing, for instance, the stomach, the intestines, or the reproductive system. Just as in the case of Vesalius's drawings, these models appear extremely vivid and their physical beauty tends to divert attention from the opened-up intestines. Most of the female bodies, for example, are shown in classic Venus-poses; while their main purpose is to display the reproductive functions of the female body, the wax models express images of the seductive goddess of love. Consequently, the aesthetic norms of external appearance outshine the realist representation of the intestines.

Lorraine Daston and Peter Galison, focusing on nineteenth-century medical representations of the body, reframe Panofsky's argument in terms of a continuous struggle between scientific objectivity and artistic subjectivity; they historicize the concept of objectivity explained by what they term "mechanical" or "noninterventionist" objectivity.[10] With the arrival of new representational technologies in the nineteenth century, scientists hoped to eliminate artistic contamination. Mechanically mediated representations were thought to be conceptually distinct from earlier attempts to produce "true to nature" depictions of the interior body. New instruments, such as photography and (later) X rays, purportedly ruled out the subjectivity of the artist, replacing it by truthful, objective imprints. Yet the introduction of mechanical inscription, as Daston and Galison convincingly show, "neither created nor terminated the debate over how to depict."[11] The substitution of the engraver by photomechanical instruments, they argue, did not eradicate interpretation; the photographer's very presence meant that images were mediated. New apparatuses mitigated the dream of perfect transparency while promoting a new image of objectivity—objectivity through mechanical reproduction.

The anatomical artifacts that Gunther Von Hagens produces reflect the historical friction between scientific accuracy and artistic or aesthetic embellishment, which he does not perceive as conflicting requirements. Each of his plastinates features a specific physiological feature—such as the musculoskeletal, digestive, or cardiovascular-respiratory system—carved out with tantalizing precision; but what attract most attention are the artistic poses in which they are sculpted. As with the drawings in Renaissance anatomical atlases, we are diverted from the abhorrence of death and the cruelty of dissection by the vivid appearance of each Ganzkörper. In line with the artistic tradition in anatomical drawings, Von Hagens's plastinates are at least as determined by artistic conventions as by scientific insights. A plastinate called *The Chess-Player*, aimed at showing the structure and functions of the nervous system, stylistically resembles Auguste Rodin's bronze *The Thinker*. Another plastinated body, entitled *The Runner*, seeks to demonstrate the workings of human kinetics; it has fluttering bits of skin and tissue attached to its limbs, to suggest the dynamics of a running man. The sculpture triggers associations with futurist art in which movement and speed were represented in new ways.

Despite the seemingly conflicting character of "true-to-nature" representation and artistic intervention, in Bodyworlds both views are explicitly conflated in single bodies. Von Hagens's anatomical bodies, like those of his historical predecessors, incorporate contemporary artistic styles and conventions —in his case, of postmodernism. Although the full-body sculptures look like direct imitations of Renaissance anatomical art, the plastinates are always modifications of earlier artistic styles, rather than straight imitations. This is particlarly illustrated by the "expanded bodies"—drawn-out plastinates that spatially expose inner organs. One such expanded plastinate is stretched lengthwise, as a result of which the body looks like a sculpted totem pole, or like the inside of a Giacometti bronze. Another plastinate is expanded in four directions, and in the three-dimensional space that is thus created, the individual parts of the body are suspended with invisible threads. A third is stretched horizontally from the diaphragm, consequently revealing the intestines. Obviously, expanded bodies are no true-to-nature representations of the body; instead, they have been spatially reconfigured into giant three-dimensional models.

If this exhibition of plastinated bodies were simply a contemporary continuation of an age-old tradition in anatomical art to relate scientific soundness with artistic aesthetics, there would have been no public outcry. But there is an important caveat to this alignment of traditional and postmodern anatomical art: Von Hagens's sculptures are not *representations* of bodies, like

Vesalius's drawings in his *Fabrica*, or Govard Bidloo's copper engravings of the interior body in *Anatomia humani corporis* (1685). Rather, the plastinates presented at Bodyworlds are *imitations of representations*, executed in modified organic material. The most prolific example illustrating this practice is Von Hagens's copy of Vesalius's muscle man: an explicit imitation of a man who carries his own skin in his hand, posing as if he has just taken off his coat. A life-size reproduction of Vesalius's muscle man was depicted on the wall behind the "look-alike" plastinated model at the Mannheim exhibition. Because the "real" body plastinate imitates a piece of art—Vesalius's drawing—object and representation seem to fuse in the sculpted body. This plastinate, at first sight, seems a "reanimation" of a representation, a wink to Renaissance artistic anatomical tradition. Postmodern literature and art, of course, are full of such gestures—playful imitations of existing styles also known as pastiche. But the alignment smoothes over an urgent question at stake here: can life—or, rather, death—imitate art? The "copy" of Vesalius's muscle man is created from an "authentic" body, which can no longer be labeled "authentic" because of its chemical modification. In our "culture of the copy," observes Hillel Schwarz, authentic and fake seem interchangeable, and their distinction is therefore obsolete.[12] Indeed, the audience may have taken umbrage at the fusion of artistic representation and organic materiality in the exhibited plastinates, yet Von Hagens's actual reversal of art-representing-body into body-representing-art is at least as disturbing.

Besides his preference for "reanimated" anatomical representations, Von Hagens also plays with the belief in mechanical objectivity. Visitors to Bodyworlds were treated to "transparent" cross-sections of the body, illustrated by the *Scheibeplastinate* or plastinated body slices. Exhibited body slices are not particularly new or shocking; cross-sectioned bodies have been showcased in anatomical museums for quite some time. Yet, if we are to believe the German anatomist-artist, his Scheibeplastinate offer "an unmediated look into the depths of the body" due to the combined techniques of cryogenic cutting and plastination. A cryogenic saw cuts up the deeply frozen body into thin slices; pieces can be cut either horizontally (coronally) or lengthwise (sagittally). The slices look like two-dimensional representations, yet Von Hagens fashions them into such a position that they ostensibly regain three-dimensionality. One of the sliced plastinates, titled *The Transparent Body*, consists of eighty-three slices some four inches apart from each other, thus forming a reclining body almost three yards long. The plastinated slices seem to provide direct access to the smallest details of the inner body—complex structures that cannot commonly be perceived by the naked eye. According to Von Hagens, the slices offer the possibility of an entirely unmediated look inside the real body, because "in today's media world in which people are increasingly informed indirectly, the need for unmediated, unadulterated originality is on the rise."[13] Just as the pioneers of photography hailed it as a mode of writing with light beams, so Von Hagens hails his technology as enabling direct inscription, eradicating all mediation between object and representation. The combined technologies of the cryogenic saw and plastination purportedly render all "subjective intervention"—inherent to representation—obsolete.

Once again, Von Hagens's techniques are not merely a prolongation of age-old representational modes. Mechanical mediation—in this case, the application of the cryogenic saw, various chemical solutions, color additives, and a plastic coating—inevitably transforms the appearance of the body. Like his nineteenth-century predecessors, the German anatomist assumes the superiority of mechanical objectivity, thus perpetuating the myth of "a scientific transparent truth"—a pure representation of the human body without the contamination of human intervention. But besides enhancing this myth, he implicitly reverses the roles of anatomical object and representation. As was the case with his imitation of Vesalius's muscle man (a reanimation of a representation), in his Scheibeplastinate he "copies" in actual flesh the most common medical-visual representation of a late-twentieth-century body: the MRI or CT scan.

The universal implementation of MRI and CT scans in medical practice has made the millimeter body slices become a common mode of representation. The introduction of magnetic resonance imaging (MRI) and computertomography (CT) in the 1980s allowed doctors to look "through" the body three-dimensionally; magnetic fields and X rays helped generate representations of cross-sections of the body, thus visualizing organic tissue and even the inside of bone material. Images resulting from MRI or CT scans—sometimes cross-sections less than a millimeter thick—actually look like slices of body parts, but of course they are "photographed" representations. These fairly recent visual technologies have familiarized the public with inferred relationships between the fragmentary, two-dimensional slices and a three-dimensional human body, even if their proper medical interpretation depends on complex insight and visual expertise on the part of the observer. Von Hagens, in turn, restyles two-dimensional representations into three-dimensional organic sculptures. Since most viewers accept the implied relationship between slices and real bodies, the claimed unmediated naturalness of plastinated body slices seems merely an extension of the MRI-induced gaze. In other words, it is precisely the familiarity of this look that distracts from the act of violence involved in cutting the body into slices, and that makes us forget that these objects are "fleshed-out" scans.

As my argument suggests, it is difficult to properly evaluate this sample of contemporary anatomical art without the perspective of art history. Like Vesalius, whose anatomical illustrations were partially based on the conventions of ancient Greek sculpture, Von Hagens looked to artistic conventions for inspiration. However, the plastinated sculptures also call for some knowledge of representation theory in order to comprehend their various signifying layers. I would argue that Von Hagens does not deploy a "true-to-nature technique" but fashions his artifacts into a "true-to-technique nature." What is remarkable about his practice is not that he reverses the order between anatomical object and representation, but that he renders the very distinction between these categories questionable. We are urged to consider Vesalius's muscle man not as a representation, but as the paper model for an animated representation; and because we are so used to seeing MRI scans as representations, we don't even flinch at seeing a body slice of a "real" cadaver. Bodies and body models, bodies and representations, seem to have become interchangeable in Bodyworlds. Plastinated organs, orthopedic cadavers, expanded corpses, and sliced body parts tell us that the anatomical body, which was already a mixed object of science and art, has also become a hybrid product of artistic models and modeled organisms. Just as the "real" tulip is now a tulip treated and perfected with chemicals to appeal to current taste, so the "real" body is now a cadaver that is surgically, chemically, and artistically modified in accordance with prevailing aesthetic standards. Before returning to the ethical implications of this technological imperative, I want to take a closer look at the settings of Von Hagens's exhibitions. . . .

THE PLASTINATION CONTROVERSY

The media debate triggered by the exhibition revolved around notions of art versus science, as the ethical acceptability of Bodyworlds was discussed almost exclusively in binary terms. The most outspoken objections came from moral theologians, who were offended by Von Hagens's desecration of human cadavers; they said they respected the scientific use of anatomical bodies, but loathed the artistic motives prominent in the plastinated figures. A second group of vocal resisters were medical scientists: anatomists and other medical specialists, who usually subject themselves to rigid protocols regulating the donation and treatment of corpses, strongly objected to Von Hagens's violation of these ethical norms by using dead bodies for frivolous purposes, *in casu* art. Directors of Europe's leading anatomical museums also responded mostly negatively to the Mannheim exhibition. They stated that plastinated bodies add no scientific or educational component to body models. Moreover, they resented Von Hagens's sensationalism and his catering to the art world; anatomical exhibitions should be in the service of science, not of art, they argued. In interviews with visitors to the exhibition, reactions ranged from indifference to strong ethical objections to the use of human cadavers for other than scientific purposes. Such dubious practices, some contended, would raise eyebrows anywhere in the world, but particularly if they took place in Germany. Inasmuch as Von Hagen's plastinates are inscribed with historical anatomical tradition, they are imbued with Germany's tainted history of scientific experiments on living and dead human bodies and the Nazi ideology of eugenics.

In response to his many and powerful foes, Von Hagens took a surprising defensive position: in all kinds of ways, he emphasized the strictly scientific nature of his work while understating its artistic value. For instance, in Mannheim he hired medical students to provide information to visitors, and to answer their questions. The donor forms distributed at the end of the tour resembled regular forms for organ donation. Countering the criticism from the medical profession, Von Hagens underscored the academic status of the Heidelberg Institut für Plastination, as well as his own scientific qualifications. The anatomist-artist claimed that he tried to liberate science from its ivory tower; unlike his peers, he considers the education of a broad audience an important asset to the discipline. The success of Bodyworlds also proved, according to Von Hagens, that ethical norms are no longer imposed from the top down—by, for instance, church authorities—but that people themselves define what they consider ethical or not. While on the one hand it is understandable that Von Hagens, for pragmatic reasons, had to capitalize on his credentials as a scientist to defy his critics, it seems nonetheless peculiar that he had to fall back on dichotomized hierarchies to defend a practice that has historically blurred those categories.

Both the vocal opposition and Von Hagens's defense reinforced the false dichotomy between art and science, implying that a different set of ethical norms or standards applies to each. I don't think that is true, however. Although the viewer of a contemporary art exhibition is expected to be more "shock proof" than the average anatomical science museum visitor, the use of severed limbs or body parts in a work of art would definitely cause unease and thus probe ethical norms. In recent years, we have witnessed various modes of artistic expression that, either deliberately or unintentionally, questioned the sanctity or integrity of human flesh. Some sensational exhibits featuring the use of organic tissue pushed the envelope of ethical permissibility in the art world.[14] It is clear that Von Hagens also walked a tightrope between artistic articulation and ethical judgment. When asked about his aesthetic-ethical sensibility, he invariably invoked the backing of historical tradition to justify his blending of scientific tools and artistic styling. Indeed, there is no denying the aesthetic refinement of his sculptures; however, the naked sensation they display is every bit as attention-grabbing as the work of some contemporary artists or, for that matter, scientists. The sensationalism, exploitation, and blatant commercialism characterizing Bodyworlds unequivocally exposed Von Hagens as a savvy businessman.

Yet neither moral outcry nor professional scorn has unveiled a more profound source of discomfort. The most disturbing aspect of Von Hagens's plastinates, in my view, is neither the transgression of art-science boundaries that purportedly fuelled the controversy, nor the resuscitation of public spectacle or display. Something that remained virtually untouched in the public debate concerning the exhibition was that plastinated cadavers prompted visitors to reconsider the status and nature of the contemporary body, both dead and alive. This body is neither natural nor artificial, but the result of biochemical and mechanical engineering: prosthetics, genetics, tissue engineering, and the like have given scientists the ability to modify life and sculpt bodies into organic forms that we once thought of as artistic ideals—models or representations. What Von Hagens does with dead bodies is very similar to what scientists do with living bodies. The reversal of body and representation also underlies the principles of cosmetic surgery; and the problem of authenticity and copy seems more urgent in genetic engineering experiments than in plastination. Perhaps most unsettling about the plastinated cadavers is their implicit statement that the very epistemological categories that guide us in making all kinds of ethical distinctions simply do not apply here. Categories such as body versus model, organic versus synthetic/prosthetic, object versus representation, fake versus real, authentic versus copied have become arbitrary or obsolete. Since we commonly ground our social norms and values in such categories, Von Hagens's anatomical art seems to elude ethical judgments.

Plastination is an illustrative symptom of postmodern culture, just as Frederick Ruysch's anatomical objects were a symptom of Vanitas art and Renaissance culture. Cadavers have become amalgams of flesh and technology, bodies that are endlessly pliable and forever manipulable, even after death. Bodies, like tulips, are no longer either real or fake, because such categories have ceased to be distinctive. Modified tulips that last longer and look absolutely perfect raise the same ethical and philosophical concerns as genetically engineered sheep and manipulated corn. By the same principle, Von Hagens's sculptures might provide an interesting critical perspective on the all-pervasive influence of technology on corporeality and the waning integrity of the flesh. Yet his defense and explanations of plastinated cadavers divulge no such critical or playful intentions. Rather than commenting on the confusion of boundaries, he reconfirms binary labels by propagating his cadavers as "real," "intact," and "authentic." While obviously sustaining the tradition of mixing scientific insights with artistic styles, he quickly retreats to the solid bastion of science to defend himself against criticism. Paradoxically, Von Hagens's denial of distinct categories of embodiment, and thus the ethical norms in which they are cemented, does not prevent him from invoking those same norms to claim the legitimacy of his practices. The artist-anatomist's plastinated cadavers seem exemplary of a culture that is "inhabited by posthumans who regard their bodies as fashion accessories rather than the ground of being."[15] The culture of the posthuman, according to Katherine Hayles, continues the liberal humanist tradition in which the body is discarded as a mere container for cognition, and the religious tradition that holds the body as a temporary vessel for the soul. Von Hagens's technique attempts to detach bodies from their living signifieds, yet they are inescapably infused with historical, local, and cultural meanings. Doctor Von Hagens is perhaps best seen as a postmodern Doctor Tulp—one who deploys medical technology to express a potentially provocative, but in actuality disturbing, commentary on our technological culture.

NOTES

1. In the exhibition's catalogue, *Körperwelten: Einblicke in den menschlichen Körper* (Heidelberg: Institut für Plastination, 1997), 217.

2. Jonathan Sawday, *The Body Emblazoned:Dissection and the Human Body in Renaissance Culture* (London: Routledge, 1975), 70.

3. Julie V. Hansen, "Resurrecting Death: Anatomical Art in the Cabinet of Dr. Frederick Ruysch," *Art Bulletin* 78, no. 4 (1996): 671.

4. Ibid., 676.

5. Ludmilla Jordanova, 1995. "Medicine and the Genres of Display," in *Visual Display: Culture Beyond Appearances*, ed. Lynne Cooke and Peter Wollen, 202–17 (Seattle: Bay Press).

6. Catalogue, *Körperwelten* (above, n. 1), 195–207, 21.

7. Donna Haraway, *Simians, Cyborgs, and Women: The Reinvention of Nature* (New York: Routledge, 1991).

8. Erwin Panofsky, "Artist, Scientist, Genius: Notes on the 'Renaissance-Dämmerung,'" in *The Renaissance: Six Essays*, ed. Wallace K. Ferguson, 123–82 (New York: Academy Library, 1953).

9. Ludmilla Jordanova, *Sexual Visions: Images of Gender in Science and Medicine between the Eighteenth and Twentieth Centuries* (Madison: University of Wisconsin Press, 1989).

10. Lorraine Daston and Peter Galison, "The Image of Objectivity," *Representations* 40 (1992): 81–128.

11. Ibid., 98.

12. Hillel Schwarz, *The Culture of the Copy: Striking Likeness, Unreasonable Facsimiles* (New York: Zone Books, 1996).

13. Catalogue, *Körperwelten*, 214.

14. For instance, Marc Quinn's model of a human head filled with his own blood; Orlan's surgical performances (1991–98) in which she refashions her own body into an organic collage of artistic representations (from the forehead of Leonardo's Mona Lisa to the chin of Botticelli's Venus); and the severed head of an executed Maori warrior auctioned at the European Fine Art Fair, 1997. The exhibition *Sensation: Young British Artists from the Saatchi Collection* (London, 1997; New York, 1999) featured some artists who produce body art, such as Mona Hatoum, Marc Quinn, and others; for an introduction to the discussion of the "ethicality" of these art works, see Norman Rosenthal, "The Blood Must Continue to Flow," in the catalogue accompanying the exhibition (London: Royal Academy of Arts, 1997).

15. N. Katherine Hayles, *How We Became Posthuman: Virtual Bodies in Cybernetics, Literature, and Informatics* (Chicago: University of Chicago Press, 1999), 5.

INVENTING THE HETEROZYGOTE: MOLECULAR BIOLOGY, RACIAL IDENTITY, AND THE NARRATIVES OF SICKLE-CELL DISEASE, TAY-SACHS, AND CYSTIC FIBROSIS

KEITH WAILOO

Historians of medicine have been interested for some time in how, where, and by whom diseases are defined (and where disease resides). They have considered, for example, the historical construction of such "diseases" as homosexuality and hysteria, as well as the historical evolution of anorexia nervosa, occupational diseases such as venereal disease, silicosis, tuberculosis, and numerous others.

Many of us also have been particularly interested in the ways in which the discourse of the medical and biological sciences shape, and are shaped by, national politics and by the politics of identity—and these issues are particularly relevant in the era of debates over research funding for AIDS and breast cancer. In our time we have seen how the biomedical sciences and the diagnostic technologies they produce—such as the HIV test—give rise to entirely new categories of persons—the HIV-positive person, an individual not yet clinically diseased. Once designated, this individual has become a social, clinical and political problem. The status of the HIV-positive person—is he or she a public danger, or "at risk" for losing health insurance, etc.—is defined by a range of moral, social, and political debates of our time. This question—the dynamic relationship between new biological disciplines and their technologies and forms of social identity—is the subject of this essay.

I consider aspects of racial, ethnic, and religious identity in post–World War II America by offering a partial history of three "racial" diseases, of the ways in which technology and questions of identity defined these diseases, and of the very different meanings of race that emerged in each case. In the process, I argue that the term "biopolitics" is, perhaps, an inadequate formulation in our attempts to understand the interaction of biological knowledge and identity. Where other scholars have suggested that the new genetic technologies of the mid-to-late twentieth century radically reconfigured questions of race, nature, and identity, this chapter offers an alternative conception of the historical interaction of biological knowledge and identity, one in which genetics provides a new template of cultural understanding on which novel interpretations of historically defined identities might be developed. As we will see, the "racial diseases" at issue in these pages are not so much new inventions of the science of genetics, they are, rather, reinterpretations that draw on particular notions of group history, identity, and memory (in addition to modern genetic technology) in order to flourish and to achieve their reality.

One disease is sickle-cell disease—since the second decade of the twentieth century labeled a disease "of Negro blood," and since the 1940s described in medicine as a characteristically "black disease"—because its incidence in African Americans is one per every 400 births. The second disease is cystic fibrosis, since the 1950s labeled a "white disease," or (according to one observer) "the most common lethal or semi-lethal genetic disease in Caucasians of (northern and) Central European origin." Cystic fibrosis has a frequency among Anglo-Americans comparable to that of sickle-cell disease in African Americans and occurs approximately once per 2,000 births across all ethnic groups. The third disorder is Tay-Sachs disease, often characterized as "a disorder prevalent among Ashkenazi Jews." Its incidence is recorded as 1 per 3,500 births according to Ashkenazi Jewish parents.

In the 1940s and 1950s, it became clear to clinicians and medical scientists that in all of these disorders, sufferers had inherited a trait from both of their parents—neither of whom had shown any outward sign of ill health. Because mechanisms of inheritance from parents to child appeared to be similar, the three diseases have often been linked in the minds of geneticists, physicians, policy makers, and others as autosomal recessive diseases. That is, children with these diseases are said to have inherited a "double dose" of the recessive trait-alleles from both parents. The parents, in the terminology of twentieth-century human genetics and public health, were labeled "heterozygotes" or (somewhat more ominously) "carriers." Not themselves sick, the parents were however responsible—jointly—for the sickness of their child. (Of course the concept of disease carriers—like the symptomless "Typhoid Mary" of the early twentieth century—was brought into the popular consciousness with the advent of the laboratory revolution in medicine. In particular, the use of bacteriological identification in medicine and public health helped detect the presence of virulent organisms in the absence of disease symptoms.) The concept of the carrier highlights how new technologies gave rise to new forms of identity, and to widespread anxieties about social interaction and new methods of surveillance.

This chapter (a history of a very different person than Typhoid Mary) might be called a history of the heterozygote, and particularly of the scientific discourse about these individuals and their social status in the United States in the late 1960s to the late 1970s. In that period significant public controversies first emerged over practices of genetic counseling, genetic testing, and mass screening for hereditary disease, with, concomitantly, an intense national refocusing on hereditary disease. Historians have variously decried and praised these developments—for their eugenic overtones, or for the aggressiveness with which national and state legislatures promoted mass screening. I will compare how particular actors in this drama (first, physicians and biological researchers, then legislators, and finally disease advocates who claimed to speak for patients themselves) portrayed the heterozygote, in cystic fibrosis (CF), Tay-Sachs (TSD), and sickle-cell disease (SCD). I am interested here in how biological stories of the heterozygote emerged from evolutionary biology, and how they were intertwined with them—contemporary narratives and anxieties about the fate of ethnic identity and racial identity. I am interested as well in how the story of the heterozygote sheds light on the changing social relations of genetics and medicine, and how in each disease the heterozygote becomes the site of community, memory, and the making of bodies.

In what follows, I will highlight particular aspects of the biological and social definition of the heterozygote in TSD and the other disorders. Both within biology and in larger communities, the heterozygote became part of discussions of group identity. The heterozygote raised questions of social surveillance. The heterozygote became the basis for discussions of the ethics of group membership—and so, for example, I will consider the role of TSD in religious rituals within Orthodox Judaism. I am also interested in the ways in which the heterozygote became a key reference point in the historical definition of race—highlighting specific historical experiences that defined group membership and defined the responsibility of group membership. These histories of the heterozygote provide, then, an important backdrop for understanding the appeal of genetic explanations of disease and the debate about ethnic difference in twenty-first-century American society. As we will see in the case of TSD, for example, the history of the heterozygote has become a key reference point in later discussions about a range of new "Jewish diseases" and in the "racialization" of Ashkenazi Jewish identity in our time.

TECHNOLOGY AND DISEASE IDENTITY

There is an important therapeutic backdrop to these diseases. All of them were obscure in the early twentieth century but became increasingly central to medical, public health, and public discourse in the last half of the century. All were discovered in the late nineteenth or early twentieth centuries, but several factors explain their rising significance after World War II. One was the decline of other childhood infectious diseases, for as tuberculosis, diphtheria, pneumonia, and other disorders disappeared, these other maladies became more visible. Particularly important in their rising visibility was the advent of penicillin (made available in World War II for soldiers and civilians); antibiotics treated the infections that often killed children with CF, SCD, and TSD. Effective antibacterial agents often treated the infections in these children, revealing their "true underlying disease" in the aftermath.

Technologies other than drugs played a role in the rising prominence of these diseases. In the 1940s and 1950s, for example, both CF and SCD were characterized as "great masqueraders," highlighting their ability to mimic other disorders. But at the same time, new diagnostic techniques migrated into clinical practice from the field of molecular biology. The growing use of electrophoresis (a late 1930s invention for chemical analysis of the composition of large molecules) in clinical diagnosis was particularly crucial to the identification of SCD and CF.

The following decades saw the increasing social prominence of these disorders. With identification and antibacterial therapy (in the case of SCD and CF) came an enlargement of the population of patients, and their growing numbers would parallel the slow transformation of disease in America—from a focus on acute infectious disorders to the brighter spotlight on chronic disease. And with longer life and the focus on chronic illness came also the focus on illness as lived experience. The Civil Rights movement in particular helped put the spotlight on SCD, which exemplified a disease of "pain and suffering" that had long been ignored in America.

It was precisely in these decades of the 1960s and 1970s that hereditary diseases (these three and others like Cooley's anemia, a disorder not unlike sickle-cell disease but prevalent in Greek and Italian Americans) became newly visible and ethnically politicized. An article in an issue of *Science* in 1972, titled "Special Treatment for Another Ethnic Disease," focused on Cooley's anemia legislation being considered by Congress and called readers' attention to "the process of transforming an obscure ethnic disease . . . into the target of a national program of research and screening." So here were newly visible diseases —partly because of new technology but also because of what these diseases symbolized for the group to which they "belonged," what they meant for group aspirations, and to social ideals.

The late 1960s and early 1970s saw the emergence of a new national politics of disease in American biomedicine. The sixties had witnessed the rising influence of particular patient constituencies (from the elderly to the CF foundation to SCD advocacy groups) and of consumer advocacy in matters of health. These groups saw diseases not merely as biopathological entities but as reflections of their political pursuit of rights and of their particular social concerns. Legislators and civic leaders responded to these social movements, even as they shaped ideas about disease to fit their own interests. A wide range of disease-specific legislation was passed in this era—from research funding for cardiovascular disease to kidney dialysis legislation, from the Sickle Cell Anemia Control Act of 1972 to the Cooley's Anemia Act later that same year, to the Genetic Disease Act of 1976.

One thing is clear. Amidst the transformations of these diseases, tensions over the allocation of research dollars, and a national politics revolving around national health insurance and health consumption, a relatively new individual emerged in public health discourse—the heterozygote.

INVENTING THE HETEROZYGOTE: SICKLE-CELL DISEASE

It was molecular biologists, evolutionary biologists, and human geneticists (and their tools) that played a leading role in characterizing the problem of hereditary disease. These scientists believed that the detection of heterozygotes was one of the most obvious and rational methods of disease control. Often (in the 1940s and 1950s) the detection of these heterozygotes or carriers occurred retrospectively. That is, after the disease CF or SCD was diagnosed in their children, only then could parents be said to be heterozygotes. But increasingly in the 1960s the production and use of electrophoresis and other diagnostic tools (coming from disciplines including molecular biology and human genetics) rendered their status "knowable" before illness appeared in their children.

The reason for creating such diagnostic tools seemed obvious to biological researchers—to estimate the supposed burden of heterozygotes on society, possibly to restrict their reproduction, and thereby to control the prevalence of disease. This was perceived as a form of preventive therapy. As many historians have written, these tools fit neatly into what many molecular biologists saw as a "new eugenics"—not, they said, anything like the coercive program of involuntary sterilization of the early twentieth century, or like efforts to eradicate the "unfit." This "new eugenics" was a medical service in the form of genetic screening and counseling for consumers. It emphasized a large degree of individual, family and community control. These diagnostic tools provided the basis for the invention of a new identity and brought narratives from evolutionary biology directly into public debates over health policy and social policy. The question emerging in the aftermath of identification of heterozygotes was simple—what to do about these people and about the hereditary diseases they might cause?

What was known about these diseases? Beginning in the late 1940s, TSD, SCD, CF, and other clinical disorders became an important area of medical focus. Clinicians observed that children diagnosed with SCD, TSD, and CF had many different clinical problems but also had a common family dilemma. The diseases were quite different, clinically speaking. SCD was a disorder characterized by sickled blood cells, joint pains, frequent infections, and high likelihood of childhood mortality. CF was characterized by severe pulmonary congestion, digestive and gastrointestinal problems, frequent infection, and (also) a high likelihood of childhood mortality. The third, TSD, was a rapidly degenerative disease of early childhood, stemming from the accumulation of lipids (fatty molecules) in particular brain cells, resulting in neurological and cognitive decline, mental retardation, cerebral seizures, loss of vision and motor control, and death between the ages of two and six. By the 1960s it was (unlike the others) almost uniformly fatal in early childhood, rather than a chronic disease with an expanding population of sufferers.

According to the rules of Mendelian inheritance, however, all three disorders had the same hereditary mechanisms. This was confirmed as clinicians took up the tools of molecular biology in the 1950s and 1960s. In all three disorders, the mating of two heterozygotes subjected their offspring to predictable odds for inheriting the disease. If such a pairing of two heterozygotes produced a child, then (if the odds played out as expected) each child had a one in four chance of inheriting the disease, a two in four (50 percent) chance of inheriting only one allele and becoming (like their parents) heterozygotes, and a one in four chance of inheriting the trait from neither parent. These mechanisms suggested the specific intervention. Molecular biologists, public health advocates, and clinicians in the 1960s and 1970s (in a time of increasing interest in the use of technologies to control reproduction) saw the heterozygote parent as a key point of intervention against disease. They spoke of these disorders, accordingly, as "hereditary diseases," emphasizing their links to family inheritance and hereditary probabilities. They did not speak of them as "genetic diseases"—a term which, at the time, implied a disorder originating in the DNA, possibly because of a mutation, possibly because of inborn errors, but not necessarily inherited.

But there were significant differences in the ways in which medical experts discussed and portrayed the heterozygote in CF, TSD and SCD. In what follows, I move back and forth between these characterizations, exploring these differences. These stories give us insight into the convergence of biological narratives (coming out of evolutionary biology and human genetics, ascendant disciplines at the time) and the politics of racial identity, ethnic heritage, group survival, and racial pride.

One finds an intense preoccupation with these themes in discussions of the sickle-cell heterozygote, and (given the similarities among diseases) a relative lack of interest in constructing parallel racial stories about cystic fibrosis. Yet, for these experts, the female CF heterozygote did hold particular interest. In the case of Tay-Sachs disease, this disorder took on a social image intricately intertwined with the problem of Jewish survival and preservation—and this unquestionably influenced images of the heterozygote. Indeed, historians and policy makers still puzzle over the question of why heterozygote screening was such a "success" in the case of TSD (embraced by American Jewish communities) and such a controversial undertaking (seen as a form of reproductive control and even "genocide") and even a failure in the case of SCD. The answer to such puzzles, I argue, resides in the meanings of the heterozygote.

The key questions about the heterozygote's status were these: If heterozygotes could be detected, should they be prevented from marrying other heterozygotes? Could mass screening of "at-risk" populations be carried out economically? Would it reduce the prevalence of the disease in the long term? How should this "hidden" identity be presented to these individuals, and to the public? Such questions were debated in scientific journals, public health meetings, religious groups, and state and national legislative hearings. Some argued that heterozygotes were not to be considered social burdens but should be understood as special individuals, to be cherished because of their difference. Whatever their character, clearly these individuals represented broader problems in family health, ethnicity, racial identity, and national character.

The prevailing view of the SCD heterozygote was that he or she was the by-product of a historical and evolutionary process. According to a study by British scientist A. C. Allison in the mid-1950s, these heterozygotes were found in high frequency in certain parts of Africa, and they seemed (based on Allison's small study and a handful of studies carried out since) to have a slight but statistically significant resistance to a strain of falciparum malaria. Allison proclaimed that the heterozygote had a "helpful defect." This speculation was handed down from one generation of researchers to another, tested only by small studies. In passing the trait to their children, African heterozygotes were said to grant their offspring a competitive advantage for survival in those regions where falciparum malaria was endemic. Where others succumbed to malaria, the sickle-cell heterozygote thrived, increasing their numbers relative to the larger African population. But if both parents passed this trait to their child, the advantage became a grave clinical problem—a disease producing high mortality in childhood. As one observer noted, "the gene causing sickle-cell trait [i.e., heterozygote status] helps to defend against malaria, but when there is no malaria [as in many urban and industrial settings], it has no use and is only disadvantageous." The disease therefore told the story of a population uniquely adapted to its native, indigenous, or original environment. But when removed from that environment, the heterozygote's favorable adaptation became a maladaptation.

By the late 1960s some physical and cultural anthropologists extended and fleshed out this biological argument. To Frank Livingstone and other anthropologists, for example, the emergence of malaria needed further scrutiny. Malaria was itself a product of African cultural evolution, Livingstone argued. He speculated that the movement from hunter-gatherer cultures to agricultural societies in certain regions resulted in the clearing of land and the creation of stagnant pools of water where mosquitoes bred and increased their population. Mosquitoes were crucial vectors for the passage of malaria to humans, and so the transformation of this entire ecosystem set the stage for the competitive advantages of the heterozygote.

As one anthropologist noted, "The new agricultural system allows expansion . . . of the population and, at the same time, is the ultimate cause of an increase in malarial parasitism." Only after malaria was thus established could the heterozygote thrive. Thus, for anthropologists the disease and the heterozygote were deeply embedded in African cultural history, their existence showing "how the rising prevalence of malaria, resulting from the changes in agricultural practice, has caused a potentially harmful characteristic to become selectively advantageous." The very genetic identity of the heterozygotes, these anthropologists argued, was shaped by the history of their culture and their human-made context.

For evolutionary biologists, the importance of SCD was due not to this cultural history but to its exemplary role as a balance polymorphism, a trait whose stability owed much to its ecological utility. By the 1960s the study of the sickle-cell heterozygote was (for molecular biologists, anthropologists, historians, and physicians) also a case in point in the nature-nurture debate, the debate over whether genes or environment were more responsible for shaping identity. Together, these experts had (by the late 1960s) constructed a rich, proud, and textured historical narrative of the heterozygote and African heritage. The SCD case became an exemplary in these disciplines—a disease that established a framework for thinking about other hereditary disorders.

THE HETEROZYGOTE AND ETHNIC IDENTITY:
TAY-SACHS DISEASE AND CYSTIC FIBROSIS

Tay-Sachs disease highlighted a very different set of associations. It was discovered in the late nineteenth century and entered the twentieth century as a mental disorder labeled as "familial idiocy" prevalent among Jews. Early on, the disease was regarded by [Bernard] Sachs (a Jewish physician working among a large Jewish population in New York City) as "almost exclusively observed among Hebrews." But in the 1940s the disorder was increasingly redefined in biochemical terms—as a lipid storage disorder of the brain with severe cognitive effects. Relative to the cases of SCD and CF, the identification of the molecular origin of this disorder came late. In the late 1960s researchers isolated the key enzyme defect (a hexosaminidase A deficiency) that resulted in the inability to remove fat from the brain. These developments from the 1940s through the 1960s led to the identification of presymptomatic cases, the possibility of widespread testing, and the possibility of heterozygote testing.

In TSD, explaining the prevalence of the disease among Ashkenazi Jews in this era brought out theories of genetic drift. Other theorists pointed to population isolation and inbreeding as key mechanisms of transmission. But others went the route of speculating about heterozygote advantage—that is, the existence of selective forces that may have encouraged the heterozygote's survival and the gene's successful transmission through many generations. These theories can best be understood as different historical narratives of group formation.

Among theories of heterozygote advantage, some speculated that large populations of Jews have lived in cities longer than most other ethnic groups. The crucial selective environment, therefore, could be the preindustrial and industrial town, where "crowd diseases" like tuberculosis were more prevalent. Perhaps, these theories ran, the heterozygote was a survivor of adverse disease conditions and had developed increased resistance to such crowd diseases. Or were there historical circumstances that had taken their toll on the heterozygote, reducing their incidence in certain populations? Such theories give insight into the historical consciousness of biological scientists (many of whom themselves identified as Jewish). Some suggested, for example, that mental ability and intelligence might explain the survival of the heterozygote. The idea here was that the scholar and the rabbi in Eastern European Jewish communities tended to marry the most desirable women and to have more children than did other men. "If true," noted one author—one of the premier authors on Jewish genetic diseases, Richard Goodman—"this would have resulted in selection for a type of rational intelligence." In this telling of the tale of hetero-zygote advantage, it was the rabbis who were responsible for the disorder.

Turning to the case of CF—and to the more amorphous group of "Cauca-sians" associated with the disease—we note some obvious contrasts. CF had only appeared as a prominent urban medical concern in the 1930s and 1940s, and by the 1950s it had gained the attention of a handful of academic special-ists. Most observers argued that "the disease is rare both in Negro and Mongo-lian populations," but they admitted that "no survey figures in such popula-tions are known." Gradually, diagnoses of the disease established its relatively high frequency in Americans of central European descent. But there were those who objected to the "racialization" of CF, noting that such a focus distracted clinicians from sound medical practice. One Washington, D.C., physician believed that "the failure to identify blacks with CF is not surprising since the disease was first studied in centers largely serving whites." By contrast, in the late 1960s there was no such rich academic discussion of the identity and ethnic heritage of the heterozygote in cystic fibrosis—researchers seemed un-interested in this dimension of the disease. There was speculation, for example, about increased resistance to influenza and cholera. But this discussion was sparse by comparison with discussions of SCD and TSD. There was, however, significant interest in heterozygote women and their fertility. Researchers rea-soned that there was good reason to believe that "a heterozygote advantage may exist in CF" and they looked to the heterozygote women for explanation.

Into the late 1960s there was, then, a certain asymmetry in the construction of a racial profile for the heterozygote in each disease. In CF, some studied what seemed to be abnormal levels of salt in CF heterozygote women and speculated that this feature may have made them more fertile, because their salt levels were more biologically welcoming to the spermatozoa. Others suggested that perhaps heterozygotes had "a greater urge to reproduce"—implying that het-erozygotes "understood," on some deeper psychological level, the likelihood of having sick offspring and sought to compensate by enhanced reproduction. One physician noted that "numerous candidates for the selective agent have been proposed ranging . . . including "the 'agreeable personality' of the moth-ers . . . but these proposals have generally not been supported." Speculating among themselves and using the evolutionary framework created by sickle-cell disease, biologists looked at CF with puzzled expression—and they constructed a problematic CF heterozygote mother to explain it. They admitted, however, that there was no valid consensus on her health status, identity, or personality.

Only in British journals like the *Lancet* and in the South African *Medical Journal*—perhaps not surprisingly in a country where "white" did evoke deeply ingrained anxieties about minority-group status—could one find richly detailed, historical, regional, and race-conscious narratives linking the CF heterozygote to historical events. One author observed that while CF was very common among white Afrikaners, it had not been found in great number among descendants of the Angola Boers. The researcher speculated that the difference between the two groups stemmed from the heroic and historical treks made by the Angola Boers from 1875 to 1905, across the Kalahari Desert, during which many died of thirst. "I suggest," he wrote, "that heterozygotes for CF, because of their impaired electrolyte regulation, died more readily during these treks, thus reducing the frequency of the deleterious gene among their descendants." There was little evidence to support such claims; nevertheless, for this author, speculations on CF heterozygotes were important primarily because they fed into questions of regional identity, ethnic heritage and pride, and distinctive, defining historical experiences—just as in the case of the sickle-cell heterozygote in the United States.

What are we to make of these highly speculative, evolutionary biological stories of the heterozygote? In sickle-cell anemia and in cystic fibrosis, we have seen how molecular biologists, clinicians, and medical experts shaped not merely scientific observations but historical narratives about the status and identity of these ambiguous persons. This was not merely idle theorizing. To understand why, let us turn to the uses of these narratives by politicians, community leaders, and heterozygotes themselves in reinforcing the sense of self, and in preserving an embattled group identity.

HETEROZYGOTES AND SOCIAL POLICY

In the United States, these narratives had particular symbolic and political importance. Let me focus here on congressional hearings surrounding sickle-cell anemia in 1971. To many grassroots critics of screening policies, it was clear that the heterozygote was not the root of the disease problem. One representative of the Black Athlete's Foundation for Research in Sickle-Cell Disease noted that "the problems of the trait carrier are mild in comparison to the agonizing pain and the terror that the sickle-cell anemia victim endures." Tennessee congressman Dan Kuykendall agreed. He noted that focusing on the trait has skewed public debate. "The word 'carrier' brings to all of us the image of 'Typhoid Mary,' " he noted, and he urged "that the words 'disease' and 'carrier' be stricken from our vocabularies when we are talking about the trait."

This white Southern Republican congressman preferred a more proud historical portrait of the heterozygote, believing that "when we talk about selling the sickle-cell anemia screening program, we must emphasize the positive aspect of being a heterozygote." The heterozygote, he argued, should be a symbol of black pride. "An individual who has the sickle-cell trait is stronger than other people are, and I wonder why we do not use some of the strengths and the positive aspects of the trait instead of emphasizing the 'disease'?" "Being a carrier," he concluded, "is not a weakness. Actually it is a historical strength . . . historically a protection from malaria." Such views echoed biological speculation and would be echoed by genetic counselors in later years. Others remained skeptical about the possibility of controlling this positive message.

How widely had stories of the heterozygote of sickle-cell anemia spread? One noted black researcher stated that "the ignorance even among experts is appalling." He juxtaposed this ignorance against his own story of the heterozygote—a story of feats of national heroism carried out by the heterozygote. He noted that "some athletes who have represented their country at the Olympic Games in Mexico and beat the world, went back to their country to be found to be sicklers on mass screening campaigns and were promptly labeled sick." In this telling, heterozygotes were national heroes, stigmatized by the screening programs of their own country; they were real achievers on the international stage, relegated to second-class status at home. Stories of the sickle-cell heterozygote had also permeated popular culture in the early 1970s, in the form of Hollywood films and made-for-TV movies.

Again, in contrast, in the case of CF, molecular biologists, counselors, geneticists, and pulmonologists watched the story of sickle-cell anemia attentively. They saw it as a paradigmatic case, drew lessons from it, but did not develop similarly complex narratives of identity, achievement, and status, at least until some years later. Most scientists and policy makers suggested that the social burden of the CF heterozygote was minimal. Only in the aftermath of sickle-cell anemia legislation did CF researchers forcefully begin to make the argument that "their" disease was "the most common inherited disease among Caucasians"—and in this atmosphere, their argument was translated into enhanced funding.

In the period from 1965 to 1975, however, an asymmetrical controversy swirled around the status of the sickle-cell heterozygote. The U.S. Air Force instituted a ban on SCD heterozygotes as pilots in the early 1970s. Of course, at the same time, there were many who celebrated and praised the achievements of the heterozygotes, noting that "parents who are concerned at either being carriers, or of their healthy children being carriers, can be assured that they are as fit, if not fitter, than average, and that they have no responsibility to reduce the number of carriers." Such celebrators of biological diversity railed against the very foundations and charities that thought they were doing a public service by educating the public: "further posters by charities about the high frequency of carriers should be more carefully worded in order to avoid the leprous connotation which is so easily conveyed . . . Heterozygote advantage is difficult to explain but unwise to destroy." By 1975 there was no scientific consensus on the social status of the heterozygote in CF and SCD. There were, indeed, many competing stories of "hereditary disease," generated by molecular biologists and human geneticists, many of them based on their concerns about parenthood, maternity, heritage, personality, regional identity, and racial identity.

CONCLUSION: FROM HEREDITARY DISEASE TO GENETIC DISEASE

Implicit in the history of the heterozygote or the genetic "carrier" of disease, then, are a host of residual questions regarding the constantly negotiated meaning of "race," the ways in which the biological sciences imagine "nature," and the politics of disease and difference among American ethnicities in post–World War II society. These case studies of "racial diseases," at the very least, should move us beyond Fanon's assertion about the powerful effects of a racialized gaze where skin and phenotypic surfaces of the body become privileged sites for the articulation of "race." Clearly, as the history of the heterozygote suggests, the epidermis does not constitute the only site at which identities are "fixed" within the body. The case of the heterozygote suggests that, with the rise of genetic technologies, different kinds of internal bioanalyses are at work in the processes by which racial identities are "assigned." At the same time, however, this study suggests that these new racial formations in biology are themselves modified, and altered by, group memories, community experiences, and constantly shifting cultural idioms and identities. As such, the heterozygote—whose origins stem from the technologies of modern genetics— becomes a cultural "work in progress." Disease identities become complex assemblages of technologies, practitioners' ideologies, and social relations. Medical technologies may help to generate historically specific forms of identity, but those technologies are always altered and reinterpreted by the users— whose notions of self, identity, suffering, memory, and cultural difference give new meanings to the technological finding.

Briefly, let me compare the discourse of "hereditary" disease that I've been describing to the discourse of "genetic" disease today. Hereditary disease was a construction that framed and informed pressing social concerns and policies, about the heterozygote, family inheritance, and in some cases gender and racial identity and the health of society. The language of "hereditary" disease focused on what parents handed to their offspring, what they owed to their social group, to society at large, and on the government's interest in their reproduction. In an era that saw the rise of genetic counseling and screening, this was the most "natural" way to describe these diseases. But this natural description was the by-product of programs, social debates, the languages and status of particular disciplines, and their constructions of disease and patient identity.

Neither of these diseases—SCD and CF—have changed their manifestations, but they are now much more frequently subsumed today under the rubric of "genetic" disease. Just as "hereditary" disease discourse reflected particular interests—those scientists, legislators, and federal agencies who sought to create and legitimate counseling programs—so too does the "genetic" disease label reflect the influence of the Human Genome Project, a new NIH funding philosophy, and a contemporary politics of disease today.

Today, genetic disease narratives are constructed not by evolutionary biologists and human geneticists but by clinical-minded medical geneticists. In other words, those biologists who sought to translate hereditary measurement into counseling and screening programs had their own definition of disease. They believed that controlling disease involved family intervention and attention to population analysis, social origins, nationality, and that it often necessitated discussing the history of race and identity. The concept of "genetic" disease relegates those concerns to a secondary status, in favor of a refocusing on the patient, an individualistic focus. "Genetic" disease also has a kind of futuristic appeal in an era when tests can be imagined to locate genes that someday, in the future, will produce disease, like the gene for Alzheimer's disease.

Our contemporary notion of "genetic" disease is, then, also a construct, the new label reflecting a new politics, a response to new technologies of genetic testing, gene therapy, and debates about their use and their meaning for identity. In many instances, this new conception is seldom environmentally and geographically contingent, never historically nuanced, and much more deterministic than suggested by the complex history of the heterozygote. And yet, as recent scholarship has suggested, genetics offers a great deal of room for maneuvering and countless possibilities for individuals, communities, and the state to reshape and redefine biological findings into potent cultural idioms about race, nature, and the politics of difference.

Anderson, Warwick. 2003. "The Third World Body." In *Companion to Medicine in the Twentieth Century*, edited by R. Cooter and J. Pickstone, 235–45. London: Routledge.

Arnold, David. 1993. *Colonizing the Body: State Medicine and Epidemic Disease in Nineteenth-Century India*. Berkeley: University of California Press.

Bachelard, Gaston. 1964. *The Poetics of Space*. Boston: Beacon Press.

Bakhtin, Mikhail. 1968. *Rabelais and His World*. Cambridge: MIT Press.

Barker, Frances. 1984. *The Tremulous Private Body: Essays in Subjection*. New York: Methuen.

Beck, Ulrich. 1992. *The Risk Society: Towards a New Modernity*. London: Sage Publications.

Blacking, John. 1977. *Anthropology of the Body*. New York: Academic Press.

Bledsoe, Caroline H. 2002. *Contingent Lives: Fertility, Time, and Aging in West Africa*. Chicago: University of Chicago Press.

Bordo, Susan. 1993. *Unbearable Weight: Feminism, Western Culture, and the Body*. Berkeley: University of California Press.

Bourdieu, Pierre. 1977. *Outline of a Theory of Practice*. New York: Cambridge University Press.

———. 1990 (1970). "The Kabyle House, Or the World Reversed." In *The Logic of Practice*, translated by Richard Nice. Stanford: Stanford University Press.

Boyarin, Jonathan. 1993. *The Ethnography of Reading*. Berkeley: University of California Press.

Bush, Barbara. 2000. "Hard Labor: Women, Childbirth, and Resistance in British Caribbean Slave Societies." In *Feminism and the Body*, edited by Londa Schiebinger, 234–62. Oxford: Oxford University Press.

Bynum, Caroline Walker. 1987. *Holy Feast and Holy Fast: The Religious Significance of Food to Medieval Women*. Berkeley: University of California Press.

Callon, Michel. 1986. "Some Elements of a Sociology of Translation: Domestication of the Scallops and the Fishermen of St. Brieux Bay." In *Power, Action and Belief: A New Sociology of Knowledge*, edited by John Law, 196–229. London: Routledge and Kegan Paul.

de Certeau, Michel. 1984. *The Practice of Everyday Life*. Berkeley: University of California Press.

Comaroff, Jean. 1985. *Body of Power, Spirit of Resistance: The Culture and History of a South African People*. Chicago: University of Chicago Press.

——. 1993. "The Diseased Heart of Africa: Medicine, Colonialism and the Black Body." In *Knowledge, Power and Practice: The Anthropology of Medicine and Everyday Life*, edited by S. Lindenbaum and M. Lock, 305–29. Berkeley: University of California Press.

Csordas, Thomas. 1994a. *Embodiment and Experience: The Existential Ground of Culture and Self*. New York: Cambridge University Press.

——. 1994b. *The Sacred Self: A Cultural Phenomenology of Charismatic Healing*. Berkeley: University of California Press.

Darwin, Charles. 1899. *The Expression of Emotion in Man and Animals*. New York: Appleton.

Derrida, Jacques. 1976. *Of Grammatology*. Baltimore: Johns Hopkins University Press.

——. 1996. *Archive Fever: A Freudian Impression*. Chicago: University of Chicago Press.

Douglas, Mary. 1966. *Purity and Danger: An Analysis of Concepts of Pollution and Taboo*. London: Routledge and Kegan Paul.

——. 1970a. *Natural Symbols: Explorations in Cosmology*. New York: Pantheon Books.

——. 1970b. "The Two Bodies." In *Natural Symbols: Explorations in Cosmology*. New York: Pantheon.

Durkheim, Emile. 1995 (1912). *The Elementary Forms of the Religious Life*. Translated by Karen E. Fields. New York: Free Press.

Elias, Norbert. 1982. *The Civilizing Process*. New York: Pantheon Books.

Farquhar, Judith. 1994. *Knowing Practice: The Clinical Encounter of Chinese Medicine*. Boulder: Westview Press.

——. 2002. *Appetites: Food and Sex in Postsocialist China*. Durham: Duke University Press.

Feldman, Allen. 1991. *Formations of Violence: The Narrative of the Body and Political Terror in Northern Ireland*. Chicago: University of Chicago Press.

Foucault, Michel. 1965. *Madness and Civilization: A History of Insanity in the Age of Reason*. New York: Pantheon Books.

——. 1970. *The Order of Things: An Archaeology of the Human Sciences*. London: Tavistock.

——. 1973. *The Birth of the Clinic: An Archaeology of Medical Perception*. New York: Pantheon Books.

——. 1977. *Discipline and Punish: The Birth of the Prison*. New York: Pantheon Books.

——. 1978. *History of Sexuality, Vol. I: An Introduction*. New York: Pantheon Books.

——. 1980. "Introduction" to *Herculine Barbin: Being the Recently Discovered Memoirs of a Nineteenth-Century French Hermaphrodite*. New York: Pantheon Books.

Gil, José. 1998. *Metamorphoses of the Body*. Minneapolis: University of Minnesota Press.

Gilman, Sander. 1985. *Difference and Pathology: Stereotypes of Sexuality, Race, and Madness*. Ithaca: Cornell University Press.

——. 1988, *Disease and Representation: Images of Illness from Madness to* AIDS. Ithaca: Cornell University Press.

Good, Byron. 1994. *Medicine, Rationality, and Experience: An Anthropological Perspective*. New York: Cambridge University Press.

Gould, Stephen Jay. 1981. *The Mismeasure of Man*. New York: Norton.

Griaule, Marcel. 1965. *Conversations with Ogotemmeli*. Oxford: Oxford University Press.

Haraway, Donna J. 1985. "A Manifesto for Cyborgs: Science, Technology, and Socialist Feminism in the 1980s." *Socialist Review* 80 (March–April).

———. 1991. *Simians, Cyborgs, and Women: The Reinvention of Nature*. New York: Routledge.

———. 1997. "Genes: Maps and Portraits of Life Itself." In *Modest_Witness@Second_Millennium. FemaleMan_Meets_OncoMouse™: Feminism and Technoscience*, 131–72. New York: Routledge.

Hebdige, Dick. 1979. *Subculture: The Meaning of Style*. London: Methuen.

Kaufman, Sharon. 1993. "Toward a Phenomenology of Boundaries in Medicine: Chronic Illness Experience in the Case of Stroke." *Medical Anthropology Quarterly* n.s. 2, no. 4: 338–54.

Keller, Evelyn Fox. 2000. *The Century of the Gene*. Cambridge, Mass.: Harvard University Press.

Kirmayer, Laurence. 1992. "The Body's Insistence on Meaning: Metaphor as Presentation and Representation in Illness Experience." *Medical Anthropology Quarterly* 6: 323–46.

Kitcher, Philip. 1996. *The Lives to Come: The Genetic Revolution and Human Possibilities*. New York: Simon and Schuster.

Kleinman, Arthur. 1988. *The Illness Narratives: Suffering, Healing, and the Human Condition*. New York: Basic Books.

Kuriyama, Shigehisa. 1999. *The Expressiveness of the Body and the Divergence of Greek and Chinese Medicine*. New York: Zone Books.

Laqueur, Thomas. 1986. "Orgasm, Generation, and the Politics of Reproductive Biology." *Representations* 14 (Spring): 1–41.

———. 1990. *Making Sex: Body and Gender from the Greeks to Freud*. Cambridge: Harvard University Press.

Latour, Bruno. 1988. *The Pasteurization of France*. Cambridge, Mass.: Harvard University Press.

———. 1993. *We Have Never Been Modern*. Cambridge, Mass.: Harvard University Press.

———, and Steve Woolgar. 1979 (1986). *Laboratory Life: The Construction of Scientific Facts*. Princeton: Princeton University Press.

de Lauretis, Teresa. 1987. *Technologies of Gender: Essays on Theory, Film, and Fiction*. Bloomington: Indiana University Press.

Law, John, and John Hassard, eds. 1999. *Actor Network Theory and After*. Malden, Mass.: Blackwell.

Lefebvre, Henri, 1991a. *The Production of Space*. Cambridge, Mass.: Blackwell.

———. 1991b. *The Critique of Everyday Life*. New York: Verso.

Lévi-Strauss, Claude. 1969a. *The Elementary Structures of Kinship*. Boston: Beacon Press.

———. 1969b. *The Raw and the Cooked*. New York: Harper and Row.

Liu, Lydia He. 1995. *Translingual Practice: Literature, National Culture, and Translated Modernity—China, 1900–1937*. Stanford: Stanford University Press.

———, ed., 1999. *Tokens of Exchange: The Problem of Translation in Global Circulations*. Durham: Duke University Press.

Lock, Margaret. 1980. *East Asian Medicine in Urban Japan: Varieties of Medical Experience.* Berkeley: University of California Press.

——. 1990. "On Being Ethnic: The Politics of Identity Breaking and Making in Canada, or Nevra on Sunday." *Culture, Medicine and Psychiatry* 14: 237–52.

——. 1993. *Encounters with Aging: Mythologies of Menopause in Japan and North America.* Berkeley: University of California Press.

——. 1993. "Cultivating the Body: Anthropology and Epistemologies of Bodily Practice and Knowledge. *Annual Reviews of Anthropology* 22: 133–55.

——. 2002. *Twice Dead: Organ Transplants and the Reinvention of Death.* Berkeley: University of California Press.

Lombroso, Cesare, and Guglielmo Ferrero. 2004 (1893). *Criminal Woman: The Prostitute, and the Normal Woman.* Durham: Duke University Press.

Low, Setha M., and Denise Lawrence-Zuñiga, eds. 2003. *The Anthropology of Space and Place: Locating Culture.* Malden, Mass.: Blackwell.

Lowe, Donald. 1982. *History of Bourgeois Perception.* Brighton, Sussex: Harvester Press.

Lukes, Steven. 1979. *Individualism.* New York: Harper and Row.

MacPherson, C. B. 1962. *The Political Theory of Possessive Individualism: Hobbes to Locke.* Oxford: Clarendon Press.

de Man, Paul. 1986. *The Resistance to Theory.* Minneapolis: University of Minnesota Press.

Marcel, Gabriel. 1997. *Le Mystère de l'être: avant-propos de Vaclav Havel.* Paris: Association Presence de Gabriel Marcel.

Martin, Emily. 1987. *The Woman in the Body: A Cultural Analysis of Reproduction.* Boston: Beacon Press.

Massumi, Brian. 1992. *A User's Guide to Capitalism and Schizophrenia: Deviations from Deleuze and Guattari.* Cambridge, Mass.: MIT Press.

——. 2002. *Parables for the Virtual: Movement, Affect, Sensation.* Durham: Duke University Press.

Mattingly, Cheryl, and Linda Garro, eds. 2000. *Narrative and the Cultural Construction of Illness and Healing.* Berkeley: University of California Press.

Mauss, Marcel. 1973 (1934). "Techniques of the Body." *Economy and Society* 2: 70–88.

——. 1985 (1935). "A Category of the Human Mind: The Notion of Person; The Notion of Self." In *The Category of the Person: Anthropology, Philosophy, History,* edited by Michael Carrithers, Steven Collins, and Steven Lukes, 1–25. New York: Cambridge University Press.

Miller, Nancy K. 1995. "Rereading as a Woman: The Body in Practice." In *French Dressing: Women, Men, and Ancien Régime Fiction,* 45–53. New York: Routledge.

Mitchell, Timothy. 2002. *Rule of Experts: Egypt, Techno-Politics, Modernity.* Berkeley: University of California Press.

Mol, Annemarie. 2002. *The Body Multiple: Ontology in Medical Practice.* Durham: Duke University Press.

Needham, Rodney, ed. 1973. *Right and Left: Essays on Dual Symbolic Classification.* Chicago: University of Chicago Press.

Nguyen, Vinh-Kim, and Karine Peschard. 2003. "Anthropology, Inequality, and Disease: A Review." *Annual Review of Anthropology* 32: 447–74.

Nichter, Mark, and Margaret Lock, eds. 2002. *New Horizons in Medical Anthropology: Essays in Honour of Charles Leslie*. New York: Routledge.

Pearce, Tola Olu. 1995. "Women's Reproductive Practices and Biomedicine: Cultural Conflicts and Transformations in Nigeria." In *Conceiving the New World Order: The Global Politics of Reproduction*, edited by F. D. Ginsburg and R. Rapp, 195–208. Berkeley: University of California Press.

Petryna, Adriana. 2002. *Life Exposed: Biological Citizens after Chernobyl*. Princeton: Princeton University Press.

Polhemus, Ted. 1978. *The Body Reader: Social Aspects of the Human Body*. New York: Pantheon.

Poovey, Mary. 1995. *Making a Social Body: British Cultural Formation, 1830–1864*. Chicago: University of Chicago Press.

Proctor, Robert. 1988. *Racial Hygiene: Medicine under the Nazis*. Cambridge, Mass.: Harvard University Press.

Scarry, Elaine. 1985. *The Body in Pain: The Making and Unmaking of the World*. New York: Oxford University Press.

Scheid, Volker. 2002. *Chinese Medicine in Contemporary China: Plurality and Synthesis*. Durham: Duke University Press.

Scheper-Hughes, Nancy, and Margaret Lock. 1987. "The Mindful Body: A Prolegomenon to an Anthropology of the Body." *Medical Anthropology Quarterly* n.s. 1, no. 1: 6–41.

Scott, Joan Wallach. 1994 (1991). "The Evidence of Experience." In *Questions of Evidence: Proof, Practice, and Persuasion Across the Disciplines*, edited by Arnold Davidson et al., 363–87. Chicago: University of Chicago Press.

Selin, Helene. 2003. *Medicine Across Cultures*. Dordrecht: Kluwer Academic.

Seremetakis, C. Nadia. 1991. *The Last Word: Women, Death, and Divination in Inner Mani*. Chicago: University of Chicago Press.

Stallybrass, Peter, and Allon White. 1986. *The Politics and Poetics of Transgression*. Ithaca: Cornell University Press.

Stoler, Ann Laura. 1995. *Race and the Education of Desire: Foucault's History of Sexuality and the Colonial Order of Things*. Durham: Duke University Press.

Strathern, Marilyn. 1992. *Reproducing the Future: Anthropology, Kinship, and the New Reproductive Technologies*. New York: Routledge.

Turner, Bryan S. 1984. *The Body and Society: Explorations in Social Theory*. New York: Basil Blackwell.

Turner, Victor. 1957. *Schism and Continuity in an African Society: A Study of Ndembu Village Life*. Manchester: Rhodes Livingstone Institute.

Valéry, Paul. 1977. *Paul Valéry, An Anthology*. Edited by James R. Lawler. Princeton: Princeton University Press.

Wacquant, Loïc. 2004. *Body and Soul: Notebooks of an Apprentice Boxer*. New York: Oxford University Press.

Weiss, Brad. 1996. *The Making and Unmaking of the Haya Lived World: Consumption, Commoditization, and Everyday Practice*. Durham: Duke University Press.

Willis, Susan. 1991. *A Primer for Daily Life*. New York: Routledge.

CITATIONS FOR TEXT SELECTIONS

I. AN EMERGENT CANON, OR PUTTING BODIES ON THE SCHOLARLY AGENDA

Friedrich Engels. 1940 (1882). "On the Part Played by Labor in the Transition from Ape to Man." In *The Dialectics of Nature*, 279–85, 288–89. New York: International Publishers. Reprinted by permission.

Robert Hertz. 1973 (1909). "The Pre-eminence of the Right Hand." In *Right and Left: Essays on Dual Symbolic Classification*, edited by R. Needham, 3–10, 13–14, 16–17, 19–21. Chicago: University of Chicago Press. © 1973 by the University of Chicago. Reprinted by permission of the University of Chicago Press.

Marcel Granet. 1973 (1933). "Right and Left in China." In *Right and Left: Essays on Dual Symbolic Classification*, edited by R. Needham, 43–50, 52–53, 54–55, 57–58. Chicago: University of Chicago Press. © 1973 by the University of Chicago. Reprinted by permission of the University of Chicago Press.

Marcel Mauss. 1973 (1935). "Techniques of the Body." *Economy and Society* 2: 70–88. Reproduced by permission of Taylor and Francis Books Ltd.

Victor Turner. 1967. "Symbols in Ndembu Ritual." In *The Forest of Symbols*, 19–26, 27–29, 39–43, 44–46. Ithaca: Cornell University Press. Reprinted by permission.

Terence S. Turner. 1993 (1980). "The Social Skin." In *Reading the Social Body*, edited by Catherine B. Burroughs and Jeffrey David Ehrenreich, 15–27, 29–39. Iowa City: University of Iowa Press. Reprinted by permission of the University of Iowa Press.

II. PHILOSOPHICAL STUDIES, OR LEARNING HOW TO THINK EMBODIMENT

Karl Marx and Friedrich Engels. 1947 (1846). *The German Ideology*, part I, 39–43, 46–60. New York: International Publishers. Reprinted by permission.

Walter Benjamin. 1978 (1966). "On the Mimetic Faculty." In *Reflections: Essays, Aphorisms, Autobiographical Writings*, edited and translated by Edmund Jephcott and Peter Demetz, 333–36. New York: Schocken Books. English translation, copyright © 1978 by Harcourt, Inc. Reprinted by permission of the publisher.

Maurice Merleau-Ponty. 1962. From *The Phenomenology of Perception*, vii–xi, xix–xxi, 52–58, 60–63, 67, 70–72. Translated by Colin Smith. London: Routledge and Kegan Paul. Reprinted by permission of Taylor and Francis Books Ltd.

Ian Hacking. 1986. "Making Up People." In *Reconstructing Individualism: Autonomy, Individuality, and the Self in Western Thought*, edited by Thomas C. Heller, Morton

Sosna, and David E. Wellbery, 222–36. Stanford: Stanford University Press. © 1986 by the Board of Trustees of the Leland Stanford Junior University.

Judith Butler. 1993. "Bodies That Matter." In *Bodies That Matter*, 27–36, 47–55. New York: Routledge. Copyright © 1993. Reproduced by permission of Routledge/Taylor and Francis Books, Inc.

Bruno Latour. 1999. From "Do You Believe in Reality?" In *Pandora's Hope: Essays on the Reality of Science Studies*, 1–10. Cambridge: Harvard University Press. Copyright © 1998 by the President and Fellows of Harvard College. Reprinted by permission of the publisher.

III. FUNDAMENTAL PROCESSES, OR DENATURALIZING THE GIVEN

E. E. Evans-Pritchard. 1963 (1940). "Time and Space." In *The Nuer*, 94–98, 100–108. Oxford: Clarendon Press. Reprinted by permission of Oxford University Press.

Caroline Walker Bynum. 1991. "Women Mystics and Eucharistic Devotion in the Thirteenth Century." Chapter 4 of *Fragmentation and Redemption: Essays on Gender and the Human Body in Medieval Religion*. New York: Zone Books. Reprinted by permission.

Kristofer M. Schipper. 1993. "On Breath." In *The Taoist Body*, 136–39. Berkeley: University of California Press. Reprinted by permission of the Regents of the University of California and the University of California Press.

Henry Abelove. 1989. "Some Speculations on the History of 'Sexual Intercourse' During the 'Long Eighteenth Century' in England." *Genders* 6: 125–30. Copyright © 1989 by the University of Texas Press. All rights reserved.

Margaret Lock. 2002. "The Worth of Human Organs." In "Human Body Parts as Therapeutic Tools: Contradictory Discourses and Transformed Subjectivities." *Qualitative Health Research* 12:10 (December): 1406–18 at 1406, 1409–11. Copyright 2006 Sage Publications. Reprinted by permission of Sage Publications.

Anna Lowenhaupt Tsing. 1993. "Meratus Embryology." In *In the Realm of the Diamond Queen*, 290–94. Princeton: Princeton University Press. © 1993 Princeton University Press. Reprinted by permission of Princeton University Press.

IV. EVERYDAY LIFE, OR EXPLORING THE BODY'S TIMES AND SPACES

Michel de Certeau. 1984. "Walking in the City." In *The Practice of Everyday Life*, translated and edited by Steven Rendall, 91–99, 106–10. Berkeley: University of California Press. Copyright © 1984 The Regents of the University of California. Reprinted by permission.

Michael Taussig. 1992. "Tactility and Distraction." In *The Nervous System*, 141–48. New York: Routledge. Copyright © 1992. Reproduced by permission of Routledge/Taylor and Francis Books, Inc.

Peter Stallybrass and Allon White. 1986. "The City: The Sewer, the Gaze, and the Contaminating Touch." In *The Politics and Poetics of Transgression*, 125–48 Ithaca: Cornell University Press. © 1986 by Peter Stallybrass and Allon White. Reprinted by permission of the publisher, Cornell University Press.

Judith Farquhar. 2002. "Medicinal Meals." In *Appetites: Food and Sex in Post-Socialist China*. Durham: Duke University Press. © 2002 Duke University Press.

Nancy K. Miller. 1995. "Rereading as a Woman: The Body in Practice." In *French Dressing: Women, Men, and Ancien Régime Fiction*, 45–53. New York: Routledge. Copyright © 1995. Reproduced by permission of Routledge/Taylor and Francis Books, Inc.

V. COLONIZED BODIES, OR ANALYZING THE MATERIALITY OF DOMINATION

Janice Boddy. 1998. "Remembering Amal: On Birth and the British in Northern Sudan." In *Pragmatic Women and Body Politics*, edited by Margaret Lock and Patricia Kaufert, 28–57. Cambridge: Cambridge University Press. Reprinted with the permission of Cambridge University Press.

Susan Pedersen. 1991. "National Bodies, Unspeakable Acts: The Sexual Politics of Colonial Policy Making." *Journal of Modern History* 63: 647–80. Reprinted by permission of the University of Chicago Press.

Stuart Cosgrove. 1988. "The Zoot Suit and Style Warfare." In Angela McRobbie, ed., *Zoot Suits and Second-Hand Dresses: An Anthology of Fashion and Music*. Boston: Unwin Hyman, 3–15, 19–20. Reprinted by permission of the author.

Patricia Leyland Kaufert and John D. O'Neil. 1995. From "*Irniktakpunga!*: Sex Determination and the Inuit Struggle for Birthing Rights in Northern Canada." In *Conceiving the New World Order: The Global Politics of Reproduction*, edited by Faye D. Ginsburg and Rayna Rapp, 59–73. Berkeley: University of California Press. Reprinted by permission of the Regents of the University of California and the University of California Press.

Jean Langford. 2002. From "Clinical Gazes." In *Fluent Bodies: Ayurvedic Remedies for Postcolonial Imbalance*, 140–47. Durham: Duke University Press. © 2002 Duke University Press.

VI. DESIRES AND IDENTITIES, OR NEGOTIATING SEX AND GENDER

John Boswell. 1980. "Intellectual Change: Men, Beasts, and 'Nature.'" In *Christianity, Social Tolerance, and Homosexuality: Gay People in Western Europe from the Beginning of the Christian Era to the 14th Century*, 315–28, 329–32. Chicago: University of Chicago Press. © 1980 by the University of Chicago. Reprinted by permission of the University of Chicago.

Gregory M. Pflugfelder. 1999. *Cartographies of Desire: Male-Male Sexuality in Japanese Discourse, 1600–1950*, 3–4, 23–27, 68–82. Berkeley: University of California Press. Copyright © 1998, The Regents of the University of California. Reprinted by permission.

Emily Martin. 1991. "The Egg and the Sperm: How Science Has Constructed a Romance Based on Stereotypical Male-Female Roles." *Signs* 16, no. 3: 485–89, 491–95, 498–501. © 1991 by the University of Chicago. Reprinted by permission of the University of Chicago Press.

Gilles Deleuze and Félix Guattari, 1977. "We Always Make Love with Worlds." From

Anti-Oedipus: Capitalism and Schizophrenia, translated by Helen Lane, Mark Seem, and Robert Hurley, 291–96. London: Continuum International Publishing Group and New York: Viking Penguin. Copyright © 1977 by Viking Penguin Inc., English language translation. Used by permission of Viking Penguin, a division of Penguin Group (USA) Inc. and The Continuum International Publishing Group London/New York.

VII. BODIES AT THE MARGIN, OR ATTENDING TO DISTRESS AND DIFFERENCE

Barbara Duden. 1991. "The Perception of the Body." From *The Woman Beneath the Skin: A Doctor's Patients in Eighteenth-Century Germany*, translated by Thomas Dunlap, 104–05, 106–12. Cambridge: Harvard University Press. Copyright © 1991 by the President and Fellows of Harvard College. Originally published as *Geschlichte Unter der Hant* by Klett Cotta in 1987. Reprinted by permission of the publisher.

Mariella Pandolfi. 1991. "Memory Within the Body: Women's Narrative and Identity in a Southern Italian Village." In *Anthropologies of Medicine*, edited by Beatrice Pfleiderer and Gilles Bibeau, 59–65. Heidelberg: Vieweg. Reprinted by permission of the author.

Nancy Scheper-Hughes. 1992. From "Embodied Lives, Somatic Culture" and "The Body as Battleground: The Madness of Nervos." In *Death Without Weeping: Mother Love and Child Death in Northeast Brazil*, 172–74, 184–92, 194–95. Berkeley: University of California Press. Reprinted by permission of the Regents of the University of California and the University of California Press.

Arthur Kleinman and Joan Kleinman. 1985. From "Somatization: The Interconnections in Chinese Society among Culture, Depressive Experiences, and the Meanings of Pain." In *Culture and Depression: Studies in the Anthropology and Cross-Cultural Psychiatry of Affect and Disorder*, edited by Arthur Kleinman and Byron Good, 429–30, 433–41, 476–79. Berkeley: University of California Press. Copyright © 1985 The Regents of the University of California. Reprinted by permission.

Alice Domurat Dreger. 2000. "Jarring Bodies: Thoughts on the Display of Unusual Anatomies." *Perspectives in Biology and Medicine* 43, no. 2: 161–72. © The Johns Hopkins University Press. Reprinted with the permission of the Johns Hopkins University Press.

VIII. CAPITALIST PRODUCTION, OR ACCOUNTING
THE COMMODIFICATION OF BODILY LIFE

E. P. Thompson. 1967. "Time, Work-Discipline, and Industrial Capitalism." In *Past and Present* 38: 56–97. Reprinted by permission of Past and Present Society and Oxford University Press.

Aihwa Ong. 1988. "The Production of Possession: Spirits and the Multinational Corporation in Malaysia." *American Ethnologist* 15, no. 1: 28–42. Reprinted by permission.

Brad Weiss. 1992. "Plastic Tooth Extraction: The Iconography of Haya Gastro-Sexual Affliction." *American Ethnologist* 19, no. 3: 538–52. Reprinted by permission.

Matthew Schmidt and Lisa Jean Moore. 1998. "Constructing a 'Good Catch,' Picking a

Winner: The Development of Technosemen and the Deconstruction of the Mono-lithic Male." In *Cyborg Babies: From Techno-Sex to Techno-Tots*, edited by Robbie Davis-Floyd and Joseph Dumit, 21–22, 25–39. New York: Routledge. Copyright © 1998. Reproduced by permission of Routledge/Taylor and Francis Books, Inc.

Margaret Lock. 2001. "Alienation of Body Parts and the Biopolitics of Immortalized Cell Lines." *Body and Society* 7, no. 2: 63–91. Copyright © Sage Publications, 2001. Reprinted by permission of Sage Publications Ltd.

IX. KNOWING SYSTEMS, OR TRACKING THE BODIES OF THE BIOSCIENCES

Shigehisa Kuriyama. 1987. "Pulse Diagnosis in the Greek and Chinese Traditions." In *History of Diagnostics*, edited by Yosio Kawakita, 43–44, 45, 49–51, 54–67. Osaka: Taniguchi Foundation. Reprinted by permission of the author.

Rayna Rapp. 1997. "Real-Time Fetus: The Role of the Sonogram in the Age of Moni-tored Reproduction." In *Cyborgs and Citadels: Anthropological Interventions into Emerging Sciences and Technologies*, edited by Gary Lee Downey and Joseph Dumit, 31–32, 33–36, 36–45. Santa Fe: School of American Research. Copyright © 1997 by the School of American Research, Santa Fe. Reprinted by permission.

Charis M. Cussins [Thompson]. 1998. "Quit Sniveling Cryo-Baby, We'll Work Out Which One's Your Mama." In *Cyborg Babies: From Techno-Sex to Techno-Tots*, edited by Robbie Davis-Floyd and Joseph Dumit, 40–66. New York: Routledge. Copyright © 1998. Reproduced by permission of Routledge/Taylor and Francis Books, Inc.

José van Dijck. 2001. "Bodyworlds: The Art of Plastinated Cadavers." *Configurations* 9: 99–126. © 2001. The Johns Hopkins University Press and Society for Literature and Science. Reprinted with permission of The Johns Hopkins University Press.

Keith Wailoo. 2003. "Inventing the Heterozygote: Molecular Biology, Racial Identity, and the Narratives of Sickle-Cell Disease." In *Race, Nature and the Politics of Differ-ence*, edited by Donald S. Moore, Jake Kosek, and Anand Pandian, 235–53. Durham: Duke University Press. © 2003 Duke University Press.

Labor, 20, 25–29; division of, 116–17, 121–24, 499–501, 505, 513

Language, 27–28, 108–9, 117, 120, 131–32, 157, 166–67, 171–73, 182, 246, 253–55, 284, 417–25, 452, 519, 538, 539

Magic, 35, 55, 132, 244, 261, 263

Marx, Karl (Marxism), 3, 10, 19, 107, 134, 167, 241, 244, 246, 266, 268–69, 386, 430–31, 489, 491, 494, 571

Materialism, 6, 8, 10–12, 15, 107–9, 114–29, 245–46, 293–94

Materiality, 1, 109–10, 164–74, 209–12, 262, 297, 556–59

Mauss, Marcel, 5, 7, 22, 23, 29, 571

Medical anthropology, 4, 9, 10, 11, 12, 20, 439

Medicalization, 364–66, 436–37, 439, 441, 457, 460–61, 467, 490, 512–29, 552, 572

Merleau-Ponty, Maurice, 3, 6, 241, 457

Microcosm. *See* Cosmology

Mimesis, 108, 130–32, 166, 171–73, 191, 205, 211–12, 262–64, 303 n.1

Mind, 2, 6, 7, 12, 178–82, 245, 264

Mind–body divide. *See* Dualism

Molecular genetics, 4, 14, 435, 493–94, 567–82, 593. *See also* Genetics; Human Genome

Naturalization, 5, 9, 21, 30, 110, 242, 386; denaturalizing critique and, 11, 110, 187, 242, 424–25, 461, 468, 557–58, 628, 639 n.1

Nature (the natural), 1, 2, 10, 11, 85–88, 91, 96–99, 101, 121, 128, 130, 183, 187–89, 193, 385–86, 389–99, 435–36, 563, 601–2, 604, 612, 623, 626–28, 640–41, 647–48, 652–53, 670–71; nature–culture divide, 11, 19, 86, 103, 187, 624, 626. *See also* Human nature

Needs, 2, 3, 119, 121

Nietzsche, Friedrich, 134, 260

Nominalism, dynamic, 109, 150–63. *See also* Social and cultural constructionism

Normality, 2, 3, 4, 8, 435

Norm, norms, 77–80, 522–25, 640; nor-malization, 5, 327, 377, 440–41, 475–85, 612–13

Objects, 7, 11–12, 23, 147–49, 167, 225, 262. *See also* Materiality

Organ transplantation, 14, 191, 224–31

Perception, 109, 136, 140–43, 147–49, 607. *See also* Sense, senses

Phenomenology, 1, 6–7, 9, 108–10, 133–38, 145–46, 182–84, 188, 457, 461

Plato, 110, 168, 171–73, 389, 451, 604–5

Politics, the political, 2, 4, 9–10, 89, 99, 165–66, 174, 232, 236, 262, 309–10, 330–45, 347–57, 374, 436, 439, 441, 461–62, 467, 471, 658, 661–62, 668–71

Practice, practices (praxis), 2, 4, 7, 11, 22, 77, 118, 123–25, 127, 242–43, 246, 251–58, 309, 532–33

Production, 220–21, 489–90; modes and forces of, 116, 121, 123–24, 127, 525–26, 544. *See also* Labor

Psychology, 6, 22–23, 55, 66–67, 143–45

Qi (*ch'i, ki*), 13, 213–16

Rabelais, François, 297–98, 313, 378–79

Race, 5, 334–46, 347–57, 493, 558, 560, 592, 594, 634–35, 638, 658–71

Religion, 33, 35, 37–39, 41, 114–15, 189, 233, 268, 389–99, 408, 410, 506–8, 514–16

Ritual, 69–82, 93–99, 102, 331–33, 337, 344–45, 527–28

Sacred–profane opposition. *See* Dualism

Science studies, 3, 4, 11–12, 176–78, 180

Sense, senses, 27–28, 76, 109, 138–40, 243, 244–45, 259–65, 273–74, 277, 284, 288, 294, 588–89, 596, 601–2, 605–6; anthropology of, 9

Sex, 65–66, 87–88, 94, 109–10, 164, 173–74, 190, 207, 217–22, 339–41, 344–46, 355, 383–87, 389, 400, 402–3, 428–32, 440–41, 521, 542–45, 614–18. *See also* Gender

MARGARET LOCK is professor of anthropology and
social studies of medicine at McGill University. She is
the author of *Twice Dead: Organ Transplants and the
Reinvention of Death* and *Encounters with Aging:
Mythologies of Menopause in Japan and North America.*

JUDITH FARQUHAR is professor of anthropology and
social sciences at the University of Chicago. She is the
author of *Appetites: Food and Sex in Postsocialist China*
and *Knowing Practice: The Clinical Encounter of Chinese
Medicine.*

Library of Congress Cataloging-in-Publication Data
Beyond the body proper : reading the anthropology of
material life / edited by Margaret Lock and Judith
Farquhar.
p. cm. — (Body, commodity, text)
Includes bibliographical references and index.
ISBN-13: 978-0-8223-3830-7 (cloth : alk. paper)
ISBN-13: 978-0-8223-3845-1 (pbk. : alk. paper)
1. Body, Human. 2. Human physiology. 3. Physical
anthropology. 4. Body, Human—Social aspects.
5. Medical anthropology. I. Lock, Margaret M.
II. Farquhar, Judith.
HM636.B48 2007
306.4—dc22   2006028687